Handbook of Research on Managerial Practices and Disruptive Innovation in Asia

Patricia Ordoñez de Pablos
University of Oviedo, Spain

Xi Zhang
Tianjin University, China

Kwok Tai Chui
The Open University of Hong Kong, Hong Kong

IGI Global
DISSEMINATOR OF KNOWLEDGE

A volume in the Advances in Business Strategy and Competitive Advantage (ABSCA) Book Series

Published in the United States of America by
IGI Global
Business Science Reference (an imprint of IGI Global)
701 E. Chocolate Avenue
Hershey PA, USA 17033
Tel: 717-533-8845
Fax: 717-533-8661
E-mail: cust@igi-global.com
Web site: http://www.igi-global.com

Library of Congress Cataloging-in-Publication Data

Names: Ordo´n~ez de Pablos, Patricia, 1975- editor. | Zhang, Xi, 1982- editor. | Chui, Kwok Tai, editor.
Title: Handbook of research on managerial practices and disruptive innovation in Asia / Patricia Ordonez de Pablos, Xi Zhang, and Kwok Tai Chui, editors.
Description: Hershey, PA : Business Science Reference, [2020] | Summary: “”This book explores emerging topics, challenges, and success factors in management, disruptive innovation, and inclusive economies in Asia”--Provided by publisher”-- Provided by publisher.
Identifiers: LCCN 2019021484 | ISBN 9781799803577 (hardcover) | ISBN 9781799803584 (paperback) | ISBN 9781799803591 (ebook)
Subjects: LCSH: Technological innovations--Economic aspects--Asia. | Management--Asia.
Classification: LCC HC415.T4 M36 2020 | DDC 658--dc23
LC record available at https://lccn.loc.gov/2019021484

This book is published in the IGI Global book series Advances in Business Strategy and Competitive Advantage (ABSCA) (ISSN: 2327-3429; eISSN: 2327-3437)

British Cataloguing in Publication Data
A Cataloguing in Publication record for this book is available from the British Library.

For electronic access to this publication, please contact: eresources@igi-global.com.

Advances in Business Strategy and Competitive Advantage (ABSCA) Book Series

Patricia Ordóñez de Pablos
Universidad de Oviedo, Spain

ISSN:2327-3429
EISSN:2327-3437

Mission

Business entities are constantly seeking new ways through which to gain advantage over their competitors and strengthen their position within the business environment. With competition at an all-time high due to technological advancements allowing for competition on a global scale, firms continue to seek new ways through which to improve and strengthen their business processes, procedures, and profitability.

The **Advances in Business Strategy and Competitive Advantage (ABSCA) Book Series** is a timely series responding to the high demand for state-of-the-art research on how business strategies are created, implemented and re-designed to meet the demands of globalized competitive markets. With a focus on local and global challenges, business opportunities and the needs of society, the **ABSCA** encourages scientific discourse on doing business and managing information technologies for the creation of sustainable competitive advantage.

Coverage

- Outsourcing
- Core Competencies
- Economies of Scale
- Differentiation Strategy
- Entrepreneurship & Innovation
- Globalization
- Value Chain
- Balanced Scorecard
- Joint Ventures
- Resource-Based Competition

IGI Global is currently accepting manuscripts for publication within this series. To submit a proposal for a volume in this series, please contact our Acquisition Editors at Acquisitions@igi-global.com or visit: http://www.igi-global.com/publish/.

The Advances in Business Strategy and Competitive Advantage (ABSCA) Book Series (ISSN 2327-3429) is published by IGI Global, 701 E. Chocolate Avenue, Hershey, PA 17033-1240, USA, www.igi-global.com. This series is composed of titles available for purchase individually; each title is edited to be contextually exclusive from any other title within the series. For pricing and ordering information please visit http://www.igi-global.com/book-series/advances-business-strategy-competitive-advantage/73672. Postmaster: Send all address changes to above address.

Titles in this Series

For a list of additional titles in this series, please visit:
https://www.igi-global.com/book-series/advances-business-strategy-competitive-advantage/73672

Theoretical and Applied Mathematics in International Business
Bryan Christiansen (Global Research Society, USA) and Fatima Shuwaikh (Université Paris Saclay, France)
Business Science Reference • copyright 2020 • 358pp • H/C (ISBN: 9781522584582) • US $225.00 (our price)

Strategic Marketing for Social Enterprises in Developing Nations
Nigel Chiweshe (University of KwaZulu-Natal, South Africa) and Debbie Ellis (University of KwaZulu-Natal, South Africa)
Business Science Reference • copyright 2019 • 373pp • H/C (ISBN: 9781522578598) • US $195.00 (our price)

Handbook of Research on Entrepreneurship, Innovation, and Internationalization
Nuno Miguel Teixeira (Polytechnic Institute of Setúbal, Portugal) Teresa Gomes da Costa (Polytechnic Institute of Setúbal, Portugal) and Inês Margarida Lisboa (Instituto Politécnico de Leiria, Portugal)
Business Science Reference • copyright 2019 • 761pp • H/C (ISBN: 9781522584797) • US $395.00 (our price)

Handbook of Research on Corporate Restructuring and Globalization
Antonio Carrizo Moreira (University of Aveiro, Portugal) and Pedro Silva (University of Aveiro, Portugal)
Business Science Reference • copyright 2019 • 408pp • H/C (ISBN: 9781522589068) • US $295.00 (our price)

Responsible, Sustainable, and Globally Aware Management in the Fourth Industrial Revolution
Ziska Fields (University of KwaZulu-Natal, South Africa) and Stefan Huesig (Chemnitz University of Technology, Germany)
Business Science Reference • copyright 2019 • 396pp • H/C (ISBN: 9781522576389) • US $225.00 (our price)

Handbook of Research on Entrepreneurial Leadership and Competitive Strategy in Family Business
José Manuel Saiz-Álvarez (Tecnológico de Monterrey, Mexico) and Jesús Manuel Palma-Ruiz (Universidad Autónoma de Chihuahua, Mexico)
Business Science Reference • copyright 2019 • 559pp • H/C (ISBN: 9781522580126) • US $345.00 (our price)

The Belt and Road Strategy in International Business and Administration
Wei Liu (University of Sydney, Australia) Zhe Zhang (University of Sydney, Australia) Jin-Xiong Chen (Wuyi University, China) and Sang-Bing Tsai (University of Electronic Science and Technology of China Zhongshan Institute, China & Capital University of Economics and Business, China & Wuyi University, China)
Business Science Reference • copyright 2019 • 273pp • H/C (ISBN: 9781522584407) • US $225.00 (our price)

IGI Global
DISSEMINATOR OF KNOWLEDGE

701 East Chocolate Avenue, Hershey, PA 17033, USA
Tel: 717-533-8845 x100 • Fax: 717-533-8661
E-Mail: cust@igi-global.com • www.igi-global.com

Editorial Advisory Board

Xinye Liu, *Tianjin University, China*
Miltiadis D. Lytras, *Effat University, Saudi Arabia*
Nivedita Mandal, *Adamas University, India*
Samrat Kumar Mukherjee, *Sikkim Manipal Institute of Technology, India*
Beatrice Orlando, *Sapienza University of Rome, Italy*
Chandra Sekhar Patro, *Gayatri Vidya Parishad College of Engineering (Autonomous), India*
Swamy Perumandla, *National Institute of Technology Warangal, India*
Preethu Rahman, *Shaanxi Normal University, China*
Jaya Rani, *Sikkim Manipal Institute of Technology, India*
Saibal Kumar Saha, *Sikkim Manipal Institute of Technology, India*
Xu Shan, *Xi'an Jiaotong University, China*
Bibeth Sharma, *Sikkim Manipal Institute of Technology, India*
Swati Sisodia, *G. D. Goenka, India*
Jia Su, *Tianjin University, China*
Xinyun Tan, *Tianjin University, China*
Mariza Tsakalerou, *Nazarbayev University, Kazakhstan*
Tao Wang, *Tianjin University, China*
Xiaoqian Wang, *Tianjin University, China*
Yufei Xia, *Tianjin University, China*
Chunfa Xu, *Tianjin University, China*
Kinza Yousfani, *Isra University, Pakistan*
Liping Yuan, *Tianjin University, China*
Cailing Zhang, *Tianjin University, China*
Xiaotong Zhang, *Tianjin University, China*
Yimin Zhang, *Tianjin University, China*
Bu Zhong, *Pennsylvania State University, USA*
Xiaodong Zhou, *Tianjin University, China*

List of Contributors

Table of Contents

Detailed Table of Contents

been mature to provide the four interactivity functions, though they are yet to be adopted by all smart TV sets so far. Our research shows that the viewer's residence was a significant moderator in the preference for the interactivity functions. Viewers from developed regions enjoyed the four functions more than those living in underdeveloped regions. Social media habits and the power usage of information and communication technology are positively associated with the preference, while need for cognition, age, and gender had little effect.

Given the fact that timber is neither traded as commodity nor is any timber-focused exchange-traded fund (ETF) available in the SAARC countries, the objectives of this chapter are to examine the plausibility of making use of the wood resources of SAARC countries in the form of ETFs, e.g. timber-focused ETF, the nature of the legal and policy frames in these countries as implemented in timber business with reference to their suitability for opening ways to construction of timber-focused ETF and the theoretical plausibility of constructing a timber-focused ETF with reference to India. It is found that in comparison with other SAARC countries, India has developed timber-based industries whose stocks are traded in the bourses and amenable to analysis for ETF construction purposes. While Bangladesh, Pakistan, Sri Lanka, and Nepal have certain legal and institutional artifacts for preservation and maintenance of forests toward gradual but by-and-large unplanned industrialization of the timber industry, Maldives and Bhutan have hardly any.

With the development of society, many industries and professions are more comprehensive and intersecting. Different industries have their own requirements for students with comprehensive backgrounds. For graduates, they may not know the skills required for various occupations, or what kind of jobs and occupations they can take based on their existing knowledge and skills, even how to acquire these abilities after they know the requirements of the jobs they want. In this chapter, authors combine the existing method to predict hot jobs with the analysis of knowledge map, aiming to achieve accurate recommendation of learning path for those who want to find a job. This chapter will help job hunters gradually master skills, and ultimately achieve the goal of optimizing resource allocation and saving social resources.

Industry 4.0 is based on the implementation of a cyber-physical system, which includes sensors, networks, computers, offering digital enhancement and well-coordinated activities. This would create a great pool of all the workforce generations, having diverse experience, agility, and different modes of working. Millennials would add more of machine learning and Generation X and Y would be the richest source of tacit and operational knowledge. Together, they would develop solutions for catering complex and networked production and aggressive logistic management, meeting the challenges of the Industry 4.0. However, the benefits of digitization and automation can be achieved, if the different generations of workforce collaborate, cooperate, and postulate together in all the business processes. Reverse mentoring is a pristine concept and ingenious method to empower learning and encourage cross-generational connections. This chapter would elaborate on the advantage of reverse mentoring in crafting Industry 4.0 more acrobatic and quick-moving.

The purpose of this chapter is to ascertain the contemporary role of social media in increased venture creation tendency along with innovative ideas. The key objective of this study is to discover the tendency of female students' innovativeness in venture creation in China. A random sampling method was used to conduct a survey in different universities in China to identify the scenario of innovativeness in venture creation. Findings from primary data collection indicated that the female students in China are highly involved with social media marketing with innovative ideas. As a result, apart from traditional marketing, society is involved with contemporary marketing where innovativeness with social media and smartphones are the key factors. Innovative ideas in venture creation may generate additional earning for people with low income in society. Future studies with mixed methodology and respondents who use different social media as a tool to innovate new venture may shed light on the undiscovered phenomenon of social media marketing in the context of the mobile phone.

Supply Chain Management (SCM) has gained importance in recent years. Innovation and technological interventions in SCM would be required to remove inefficiencies. It has become imperative for firms to undertake new innovations in SCM to remain competitive. This chapter focuses on physical and digital

Human Resource Management plays an essential role for attaining organizational goals. Nowadays, practitioners, researchers, and academicians around the world are emphasizing to transform and reshape the practice of human resource. However, very few research works have been done in the area of Transformational Human Resource Management (T-HRM). Hence, the aim of this chapter is to propose an integrated framework of T-HRM and organizational efficiency. In light of that, this study has proposed potential factors of the T-HRM. Secondly, this study presented positive effect of the factors of T-HRM on organizational efficiency management. Concept of knowledge management has introduced as a potential mediator, and ICT and organizational alignment has presented as a potential moderator of this study. Finally, knowledge of this study will provide better insights on T-HRM for ensuring organizational efficiency.

Floriculture based economy is slowly increasing throughout the world. Orchids, which were once found only in the wild are now being widely cultivated and being sold in the market. In this chapter, an attempt has been made to understand the inter-relational dynamics of different factors impacting the cymbidium orchids-based economy in Sikkim, India using DEMATAL. A total of 14 factors were identified in this regard and based on expert opinion, these factors were rated and analysed using DEMATAL. The cause and effect relationship of the different factors was established in the process. It was found that policy, technology, e-commerce, floricultural parks, certification and infrastructure are the causes while cold chain, quality, pricing, promotion, market development, product development, entrepreneurs, and farming are the effects.

Employee welfare is a prerequisite element for the success and growth of any form of organisation. The provision of welfare facilities improves the relations among the employees and the management of an organisation. These provisions boost the competence levels and value of the employees. The balance between employees' quality of life at the workplace and home is vital, as employees are the pillars of any organisation. The central aim of any organisation in adopting the welfare schemes is to secure the workforce by providing a proper work environment and minimising its hazardous effect on the employees' work life. The basic purpose of employee welfare is to enrich the lives of employees and to keep them

Preface

INTRODUCTION

Managerial practices are diverged among companies and industries attributable to different beliefs and culture but are converged towards to goal of smart city: Increase the growth, profitability and productivity. Disruptive innovation aims at bombarding people's mind to execute management policies in a new and exciting way of thinking. Computational methods including data analytics and machine learning are the emerging trend of research in managerial practices. This encourages management to make scientifically proven changes for better management but discourages management from making decisions purely on whim.

The book explores the state of the art, new and emerging topics and future directions in management, disruptive innovation and inclusive economies in Asia, with a special focus on case studies and empirical studies.

The book aims to be an international platform to bring together academics, researchers, lecturers, decision makers, policy makers, and practitioners to share new theories, research findings, and case studies, to enhance understanding and collaboration in business, management and technology in Asia now and in the future.

TARGET AUDIENCE

Professors in academia, deans, heads of departments, director of masters, students (undergraduate and postgraduate level), policy makers, corporate heads of firms, senior general managers, managing directors, information technology directors and managers, libraries, etc. with interest in disruptive innovation and management in Asia.

CONTENTS OF THE BOOK

The book is formed by an outstanding collection of 19 chapters written by experts from several Asian countries. A few chapters are extended versions of best papers published in International Journal of Asian Business and Information Management in previous years. The rest of the chapters are new material from leading experts on disruptive innovation and management in Asia.

Chapter 1, titled "The Utilization of Web 2.0 for Knowledge Sharing: The Case Tertiary Education in Brunei Darussalam" (authors: Nurul Afiqah Nor Amin, Mohammad Nabil Almunawar, Amy Suliza Hasnan, and Nurul Nazira Besar Besar) examines "the current knowledge creation and sharing processes in Brunei Darussalam's tertiary education. It explains the preferences, benefits, and barriers to knowledge creation and sharing processes in tertiary education. A descriptive research method is used, in which a quantitative approach was selected to collect data. This study revealed that most of the respondents highly utilize emails, knowledge system that is provided by the host tertiary institutes and instant text messaging platforms. The benefits of using Web 2.0 due to its flexibility and ease of use. Due to these benefits, Web 2.0 technologies are used for communication, collaboration, and documentation purposes. However, there are some concerns in using Web 2.0 tools for knowledge sharing, mainly in term of privacy issues and reliability of information and knowledge shared due to its high risk of collaborators".

Chapter 2 titled "An Empirical Note on Health Vulnerability and Health Information Digital Divide: A Study of Indian Patients" (authors: Jaya Rani, Ajeya Jha, Jitendra Kumar, Samrat Kumar Mukherjee, and Saibal Kumar Saha) states that "it is important to know who are these people who actively search for online healthcare information and also those who are unable to do so. In all, 754 respondents participated in the survey. The variables selected from literature survey and exploratory study are Health Information igital Divide , Income, Having E-mail id, access to internet, geographical location, Education, family-type, age and gender. As the data is categorical the significance of difference has been calculated using Chi-square test. Later discriminant analysis was conducted to predict patients who make online health information search and the ones who do not. Using discriminant analysis 94.5 percent patients who make online health information search could be correctly predicted. Prediction is 99.7% for the patients who do not indulge in online health information search".

Chapter 3 titled "Decoding Cool in Indian Context: Meaning, Value, and Brands" (authors: Harsh V. Verma and Ekta Duggal) analyzes "the meaning of cool by tracing its genesis in theological texts and popular culture. Cool has connections with theological discourses of religions like Hinduism, Stoicism and Buddhism. Buddhism's fundamental tenet is cultivation of equanimity. Hinduism's sacred text Bhagavat Gita exhorts development of control over senses. Stoicism's core principle is 'Ataraxia' or indifference. Cool in popular culture originated from the sites of slavery by people who cultivated an attitude to calmness as means to survive in punishing conditions. This study assumed cool as a signifying system and sought to uncover the hidden meaning it stands for. It found that cool stands for four human facets. These are composure, paradox, good and cheeky".

Chapter 4 titled "Analysis of Social Value of TV Dramas Based on Audience Comments" (authors: Xinye Liu, Xiaotong Zhang, Tao Wang, Kun Cheng, Shangbing Jiao, Diqing Liu, Jia Su, and Xiaoqian Wang) studies "the social value of the TV drama Entrepreneurial Age through the mining of the audience's comments, so as to provide reference for the TV drama producers in topic selection, casting and script design. Design/methodology/approach: The research is based on a three-step approach including data crawling, two-dimension data tags and the random forest algorithm design. Findings: This paper finds that there are three factors related to demand of TV drama:1) the appearance and acting skill of actors; 2) the closeness between TV plays and real life; 3) whether the topic of TV plays has high attention. Value: Based on the big data of audience comments, this paper explores the factors that influence the number of TV plays. It provides an important reference for TV drama producers on how to design the plot of TV drama, how to choose actors and how to create topics".

Chapter 5 titled "Disruption in HR Through Inclusive Emotional Culture: Key to Sustainable Growth in Indian Businesses" (authors: Nitu Ghosh) suggests that "diversity in workforce has demanded the need to develop inclusive culture that suffices the needs, interests and emotions of all equally. High performing organizations have incorporated emotions in their mission, creating an emotional culture that accentuates emotional intelligence of their talent force rather than suppression of emotions. The chapter explains how and why organisational cultures are transforming into inclusive emotional cultures. Through an analysis of cases of Indian multinationals, the antecedents and outcomes of Inclusive emotional culture is described through Inclusive Emotional Culture Framework model. Views of HR managers is expressed through content analysis qualitative technique. The chapter provides key to sustainable development of Indian businesses amidst the doldrums of global business environment through inclusive emotional culture integrating values of long-term sustenance and development".

Chapter 6 is titled "Employee Experience Design: An Innovation for Sustainable Human Capital Management Practice" (authors: Nitu Ghosh and Urmila Itam). The philosophy of employee experience has been advocated as an effective tool to achieve the highest level of employee engagement in a business environment characterised by increased dynamism in workforce demography, psychography, competencies and expectations. Employee experience encapsulates the overall journey of an employee in an organisation from hire to exit, encompassing their experiences, what they feel, do and achieve from the company. It is a comprehensive view of the relationship shared and experienced between the individual and the organisation right from the application stage to stage they join the alumni group after their exit. The chapter explores this new philosophy in current Human Capital management practices that aims at revamping the HR practices and policies so as to create an experience that provides the highest level of engagement. Based on content analysis qualitative survey of various HR managers, opinions and new age thoughts in HRM practices has been highlighted in this chapter, giving an innovative dimension to HRM.

Chapter 7 titled "From Entertainment Device to IoT Terminal: Smart TV Helps Define the Future Living in Smart Home" (authors: Bu Zhong and Fan Yang) states that "following the trend of home appliances and devices being growingly interconnected into the Internet of Things (IoT) system, smart TV helps define the future of human living as part of the smart home IoT system. This research explores viewers' perceptions toward four emerging interactivity functions of smart TV: the interaction between viewers and TV functions, between viewers and friends, between viewers and programming, and viewers' interaction with products appearing in TV shows. The technology has been mature to provide the four interactivity functions, though they are yet to be adopted by all smart TV sets so far. Our research shows that the viewer's residence was a significant moderator in the preference for the interactivity functions. Viewers from developed regions enjoyed the four functions more than those living in underdeveloped regions. Social media habits and the power usage of information and communication technology are positively associated with the preference, while need for cognition, age and gender had little effect".

Chapter 8 titled "Is Construction of Timber-Focused Exchange Traded Fund Plausible in SAARC Countries With Reference to India?" (authors: Nivedita Mandal and Rituparna Das) proposes that "given the fact that timber is neither traded as commodity nor is any timber focused exchange-traded fund (ETF) available in the SAARC countries the objectives of this work are to examine the plausibility of making use of the wood resources of SAARC countries in the form of ETFs, e.g. timber-focused ETF, the nature of the legal and policy frames in these countries as implemented in timber business with reference to their suitability for opening ways to construction of timber-focused ETF and the theoretical plausibility of constructing a timber-focused ETF with reference to India. It is found that in comparison with other SAARC countries India has developed timber-based industries whose stocks are traded in the bourses and amenable to analysis for ETF construction purposes. While Bangladesh, Pakistan, Srilanka and Nepal have certain legal and institutional artifacts for preservation and maintenance of forests toward gradual but by and large unplanned industrialization of timber industry, Maldives and Bhutan have hardly any".

Chapter 9 titled "Learning Path Recommendation Method Based on Knowledge Map" (authors: Xiaodong Zhou, Yi Li, Liping Yuan, Gaofeng Ma, Xinyun Tan, Kaihua Zhang, Lijiao Gong, and Boxiang Jia) indicates that "with the development of society, many industries and professions are more comprehensive and intersecting. Different industries have their own requirements for students with comprehensive backgrounds. For graduates, they may not know the skills required for various occupations, or what kind of jobs and occupations they can take based on their existing knowledge and skills, even how to acquire these abilities after they know the requirements of the jobs they want. In this paper, the authors combine the existing method to predict hot jobs with the analysis of knowledge map, aiming to achieve accurate recommendation of learning path for those who want to find a job. This article will help job hunters gradually master skills, and ultimately achieve the goal of optimizing resource allocation and saving social resources".

Chapter 10 titled "Reverse Mentoring the Editing Edge in Management 4.0" (authors: Swati Sisodia and Neetima Agarwal) states that "Industry 4.0 is based on the implementation of a cyber-physical system, which includes sensors, networks, computers, offering digital enhancement and well-coordinated activities. This would create a great pool of all the workforce generations, having diverse experience, agility and different modes of working. Millennials would add more of machine learning and Generation X and Y would be the richest source of tacit and operational knowledge. Together, they would develop solutions for catering complex and networked production and aggressive logistic management, meeting the challenges of the Industry 4.0. However, the benefits of digitization and automation can be achieved, if the different generation of workforce collaborates, cooperate and postulate together in all the business processes. Reverse mentoring is a pristine concept and ingenious method to empower learning and encourage cross-generational connections. This chapter would elaborate on the advantage of reverse mentoring in crafting Industry 4.0 more acrobatic and quick moving".

Chapter 11 titled "Social Media and Increased Venture Creation Tendency With Innovative Ideas: The Case of Female Students in Asia" (authors: Syed Far Abid Hossain, Xu Shan, Mohammad Musa, and Preethu Rahman) focuses on "the contemporary role of social media in increased venture creation tendency along with innovative ideas. The key objective of this study is to discover the tendency of female students' innovativeness in venture creation in China. A random sampling method was used to conduct a survey in different universities in China to identify the scenario of innovativeness in venture creation. Findings from primary data collection indicated that the female students in China are highly involved with social media marketing with innovative ideas. As a result, apart from traditional marketing, society is involved with contemporary marketing where innovativeness with social media and smartphones are the key factors. Innovative ideas in venture creation may generate additional earning for people with low income in society. Future studies with mixed methodology and respondents who use different social media as a tool to innovate new venture may shed light on the undiscovered phenomenon of social media marketing in the context of the mobile phone.

Chapter 12 titled "Study of Technology-Based Innovations in Supply Chain Management Function of Indian Firms: Strategic Imperatives" (authors: Som Sekhar Bhattacharyya, Bibhash Laik, and Raunak Jaiswal) states that "supply Chain Management (SCM) has gained importance in recent years. Innovation and technological interventions in SCM would be required to remove inefficiencies. It has become imperative for firms to undertake new innovations in SCM to remain competitive. This paper focussed on physical and digital innovation in Indian market context in the context of SCM. The authors explored the strategic imperative of technology based SCM innovation by performing detailed literature review regarding new automated technological innovations in SCM to understand the new set of business gains to be incurred from SCM. The authors then carried out through a semi-structured questionnaire, in-depth personal interviews of the 24 SCM experts the study. Thematic content analysis was done. The main finding of the study was that physical innovation in SCM has occurred at a slower pace as compared to digital innovation. Digital innovation was perceived to be helping firms' more than physical innovation in SCM. The major challenge has been regarding the integration of new system with the existing SCM system".

Chapter 13 titled "The Role of Digital Economies in the Development and Growth in Asian Business Models" (authors: Kinza Yousfani and Farhana Khowaja) examines "the development of information and communication technology across Asian economies. Digital technology is the presentation of information in bits, that emphases on digital technology which covers all business, financial, social, cultural events that is supported by the web and other digital communication technologies. Technology has minimized the cost of storage, and flow of information. In the most recent 15 years, digital technology affects human lives and examines how digital technology changes economic activity. There are three principal segments e-business, e-business framework, and e-commerce. The digital economy is known as the web economy because of its dependence on the network. Modern technologies, cloud computing, mobile app, and social media influence the business landscape, reshaping the idea of work, boundaries of enterprises and the obligations of business pioneers. Thus, the digital economy features the opportunity for organizations and people to utilize to execute existing tasks on the PC frequently as before".

Chapter 14 titled "The Role of Total Quality Management, Entrepreneurial Orientation, Export Market Orientation, and Brand Orientation in Firm Export Performance: Resources and Export Performance of China Automobile" (authors: Muhammad Imran and Jawad Abbas) states that "during the last four years, the China automobile industry experienced a 49% drop in sale in the domestic and international market. This situation is attaining the attention of companies' owners and Government of China to explore the factors which could help them to overcome the issues relating to sale, reputation and brand image. Nonetheless, the investigation of company export performance factors in automobile sector of China has largely been ignored. However, the present study conducted the literature review on factors of firm export performance. Therefore, the conceptual framework has been developed to identify the factors of firm export performance such as total quality management (TQM), entrepreneurial orientation (EO), export market orientation (EMO), brand orientation (BO) based on resource base based theory. The future studies should validate the proposed research framework empirically in the context of China automobile industry".

Chapter 15 titled "Transformational Human Resource Management: Crafting Organizational Efficiency Management" (authors: Ikramul Hasan, Nazmul Islam, and Mohammad Ashraful Ferdous Chowdhury) affirms that "Human Resource Management plays an essential role for attaining organizational goals. Now-a-days practitioners, researchers and academicians around the world are emphasizing to transform and reshape the practice of human resource. However, very few research works have been done in the area of Transformational Human Resource Management (T-HRM). Hence, aim of this chapter is to propose an integrated framework of T-HRM and organizational efficiency. In light of that, first, this study has proposed potential factors of the T-HRM. Secondly, this study presented positive effect of the factors of T-HRM on organizational efficiency management. Concept of knowledge management has introduced as a potential mediator as well as ICT and organizational alignment has presented as a potential moderator of this study. Finally, knowledge of this study will provide better insights on T-HRM for ensuring organizational efficiency".

Chapter 16 titled "Understanding Inter-Relational Dynamics of Different Factors Impacting the Cymbidium Orchids Based Economy in Sikkim, India Using DEMATAL" (authors: Bibeth Sharma, Saibal Kumar Saha, and Ajeya Jha) states that "floriculture based economy is slowly increasing throughout the world. Orchids, which were once found only in the wild are now being widely cultivated and being sold in the market. In this study an attempt has been made to understand the inter-relational dynamics of different factors impacting the cymbidium orchids based economy in Sikkim, India using DEMATAL. A total of 14 factors were identified in this regard and based on expert opinion these factors were rated and analysed using DEMATAL. The cause and effect relationship of the different factors was established in the process. It was found that policy, technology, e-commerce, floricultural parks, certification and infrastructure are the causes while cold chain, quality, pricing, promotion, market development, product development, entrepreneurs, and farming are the effects".

Chapter 17 titled "Employee Welfare Provisions: An Administration Tool to Enhance Employee Efficiency" (authors: Chandra Sekhar Patro) proposes that "employee welfare is a prerequisite element for the success and growth of any form of organisation. The provision of welfare facilities improves the relations among the employees and the management of an organisation. These provisions boost the competence levels and value of the employees. The balance between employees' quality life at the workplace as well as home is vital, as they are the pillars of any organisation. The central aim of any organisation in adopting the welfare schemes is to secure the workforce by providing a proper work environment and minimising its hazardous effect on the employees work life. The basic propose of employee welfare is to enrich the lives of employees and to keep them happy and conducted. The provision of employees' welfare may be regarded as a wise investment as these would bring a profitable return in the form of greater efficiency. The chapter focuses on determining the various employee welfare provisions adopted by different private organisations and its influence on the employee's satisfaction and effectiveness".

Chapter 18 titled "Sustainoprenuership: A New World Order" (authors: Jawad Iqbal and Shakeela Kousar) states that "the current millennium is characterize with several global problems including unemployment, poverty, hunger, starvation, social exclusion and environmental degradation. These social, economic and environmental trends require the changes in current world order to provide such global mechanism where basic needs of each person will be satisfied and each person will have equal rights to resources and opportunity to utilize his full human potential. Entrepreneurship is an important driver that can transform a society towards a more sustainable future. Sustainopreneurship has emerged as a mutual product of sustainability and entrepreneurship. Nations have included the entrepreneurship in their developmental agenda for achieving the sustainable development goals. The purpose of this chapter is to briefly discuss the concept of sustainable entrepreneurship, its definition, evolution of the concept, transition from sustainable development to Sustainopreneurship as a mechanism for disruptive innovations".

Finally the last chapter of the book, Chapter 19 titled "Crime Hotspot Prediction Using Big Data in China" (authors: Chunfa Xu, Xiaoyang Hu, Anqi Yang, Yimin Zhang, Cailing Zhang, Yufei Xia and Yanan Cao) explores "the utilization of big data and machine learning to predict crime is feasible in China. The researchers introduce five new machine learning algorithms into the field of crime prediction, and compare them with four methods widely used in previous research. Using a weekly dataset in 213 street-level cells of Shanghai from April 2017 to March 2018, the researchers find new methods work better in predicting whether a specific cell will be a crime hotspot in next week. Five among nine methods can predict crime with more than 90 percent accuracy. These findings provide a scientific reference for urban safety protection. The research adds some significant evidence to a theoretical literature emphasizing that big data could predict crime".

We would like to say thank you to the Editorial Advisory Board of the book and the pool of reviewers for their time and effort providing valuable advice and review comments for authors of chapters, respectively. Additionally, we would like to acknowledge the continuous support and help of IGI-Global staff, from Jan Travers to Courtney Tychinski, for their encouragement during the development of the book.

Patricia Ordóñez de Pablos
The University of Oviedo, Spain

Xi Zhang
Tianjin University, China

Kwok Tai Chui
The Open University of Hong Kong, Hong Kong

Chapter 1
The Utilization of Web 2.0 for Knowledge Sharing:
The Case of Tertiary Education in Brunei Darussalam

Nurul Afiqah Nor Amin
Universiti Brunei Darussalam, Brunei

Mohammad Nabil Almunawar
Universiti Brunei Darussalam, Brunei

Amy Suliza Hasnan
Universiti Brunei Darussalam, Brunei

Nurul Nazira Besar
Universiti Brunei Darussalam, Brunei

ABSTRACT

This chapter assesses the current knowledge creation and sharing processes in Brunei Darussalam's tertiary education. The chapter explains the preferences, benefits, and barriers to knowledge creation and sharing processes in tertiary education. A descriptive research method is used, in which a quantitative approach was selected to collect data. This study revealed that most of the respondents highly utilize emails, learning management system, knowledge system that is provided by the host tertiary institutes and instant text messaging platforms. The benefits of using Web 2.0 are its flexibility and ease of use. Due to these benefits, knowledge sharing utilizing Web 2.0 technologies are used for communication, collaboration, and documentation purposes. However, there are some concerns in using Web 2.0 tools for knowledge sharing, mainly in term of privacy issues and reliability of information and knowledge shared due to its high risk of collaborators.

DOI: 10.4018/978-1-7998-0357-7.ch001

INTRODUCTION

Information and Communication Technology (ICT), the Internet, and its Web 2.0 technologies have transformed a variety of fields (Kalantzis-Cope, 2011). Especially in a learning environment, it creates a new experience for many people of all ages, no matter their whereabouts (Pieri & Diamantini, 2014). Web 2.0 was first coined by DiNucci (1999) and popularized by Tim O'Reilly and Dale Dougherty in 2004 as a second generation internet service (O'Reilly, 2007). Web 2.0 comprises of tools or applications which allow individual and collective publishing and sharing multimedia information in the form of texts, images, audios, and videos. Also, it includes the formation and maintenance of online social networks (Bennett, Bishop, Dalgarno, Waycott, & Kennedy, 2012). Examples of Web 2.0 technologies are Wiki, blogs, Google Docs, Google+, Instagram, Facebook, Twitters, and instant text messaging applications such as WhatsApp, Facebook Messenger, Telegram, WeChat, etc.

Web 2.0 technologies are prevalent and globally adopted by many people, and these technologies are now becoming a part of the everyday lives of many people, including in tertiary education institutions. As the learning environment in tertiary education involves several individuals such as lecturers, professors, students, and administrations, Web 2.0 has become a widespread and knowledge sharing tools. Despite its popularity, some concerns need to be adequately addressed. Our main objective is to answer the following research question: how Web 2.0 influence activities in sharing and knowledge aspect? With this objective, this chapter discusses the preferences of Web 2.0 technologies among the individuals in tertiary education, its benefits, and barriers, using Brunei Darussalam tertiary education as the research setting. The rest of this chapter is organized as follows. The next section is the literature review followed by the methodology. We then discuss the findings. The last parts of the chapter are the conclusion, limitation of the study, and recommendation for future studies.

LITERATURE REVIEW

The literature review of this study encompassed of four parts which started with the definition of knowledge, knowledge creation and sharing and its role in tertiary education, the benefits and barriers of Web 2.0 as knowledge creation and sharing platform, examples of Web 2.0 technologies and the ICT initiatives in Brunei Darussalam education.

Knowledge Creation and Sharing

"Knowledge creation and sharing are defined as generating, storing and sharing of knowledge for the benefit of the organization and its individuals to ensure comprehensive and understandable management initiatives and procedures in the organization" (Bell, 2001, p. 49). Knowledge can be in terms of information, skill, experience, and intellect. In general, it can be classified into two forms, explicit and tacit knowledge. Explicit knowledge is the knowledge that is codified, and that can be stored digitally. Tacit knowledge is information that embeds in a person such as experiences, emotions, or skills (Choi & Lee, 2002; Huang & Liaw, 2004). The continuous knowledge creation between these modes affects and creates knowledge. The transformation processes depicted in Figure 1 are explained below.

1. Externalization is a process to justify tacit knowledge and express its concept and formal models. Therefore it involves the conversion of tacit into explicit knowledge. Examples of this process are commentary, narration, or write instruction manuals.
2. The combination is a process that enables explicit knowledge collected from inside or outside the organization to be combined, edited, integrated processed to form new knowledge and disseminated it to all individuals and members using communication networks. For example, course materials and syllabus are shared among teachers and students.
3. Internalization is a process when knowledge is created and shared through an online learning community and delivers by an individual. This involves internalizing explicit knowledge into tacit knowledge. Example of internalization is reading course materials and having a test online.
4. Socialization is a process that involves social and cultural activities, such as discussion and informal meetings. Therefore socialization is sharing experiences from tacit knowledge to tacit knowledge.

As illustrated in Figure 1, the knowledge creation process can be perceived as a continuous and cumulative learning process (Bhatt, 2000). With this process, it leads to the improvement of skill development and academic performance (Zaqout & Abbas, 2012). Most importantly, it leads to the creation of new knowledge and ideas (ibid). Knowledge was once created and shared using paper-based (hardcopies) (Myrberg & Wiberg, 2015; Noyes & Garland, 2008). Although paper-based knowledge is still relevant, some studies (such as Myrberg & Wiberg, 2015; Noyes & Garland, 2008) stated that cognitive workload and memory measures are replaced by the use of a computer. A knowledge-based organization, especially in the education sector, has made learning through knowledge creation and sharing, become a crucial factor as it involves various stakeholders such as students, administration, and academics (Marić, 2013). In line with this statement, Charband & Jafari Navimipour (2018) mentioned that knowledge sharing is particularly crucial in educational and learning institutions for its role to provide, manage and blend knowledge which in turn help the education sector to increase competitive advantage, enhance innovation and optimization of learning effect. Thus, ICT has become an essential role in supporting and facilitating

Figure 1. Knowledge creation and sharing process framework of Huang and Liaw (2004)

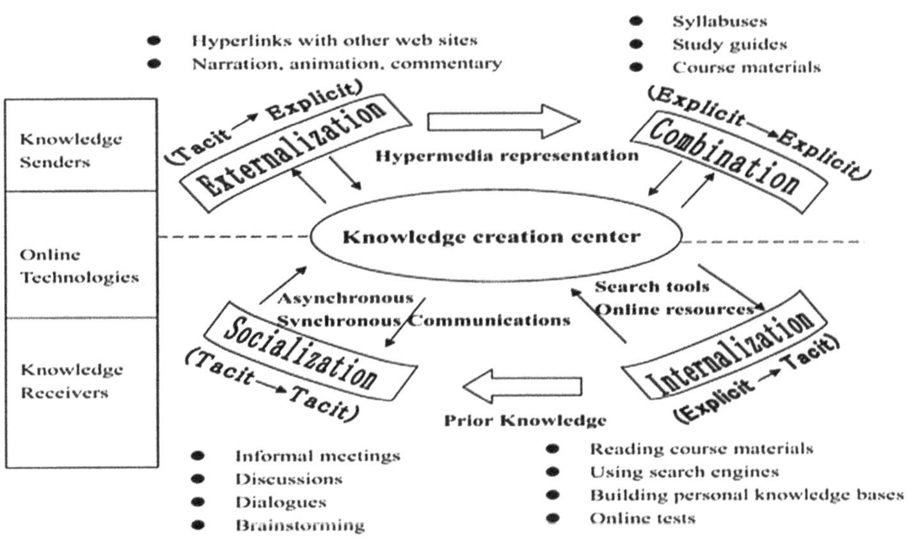

knowledge creation and sharing in organizations to which it is easier to acquire, store, or disseminate knowledge (López, Peón, & Ordás, 2009). Individual, organizational, and technology that exist affect knowledge sharing practice (Abdelwhab Ali, Panneer Selvam, Paris, & Gunasekaran, 2019). Therefore, it is vital to leverage ICT in a virtual knowledge-sharing community to facilitate the success of knowledge initiative (Ardichvili, Page, & Wentling, 2003) in terms of knowledge, information and ideas sharing, exchange and communication (Kalema, Motsi, & Motjolopane, 2017). One of the technologies used to cater to these activities is Web 2.0. The detail explanations of Web 2.0 as a form of knowledge creation and sharing platform are explained in the next section.

Web 2.0 as a Knowledge Sharing Platform

Web 2.0 is a new concept of the web which allows easy creation of websites that improve the sharing of knowledge. According to Patrick & Dotsika (2007), social software is seen as one of the components of Web 2.0. Social software of Web 2.0 is also more collaborative, interactive, and dynamic. Boutin (2006) described Web 2.0 as "a platform for interacting with content, reflecting a set of online tools that enables the aggregation and remixing of content, through interfaces that combine content from different sources in a manner not possible through a single domain." Also, Macaskill & Owen (2006) added that Web 2.0 allows users to gain access, contribute, tag, annotate and bookmark web-mediated content in various formats such as text, video, audio, pictures, and graphs. Therefore, Web 2.0 technologies enable content sharing and creation, collaboration, and communication among users all over the world.

Benefits of Using Web 2.0

The existence of Web 2.0 technologies has inspired many individuals in tertiary education to consider using Web 2.0 as a knowledge sharing platform. It is no wonder the Web 2.0 use in educational institutions has been surprisingly increasing. As stated by Hargadon (2008), Web 2.0 is the future of education. Also, Web 2.0 technologies have an emerging role in transforming teaching and learning (Levine & Alexander, 2008). Other than as educational tools, Web 2.0 can be used as multi-purpose tools (Virkus & Bamigbola, 2011). For instance, Kulakli & Mahony (2014) has stated the growing interests of Web 2.0 technologies help education institutions to adapt Web 2.0 as supplementary communication and collaboration medium by promoting resource sharing among individuals.

As stated above, collaboration is one of the benefits of using Web 2.0, which enables knowledge to be distributed efficiently and effectively. As such, Bennett et al., (2012) stated the Web 2.0 technologies could develop students' skills that are gained through participation; user-generated content, and collaboration. In detail, research done by Ward, Moule, & Lockyer (2008) showed that the majority of students have a positive experience with collaborative writings by commenting and editing group work. A similar finding by Zeeng, Robbie, Adams, & Hutchison (2009) has found that Web 2.0 enables a global classroom to work independently on a collaborative activity despite different time zones. Due to its collaboration factor, knowledge sharing using the platform promotes knowledge sharing culture that fosters effective communication and increases involvement in research activities (Naeem, 2019). It has also been suggested that Web 2.0 technologies are widely used to improve communication, outreach services, and knowledge sharing in educational institutions (Balubaid, 2013).

Another significant benefit that can be gained from the adoption of Web 2.0 is the flexibility and ease of use of technology. According to Echeng, Usoro, Echeng, & Usoro (2016), Web 2.0 used for learning activities are reasonably accessible and do not require extensive training effort to serve as motivation to users. Similarly, An, Aworuwa, Ballard, & Williams (2009) mentioned that it eliminates time constraints in the learning environment. It is also easier to find information for various study and job-related tasks using Web 2.0, as it is easy to learn and use where minimal technical problems were encountered (Popescu, 2010). Similar research (Singh, 2018; Tyagi, 2012) explained that Web 2.0 has up-to-date information on a related topic of interest. Apart from benefitting students and lecturers, the adoption of Web 2.0 can improve the quality of library services. For instance, it provides a more interactive and user-oriented reference services and, shared news and promoted library services in relatively less time (Tandi Lwoga, 2014).

Studies have also demonstrated that the utilization of Web 2.0 is associated with constructive behavior. For example, it enhances students' positive learning attitude and well-being development (Tian, Yu, Vogel, & Kwok, 2011). Web 2.0 can also help build a sense of community by increasing interaction and communication among the instructor, students, and other people (An et al., 2009). London (2012) stated that Web 2.0 technologies could promote generative learning processes in which team members are open to new ideas, explore new ways of interacting, and apply their learning. Ultimately, Almekhlafi & Abulibdeh (2018) stated that Web 2.0 used in the education sector could improve students' performance. Hence, Web 2.0 technologies serve as a platform that encourages individuals to share their thoughts and experiences while at the same time, it leverages behavior improvements at all levels.

Barriers of Using Web 2.0

There are several barriers associated with Web 2.0 usage in organizations. Organizations face a challenging task of deciding the type of ICT solutions to deploy in support of their knowledge management initiatives. This is because, according to Shannon & Weaver (1949), knowledge sharing in information technology sometimes undergoes problems in its technical, semantic, and effectiveness aspects. In the tertiary education context, Sandnes, Jian, Hagen, & Talberg (2007) reported that submitting coursework through the learning management system is complicated.

Moreover, according to Pedro, Soares, Matos, & Santos (2008), the learning management platform mainly concerns technical features such as slow access, limitations in space available for uploading files and difficulties in managing the platform and course editing. When the tools used are new and still underdeveloped (Bennet et al., 2012), it can lead to people in the organization not willing to engage with the tool due to resistance in learning it (Ward et al., 2008). Hence, the purpose of Web 2.0 needs to be embraced as a means of innovation, open communication, collaboration, and user-generated content. If they can achieve this, it may contribute to a friendly working culture as well as to be more open to face any changes (Sivarajah, Irani, & Weerakkody, 2015).

Adding to that, in some organizations, in order to be in line with the implementation of technologies that incorporated Web 2.0, existing staff are required to undergo education and training. Lack of professional training that includes the use of Web 2.0 tools such as microblogging, creation and sharing an audio file, and photo sharing and content development, an audio file, and photo sharing and content development may hinder academicians from using Web 2.0 (Almekhlafi & Abulibdeh, 2018). However, undergoing the training process can be time-consuming (An et al., 2009). Similar finding by Tandi Lwoga (2014) stated that the implementation of Web 2.0 technologies is related to infrastructure readiness and a shortage of trained staff. This can lead to several problems, such as misleading perceptions, tangents, and poor decisions (London, 2012). Hence, readiness is considered as a barrier for Web 2.0 usages as it takes time away from learning subject matter content, particularly in the education institutions.

The open nature of Web 2.0 is another main issue that brings out significant challenges over the information shared. It is considered as a security threat in any organizations where privacy and confidentiality are prone to be exposed or leaked. A study done by Abdelwhab Ali et al., (2019) mentioned that an individual is more confident to share knowledge with their colleagues when they feel that it is expected in the organization to have knowledge sharing behavior. However, in educational institutions, this feature of Web 2.0 cause uneasiness to many students, and this has led to their reluctance to participate in class activities that utilize Web 2.0 technologies. This statement has been proved in a study done by Smith and Lambert (2014) that stated there is a possibility of faculty members spying or breaching on their "virtual" identity, hence, affecting student-staff relationships boundary.

Similarly, students tend to inherent risks of reducing the education structure since it is hard to balance between the powers of academics and students (Ward et al., 2008). This concept is further addressed by Bateman and Willems (2012) whereby the blurring of professional and personal relationships based on the notion of ''friending" on social media. Furthermore, students are afraid of their personal information to be misused (Singh, 2018)

There are also concerns over reliability, accuracy, and authority that arise from utilizing Web 2.0. The search engines, such as Google also make students have uncoordinated and often irrelevant and bring to an "information overload" (Pieri & Diamantini, 2014). This also can cause issues such as authorship whereby Web 2.0 may cause claimed of work by anyone in intellectual contributions (Noss et al., 2008). This indicates thatthe more information is shown, the more information must be managed.

Example of Web 2.0 Technologies

Web 2.0 technologies allow users to create, describe, post, search, collaborate, share, and communicate online content in the multimedia information. The following are some examples of popular Web 2.0 technologies.

E-mails

E-mail and mailing lists is a typology of Web 2.0 technologies for communication (Fuchs, 2008). Features of the email include replying, sending, or forwarding the email to other users (Baines, 1992). Some examples of emails are Gmail, Hotmail, Mail, Yahoo mail. Research by Burton (1994) provided an in-depth analysis of the literature regarding the advantages of using email. The author stated that email enables one to distribute information that can be sent across vast distances, regardless of the time. In the academic context, the use of email in student-lecturer interactions has become increasingly popular over the last 20 years. Perhaps, this is because email enables users in academia to send and receive academic information, exchange research idea, and generally facilitate cheap and effective communication (Akinseinde & Adomi, 2004). Sun (2011) stated that email is the most used form of communication by students. In the same vein, McNeill, Ming Diao, & Gosper (2011) mentioned that in their ethnographic research, students use emails for both communication and collaboration purposes. Also, email increases the interpersonal relationship between students and lecturers (Sheer & Fung, 2007). This has make e-mail as an efficient platform for knowledge sharing (Kalema et al., 2017).

Google+

Google+ is a multilingual social networking and identity tool owned and operated by Google Inc. Google+ as described by Google is a social layer which consisted of not just a single site but rather an all-encompassing layer which covers many of its online properties such as Google Drive, Gmail, Google Play, Google maps, etc. With Google+, users can selectively share their network with specific circles of people or groups. The ability to communicate with groups of people at a time has great possibilities for professional associations and informal groups, hence ensuring that the platform is secure and provide privacy and smooth interaction (Elson Anderson & Still, 2011). One of its online property, Google Docs foster collaboration among individuals, to which it enables different users to edit a document. Thus, Google Docs can be used as a platform to communicate in real time where sessions and paperwork can be shared (Carlisle, Hays, Pribesh, & Wood, 2017). This can improve overall writing quality by adding informational elements to support the organization of their writing, and revisions were mostly made at the sentence level (Semeraro & Moore, 2016). Google also remains as an essential research tool and offers access to works in the public domain in a variety of languages and fields of study (Leonardo, 2012). This has made Google scholar to help researchers in their current and future research and also to expand their scholarly networking for collaborations (Zientek, Werner, Campuzano, & Nimon, 2018).

Facebook

Facebook is created by Mark Zuckerberg in February 2004 as a social networking website for college students to stay in touch with each other. Years after, the website became open to public use, which leads to an increase in users using the website for academic purposes (Kosik, 2007). Although Facebook is not explicitly created for an educational platform, it offers opportunities to collaborate and learn outside of the classroom (Towner & Lego Muñoz, 2011). This notion is supported by González-Ramírez, Gascó, & Llopis Taverner (2015) that stated students that are registered on the subject's Facebook web page are satisfied with their experience in using Facebook as a learning platform. Perhaps, this is because Tyagi (2012) showed that most of the faculty members used Facebook to engage in collaborative and cooperative work. As such, the platform allows its users to create groups for academic courses in which users can discuss and post pictures, videos, and notes. This enables users to reflect upon prior knowledge, capture new experiences, and provide feedback (Chan, Chu, Lee, Chan, & Leung, 2013). Other than that, Tiruwa, Yadav, and Suri (2018) stated that the sharing capability of Facebook helps in self-learning and going through the shared material makes it an good for learning and creating a better team-building environment. Similarly, Bateman and Willems (2012) mentioned that Facebook has the potential for geographically dispersed learners such as peer teaching and resource sharing. Due to this, research by Singh (2018) indicated that social networking sites have received prominent responses among students for its up-to-date information, academic and research, and socializing. Furthermore, Facebook is also deemed to be acceptable in higher education (Costa, Alvelos, & Teixeira, 2019)

YouTube

YouTube is a video sharing website created by former PayPal employees in February 2005. With YouTube, users can upload, view, and share videos. Users can also leave comments in other videos. Academically, YouTube has served as a useful online tool to share and demonstrate video for learning purposes. Cem

(2009) also agreed that for academic purposes, YouTube served as an excellent platform for knowledge sharing between academic departments and students where information about an academic department such as location, module registration method of a university or college can be shared in YouTube. Lectures can also be delivered to students by uploading lectures videos for students to access. Other than that, students are aware of the importance of video sharing platforms in supporting their learning activities (Almobarraz, 2018; Costa et al., 2019). It also increases their knowledge when they watch videos that are linked to the subjects that they learn (Almobarraz, 2018). Through this, it harnesses the collaborative spirit between contributors and consumers of educational videos in a way that will enhance participatory culture and self-directed learning (Madden, Ruthven, & McMenemy, 2013). Ultimately, this will benefit educational institutions in terms of educational content they need to provide as they move towards the direction of building collaborative digital learning platforms to enable users to participate in lifelong self-learning and education.

Twitter

Twitter is an online news, and social networking service with users post and interact with message restricted to 140 characters. User access Twitter through its website interface, SMS, or a mobile device app. Despite its limitation on characters, it has been used as a training tool for higher education students. As stated by West & Kimmons (2019), Twitter is a microblogging platform that can support teaching and learn, such as professional development, lifelong learning, and informal educational activities. This has been proved by Hull and Dodd (2017) that reported positive learning in using Twitter in the classroom. As such, it is used to improve students reflective, critical judgment, and information selection skills and at the same time, learning to save time for specific tasks (Ricoy & Feliz, 2016). Jones (2011) mentioned that Twitter helped students to improve their spoken and written arguments, and deepen their understanding of challenging texts. Twitter also has a hashtag function that enables a user to click a topic of discussion. Through this, both the lecturers and students can feel more bonded to the topic of discussion, and by default, more connected to the course and its instructor (Chamberlin & Lehmann, 2011). Also, lecturers can tweet extra resources to their students while others are sharing their courses' backchannels with the entire world (ibid). Twitter has also become a teaching-supporting tool in avoiding time-consuming difficulties in face to face learning academic activities (Cohen & Duchan, 2012). This means that users can receive immediate and frequent course information, asking questions to mentor, update course assignments and to share helpful information from outside the textbook with their classmates and mentors (Bista, 2015). Interestingly, Twitter is also related to providing support for students during a problematic module whereby it facilitated communication, relieving anxieties, and raised morale (Hennessy, Kirkpatrick, Smith, & Border, 2016).

Instant Text Messages Platforms Such as WhatsApp

Instant text messaging technology is a type of online chat that offers real-time text transmission. As stated previously, some example of instant text messaging platforms is WhatsApp, Facebook Messenger, WeChat, etc. These instant text messaging platforms are an application for smartphones, but can also be accessible through websites. It enables the user to make voice calls, video calls, send text messages, images, GIF, videos, documents, user location, and audio files to other users. Using instant text messaging has been proved to benefit the learning platform. As such, interactivity, knowledge sharing, sense of presence, collaboration, and ubiquity are among the benefits that can be gained from using these platforms for education (Klein, Junior, Mattiello da Silva, Barbosa, & Baldasso, 2018). It also creates social bonding between students and lecturers by allowing students to clarify the ambiguity of concepts learned in class and create a sense of connection with their lecturers which leads to better student engagement (Nkhoma et al., 2018). It also helps the student to overcome their shyness and language barriers. It has been suggested by Sundararajan, Sheehan, and Gilbert (2013) that the students did a better job with instant messaging. With this support, one example of an instant text messaging platform, WhatsApp, has shown that its role in facilitating knowledge sharing in education has increased (Barhoumi, 2015). WhatsApp is used after school hours to enable quick transference of links to study materials (Bouhnik & Deshen, 2014). Students also use WhatsApp to practice language and have opportunities for them to relate their opinions to others (Fathy, Said, & Fattah, 2015). The author further stated that WhatsApp enables students-teachers comprehensive relationship that makes students learn better. Similar finding by Sayan (2016) has stated that using WhatsApp offers external activity around student-centered learning for exam preparation, hence has a positive relationship with students achievement.

Learning Management Systems (LMS)

LMS is a technological system that creates online courses. The current LMS incorporate Web 2.0 technologies (Holmes & Prieto-Rodriguez, 2018) which can be used as a means to access content, evaluate and give feedback to students while at the same time promoting teacher-student and student-student communication (Porter, 2013) via message boards, forums, chats, video-conferences (Sánchez & Hueros, 2010). Thus, LMS has become an essential technological development in higher education (ibid). This shows that LMS is a basic and easy to use the platform (Dash, 2019). LMS has proved to have many benefits, such as providing more accessible communication, collaboration, and transfer of information (Al-Busaidi & Al-Shihi, 2012). Furthermore, LMS provides an online education whereby it enables the control and the monitoring of those interactions through the tracking of postings and emails with useful feedback and monitor performance (Wieser & Seeler, 2018). For this reason, LMS has become the most accepted technology for education purposes (Costa et al., 2019).

ICT Initiatives in Brunei Darussalam Education

The first basic structure of the education system in Brunei Darussalam was mandated in the first National Development Plan (1954-1959) encompassed of primary education, lower secondary and upper secondary. A more comprehensive education that involved tertiary education developed with the implemented of Bilingual Education Policy in 1985 whereby education started with primary level, secondary level, pre-university and post-secondary training or vocational and technical education. Universiti Brunei Darussalam, Universiti Technology Brunei, technical and engineering schools, Wasan Vocational School, Universiti Islam Sultan Sharif Ali, Brunei Polytechnic are the tertiary educational institutions are the established institutions that provide post-secondary and tertiary education in Brunei.

Computer Studies and Design and Technology were introduced at the secondary level of education in 1993 and 2002, respectively. Some educational services were even upgraded by incorporating teaching and learning of ICT across the curriculum by providing e-education such as Edunet, E-learning, Education Information, and Digital Library and Human Capacity Building under the Ministry of Education (Oxford Business Group, 2011). It was stated by Joia (2000) that the usage of ICT in teaching and learning activities would produce individuals that can be counted as intellectual capacity for the society. This is in line with research done by Lubis, Yunus, Lampoh, & Ishak (2010) that stated modern technology is encouraged as it helps towards the development of positive thinking and can innovate and to trigger the driver for self-improvement for Brunei teachers and students.

Apart from that, the benefit of integrating ICT also includes the knowledge and familiarity of new technologies as an important dimension of employability in the information society, and the usage of ICT will improve the quality of educational experience by providing a favorable environment for learning (Seyal, 2012). Due to the aid of ICT can provide, tertiary educations have looked to a variety of solutions for managing the learning, by assimilating learning management systems within their educational program. As such, the Universiti Brunei Darussalam uses a learning management system called Canvas which allows students to send and receive documents that include assignments and test paper for their courses, and mainly to track and manage their courses online. Another example is Polytechnic, and University Technology Brunei is both using Moodle that serves as an open source platform for educators to develop and manage courses online and also for all kind of collaborative activities.

METHODOLOGY

The descriptive research method was chosen for this research as it can provide information about the attitudes, behavior, or other characteristics of a particular group (Jackson, 2009). In descriptive studies, there are two general methods of conducting research which is qualitative or quantitative. For this research, the quantitative approach was selected in order to get precise data on the use of the Web 2.0-based knowledge sharing platform in tertiary education. Questionnaires are employed as the research instrument which aims at gathering primary source data from numbers of samples selected for the study. Data gathered from the survey were analyzed descriptively using Microsoft Excel.

Literature Survey

A literature survey is a method by which it provides guides or helps the researcher to define or identify a problem. The literature survey uses secondary data from books, journals, and online journals in order to conduct a literature review.

Population, Sampling, and Sample Design

A sampling method is a method of choosing a group of peoples from the entire target population in order to obtain the required information. For this research, a non-probability sampling method was invoked as it is suitable for a widely dispersed population, such as tertiary level students. The target population for this research is limited to tertiary level students and lecturers. For the sample design, a questionnaire was designed and distributed by using an online survey through www.google.com/forms. By using the online survey method, data can be collected easily and on time.

Sample Size

Sample size will determine the clarity and consistency of research; hence, larger sample size will be more representative of a population. As this research is preliminary research, a sample size of 100 tertiary students, education officers and lecturers were used. The main aim of conducting the survey is to examine the sharing and knowledge aspects of using Web 2.0 between students and lecturers in tertiary education.

FINDINGS AND DISCUSSIONS

Respondents' Demographic Information

There were 100 questionnaires distributed, but only 60 proper responses were received (60% response rate). Five (5) types of demographic information have been gathered from respondents. Results of demographic information collected in this research are shown in Table 1.

Most of the respondents are female (77%), while the remaining 23% of the respondents were male. There are 85% of the respondents in the range of age between 21-29, 10% are between 15- 20 years old, 5% are between 30 – 39 years old and none of the respondents are 40 years old above.

In term of education, qualifications of the respondents are 64% bachelor degree holder, 28% are a master degree holder, 8% are a higher national diploma (HND) and diploma holder, and none of the respondents is a Ph.D. holder. We managed to get 92% respondents from students, 3% from the lecturer, and 5% from officers that are working in tertiary education. Most of the respondents were from Universiti Brunei Darussalam (UBD), and the rest were from other tertiary education in Brunei Darussalam. Table 1 summarized the demography data of the respondents.

Knowledge Creation and Sharing Using Web 2.0

Based on our analysis, most respondents agree that Web 2.0 is used to update and review information, while at the same time, facilitating communication between lecturers and students. This also shows that the respondents agree that the current Web 2.0 system supports information and knowledge sharing. However, 50% of the respondents are neither feeling satisfied or unsatisfied with the current quality use of Web 2.0 provided by their institutions. This finding was supported by López et al. (2009) and Ardichvili et al. (2003) that stated the need for leveraging knowledge management in an organization is an important mechanism to facilitate and support interaction among individuals in organizations.

More than 40% of respondents agree with sharing skills and information with their fellow students' colleague and lecturers in the same faculty and other faculties. While less than 10% disagree with this point, majority of the respondents involved (more than 50%) in this survey felt neutral about this point. This has shown that knowledge, such as information and skills, are indeed included as knowledge creation and sharing. In terms of its usage, there were more than 30% of the respondents who agree that Web 2.0 distributes information across faculties. This result support Huang and Liaw (2004), whereby in the process of knowledge sharing, knowledge is internalized when knowledge is created and shared through an online learning community and delivers by an individual. The concept of combination is also used to enable knowledge to be spread to all individual, whether it is from students to students, students to lecturers, or vice versa using communication networks. This agrees with research done by López et al.

Table 1. Summary of demographic information (n=60)

No	Demographic	Frequency	Percentage (%)
1	*Gender*		
	Female	46	23
	Male	14	77
2	*Highest Qualification*		
	PhD	0	0
	Master Degree	17	28
	Degree	38	64
	HND, Diploma	4	8
3	*Current Position*		
	Student	54	92
	Lecturer	2	3
	Officer	3	5
4	*College/ University/ Institution*		
	Technical School	0	0
	Vocational School	0	0
	ITB	3	5
	UBD	49	82
	UNISSA	2	3
	Polytechnic	0	0
	IGS	3	5
	Laksamana College	0	0
	Cosmopolitian College	0	0
	Other	3	5
5	*Age Group*		
	15 - 20	6	5
	21 - 29	51	85
	30 - 39	3	10
	40 above	0	0

(2009) that stated technology had become an important role in acquiring and disseminating knowledge. Also, most of the respondents (less than 70%) agree than Web 2.0 is used to facilitate communication. This agrees with previous research (such as Balubaid, 2013; Bennett et al., 2012; Kulakli & Mahony, 2014; Naeem, 2019) to which Web 2.0 is used as a supplementary communication and collaboration medium in the tertiary education.

Figure 2. Knowledge creation and sharing using Web 2.0

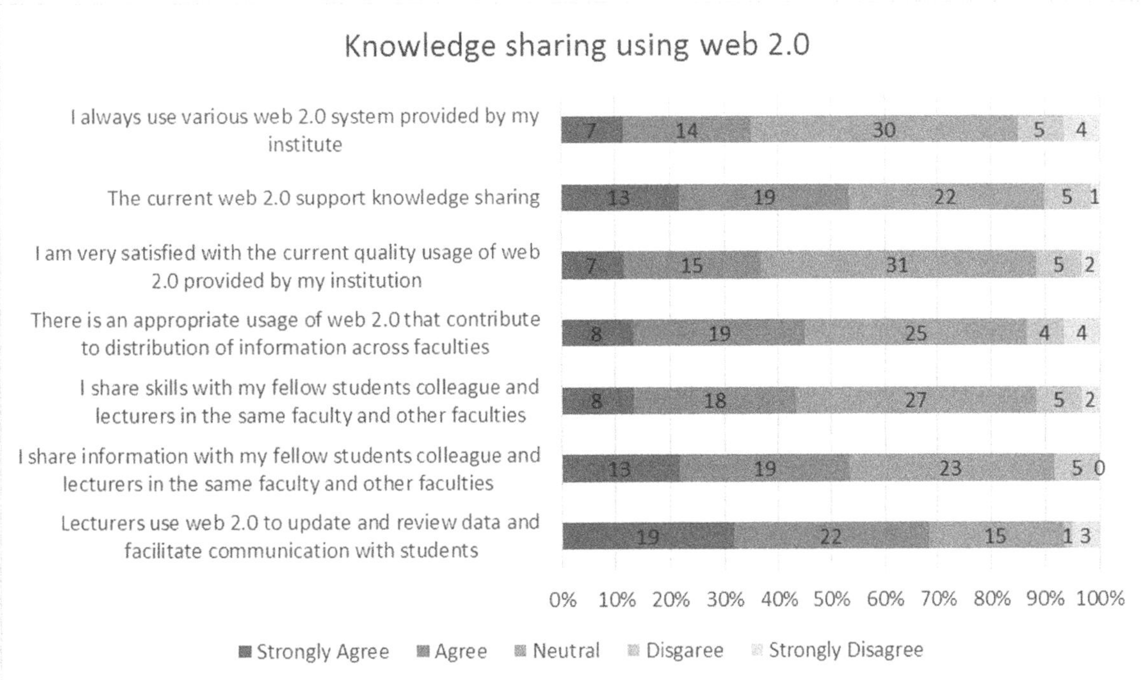

Web 2.0 Knowledge Sharing Platform Preferences

Two questions regarding knowledge sharing platform were asked to the respondents in order to find out which sharing activities and platform were mostly preferred by them. These are shown and summarized in Figure 3 and Figure 4.

Figure 3 shows the preferred knowledge sharing activities ranked by the respondents using a Likert scale of 1 (strongly disagree) to 5 (strongly agree). Based on the graphs shown above, most respondents preferred to use Web 2.0 as a place to document knowledge and practices, exchange of work, forward useful information, and to participate in an online discussion.

Most people would prefer to use Web 2.0 as a platform to document knowledge and practices, exchange of work, and forward useful information. This is highlighted by several studies to which Web 2.0 technologies such as Facebook can be used to share materials (Tiruwa, Yadav, and Suri, 2018) and collaborative and cooperative work (Tyagi, 2012), Youtube for sharing videos that are related to the subjects that they learn (Almobarraz, 2018) and Google for research purposes (Zientek et al., 2018) and collaboration of working paper (Carlisle et al., 2017; Semeraro & Moore, 2016). The ease to forward useful information are also highlighted by various studies such as WhatsApp for its interactivity and knowledge sharing such as links of study materials (Bouhnik & Deshen, 2014; Klein et al., 2018) and Twitter to 'tweet' extra resources (Chamberlin & Lehmann, 2011).

Figure 4 summarizes the results of the response against the questions about the preferred online applications according to a Likert scale of 1 (strongly disagree) to 5 (strongly agree). The questions are directly and indirectly inter-relate to Web 2.0 technologies specifically Web board, emails, instant messaging applications such as WhatsApp, Social Networking Sites such as Facebook, Twitter and In-

Figure 3. Preferred knowledge creation and sharing activities

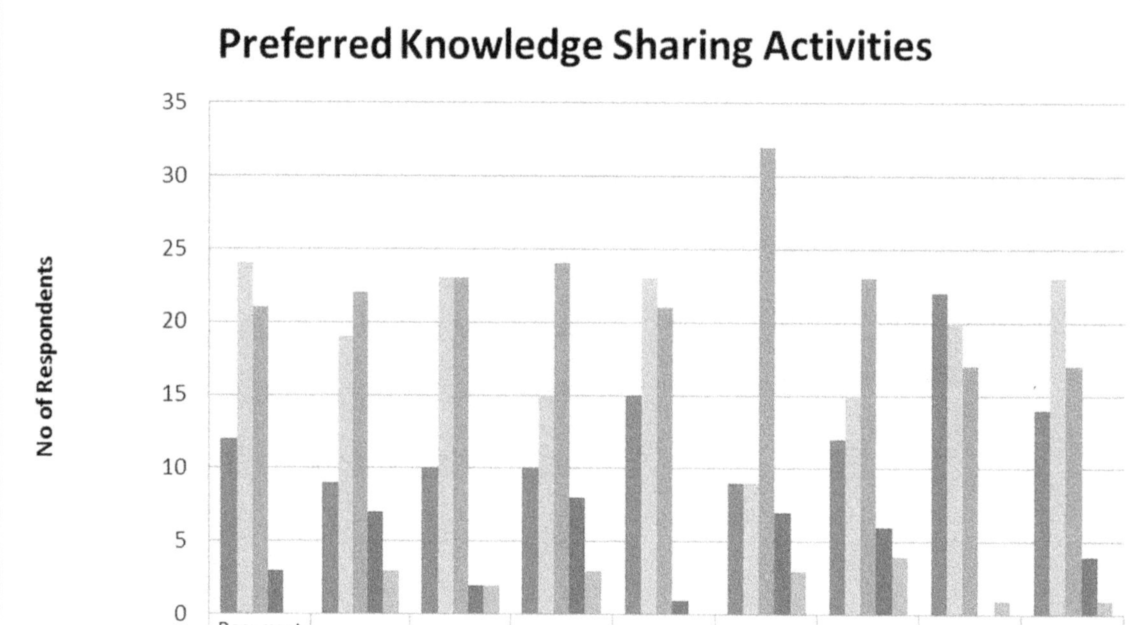

	Documentation of knowledge and practives in websites	Online Story Telling	Crossgroup learning	Online Mentoring	Exchange of Work	Writing posting on weblog	Writing and posting on journal or newslettee	Forwarding useful information	Participating in online discussion
Strongly Agree	12	9	10	10	15	9	12	22	14
Agree	24	19	23	15	23	9	15	20	23
Neutral	21	22	23	24	21	32	23	17	17
Disagree	3	7	2	8	1	7	6	0	4
Strongly Disagree	0	3	2	3	0	3	4	1	1

stagram, and Web 2.0 knowledge sharing platform provided by respective institutions such as Learning Management System, Canvas, and Edmodo.

Based on Figure 4 shown above, email was mostly preferred by the respondents for Web 2.0 technologies. The findings of the current study are consistent with those of McNeill et al. (2011) and Sun (2011) who stated email is the most used form of Web 2.0 technology, thus making it as an efficient platform for knowledge sharing (Kalema et al., 2017).

Besides email, most respondents prefer Web 2.0 knowledge sharing platforms that are provided by their respective universities or institutions. The main reason behind this is that the Web 2.0 knowledge sharing platforms are the formal system that was required by the universities. Tertiary education has looked into a variety of solutions for assimilating learning into the institute management system to favor the environment for learning (Seyal, 2012). Thus institutions have made it a mandate to use the learning management system as part of the knowledge creation and sharing. For example, Universiti Brunei Darussalam (UBD) requires their lecturers and students to fully utilize the UBD Canvas which is a Web 2.0 learning management system where students and lecturers can submit and share slides and notes,

Figure 4. Preferred Web 2.0 technologies

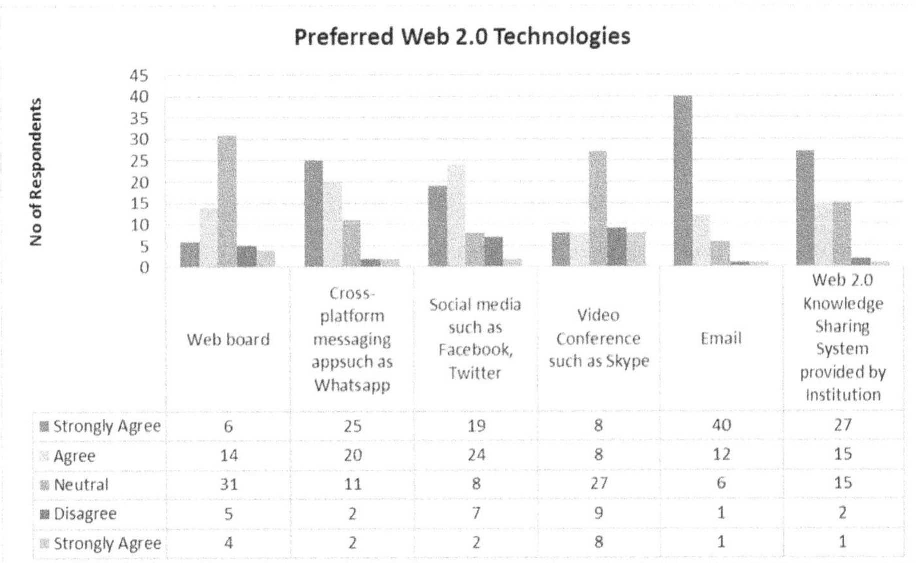

gives grades, and have a discussion among each other. This finding is by a study done by (Costa et al., 2019) that revealed LMS as the most accepted platform in an educational context.

A large number of respondents are also seen to prefer cross-platform messaging and social medias such as Facebook and WhatsApp, which shows that this form of Web 2.0 supplement class learning. Hence this confirms that social media and cross-platform messaging can cater to collaborative and cooperative work (Bista, 2015; Bouhnik & Deshen, 2014; Tyagi, 2012) teaching-supporting tool (Cohen & Duchan, 2012; Fathy et al., 2015) and most importantly it has improved positive student learning (Sayan, 2016).

Benefits of Knowledge Sharing Using Web 2.0

Respondents were asked with two questions regarding the benefits of knowledge sharing using Web 2.0 and the barriers of knowledge sharing using web 2.0 based on individual, technology, and knowledge sharing capabilities. The result findings are shown in Figure 5.

Figure 5 shows the results on the benefits of knowledge sharing using Web 2.0, which was measured according to the Likert Scale of 1 (strongly disagree) to 5 (strongly agree). Based on the findings, most respondents (more than 80%) agreed that the most important benefit of knowledge sharing using Web 2.0 is due to the fact that it increases flexibility and ease of use. This finding is supported with several previous findings (such as: Echeng et al., 2016; Naeem, 2019; Popescu, 2010; Tyagi, 2012) that found Web 2.0 provides a more flexible learning in terms of interaction, communication and collaboration as a whole. Moreover, the features of web 2.0 offer convenience in such a way that people with minimal web skills can also adapt these tools to knowledge sharing.

Figure 5. Benefits of knowledge creation and sharing using Web 2.0

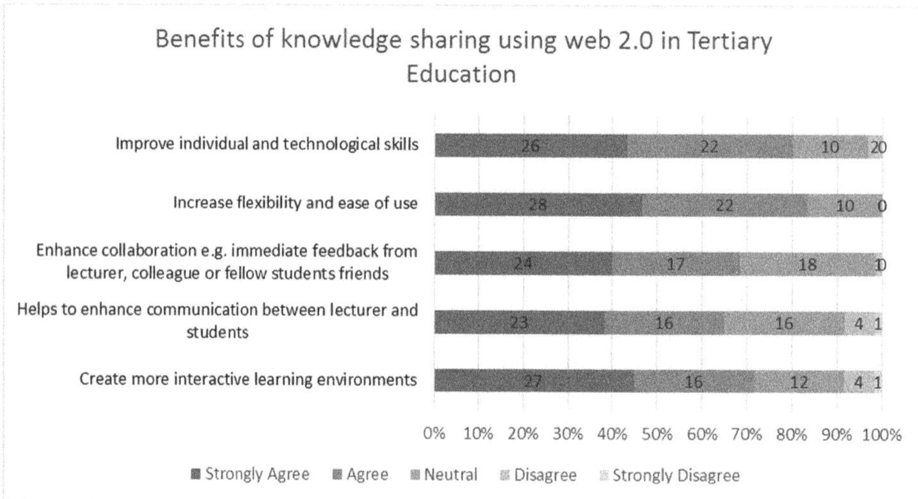

Barriers of Knowledge Sharing Using Web 2.0

Figure 6 illustrates the results on the barriers of knowledge sharing using web 2.0 based on individual views using a Likert scale of 1 (strongly disagree) to 5 (strongly agree). As can be seen from the results, the majority of the respondents agreed that the major barrier is the fear of accidentally sharing confidential information. According to Singh (2018), the fear of publishing or sharing confidential information is one of the factors that could hinder the use of Web 2.0 technologies in organizations and thus, this shows that it is significant to our result findings. Moreover, as supported by An et al. (2009), most students

Figure 6. Barriers of knowledge creation and sharing using Web 2.0: individual

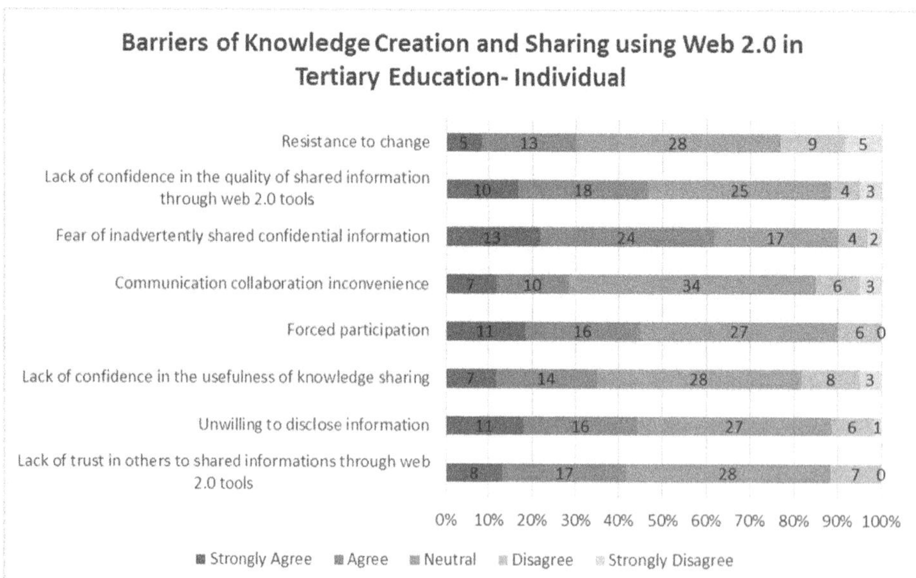

are very uncomfortable with openness and are reluctant to share confidential information via Web 2.0 technologies. The other barriers that are listed show that most of the respondents have no opinion on them. With this result, this study only takes into account that the barrier of knowledge sharing using Web 2.0 in this research setting is only the fear of sharing their confidential information.

Figure 7 shows the results on the barriers of knowledge sharing using Web 2.0 in terms of technology, and it was measured using a Likert scale of 1 (strongly disagree) to 5 (strongly agree). From the findings, it shows that two important barriers in utilizing Web 2.0 technologies are the problem of inadequate hardware and software facilities as well as due to outdated and unreliable information on the website. Some Web 2.0 technologies might be having technical glitches or might not work well with current course management systems (An et al., 2009) or even the system has its limitation (Bennett et al., 2012; Pedro et al., 2008). Hence it prevents students from accessing or even use the system that incorporated Web 2.0.

Figure 8 shows the results on the barriers of knowledge sharing using Web 2.0 in terms of knowledge sharing according to a Likert scale of 1 (strongly disagree) to 5 (strongly agree). The majority of respondents agreed that the barrier in knowledge sharing is the tendency to share irrelevant information when using an online forum. This finding is in agreement with Smith and Lambert (2014), and Ward et al. (2008) that stated that infringing their virtual self is vulnerable. Furthermore, over half of those surveyed reported that they lack confidence in the usefulness of knowledge sharing. These might be because of their concerns of the accuracy of the information in the Web 2.0 platform (Noss et al., 2008), For instance, it is difficult to distinguish between fact and fiction in Wikipedia.

CONCLUSION

In this paper, the utilization of Web 2.0 for knowledge creation and sharing processes has been investigated, in terms of its benefits, barriers, and preferences. The results showed that the role of Web 2.0 in facilitation information and knowledge sharing in tertiary education is crucial. This study has indicated that Web 2.0 platforms such as emails, LMS, and social media have become a crucial platform

Figure 7. Barriers of Knowledge Creation and Sharing using web 2.0 – technology

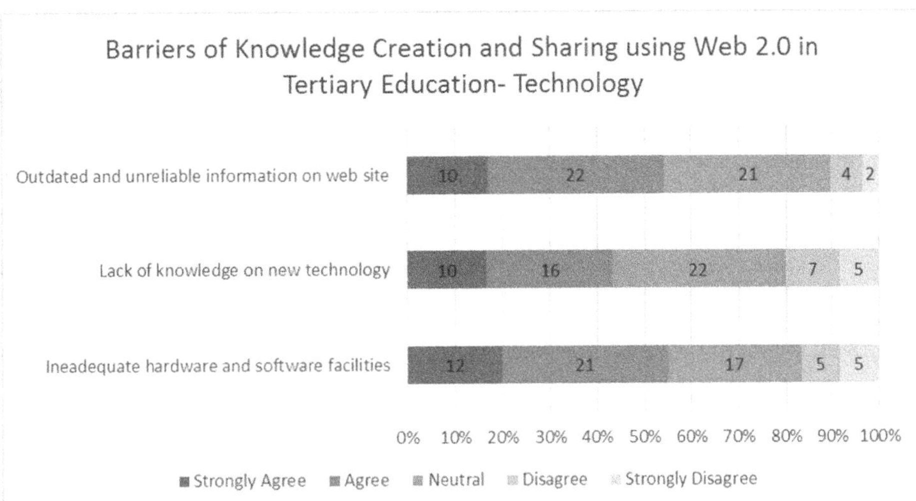

Figure 8. Barriers to knowledge sharing using web 2.0 – knowledge sharing

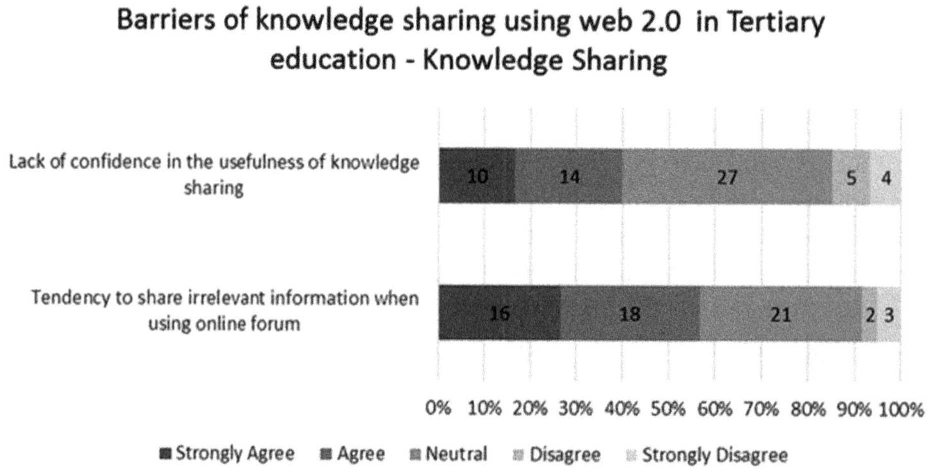

for knowledge sharing processes such as information sharing and collaboration purposes to facilitate learning. This is mainly due to its flexibility and ease of use that makes it convenient for individuals in tertiary education to utilize it. However, there are some notable barriers to knowledge sharing using Web 2.0 based on the results, which are mainly the fear inadvertently shared confidential information over Web 2.0. On the technological side, barriers to knowledge sharing using Web 2.0 are the problem of inadequate hardware and software facilities as well as due to outdated and unreliable information on websites. Finally, the barrier of using Web 2.0 in information and knowledge sharing includes the tendency to share irrelevant information when using an online forum, which creates some concerns.

LIMITATIONS OF THE STUDY AND FURTHER RESEARCH

The research is exploratory research. There are some limitations that should be taken into consideration, which are as follows:

1. The limitations are related to the small sample size (60 out of 100 respondents) with the majority of the respondents are students from Universiti Brunei Darussalam which is too small and limited to represent the population of students, education officers and lecturers of tertiary education in Brunei Darussalam.
2. Due to the majority of the respondents were students, the results of this study might be imbalance hence lead to bias.
3. Due to time constraint, only small numbers of respondents were able to be attracted to participate.

As such, follow-up research which a much larger number of respondents need to be conducted. Also, research on information and knowledge sharing using Web 2.0 has progressed day by day and logically, the area will be developed more in the near future. Due to this, the research shows potential in a research opportunity, not just in the area of education or specifically tertiary education, but also other sectors such as organizations, the public and private sector which can contribute to the subject of knowledge sharing using Web 2.0. Further studies might be useful, for example, finding trends in the use of Web 2.0 as a knowledge sharing platform over recent decades.

REFERENCES

Abdelwhab Ali, A., Panneer Selvam, D. D. D., Paris, L., & Gunasekaran, A. (2019). Key factors influencing knowledge sharing practices and its relationship with organizational performance within the oil and gas industry. *Journal of Knowledge Management*. doi:10.1108/JKM-06-2018-0394

Akinseinde, S. I., & Adomi, E. E. (2004). E-Mail Usage by Technical Education Students in Nigerian Universities. *Library Hi Tech News*, *21*(9), 15–20. doi:10.1108/07419050410572762

Al-Busaidi, K. A., & Al-Shihi, H. (2012). Key factors to instructors' satisfaction of learning management systems in blended learning. *Journal of Computing in Higher Education*, *24*(1), 18–39. doi:10.100712528-011-9051-x

Almekhlafi, A. G., & Abulibdeh, E. S. A. (2018). K-12 teachers' perceptions of Web 2.0 applications in the United Arab Emirates? *Interactive Technology and Smart Education*, *15*(3), 238–261. doi:10.1108/ITSE-11-2017-0060

Almobarraz, A. (2018). Utilization of YouTube as an information resource to support university courses. *The Electronic Library*, *36*(1), 71–81. doi:10.1108/EL-04-2016-0087

An, Y.-J., Aworuwa, B., Ballard, G., & Williams, K. (2009). Teaching with Web 2.0 Technologies: Benefits, Barriers and Best Practices. In M. Simonson (Ed.), *Proceedings of Annual Convention of the Association for Educational Communications and Technology*. Florida: Nova Southeastern University. Retrieved from https://members.aect.org/pdf/Proceedings/proceedings09/2009/09_1.pdf

Ardichvili, A., Page, V., & Wentling, T. (2003). Motivation and barriers to participation in virtual knowledge-sharing communities of practice. *Journal of Knowledge Management*, *7*(1), 64–77. doi:10.1108/13673270310463626

Baines, A. (1992). Electronic mail. *Work Study*, *41*(5), 24–25. doi:10.1108/EUM0000000002678

Balubaid, M. A. (2013). Using Web 2.0 Technology to Enhance Knowledge Sharing in an Academic Department. *Procedia: Social and Behavioral Sciences*, *102*, 406–420. doi:10.1016/j.sbspro.2013.10.756

Barhoumi, C. (2015). The Effectiveness of WhatsApp Mobile Learning Activities Guided by Activity Theory on Students' Knowledge Management. *Contemporary Educational Technology, 6*. Retrieved from https://files.eric.ed.gov/fulltext/EJ1105764.pdf

Bateman, D., & Willems, J. (2012). Facing off: Facebook and Higher Education. In D. Bateman, & J. Willems (Eds.), *Misbehavior Online in Higher Education (Cutting-edge Technologies in Higher Education* (Vol. 5, pp. 53–79). Emerald Group Publishing Limited; doi:10.1108/S2044-9968(2012)0000005007

Bell, D. (2001). *An Introduction to Cyberspace*. London, UK: Routledge.

Bennett, S., Bishop, A., Dalgarno, B., Waycott, J., & Kennedy, G. (2012). Implementing Web 2.0 technologies in higher education: A collective case study. *Computers & Education*, *59*(2), 524–534. doi:10.1016/j.compedu.2011.12.022

Bista, K. (2015). Is Twitter an effective pedagogical tool in higher education? Perspectives of education graduate students. *The Journal of Scholarship of Teaching and Learning*, *15*(2), 83–102. doi:10.14434/josotl.v15i2.12825

Bouhnik, D., & Deshen, M. (2014). WhatsApp Goes to School: Mobile Instant Messaging between Teachers and Students. *Journal of Information Technology Education: Research*, *13*, 217–231. doi:10.28945/2051

Boutin, P. (2006). *Web 2.0: The new Internet doesn't live up to its name.* Retrieved from https://slate.com/technology/2006/03/web-2-0-doesn-t-live-up-to-its-name.html

Burton, P. F. (1994). Electronic Mail as an Academic Discussion Forum. *The Journal of Documentation*, *50*(2), 99–110. doi:10.1108/eb026926

Carlisle, R. M., Hays, D. G., Pribesh, S. L., & Wood, C. T. (2017). Educational Technology and Distance Supervision in Counselor Education. *Counselor Education and Supervision*, *56*(1), 33–49. doi:10.1002/ceas.12058

Cem, B. (2009). Long Live, YouTube: L2 Stories about YouTube in Language Learning. In *E-proceedings of the International Online Language Conference.*

Chamberlin, L., & Lehmann, K. (2011). Twitter in higher education. In C. Wankel (Ed.), Educating Educators with Social Media (Cutting-edge Technologies in Higher Education, Volume 1) (pp. 375–391). Emerald Group Publishing Limited. doi:10.1108/S2044-9968(2011)0000001021

Chan, R. C. H., Chu, S. K. W., Lee, C. W. Y., Chan, B. K. T., & Leung, C. K. (2013). Knowledge management using social media: A comparative study between blogs and Facebook. *Proceedings of the American Society for Information Science and Technology*, *50*(1), 1–9. doi:10.1002/meet.14505001069

Charband, Y., & Jafari Navimipour, N. (2018). Knowledge sharing mechanisms in the education. *Kybernetes*, *47*(7), 1456–1490. doi:10.1108/K-06-2017-0227

Choi, B., & Lee, H. (2002). Knowledge management strategy and its link to knowledge creation process. *Expert Systems with Applications*, *23*(3), 173–187. doi:10.1016/S0957-4174(02)00038-6

Cohen, A., & Duchan, G. (2012). The usage characteristics of Twitter in the learning process. *Interdisciplinary Journal of E-Learning and Learning Objects, 8*. Retrieved from http://www.openu.ac.il/research_center/chais2011/papers.html

Costa, C., Alvelos, H., & Teixeira, L. (2019). Investigating the Use and Acceptance of Technologies by Professors in a Higher Education Institution. *International Journal of Online Pedagogy and Course Design*, *9*(2), 1–20. doi:10.4018/IJOPCD.2019040101

Dash, S. (2019). Google classroom as a learning management system to teach biochemistry in a medical school. *Biochemistry and Molecular Biology Education*, bmb.21246. doi:10.1002/bmb.21246

Echeng, R., Usoro, A., Echeng, R., & Usoro, A. (2016). Enhancing the use of Web 2.0 Technologies in Higher Education: Students' and Lectures' Views. *Journal of International Technology and Information Management*, *25*(1), 89–106. Retrieved from http://scholarworks.lib.csusb.edu/jitimhttp://scholarworks.lib.csusb.edu/jitim/vol25/iss1/6

Elson Anderson, K., & Still, J. M. (2011). An introduction to Google Plus. *Library Hi Tech News*, *28*(8), 7–10. doi:10.1108/07419051111187842

Fathy, S., Said, E., & Fattah, A. (2015). The Effectiveness of Using WhatsApp Messenger as One of Mobile Learning Techniques to Develop Students' Writing Skills. *Journal of Education and Practice*, *6*(32), 115–127.

Fuchs, C. (2008). *Internet and Society: Social theory in the information age*. New York, NY: Routledge.

González-Ramírez, R., Gascó, J. L., & Llopis Taverner, J. (2015). Facebook in teaching: Strengths and weaknesses. *International Journal of Information and Learning Technology*, *32*(1), 65–78. doi:10.1108/IJILT-09-2014-0021

Hargadon, S. (2008). *Web 2.0 Is the Future of Education*. Retrieved from http://www.stevehargadon.com/2008/03/web-20-is-future-of-education.html

Hennessy, C. M., Kirkpatrick, E., Smith, C. F., & Border, S. (2016). Social media and anatomy education: Using twitter to enhance the student learning experience in anatomy. *Anatomical Sciences Education*, *9*(6), 505–515. doi:10.1002/ase.1610 PMID:27059811

Holmes, K., & Prieto-Rodriguez, E. (2018). Student and Staff Perceptions of a Learning Management System for Blended Learning in Teacher Education. *Australian Journal of Teacher Education*, *43*(3), 21–34. doi:10.14221/ajte.2018v43n3.2

Huang, H.-M., & Liaw, S.-S. (2004). The Framework of Knowledge Creation for Online Learning Environments. *Canadian Journal of Learning and Technology*, *30*(1). doi:10.21432/T26G6Z

Hull, K., & Dodd, J. E. (2017). Faculty use of Twitter in higher education teaching. *Journal of Applied Research in Higher Education*, *9*(1), 91–104. doi:10.1108/JARHE-05-2015-0038

Joia, L. A. (2000). Using intellectual capital to evaluate educational technology projects. *Journal of Intellectual Capital*, *1*(4), 341–356. doi:10.1108/14691930010359243

Jones, A. (2011). How Twitter Saved My Literature Class: A Case Study with Discussion. In C. Wankel (Ed.), Teaching Arts and Science with the New Social Media (Cutting-edge Technologies in Higher Education, Volume 3) (pp. 91–105). Emerald Group Publishing Limited. doi:10.1108/S2044-9968(2011)0000003008

Kalantzis-Cope, P. (2011). Properties of Technology. In V. Kalantzis-Cope, & K. Gherab-Martin (Eds.), *Emerging Digital Spaces in Contemporary Society: Properties of Technology* (pp. 3–9). London, UK: Palgrave Macmillan.

Kalema, B. M. M., Motsi, L., & Motjolopane, I. M. (2017). Utilizing IT to Enhance Knowledge Sharing for School Educators in Developing Countries. *The Electronic Journal on Information Systems in Developing Countries*, *73*(1), 1–22. doi:10.1002/j.1681-4835.2016.tb00533.x

Klein, A. Z., Junior, J. C., Mattiello da Silva, J. V. V. M., Barbosa, J. L. V., & Baldasso, L. (2018). The Educational Affordances of Mobile Instant Messaging (MIM). *International Journal of Distance Education Technologies*, *16*(2), 51–64. doi:10.4018/IJDET.2018040104

Kosik, A. (2007). The implications of Facebook, sharing the Commonwealth: Critical issues in higher education. *Critical Issues in Higher Education*, 9–10.

Kulakli, A., & Mahony, S. (2014). Knowledge Creation and Sharing with Web 2.0 Tools for Teaching and Learning Roles in So-called University 2.0. *Procedia: Social and Behavioral Sciences*, *150*, 648–657. doi:10.1016/j.sbspro.2014.09.084

Leonardo, D. (2012). Google Books: Primary sources in the public domain. *Collection Building*, *31*(3), 103–107. doi:10.1108/01604951211243498

Levine, A., & Alexander, B. (2008). Web 2.0 Storytelling: Emergence of a New Genre. *EDUCAUSE Review*, *43*(6). Retrieved from https://er.educause.edu/articles/2008/10/web-20-storytelling-emergence-of-a-new-genre

London, M. (2012). Generative team learning in Web 2.0 environments. *Journal of Management Development*, *32*(1), 73–95. doi:10.1108/02621711311287035

López, S. P., Peón, J. M. M., & Ordás, C. J. V. (2009). *Information Technology as an Enabler Of Knowledge Management: An Empirical Analysis* (pp. 111–129). Boston, MA: Springer; doi:10.1007/978-1-4419-0011-1_8

Lubis, M. A., Yunus, M. M., Lampoh, A. A., & Ishak, N. M. (2010). The use of ICT in teaching islamic subjects in Brunei Darussalam. In *International Conference on Education and Educational Technologies* (pp. 212–217). Iwate. Retrieved from https://ukm.pure.elsevier.com/en/publications/the-use-of-ict-in-teaching-islamic-subjects-in-brunei-darussalam

Macaskill, W., & Owen, D. (2006). Web 2.0 to go. In *Proceedings of LIANZA Conference*. Wellington, New Zealand.

Madden, A., Ruthven, I., & McMenemy, D. (2013). A classification scheme for content analyses of YouTube video comments. *The Journal of Documentation*, *69*(5), 693–714. doi:10.1108/JD-06-2012-0078

Marić, I. (2013). Stakeholder analisys of higher education institutions. *Interdisciplinary Description of Complex Systems*, *11*(2), 217–226. doi:10.7906/indecs.11.2.4

McNeill, M., Ming Diao, M., & Gosper, M. (2011). Student uses of technology in learning: Two lenses. *Interactive Technology and Smart Education*, *8*(1), 5–17. doi:10.1108/17415651111125478

Myrberg, C., & Wiberg, N. (2015). Screen vs. paper: What is the difference for reading and learning? *Insights the UKSG Journal*, *28*(2), 49–54. doi:10.1629/uksg.236

Naeem, M. (2019). Uncovering the role of social media and cross-platform applications as tools for knowledge sharing. *VINE Journal of Information and Knowledge Management Systems*, VJIKMS. doi:10.1108/VJIKMS-01-2019-0001

Nkhoma, C. A., Thomas, S., Nkhoma, M. Z., Sriratanaviriyakul, N., Truong, T. H., & Vo, H. X. (2018). Measuring the impact of out-of-class communication through instant messaging. *Education + Training, 60*(4), 318–334. doi:10.1108/ET-12-2017-0196

Noss, R. T.-T., Selwyn, N., Crook, C. B., Carr, D., Carmichael, P., & Laurillard, D. (2008). Education 2.0? Designing the web for teaching and learning: A Commentary by the Technology Enhanced Learning phase of the Teaching and Learning Research Programme. Retrieved from https://www.semanticscholar.org/paper/Education-2.0-Designing-the-web-for-teaching-and-A-Noss-Tlrp-Tel/41c6c7f9316a30c422c447a618dd777b41edbb9d

Noyes, J. M., & Garland, K. J. (2008). Computer- vs. paper-based tasks: Are they equivalent? *Ergonomics, 51*(9), 1352–1375. doi:10.1080/00140130802170387 PMID:18802819

O'Reilly, T. (2007). What Is Web 2.0: Design Patterns and Business Models for the Next Generation of Software. *International Journal of Digital Economics, 65*, 17–37. Retrieved from https://mpra.ub.uni-muenchen.de/4580/

Oxford Business Group. (2011). *The report Brunei Darussalam 2011.*

Patrick, K., & Dotsika, F. (2007). Knowledge sharing: Developing from within. *The Learning Organization, 14*(5), 395–406. doi:10.1108/09696470710762628

Pedro, N., Soares, F., Matos, F., & Santos, M. (2008). *The Use of Learning Management Platforms in School Context - a National Study.* Retrieved from http://repositorio.ul.pt/bitstream/10451/5305/1/The_use_of_learning_manegement_national_study_english_version_.pdf

Pieri, M., & Diamantini, D. (2014). An E-learning Web 2.0 Experience. *Procedia: Social and Behavioral Sciences, 116*, 1217–1221. doi:10.1016/j.sbspro.2014.01.371

Popescu, E. (2010). Students' Acceptance of Web 2.0 Technologies in Higher Education: Findings from a Survey in a Romanian University. In 2010 Workshops on Database and Expert Systems Applications (pp. 92–96). IEEE. doi:10.1109/DEXA.2010.38

Porter, G. W. (2013). Free choice of learning management systems. *Interactive Technology and Smart Education, 10*(2), 84–94. doi:10.1108/ITSE-07-2012-0019

Ricoy, M.-C., & Feliz, T. (2016). Twitter as a Learning Community in Higher Education. *Journal of Educational Technology & Society*. International Forum of Educational Technology & Society. doi:10.2307/jeductechsoci.19.1.237

Sánchez, R. A., & Hueros, A. D. (2010). Motivational factors that influence the acceptance of Moodle using TAM. *Computers in Human Behavior, 26*(6), 1632–1640. doi:10.1016/j.chb.2010.06.011

Sandnes, F. E., Jian, H.-L., Hagen, S., & Talberg, O. (2007). Student Evaluation of the Learning Management System Fronter From an HCI Perspective. In *International Conference on Engineering Education – ICEE 2007*. Coimbra, Protugal. Retrieved from http://icee2007.dei.uc.pt/proceedings/papers/86.pdf

Sayan, H. (2016). Affecting Higher Students Learning Activity by Using Whatsapp. *European Journal of Research and Reflection in Educational Sciences, 4*(3), 88–93. Retrieved from www.idpublications.org

Semeraro, J., & Moore, N. S. (2016). The Use of Google Docs Technology to Support Peer Revision. In E. Ortlieb, E. H. J. Cheek, & W. Verlaan (Eds.), Writing Instruction to Support Literacy Success (Literacy Research, Practice and Evaluation, Volume 7) (pp. 203–220). Emerald Group Publishing Limited. doi:10.1108/S2048-045820160000007013

Seyal, A. H. (2012). A preliminary study of school administrators' use of information and communication technologies: Bruneian perspective. *International Journal of Education and Development using Information and Communication Technology (IJEDICT)* (Vol. 8). Retrieved from www.bit.gov.bn

Shannon, C. E., & Weaver, W. (1949). *The mathematical theory of communication.* Urbana, IL: University of Illinois Press.

Sheer, V. C., & Fung, T. K. (2007). Can Email Communication Enhance Professor-Student Relationship and Student Evaluation of Professor?: Some Empirical Evidence. *Journal of Educational Computing Research, 37*(3), 289–306. doi:10.2190/EC.37.3.d

Singh, M. (2018). Adoption of Web 2.0 tools in higher education in India: A study. *Global Knowledge. Memory and Communication, 67*(4/5), 297–315. doi:10.1108/GKMC-10-2017-0084

Sivarajah, U., Irani, Z., & Weerakkody, V. (2015). Evaluating the use and impact of Web 2.0 technologies in local government. *Government Information Quarterly, 32*(4), 473–487. doi:10.1016/j.giq.2015.06.004

Smith, T., & Lambert, R. (2014). A systematic review investigating the use of Twitter and Facebook in university-based healthcare education. *Health Education, 114*(5), 347–366. doi:10.1108/HE-07-2013-0030

Sun, S. (2011). The Internet Effects on Students Communication at Zhengzhou Institute of Aeronautical Industry Management. In S. Lin, & X. Huang (Eds.), *Advances in Computer Science, Environment, Ecoinformatics, and Education. CSEE 2011* (pp. 418–422). Berlin, Germany: Springer; doi:10.1007/978-3-642-23357-9_74

Sundararajan, B., Sheehan, L., & Gilbert, S. (2013). Mediated Discourse in Higher Ed Classrooms Using Text Messaging. In C. Wankel (Ed.), Increasing Student Engagement and Retention Using Classroom Technologies: Classroom Response Systems and Mediated Discourse Technologies (Cutting-edge Technologies in Higher Education, Volume 6 Part E) (pp. 199–232). Emerald Group Publishing. doi:10.1108/S2044-9968(2013)000006E010

Tandi Lwoga, E. (2014). Integrating Web 2.0 into an academic library in Tanzania. *The Electronic Library, 32*(2), 183–202. doi:10.1108/EL-06-2012-0058

Tian, S. W., Yu, A. Y., Vogel, D., & Kwok, R. C. W. (2011). The impact of online social networking on learning: A social integration perspective. *International Journal of Networking and Virtual Organisations, 8*(3), 264–280. doi:10.1504/IJNVO.2011.039999

Tiruwa, A., Yadav, R., & Suri, P. K. (2018). Modelling Facebook usage for collaborative learning in higher education. *Journal of Applied Research in Higher Education, 10*(3), 357–379. doi:10.1108/JARHE-08-2017-0088

Towner, T. L., & Lego Muñoz, C. (2011). Facebook and education: A classroom connection? In C. Wankel (Ed.), *Educating Educators with Social Media (Cutting-edge Technologies in Higher Education* (Vol. 1, pp. 33–57). Emerald Group Publishing Limited. doi:10.1108/S2044-9968(2011)0000001005

Tyagi, S. (2012). Adoption of Web 2.0 technology in higher education: A case study of universities in National Capital Region, India. [IJEDICT]. *International Journal of Education and Development Using Information and Communication Technology*, *8*(2), 28–43. Retrieved from https://files.eric.ed.gov/fulltext/EJ1084132.pdf

Virkus, S., & Bamigbola, A. A. (2011). Students' conceptions and experiences of Web 2.0 tools. *New Library World*, *112*(11/12), 479–489. doi:10.1108/03074801111190473

Ward, R., Moule, P., & Lockyer, L. (2008). Adoption of Web 2.0 technologies in education for health professionals in the UK. Where are we and why? In *7th European Conference on E-Learning, Agia Napa, Cyprus*. Retrieved from http://eprints.uwe.ac.uk/65/

West, D. M., & Kimmons, R. (2019). Twitter in Education. In *The International Encyclopedia of Media Literacy* (pp. 1–6). Hoboken, NJ: Wiley; doi:10.1002/9781118978238.ieml0237

Wieser, D., & Seeler, J.-M. (2018). Online, Not Distance Education: The Merits of Collaborative Learning in Online Education. In A. Altmann, B. Ebersberger, C. Mössenlechner, & D. Wieser (Eds.), *The Disruptive Power of Online Education* (pp. 125–146). Emerald Publishing Limited. doi:10.1108/978-1-78754-325-620181008

Zaqout, F., & Abbas, M. (2012). Towards a model for understanding the influence of the factors that stimulate university students' engagement and performance in knowledge sharing. *Library Review*, *61*(5), 345–361. doi:10.1108/00242531211280478

Zeeng, L., Robbie, D., Adams, K. M., & Hutchison, C. (2009). Where's my class? Using Web 2.0 for collaboration in a design environment. In *Ascilite*. Auckland, New Zealand. Retrieved from http://www.ascilite.org/conferences/auckland09/procs/zeeng.pdf

Zientek, L. R., Werner, J. M., Campuzano, M. V., & Nimon, K. (2018). The Use of Google Scholar for Research and Research Dissemination. *New Horizons in Adult Education and Human Resource Development*, *30*(1), 39–46. doi:10.1002/nha3.20209

KEY TERMS AND DEFINITIONS

Knowledge Sharing: Knowledge sharing is various activities that are used to exchange knowledge among participants.

Social Media: Social media is websites of web-based tools that allow people to have social networking in which they communicate, interact, collaborate, by creating and sharing information and knowledge.

Tertiary Education: Tertiary education is the next level of education for those who complete their secondary education, which normally conducted at universities and colleges.

Web 2.0: The second stage development of the World Wide Web, characterized by the change from static web pages to dynamic or user-generated content and the growth of social networking.

Chapter 2
An Empirical Note on Health Vulnerability and Health Information Digital Divide:
A Study of Indian Patients

Jaya Rani
Sikkim Manipal University, India

Ajeya Jha
https://orcid.org/0000-0003-0491-5008
Sikkim Manipal University, India

Jitendra Kumar
Sikkim Manipal University, India

Samrat Kumar Mukherjee
Sikkim Manipal University, India

Saibal Kumar Saha
Sikkim Manipal University, India

ABSTRACT

Availability of healthcare information on the Internet has made it possible for patients or their relatives to search for such information. Considering the delicate nature of such information as well as its great need felt by the society, it is important to know who are these people who actively search for online healthcare information and also those who are unable to do so. In all, 754 respondents participated in the survey. The variables selected from literature survey and exploratory study are Health Information Digital Divide, Income, Having E-mail id, access to Internet, geographical location, Education, family-type, age, and gender. As the data is categorical, the significance of difference has been calculated using Chi-square test. Later discriminant analysis was conducted to predict patients who make online health information searches and the ones who do not. Using discriminant analysis, 94.5 percent of patients who make online health information searches could be correctly predicted. Prediction is 99.7% for the patients who do not indulge in online health information search.

DOI: 10.4018/978-1-7998-0357-7.ch002

INTRODUCTION

Human societies across the globe have had a great divide between the rich and the poor – "Haves" and Have-nots". Earlier the divide was rooted in land – those who possessed it and those who were landless. Natural resources, industry, technology, language consolidated the divide in favour of "Haves". It is a human endeavor to bridge this gap. Arrival of information technology has once again emerged as another factor in the divide. A digital divide implies socio-economic inequality in access to, use of, or impact of information and communication technologies (ICT).

Digital divide is conceptualized on the basis of attributes that distinguish to describe the divide. Most common happen to be income, education, age, geographic location, race, gender, age, skills, awareness, political, cultural and psychological attitudes and motivation (Hilber & Martin, 2011; Mossberger et al., 2006; Lawton & Tait, 2007; Wan, 2001; Guillen, 2005; Ernest et al., 2004; Carr, 2007; Wilson, 2003; Slate et al., 2002; Losh, 2004), economic (McLaren & Zappalà, 2002), political (Hung, 2004) and cultural (Press et al., 2003). Further digital divide is a dynamic phenomenon in a sense that as a gap closes, the other gaps open up (van Dijk, 2002; van Dijk & Hacker, 2003). As the access gap is being narrowed skill and usage gaps have cropped up.

Healthcare is an important concern for any society. Any divide across various groups for healthcare are to be discouraged as right to life and right to healthy life are fundamental in the context to WHO vision as well as the constitution of most nations. This paper has been envisioned to understand the extent of digital divide for healthcare information in the context of India. The diverse nature of divide needs to be established and understood if any curative measures are to be designed. The paper additionally also attempts to measure the health care vulnerability of various demographic groups to bring out the stark need for action in this respect.

LITERATURE REVIEW

Healthcare, along with agriculture, remains the core concerns for human societies as these are most elemental to life itself. Consequently, hence how the digital divide affects health care is an important question. Further Information is considered a vital resource in health condition improvement because it is "the first step to every healthy choice" (Gann, 1986) and it is "on a continuum between health education and health promotion" (Rolinson, 1998).

Healthcare has been affected because of the digital revolution we have experienced in last 3 decades. Before the advent of the Internet, healthcare providers were the sole and legally authorized individuals with access to gain, interpret, analyze and convey health information. This naturally resulted in them having exalted status in their relations with patients or health consumers (Giddens, 1991; Goldsmith, 2000; Hardey, 1999). Internet has at once opened the floodgates of information at the disposal of patients and common men. It, therefore needs to emphasis that, healthcare consumers seeking online information do not portray merely a change in transition in the way they seek information, or just an opportunity to avail what they could not earlier, but a dynamic shift in their empowerment vis-avis their own health. (Cotten, 2001). They have far greater ability to actively understand and influence their health status (D'Alessandro & Dosa, 2001), assume more responsibility for their own health, and participate in health care decisions (Anderson et al., 2003). This ability, though, has brought forth certain misgivings (Henwood et al., 2003) yet is being hailed as a right to information and physicians are being implored to treat patients as individuals and not as treatment opportunities' (Gann & Needham, 1992). Even Information professionals, advocate empowerment of people through accessible health information (Calvano, 1996). Health information is promoted as essential in healthcare as it is believed to provide both direction and rationale for guiding strategic health behaviors, treatments, and decisions (Kreps, 2001).

More and more patients are now accessing Internet increasingly for finding answers to their queries on health issues – disease state, symptoms, precautions and medication options as also are able to self-diagnose, investigate their conditions, share experiences with other patients and care-givers, and gain knowledge about treatment options. Access to healthcare information has been generally acknowledged as a positive development despite the traditional and contemporary misgivings in this regard and consequently it is important to identify the manner and extent of digital divide in accessing online information on healthcare by the patients. Demographic factors have been found to be critical by various researchers who identify age, gender, geographical location, income, education and language as the prime barriers that perpetuate digital divide across the globe. Influence of digital divide in the general population has been a focused area of research. (Cotten & Gupta, 2004).

Van Dijk and Hacker (2003) explains the reason for digital divide as in terms of access to digital resources being a multi-faceted phenomenon comprising mainly of four factors namely psychological, material, skills and usage, that regulate access.

In India Internet today is the foremost source of uninterrupted and instant information. (Cotten & Gupta, 2004). It is economical and is available widely across the country and hence in India also, a large majority of people use net regularly and they also search health related information available online. As in the developed countries the developing countries people too are searching out their health-care related queries by consulting doctors. In recent years Indians have also developed greater awareness for their healthcare. Due to the emergence of Internet the health information seeking behavior of the patients has totally altered. Regular online search is becoming common and accepted. Since it is effortless to use, hence very frequently people visit online and search information related to health issues. People search different health related information such as on prescription drugs, alternative medicines, nutrition, exercise, medical conditions, current health topics, illegal drugs etc. (Baker et al., 2003; Escoffery et al., 2005; Fox, 2006; Taha et al., 2009).

Technically direct to consumer promotion is legal in only two of the nations: U.S. and New Zealand. There also we find strong voices of protests against such a policy. The Drugs and Magic Remedies (Objectionable Advertisements) Act, 1954 of India prohibits any kind of promotion of prescription drugs in India. From the literature survey we find that most such studies have been undertaken in developed region of the world - USA, Europe, UK, Canada and Australia. Other regions of the world have remained untouched in this respect. In view of this the study was undertaken with the purpose to identify the common and India-specific factors resulting in health-information digital divide.

The paper also dwells upon the vulnerability assessment of groups and match it against their online behavior in the context of healthcare information.

- **Objectives of the Study:** The major objective of the study is to study the impact of demographical variables on online health search behaviour of the patients in India. Yet another objective is to assess the health vulnerability of certain demographic groups.
- **Hypotheses**: Keeping the above objectives in view the following hypothetic framework was proposed:

Ho1: Demographical variables determine the health information digital divide India.

The demographic variables included in this study are Income, Age, Email, and Internet availability, Location, Education, English Knowledge, Gender and Family-type.

These variables have been arrived at on the basis of an exploratory study whereby patients/their relatives were interviewed for their propensity to search online information for health-related issues. These indicators appear relevant in the present case because:

- Income is an important indicator determining access to computer and communication related technologies and hence search for online information for health-related issues
- Younger generation is more comfortable with Internet and related information and hence age appears to be an important factor
- Those having and email-id will be having a greater inclination to surf Internet and vice-versa and hence having email could be a critical indicator to identify patients surfing online information
- Availability of Internet on a continual basis too is a fundamental requirement of surfing Internet for health-related information.
- In India Internet is mostly an urban and to some extent a suburban phenomenon and hence geographical location is considered an essential condition for online health-care search.
- Net-surfing is a behavior mostly associated with people having a basic education. Unlike west India still harbours a large population that is bereft of any education and hence education levels have been incorporated in this study.
- Internet (at least in Indian context) is overwhelmingly English-based. In India English is widespread yet is understood by less than 20% of the population hence it is being assumed that People not comfortable in English do not look for online healthcare information.
- Genders differ in healthcare needs as well as in access to Internet and other social issues and hence gender too has been included in this study.
- India traditionally has been a nation of joint-families. In last few decades, however, because of socio-economic reasons nuclear families too have emerged quickly. It is being assumed that this two sets of families differ in their online healthcare search behavior.

Ho2: There is no significant health vulnerability across the demographic variables. The demographic factors selected include age, gender, income and geographic location

Scope of Study

The study is applicable in Pharmaceutical industry only, on Indian Patients and limited to health information digital divide and health vulnerability assessment.

METHODOLOGY

Nature of Research

The present research is exploratory and empirical in nature with descriptive statistics based on the data on the belief expressed by patients/their relatives on the harmful impact of direct-to-consumer promotion of diseases and drugs.

Research Design

The research-design for the research work is conclusive. To arrive at conclusions descriptive approach has been used

Sample

The sample respondents of this first objective consist of patients with varying age, gender income, location, education, English proficiency, family type, email-id and access to information. For the second objective sample respondent comprises of physicians and other healthcare professionals such as nurses and pharmacists. Sampling method used was cluster based Random sampling method.

Sampling Frame

In order to draw a representative sampling it is assumed that patients and relatives of patients belonging to different sized cities may have difference in beliefs and hence, we have drawn proportionate samples from metropolitan cities with population above 10 millions, cities with population between 2-5 million and from cities having population below 2 million. That's why it is clustered sampling method. The cities have been decided upon by random sampling. The patients and relatives have been identified by visiting randomly selected hospitals. 200 each respondents are from cities with population above 10 million (Bengaluru and Kolkata), 80 each respondents are from cities with population 2-5 million (Raipur and Patna), 60 each from cities below 2 million (Siliguri, Jamshedpur, Gangtok, and Guwahati).

Sample Size

In all a sample size of 800 respondents was arrived at. This sample was considered statistically relevant for the study. Out of these 46 declined to respond. For second survey a sample size of 300 was decided. To touch this figure we needed to approach 331 respondents of which 31 did not respond.

Tools

The tool is simple and based on categorical data on Income (in four categories), Age (In 3 categories), Email (Yes/No), Internet accessibility (Yes No), Location (Urban/Suburban/Rural), Education, English proficiency (Yes/No), Gender (Male/Female) and Family-type (Nuclear/Joint-extended). The tool comprised of seeking belief on the health vulnerability of nine demographic groups on a scale of 1 to 10 with 10 implying highest possible vulnerability.

Data Analysis

The test statistics was checked and found to follow normal distribution. For the hypothesis testing the confidence limit is set at 95%.

Hypotheses Testing

The Hypotheses testing has been done using Chi-square-test. The chi-square (I) test is used to determine whether there is a significant difference between the expected frequencies and the observed frequencies in one or more categories. The Alternate Hypothesis is accepted if calculated Chi-square value is more than the tabulated one and Significance value is less than 0.05. For the second part, hypothesis testing is based upon paired sample t-test and ANOVA.

RESULT AND DISCUSSION

Result and discussion comprises of two parts to cater to the two objectives we have stared with.

Digital Divide: Cross-Tabulation

Income

From Table 1 we find that percentage of respondents responding positively to the question whether or not they search net for health information are 2%, 11%, 77% and 78% for income groups below Rs. 10,000/month, 10000 to 19999, 20000 29999 and above 30,000 respectively. From this it is obvious that seeking online healthcare information is strongly a function of income with higher income groups responding positively to it and lower income group responding relatively negatively to it. This is further confirmed by the corresponding pearson chi square value of 386.142 (Table 2) which is higher than the tabulated Chi square value. We, therefore, reject the null hypothesis. This is confirmed also by the consequent significance which is almost zero - much below 0.05. In other words, we can say that Income has an effect on online health information search behaviour. Higher income group display a distinctly high tendency than their lower income groups counterparts for searching online health–related information.

Our study identifies income as an important dimension of health digital divide. This is in consonance of the conclusion of various other researchers who have studied digital divide in general or in the context of health information (Peizer, 2000; Garcia, 2000). In fact, many researchers insist that income is the sole determinant of health digital divide as all other divides too are a product of income or lack of it. Lack of income results in an inability to buy ICT products. This is an alarming scenario as almost 50% of the global population survives on less than $2.00 a day (Chanda, 2000). This gap is narrowing as more poor are buying ICT products than the richer ones. However, it may take many decades to narrow it down to insignificance although the quality technology may remain an issue for a longer time to come. Senecal et al., (2018) recently found no significant correlation between socioeconomic environment and digital health search. This could be because at least in certain societies Internet penetration is exceptionally high.

Table 1. Cross tabulation variables x information search

S. No.	Variables		Information Search	
			Yes	No
1	Income	Below 10000	2% (3)	98% (163)
		10000 19999	11% (25)	89% (194)
		20000 to 29999	77% (162	23% (49)
		30000 and above	78% (124)	22% (34)
2	Age	18-30	39% (88)	61% (135)
		31-50	42% (133)	58% (180)
		50 and above	43% (93)	57% (125)
3	Email	No	4% (16)	96% (409)
		Yes	91% (298)	9% (31)
4	Internet availability	No	3% (11)	97% (403)
		Yes	89% (303)	1% (37)
5	Location	Rural	1% (4)	99% (343)
		Suburban	74% (234)	26% (84)
		Urban	85% (76)	15% (13)
6	Education	Illiterate	2% (3)	98% (158)
		primary	11% (25)	89% (194)
		High School	75% (164)	25% (55)
		College Degree	79% (122)	21% (33)
7	English	No	0.2%(1)	100% (416)
		Yes	93% (313)	7% (24)
8	Family-type	Nuclear	42% (148)	58% (207)
		Joint	42% (166)	58% (233)
9	Gender	Male	47% (166)	53% (188)
		Female	37% (148)	63% (252)

Age

From Table 1 we find that percentage of respondents responding positively to the question whether or not they search net for health information are 39%, 42% and 43% for age groups below 15-30, 31-50 and above 50 respectively. This implies that seeking online healthcare information is not dependent on age. As all the age groups practically have same percentage respondents responding positively and negatively. This is further confirmed by the corresponding low pearson chi square value of 0.622 (Table 2) which is lower than the tabulated Chi square value. We, therefore, accept the null hypothesis. This is confirmed also by the consequent significance which is 0.624 – far above 0.05. This brings forth the inference that age has no bearing on online health information search behavior.

Age has been identified as a key contributing factor to health digital divide (Swartz et al., 2004; Fang et al., 2018). He and other researchers categorize health consumers in three broad groups the youth, the

Table 2. Chi square test value of variables x information search

Sl.	Variable	Chi-Square Tests		Asymp. Sig. (2-sided)		Null hypothesis
		Pearson	Likelihood Ratio	Pearson	Likelihood Ratio	
1	Income	386.142[a]	445.283	.000	.000	Rejected
2	Age	.622[a]	.624	.733	.732	Accepted
3	Email	575.097[a]	682.346	.000	.000	Rejected
4	Internet availability	574.256[a]	688.638	.000	.000	Rejected
5	Location	437.702[a]	539.239	.000	.000	Rejected
6	Education	374.393[a]	431.368	.000	.000	Rejected
7	English	658.170[a]	836.981	.000	.000	Rejected
8	Family-type	.622[a]	.624	.733	.732	Accepted
9	Gender	7.563[a]	7.567	.006	.006	Rejected

middle-aged and old. They report middle aged as the most actively online-information seekers perhaps because of their own and their dependents health-issues and access to Internet. Shelia & Gupta (2004), Witry et al. (2018) identify the old age group as the worst victims of health digital divide because they need it the most and yet have no access to it because of low income and skills. Similarly Levy et al. (2015) reported that not even a third of Americans over 65 years of age use the Web for health information they further conclude that percentage of those with low health literacy or ability to navigate the health-care system lies below 10. Our results differ from these researchers primarily because of exponential growth of Internet users in last few years. These very researchers had predicted that because of this exponential growth the age gap will diminish and may disappear altogether (Judy & D'Amico, 1997; Shelia & Gupta, 2004). Their views appear to have come true as our study does not reflect any health information age divide. At the hindsight it is felt that with perhaps age group above 65 years of age the proclivity to search online health information would be much lesser and it has already been pointed out that it is this group of people who perhaps depend upon online information the most. Yoon et al. (2018) in fact finds old age people using health information more frequently. Later studies, however, continue to find that older people find seeking health information still challenging. Fox & Connolly et al. (2018) reports that lower adaptability and high-risk perception as challenges that keep old people on the negative side of the digital divide. Nancy et al. (2018) old age people still lack access to technology and are less confident about using the same. Witry et al. (2019) identifies frailty also as a challenge by aged people in seeking health information. Francis et al. (2019) concludes that if old age people embrace m-health, they would be in a better position to experience better quality of life. It has also been found that health information search by old people enhances positive health perception and medical satisfaction. Bluethmann et al. (2018) advise old adults to use online information more often to gain confidence and for better health management. Younger people on the other hand continue to use online information for health management (Russell et al. (2019), Bol et al. (2018)

Email-ID

We have already indicated that health information digital divide is not merely a function of access to technology but also of technological skills too. One measure of technological skills is whether or not a person has an email-id. From the table 01 we find that percentage of respondents having email-id and responding positively to health information online search are 91% against only 4% not having email-id. It is obvious from the figures that people having email-id (signifying technological skill) are overwhelmingly more likely to search for online health information. This is further reflected in the high calculated pearson chi-square value (557.97) and which is extraordinarily higher than the tabulated chi-square value. The consequent p-value (almost zero) further substantiates it.

Having a personal email-id reflects a certain threshold level of Internet skills. Skills have already been identified as a crucial factor in health information digital divide (Bucy & Newhagen, 2004). Access to ICTs and the Internet is does not automatically translate into accessing the content (Van Dijk, 2004). We find that individuals with email-id are distinctly more prolific searchers of online health information than their non-email-id counterparts. Once again this could be because email-id, apart from reflecting required Internet skills also is a product of higher income and education. In the hindsight we feel we should have included social networking site membership because it is emerging that many people may have such a membership even as they have no email-id. Despite this flaw we still feel our conclusions are valid as the underlying logic remains sound.

Internet Availability

From Table 1 we find that only 3% of the respondents having no Internet availability search online health information and which is extremely low in comparison to the ones who have access to Internet (89%). This stark distinction is well reflected in the corresponding chi-square value which is as high as 574.256 – an unambiguous conclusion that online health information search is dependent on access to Internet. This is further confirmed by the corresponding p-value which is zero for all practical purpose.

Internet skills have to go along with Internet access too. Access have been identified as an essential factor of health information digital divide (Loader & Keeble, 2004; Chinn & Fairlie, 2004; Caspary & O'Connor, 2003). It is said to be the first level of digital divide. Our study reveals that indeed it is and that people with access search net for health information more actively than the ones with no access.

Geographical Location

Is there a rural urban divide in online health information search behavior? Table 1 indicates that merely 1% rural, 74% suburban and 85% urban dwellers respond positively to searching online health information. This again translates into unmistakable implication that a stark digital location divide does exist in this respect. The corresponding extremely high chi-square value (437.7) validates this further as also by the resultant low p-value which again effectively is zero implying that location is a significant factor in perpetuating health information digital divide.

We found a considerable difference between online health information search behaviour amongst urban, suburban and rural dwellers (Guillen & Suárez, 2005; Mossberger et al., 2006; Hilbert, 2011). The differences exist because on average rural people are less educated, have less income and have less access to Internet facility Hale (2010). In the context of India it is generally known that people living in rural regions are less educated. The reason could be that as farming and other rural based occupations do not require higher education. Also consequently, most higher learning centers in India are in urban setting. People in rural regions have considerably less income. Further use of English, in India once again is an urban phenomenon and hence most of the rural dwellers may not be able to benefit from online health information even if somehow accessible to them. Finally, there is an enormous infrastructural gap in rural and urban India with Internet penetration representing perhaps the most critical one. Greenberg et al. (2018) found significant difference between the usages of online health information between rural and urban residents, primarily because technology is urban centric even now. Bakken et al. (2019) opines that rural residents continue to lag behind their urban counterparts even if other factors such as age, income, education and accessibility are suppressed.

Education

Does education affect inclination to search for online health information? It appears to be a rational assumption. Table 1 provides us with the answer. 2% illiterate, 11% primary education holders, 75% high school pass respondents, and 79% of the college degree holders report to search for online health information. The figures in Table 1 and Table 2 project a distinct interpretation that education is a critical factor in digital health information divide. A chi-square value of 374.93 clearly establishes such an inference as also the consequent p-value that is zero for all practical purposes. 2% illiterate and 11% primary pass respondents saying "yes" to seeking online health information belong to relatively higher income group and who take help of family members to gain online information about their health predicaments.

We have arrived at the conclusion that education is a significant denominator of health information digital divide. It stands to reason as we have already agreed that digital divide has two components – access and skill. Education affects both the components and as higher education relates to higher income which in turn leads to higher access. Similarly, higher education also means higher online skills too. Education is closely linked to income and hence we may say that this too is a derived divide – a product of income-push divide. Other researchers also have reported similar conclusion that it is people with higher education who primarily search online information (Hoffman & Novak, 1998; Lach, 2000; Fattah, 2000). Higher education not only results in higher income (to make ICT accessible) but also means higher online search skills and better understanding of medical and health issues. Better search skills lead them to better websites and the consequent better satisfaction. Better understanding of health-related concepts and terminologies also make their efforts more fruitful and encourage future visits.

Language

As most health information websites are in English it was natural on our part to investigate this aspect of health information digital divide. Table 1 and 2 present an interesting picture in this respect. Only 0.2% respondents who are not conversant with English still claim to search for online health information as against 93% who are conversant with English. Obviously English emerges as the most vital component of health information digital divide as is further evinced by highest chi-square value (658.17) of this study and near zero resultant p-value.

The study concludes that knowledge of English provides a distinct advantage in online health search behavior. This is understandable, because, more than 90% health care websites are in English (with hardly any being in other Indian languages). In India hardly 10% population is conversant with English and fluent speakers hardly about 2%. Other researchers have found language as a critical determinant of the health information digital divide. In fact in our study Language (English) emerges as the most critical factor governing health information digital divide (With 93% English speakers responding positively to searching online health information as against just 0.2% non-English speakers with the corresponding calculated chi square value being as high as 658.17) . The most enlightening example of digital language divide is perhaps Europe where income and education are not much an issue yet it is non-English speakers who are on the disadvantaged end of the health information divide (Black, 1999). It is further explained that non-English speakers in the developed world include two, sometimes overlapping groups: the poor, and the aged (Black, 1999).

Family Type

Social structure in India accommodates two major family types – nuclear (Mother-father – children unit) and extended (aged parents living with nuclear family of their children) or joint family (where two or more brothers along with their nuclear family share the same household. Differences with respect to online health search behavior are expected. Table1 and 2, however, reveal that no differences exist between the two types of families in this context. In both nuclear and joint families, the percentage of online health information search is undertaken by identical (42%) of the respondents. The consequent Chi-square value is just 0.622 – much below the tabulated chi-square value. Consequential p-value is 0.624 – much above 0.05.

This study analyzed hitherto untouched variable namely family-type also for any kind of digital divide. Families were identified as nuclear and joint/extended. Underlying hypothesis was that nuclear families by virtue of higher education and upward mobility will have higher inclination for online health search. We, however, could not find any difference in their propensity in this context. The reason could be that nuclear families exist in lower strata of the society also because of economic compulsions to migrate and live as a nuclear family. Other reason could be that joint families perhaps have higher need for health information because of the presence of old-age people within the family structure.

Gender

Do men and women differ in their propensity for online health information search? We find the answer in Table 1 and Table 2. Percentage of women responding in affirmative to online health information search is 37% as against 47% that of men. It, therefore, may be concluded that indeed the differences exist, even though they are clearly not as pronounced as that for Income, Internet availability, education, and familiarity with English. The related chi-square value is 7.563 which though is low, is still higher than the tabulated chi-square value and hence it is confirmed that there is a significant health information gender divide.

In literature we come across two divergent views on gender in the context of health digital divide. One considers it to be a significant factor with women being critically disadvantaged, particularly in the developing countries (Moolman et al., 2007; Hafkin et al., 2001; Ono & Zavodny, 2007; Clauss et al., 2013; Gil et al., 2010; Melhem et al., 2009) Others health digital divide to be independent of gender (Hern et al., 1998; Miller, 2001, Brodie et al., 2000; Pandey et al., 2003) with some finding that women being more active searchers. In our case we have found that difference is significant. This echoes the conclusion of many such studies of this nature conducted in developing countries. Women reportedly constitute less than 25% of Internet users in Africa, 22% in Asia, 38% in Latin America, and a meagre 6% in the Middle East. Similarly, they make up for below 10%, 20% and 25% of Internet users in Guinea and Djibouti, Nepal and India respectively (Hafkin, 2001). In fact, the gender digital divide is one of the most significant inequalities amplified by the digital revolution (Moolman, 2007). Non-users reportedly are disproportionately female apart from including older, less educated and poor. Further this divide is across all ages. Problem also needs to be understood in the context of numbers. An estimated 80% live in developing countries and they often suffer far greater gender-related inequity than women in developed world (Hilbert, 2011)

What could be the reason for gender-based health digital divide? Initial understanding of the gender digital divide was thought to be deficit-focused, lately, however, it is being increasingly acknowledged that the divide is also of obstacles to Internet use. It is suggested that pre-existing socio-economic measures of inequality predict inequality in IT access and usage also (Ono & Zavodny, 2007). Women are poorer, less educated, overworked and are socially deprived/disadvantaged. Gil et al. (2010) mentions four barriers impeding women's access to and use of ICTs - Exclusion from technology education and design; limited free time; social norms favoring men; and financial and/or institutional constraints. Technology still is considered primarily a male domain. Melhem et al. (2014) insist that women and girls have little access to scientific and technical education specifically and to education in general compared to their male counterparts. India reportedly has 51% of women literacy compared to 75% men. In the absence of this elementary skill, the Internet and its benefits remain beyond their realms.

Additionally, social norms also hinder women's ability to access net. Internet cafes, in developing countries remain inaccessible to women as they cannot leave home for socio-religious and cultural reasons. Further such public places are intimidating for women because of technological expertise, low self-esteem and their belief that they are socially unwelcome (Clauss et al., 2013). One in five women in India and Egypt believe that the Internet is not appropriate for them, or that their families would disapprove, and that engaging online would not be beneficial for them.

Women also reportedly have Limited free time (Ono & Zavodny, 2007; Pande, 2012) and they often have no control on finances or have adequate personal income to purchase products or pay Internet service providers for monthly access (Melhem, 2014).

There are other studies that claim that women face no divide in access to Internet. A study in West Arica (Benin, Burkina, Faso, Cameroon, Mali, Mauritania, and Senegal) found that no gender gap exists in connectivity or usage among young women educated to secondary school level and beyond (Hafkin, 2007). Highly educated women reportedly access and use the Internet as much as men do it therefore is being argues that with education and the means to do so, women will use Internet as much as men do.

The divergent views are accounted by the fact that in most societies digital divide exists because of fundamental divide in terms of socio-economic structures. Women are less educated, have less free time, have less access to Internet and have less skills in the developing countries. However, when these anomalies are overcome women search online health information to an equal extent if not more. Even in studies where women emerged as victims of digital divide because fewer women lack education, employment and income. The moment these variables are controlled; women are found to be equally or even more active users of digital technologies than men. Education, therefore, emerges as a powerful means to overcome gender gap in Internet access and use (Polat, 2012)

Most studies in the developed world in fact now report that that it is women who lead men in using the Internet for health information (Hern et al., 1998; Miller, 2001). This appears to be taking place across the age as Brodie et al. (2000) has reported that girls are more likely than boys to seek health information. Acquavita et al. (2019) finds that African women prefer mobile phones as this provides greater privacy. Zimmerman (2018) reports than women from low income group use Internet to seek health information primarily for reproductive health. This may reflect perhaps grater need on part of females for such information. Pandey et al. (2003) explain this phenomenon by suggesting as Internet now is a fundamental part of developed human societies that health-conscious women use Internet pro-actively.

In fact, digital technologies have the potentially to enable women and other marginalized groups to prevail over historic inequalities by creating employment opportunities and an increase income, apart from availing cost-effective health care and education (Hilbert, 2011a).

Discriminant Analysis

Since a number of factors such as Income, email, Internet availability, location, education, knowledge of English, and gender have been found to have close bearing on whether or not a sample unit (patient) makes online health information search the next logical step is to determine if it is possible to predict whether or not a person will search for online health-information. This was achieved using discriminant analysis. Following are the results:

We get an Eigen value to be 6.926 which is very high. Such a high value is associated with strong function. Thus it may be interpreted that it is worthwhile to use discriminant analysis in this instance. The value of corresponding Wilk's Lambda is 0.126. Wilk's Lambda is the ratio of the within-groups sum of squares to the total sum of squares. This is the proportion of the total variance in the discriminant scores not explained by the difference among groups. A small Lambda indicates that the difference is significant as is in this case. Chi-square is also the measure of whether the two levels of function significantly differ from each other based on the discriminant function. A high value (In this case it is 1548.43 which is extremely high) indicates that the function discriminates well. The associated significance value indicates whether the difference is significant. In our case this is further confirmed as the significance value is 0 – much below 0.05.

The Table 3 on structure matrix provides us with function values of variables (Income, email, Internet availability, location, education, knowledge of English, and gender) that are being utilized to predict whether or not a person surfs net for healthcare information. From the table we find that most discriminating variable in health information divide is Language (0.996) followed by Email (0.681), Internet accessibility (0.679, Geographical location (0.384), Income (0.330), Education (0.324, gender (0.038, and age (0.009)

Finally, we arrive at the classification results as given in the Table 4. Out of total 440 cases that do not make online search for health information, as many as 416 could be predicted correctly this amounts to 94.5% success and is exceptionally high. In case of those who make online search for health information, out of 314 just one case could not be predicted. The corresponding percentage of the correct prediction of patients who search for healthcare information is 99.7 which is near perfect prediction. It may therefore be concluded that it is possible to predict the behavior of online search for health information of patients in India as it follows a rigid pattern along the identified variables.

Health Vulnerability

The details of the health vulnerability have been provided in Table 5.

Table 3. Structure matrix

	Function 1
Language	.996
Email	.681
Internet availability	.679
Location	.384
Income	.330
Education	.324
Gender	-.038
Age	.009

Table 4. Discriminant analysis classification results

			No	Yes	
Original	Count	No	416	24	440
		Yes	1	313	314
	%	No	94.5	5.5	100.0
		Yes	.3	99.7	100.0

Gender

The mean vulnerability for males and females has been found to be 4.96 and 5.45. Women, therefore have been identified as more vulnerable than men. This gets confirmed by the corresponding t-value which stands at 3.476 and implies that difference is significant. Our Null hypothesis stands rejected in this instance. How accurate is this picture? The correlation coefficient (R) provides interesting insight

Table 5. Health vulnerability assessment

Factors	Mean	S.D.	t/F value	p-value	R value
Male	4.96	1.72	3.476	0.001	-0.47
Female	5.45	1.69			
Youth	3.57	1.54	93.53	0	0.67*
Middle-aged	4.68	1.78			
old age	6.13	1.49			
Rural	4.81	1.58	1.19	0.198	-0.749
Urban	4.63	1.6			
High Income	4.95	1.63	0.512	0.609	-0.471
Low Income	4.72	1.52			
mean of 3 bivariate R value					

as it has a value of -0.47. A negative correlation implies inverse relationship. It appears that people who rate women as more vulnerable rate men as less vulnerable and others who rate women as less vulnerable believe men are more vulnerable. Further questioning reveals that many respondents feel women are more vulnerable because they are physically less strong and also because they bear the major brunt of reproductive responsibility. Many on the other hand believe that men are more vulnerable as they are biologically weaker. Men are known to be more vulnerable to cardiovascular and many other diseases and also their life is marked with greater risks. More men than women die in road accidents. This variability of research findings find resonance in literature also. Female gender is found to have increased likelihood of the frailty condition (Bolina et al., 2019). Women were found to be most vulnerable group in Afghanistan (Trani et al., 2010). Gender has also been found to be statistically independent of health vulnerability. Other studies find men to have higher vulnerability (Bila & Egrot, 2009). This issue could be perceptual also as women consistently rate their health worse than those of men (del Mar García-Calvente et al., 2012). It appears women in younger age group are more vulnerable due to the reproductive burden they shoulder. Men in older age are more vulnerable because of socio-biological issues and perhaps also because they are exposed to harsher environment because of their professional responsibilities.

Age

As we age the vulnerability increases. This appears to be the conclusion of the result. Mean for young, middle aged and old age group is 3.57, 4.69, and 6.13 respectively. Corresponding F value (93.53) and p-value (near zero) confirms the rejection of null hypothesis again. This appears to be true as the mean of 3 bivariate R values is 0.67 indicating a consistent belief in this respect by the respondents. Old people suffer greater health vulnerability because old age diseases are very common and chronic as well. This view is well supported by the available literature (Narushima et al., 2018; Walker & Peterson, 2018; Wiles et al., 2019). Old people are more vulnerable and at the same time are on the negative side of the digital divide. They, therefore, are more vulnerable than they seem to. Healthcare policies need to be built specifically for them. Specially as many studies have shown strong linkages between healthcare learning by old age people and positive health as well as with positive health perceptions (Dolan et al., 2012; Feinstein & Hammond et al., 2004; Field, 2009; Hammond & Feinstein, 2006; Manninen et al., 2014; Schuller et al., 2004; Bailey et al., 2010; Jang et al., 2018).

Geographic Location

Healthcare vulnerability across geographic location appears to be almost equal. The rural and urban mean are 4.81 and 4.63 respectively. There is no significant difference as is indicated by the t-value (1.19) and the p-value (0.198. Yet again the R-value (-0.749) reflects an inverse relationship. People that people who regard urban-dwellers as more vulnerable rate rural-dwellers as less vulnerable and others who rate urban-dwellers as less vulnerable believe rural-dwellers are more vulnerable. Further investigation reveals that a set of respondents believe that urban environment and life-style is hazardous to human health. Another set of respondents consider rural-dwellers as more vulnerable because of unavailability and inaccessibility of healthcare services. We find agreement in available literature on this issue. Unavailability and inaccessibility of healthcare services is a major challenge to rural dwellers (George et al., 2018; Burrow et al., 2018; Gonzalez et al., 2018; Easterday et al., 2019; Ballesteros & Roberts, 2018) Low income and low education could enhance the vulnerability of rural people.

Income

Both high- and low-income group appear to be equally vulnerable as is evinced from the mean values of vulnerability (4.95 and 4.72). Again as the R-value is negative it implies a major difference in opinion. Rich and poor both are considered more vulnerable by the two differing sections of respondents. Recent onslaught of life-style diseases such as diabetes, hyper-tension and depression are more rich-centric. Poor on the other hand are exposed to greater risk in their day-to-day life. Poor people further have less access to healthcare services including healthcare information. Poverty remains a major factor in health vulnerability.

CONCLUSION

The study concludes that variables that affect the propensity to search Internet for healthcare information are Language, Income, and Having E-mail id, availability of Internet, geographical location, Education and gender. Income is an obvious indicator as digital divide is primarily its function. Higher income group people have access to online information whereas lower income group are denied it. Having an email id is a positive indicator in identifying online health searchers this because one of the first activity undertaken by individuals having access to Internet is to use email facility. In other words, email id confirms that a person has access to Internet. Easy and frequent access to Internet too is an evident indicator to recognize online health-care searchers. Urban people, by virtue of having easy access to online facility too have emerged as important indicator in this respect. Rural people by contrast do not show much inclination to search online health-care information. Education too is a primary pointer of digital divide. We find in our study too that online health searchers generally have higher levels of education than the lower ones. Men have shown slightly higher propensity to search for online health-care information than the women. The reason could be generally higher accessibility to resources that men enjoy in India. This is further confirmed when we compare the gender on access to Internet availability. 50% men respond yes to having Internet access as against only 40% women in this study.

It was assumed that age could be an important determinant of online healthcare search behavior because of greater comfort with Internet that younger generation enjoys comparatively. Our study, however, discounts such a hypothesis possibly because older people have greater need to search for healthcare information. This needs to be confirmed through another study. Similarly type of family, whether joint or nuclear, too does not determine the propensity for online healthcare search behavior.

The discriminant analysis results are also extremely favorable and they suggest that predicting a patient who makes online health information search possible with high degree of precision and accuracy on the basis of selected demographic variables.

Our study, therefore, succeeds in identifying important indicators for tendency of individuals to search for online healthcare information in Indian context. This information may be useful for Governments who have overwhelming responsibility to ensure that health information reaches the ones who have been beyond its purview in India. It could be useful for companies that market pharmaceutical products to fine tune informational needs of people in general to enhance overall well-being of the patients.

REFERENCES

Acquavita, S. P., Krummel, D. A., Talks, A., Cobb, A., & McClure, E. (2019). Assessing the digital divide among low-income perinatal women: Opportunities for provision of health information and counseling. *Telemedicine Journal and e-Health*, *25*(1), 48–54. doi:10.1089/tmj.2017.0292 PMID:29708865

Antonio, A., & Tuffley, D. (2014). The gender digital divide in developing countries. *Future Internet*, *6*(4), 673–687. doi:10.3390/fi6040673

Bailey, N., Breen, J., & Ward, M. (2010). *Community Education: More than Just a Course. Exploring the Outcomes and Impact of Department of Education and Skills Funded Community Education*. AONTAS The National Adult Learning Organisation. 2nd Floor 83-87 Main Street [Ireland.]. *Ranelagh, Dublin*, *6*, D6.

Baker, L., Wagner, T. H., Singer, S., & Bundorf, M. K. (2003). Use of the Internet and e-mail for health care information: Results from a national survey. *Journal of the American Medical Association*, *289*(18), 2400–2406. doi:10.1001/jama.289.18.2400 PMID:12746364

Bakken, S., Marden, S., Arteaga, S. S., Grossman, L., Keselman, A., Le, P. T., ... Das, R. (2019). Behavioral Interventions Using Consumer Information Technology as Tools to Advance Health Equity. *American Journal of Public Health*, *109*(S1), S79–S85. doi:10.2105/AJPH.2018.304646 PMID:30699018

Ballesteros, G., & Roberts, L. (2018). A Closer Look at the Peruvian Healthcare System and the Patient Medical Treatment Experience.

Beazley, M., & Smith, M. (1999). Record of the DTI/social exclusion policy action team 15 visit to the Sparkbrook, Sparkhill and Tyseley Area Regeneration Initiative (SSTARI) Birmingham. Information technology and black and ethnic minority communities. School of Public Policy, University of Birmingham, in conjunction with Sparbrook, Sparkhill and Tyseley Area Regeneration Initiative, Birmingham.

Bila, B., & Egrot, M. (2009). Gender asymmetry in healthcare-facility attendance of people living with HIV/AIDS in Burkina Faso. *Social Science & Medicine*, *69*(6), 854–861. doi:10.1016/j.socscimed.2009.05.035 PMID:19539415

Black, J. (1999). *Information rich, information poor: bridging the digital divide*. International Institute for Communication and Development.

Bluethmann, S. M., Coa, K. I., Alfano, C. M., & Hesse, B. W. (2018). Electronic health information exchange opportunities for self-management of care: Responses from older adults with and without cancer history in the United States. *Current Oncology Reports*, *20*(4), 30. doi:10.100711912-018-0674-1 PMID:29572671

Bol, N., Helberger, N., & Weert, J. C. (2018). Differences in mobile health app use: A source of new digital inequalities? *The Information Society*, *34*(3), 183–193. doi:10.1080/01972243.2018.1438550

Brodie, M., Flournoy, R. E., Altman, D. E., Blendon, R. J., Benson, J. M., & Rosenbaum, M. D. (2000). Health Information, The Internet, And The Digital Divide: Despite recent improvements, Americans' access to the Internet—and to the growing body of health information there—remains uneven. *Health Affairs*, *19*(6), 255–265. doi:10.1377/hlthaff.19.6.255 PMID:11192412

Bucy, E. P., & Newhagen, J. E. (Eds.). (2004). Media access: Social and psychological dimensions of new technology use. London, UK: Psychology Press.

Burrow, S., Goldberg, L., Searle, J., & Aston, M. (2018). Vulnerability, harm, and compromised ethics revealed by the experiences of queer birthing women in rural healthcare. *Journal of Bioethical Inquiry, 15*(4), 511–524. doi:10.100711673-018-9882-5 PMID:30402734

Calvano, M., & Needham, G. (1996). Public empowerment through accessible health information. *Bulletin of the Medical Library Association, 84*(2), 253. PMID:8826633

Carr, D. (2007). The global digital divide. *contexts, 6*(3), 58.

Caspary, G., & O'Connor, D. (2003). Providing Low-Cost Information Technology Access to Rural Communities in Developing Countries: What Works? What Pays? OECD Development Centre Working Paper No. 229 (Formerly Webdoc No. 17). *OECD Publishing (NJ1).*

Chanda, N. (2000). Asian innovation awards: The digital divide. *Far Eastern Economic Review, 163*(42), 50-53.

Chinn, M. D., & Fairlie, R. W. (2004). The determinants of the global digital divide. *Choice (Chicago, Ill.), 42*, 7–8.

Clauss, K., Wakahiu, J., & Salvaterra, M. (2013). Using technology to educate women religious in Africa. *Am. Int. J. Soc. Sci, 2*, 29–37.

Cotten, S. R. (2001). Implications of Internet technology for medical sociology in the new millennium. *Sociological Spectrum, 21*(3), 319–340. doi:10.1080/027321701300202019

Cotten, S. R., & Gupta, S. S. (2004). Characteristics of online and offline health information seekers and factors that discriminate between them. *Social Science & Medicine, 59*(9), 1795–1806. doi:10.1016/j.socscimed.2004.02.020 PMID:15312915

D'alessandro, D. M., & Dosa, N. P. (2001). Empowering children and families with information technology. *Archives of Pediatrics & Adolescent Medicine, 155*(10), 1131–1136. doi:10.1001/archpedi.155.10.1131 PMID:11576008

del Mar García-Calvente, M., Hidalgo-Ruzzante, N., del Río-Lozano, M., Marcos-Marcos, J., Martínez-Morante, E., Maroto-Navarro, G., ... Gil-García, E. (2012). Exhausted women, tough men: A qualitative study on gender differences in health, vulnerability and coping with illness in Spain. *Sociology of Health & Illness, 34*(6), 911–926. doi:10.1111/j.1467-9566.2011.01440.x PMID:22443288

Dolan, P., Fujiwara, D., & Metcalfe, R. (2012). Review and update of research into the wider benefits of adult learning. *BIS Research Paper,(90)*, 1-47.

Easterday, A., Driscoll, D., & Ramaswamy, S. (2019). Rural homelessness: Its effect on healthcare access, healthcare outcomes, mobility, and perspectives of novel technologies. *Journal of Social Distress and the Homeless, 28*(1), 56–64. doi:10.1080/10530789.2019.1567978

Ernest III, J., Wilson, E. J., & Wilson III III, E. J. (2004). *The information revolution and developing countries*. Cambridge, MA: MIT Press.

Escoffery, C., Miner, K. R., Adame, D. D., Butler, S., McCormick, L., & Mendell, E. (2005). Internet use for health information among college students. *Journal of American College Health, 53*(4), 183–188. doi:10.3200/JACH.53.4.183-188 PMID:15663067

Fang, M. L., Canham, S. L., Battersby, L., Sixsmith, J., Wada, M., & Sixsmith, A. (2018). Exploring privilege in the digital divide: Implications for theory, policy, and practice. *The Gerontologist, 59*(1), e1–e15. PMID:29750241

Fattah, H. (2000). Politics or real problem? *Technology Marketing Intelligence, 20*(9), 83–87.

Feinstein*, L., & Hammond, C. (2004). The contribution of adult learning to health and social capital. *Oxford Review of Education, 30*(2), 199-221.

Fernandes Bolina, A., Rodrigues, R. A. P., Tavares, D. M. D. S., & Haas, V. J. (2019). Factors associated with the social, individual and programmatic vulnerability of older adults living at home. *Revista da Escola de Enfermagem da U S P.*, 53. PMID:30810627

Fernandes Bolina, A., Rodrigues, R. A. P., Tavares, D. M. D. S., & Haas, V. J. (2019). Factors associated with the social, individual and programmatic vulnerability of older adults living at home. *Revista da Escola de Enfermagem da U S P.*, 53. PMID:30810627

Field, J. (2009). *Well-being and happiness.* National Institute of Adult Continuing Education.

Fox, G., & Connolly, R. (2018). Mobile health technology adoption across generations: Narrowing the digital divide. *Information Systems Journal, 28*(6), 995–1019. doi:10.1111/isj.12179

Fox, S. (2006). Online Health Search 2006. Pew Internet and American Life Project. Retrieved from http://www. pewInternet. org/pdfs/PIP_Online_Health_2006. pdf

Francis, J., Ball, C., Kadylak, T., & Cotten, S. R. (2019). Aging in the Digital Age: Conceptualizing Technology Adoption and Digital Inequalities. In *Ageing and Digital Technology* (pp. 35–49). Singapore: Springer. doi:10.1007/978-981-13-3693-5_3

Gann, B., & Needham, G. (Eds.). (1992). *Promoting Choice: Consumer Health Information in the 1990's.* Consumer Health Information Consortium.

Gann, R. (1986). The health information handbook: resources for self care. Abingdon-on-Thames, UK: Gower Pub.

George, S., Daniels, K., & Fioratou, E. (2018). A qualitative study into the perceived barriers of accessing healthcare among a vulnerable population involved with a community centre in Romania. *International Journal for Equity in Health, 17*(1), 41. doi:10.118612939-018-0753-9 PMID:29615036

Giddens, A. (1991). *Modernity and self-identity.* Cambridge, UK: Polity Press.

Gill, K., Brooks, K., McDougall, I., & Patel, P. K., (2010) A. Bridging the Gender Divide: How Technology Can Advance Women Economically; International Centre for Research on Women: Washington, DC.

Goldsmith, J. (2000). How will the Internet change our health system? *Health Affairs, 19*(1), 148–156. doi:10.1377/hlthaff.19.1.148 PMID:10645081

Gonzalez, K. M., Shaughnessy, M. J., Kabigting, E. N. R., Tomasulo West, D., Callari Robinson, J. F., Chen, Q., & Stewart Fahs, P. (2018). A Systematic Review of the Health of Vulnerable Populations within US Rural Societies. *Online Journal of Rural Nursing and Health Care: the Official Journal of the Rural Nurse Organization*, *18*(1), 112–147. doi:10.14574/ojrnhc.v18i1.507

Gordon, N. P., & Hornbrook, M. C. (2018). Older adults' readiness to engage with eHealth patient education and self-care resources: A cross-sectional survey. *BMC Health Services Research*, *18*(1), 220. doi:10.118612913-018-2986-0 PMID:29587721

Greenberg, A. J., Haney, D., Blake, K. D., Moser, R. P., & Hesse, B. W. (2018). Differences in access to and use of electronic personal health information between rural and urban residents in the United States. *The Journal of Rural Health*, *34*, s30–s38. doi:10.1111/jrh.12228 PMID:28075508

Guillén, M. F., & Suárez, S. L. (2005). Explaining the global digital divide: Economic, political and sociological drivers of cross-national Internet use. *Social Forces*, *84*(2), 681–708. doi:10.1353of.2006.0015

Hafkin, N., & Huyer, S. (2007). Women and gender in ICT statistics and indicators for development. *Information Technologies and International Development*, *4*(2), 25–41. doi:10.1162/itid.2008.00006

Hafkin, N., & Taggart, N. (2001). Gender, information technology, and developing countries: An analytic study (pp. 42-48). Office of Women in Development, Bureau for Global Programs, Field Support and Research, United States Agency for International Development.

Hammond, C., & Feinstein, L. (2006). *Are those who flourished at school healthier adults? What role for adult education?* (Wider Benefits of Learning Research Report No. 17). Centre for Research on the Wider Benefits of Learning.

Hardey, M. (1999). Doctor in the house: The Internet as a source of lay health knowledge and the challenge to expertise. *Sociology of Health & Illness*, *21*(6), 820–835. doi:10.1111/1467-9566.00185

Henwood, F., Wyatt, S., Hart, A., & Smith, J. (2003). 'Ignorance is bliss sometimes': Constraints on the emergence of the 'informed patient'in the changing landscapes of health information. *Sociology of Health & Illness*, *25*(6), 589–607. doi:10.1111/1467-9566.00360 PMID:12919447

Henwood, F., Wyatt, S., Hart, A., & Smith, J. (2003). 'Ignorance is bliss sometimes': Constraints on the emergence of the 'informed patient'in the changing landscapes of health information. *Sociology of Health & Illness*, *25*(6), 589–607. doi:10.1111/1467-9566.00360 PMID:12919447

Hern, M. J., Weitkamp, T., Hillard, P. J. A., Trigg, J., & Guard, R. (1998). Promoting women's health via the World Wide Web. *Journal of Obstetric, Gynecologic, and Neonatal Nursing*, *27*(6), 606–610. doi:10.1111/j.1552-6909.1998.tb02629.x PMID:9836154

Hilbert, M. (2011). The end justifies the definition: The manifold outlooks on the digital divide and their practical usefulness for policy-making. *Telecommunications Policy*, *35*(8), 715–736. doi:10.1016/j.telpol.2011.06.012

Hilbert, M. (2011 a). Digital gender divide or technologically empowered women in developing countries? A typical case of lies, damned lies and statistics. *Women's Studies International Forum*, *34*(6), 479–489. doi:10.1016/j.wsif.2011.07.001

Hoffman, D. L., & Novak, T. P. (1998). Bridging the Digital Divide: The Impact of Race on Computer Access and Internet Use.

Hung, C. L. (2004). The political economy of the digital divide in Taiwan.

Jang, Y., Park, N. S., Yoon, H., Huang, Y. C., Rhee, M. K., Chiriboga, D. A., & Kim, M. T. (2018). The risk typology of healthcare access and its association with unmet healthcare needs in Asian Americans. *Health & Social Care in the Community*, *26*(1), 72–79. doi:10.1111/hsc.12463 PMID:28620950

Judy, R. W., & d'Amico, C. (1997). *Workforce 2020: Work and workers in the 21st century*. Hudson Institute, Herman Kahn Center, PO Box 26-919, Indianapolis, IN 46226; tele.

Kalichman, S. C., Weinhardt, L., Benotsch, E., & Cherry, C. (2002). Closing the digital divide in HIV/AIDS care: Development of a theory-based intervention to increase Internet access. *AIDS Care*, *14*(4), 523–537. doi:10.1080/09540120208629670 PMID:12204154

Kreps, G. L. (2001). The evolution and advancement of health communication inquiry. *Annals of the International Communication Association*, *24*(1), 231–253. doi:10.1080/23808985.2001.11678988

Lach, J. (2000). Crossing the digital divide. *American Demographics*, *22*(6), 9–12.

Levy, H., Janke, A. T., & Langa, K. M. (2015). Health literacy and the digital divide among older Americans. *Journal of General Internal Medicine*, *30*(3), 284–289. doi:10.100711606-014-3069-5 PMID:25387437

Loader, B., & Keeble, L. (2004). Challenging the digital divide?: a literature review of community informatics initiatives. York, UK: Joseph Rowntree Foundation.

López, L., Green, A. R., Tan-McGrory, A., King, R. S., & Betancourt, J. R. (2011). Bridging the digital divide in health care: The role of health information technology in addressing racial and ethnic disparities. *Joint Commission Journal on Quality and Patient Safety*, *37*(10), 437–445. doi:10.1016/S1553-7250(11)37055-9 PMID:22013816

Losh, S. C. (2004). Gender, educational, and occupational digital gaps 1983-2002. *Social Science Computer Review*, *22*(2), 152–166. doi:10.1177/0894439303262557

Manninen, J., Sgier, I., Fleige, M., Thöne-Geyer, B., Kil, M., Možina, E., & Diez, J. (2014). *Benefits of lifelong learning in Europe: Main results of the BeLL-project*. German Institute for Adult Education DIE.

McLaren, J., & Zappala, G. (2002). The'digital divide'among financially disadvantaged families in Australia. *First Monday*, *7*(11). doi:10.5210/fm.v7i11.1003

Melhem, S., Morell, C., & Tandon, N. (2009). Information and communication technologies for women's socio-economic empowerment. Washington, DC: The World Bank. doi:10.1596/978-0-8213-8133-5

Miller, J. D. (2001). Who is using the web for science and health information? *Science Communication*, *22*(3), 256–273. doi:10.1177/1075547001022003003

Moolman, J., Primo, N., & Shackleton, S. (2007). Taking a byte of technology: Women and ICTs. *Empower. Women Gend. Equity*, *21*, 4–14.

Mossberger, K., Tolbert, C. J., & Gilbert, M. (2006). Race, place, and information technology. *Urban Affairs Review*, *41*(5), 583–620. doi:10.1177/1078087405283511

Narushima, M., Liu, J., & Diestelkamp, N. (2018). Lifelong learning in active ageing discourse: It's conserving effect on wellbeing, health and vulnerability. *Ageing and Society*, *38*(4), 651–675. doi:10.1017/S0144686X16001136 PMID:29551843

Ono, H., & Zavodny, M. (2007). Digital inequality: A five country comparison using microdata. *Social Science Research*, *36*(3), 1135–1155. doi:10.1016/j.ssresearch.2006.09.001

Pande, R. (Ed.). (2012). Globalization, Technology Diffusion and Gender Disparity: Social Impacts of ICTs: Social Impacts of ICTs. Hershey, PA: IGI Global. doi:10.4018/978-1-4666-0020-1

Pandey, S. K., Hart, J. J., & Tiwary, S. (2003). Women's health and the Internet: Understanding emerging trends and implications. *Social Science & Medicine*, *56*(1), 179–191. doi:10.1016/S0277-9536(02)00019-9 PMID:12435560

Peizer, J. (2000). What do we mean when we say 'digital divide?'. *Retrieved December 2*, 2000.

Polat, R. K. (2012). Digital exclusion in Turkey: A policy perspective. *Government Information Quarterly*, *29*(4), 589–596. doi:10.1016/j.giq.2012.03.002

Press, L., Foster, W., Wolcott, P., & McHenry, W. (2003). The Internet in India and China. *Information Technologies & International Development, 1*(1), pp-41.

Rolinson, J. (1996). *Health information for the teenage years: what do they want to know? Department of Information and Library Studies*. Loughborough University, UK.

Russell, E., Lloyd-Houldey, A., Memon, A., & Yarker, J. (2019). Factors influencing uptake and use of a new health information app for young people. *Journal of Technology in Human Services*, 1–19.

Schuller, T., Preston, J., Hammond, C., Brassett-Grundy, A., & Bynner, J. (2004). The benefits of learning: The impact of education on health, family life and social capital. Abingdon-on-Thames, UK: Routledge. doi:10.4324/9780203390818

Senecal, C., Widmer, R. J., Bailey, K., Lerman, L. O., & Lerman, A. (2018). Usage of a Digital Health Workplace Intervention Based on Socioeconomic Environment and Race: Retrospective Secondary Cross-Sectional Study. *Journal of Medical Internet Research*, *20*(4), e145. doi:10.2196/jmir.8819 PMID:29685862

Shelia, R. C., & Gupta, S. S. (2004). Characteristics of online and offline health information seekers and factors that discriminate between them. *Social Science & Medicine*, *59*(9), 1795–1806. doi:10.1016/j.socscimed.2004.02.020 PMID:15312915

Slate, J. R., Manuel, M., & Jr., K. H. B. (2002). The "Digital Divide": Hispanic college students' views of educational uses of the Internet. *Assessment & Evaluation in Higher Education*, *27*(1), 75–93. doi:10.1080/02602930120105081

Swartz, S. H., Cowan, T. M., & Batista, I. A. (2004). Using claims data to examine patients using practice-based Internet communication: Is there a clinical digital divide? *Journal of Medical Internet Research, 6*(1), e1. doi:10.2196/jmir.6.1.e1 PMID:15111267

Taha, J., Sharit, J., & Czaja, S. (2009). Use of and satisfaction with sources of health information among older Internet users and nonusers. *The Gerontologist, 49*(5), 663–673. doi:10.1093/geront/gnp058 PMID:19741112

Tait, L. (2007). 15 Years of Chinese Internet Usage in 13 Pretty Graphs. NanjingMarketingGroup.com. Beijing, China: CNNIC.

Trani, J. F., Bakhshi, P., Noor, A. A., Lopez, D., & Mashkoor, A. (2010). Poverty, vulnerability, and provision of healthcare in Afghanistan. *Social Science & Medicine, 70*(11), 1745–1755. doi:10.1016/j.socscimed.2010.02.007 PMID:20359809

Van Dijk, J. (2004). Divides in succession: Possession, skills, and use of new media for societal participation. *Media access: Social and psychological dimensions of new technology use*, 233-254.

Van Dijk, J., & Hacker, K. (2003). The digital divide as a complex and dynamic phenomenon. *Inf. Soc. Int. J., 19*(4), 315–326. doi:10.1080/01972240309487

Van Dijk, J. A. G. M. (2002). A framework for digital divide research. *Electronic Journal of Communication, 12*(1), 2.

Walker, C., & Peterson, C. L. (2018). A sociological approach to resilience in health and illness. *Journal of Evaluation in Clinical Practice, 24*(6), 1285–1290. doi:10.1111/jep.12955 PMID:29901240

Wang, W. (2001). Impact of ICTs on farm households in China. *ZEF of University Bonn.*

Wiles, J., Miskelly, P., Stewart, O., Kerse, N., Rolleston, A., & Gott, M. (2019). Challenged but not threatened: Managing health in advanced age. *Social Science & Medicine, 227*, 104–110. doi:10.1016/j.socscimed.2018.06.018 PMID:29941204

Wilson, K. R., Wallin, J. S., & Reiser, C. (2003). Social stratification and the digital divide. *Social Science Computer Review, 21*(2), 133–143. doi:10.1177/0894439303021002001

Witry, M., Comellas, A., Simmering, J., & Polgreen, P. (2018). The Association between Technology Use and Health Status in a Chronic Obstructive Pulmonary Disease Cohort: Multi-Method Study. *Journal of Medical Internet Research, 20*(4), e125. doi:10.2196/jmir.9382 PMID:29610113

Yoon, H., Jang, Y., Vaughan, P. W., & Garcia, M. (2018). Older adults' Internet use for health information: Digital divide by race/ethnicity and socioeconomic status. *Journal of Applied Gerontology.* PMID:29661052

Zimmerman, M. S. (2018). Assessing the reproductive health-related information-seeking behavior of low-income women: Describing a two-step information-seeking process. *Journal of Health Communication, 23*(1), 72–79. doi:10.1080/10810730.2017.1411996 PMID:29265926

Chapter 3
Decoding Cool in Indian Context:
Meaning, Value, and Brands

Harsh V. Verma
University of Delhi, India

Ekta Duggal
Motilal Nehru College, India

ABSTRACT

The term cool is both nebulous and valuable. It contains some magical property which transforms perceived value of anything to which it is added as prefix. Therefore, marketers want to make their brands cool for superior outcomes. But the challenge lies in apprehending what cool actually stands for and how to use it in brand building. This paper explores meaning of cool by tracing its genesis in theological texts and popular culture. Cool has connections with theological discourses of religions like Hinduism, Stoicism, and Buddhism. Buddhism's fundamental tenet is cultivation of equanimity. Hinduism's sacred text Bhagavat Gita exhorts development of control over senses. Stoicism's core principle is 'Ataraxia' or indifference. Cool in popular culture originated from the sites of slavery by people who cultivated an attitude to calmness as means to survive in punishing conditions. This study assumed cool as a signifying system and sought to uncover the hidden meaning for which it stands. It found that cool stands for four human facets: composure, paradox, good, and cheeky.

INTRODUCTION

A multi- level marketing company has made an entry into premium skin care segment with a new brand. The brand is being aggressively promoted on television. But unlike many cosmetic brands which employ 'ingredient-benefit' strategy this brand exclusively focuses on a particular user personality. One of the ads of the brands goes as follows: the setting is that of a fashion show and models in their high precision choreographed walk on the ramp are appearing one after the other. And then emerges a gorgeous model

DOI: 10.4018/978-1-7998-0357-7.ch003

and camera zooms to show her meticulous and confident walk down the stage. She is probably the most admirable of all for the confidence and control of the moment and then there happens the unexpected. Her stilettos go loose and she tumbles. She smiles, collects her stilettos and holds them in her hands, claps, and stands up only to continue her walk up to the end of the ramp unruffled and perfectly at ease. The communication refrains from making elaborate description of product and only introduces the name and who it is meant for. The brand seeks to connect with women, who can laugh at their own self. That is a person who can take imperfection, failures and stress with a characteristic ease.

In another case, a soft drink brand seeks to connect with its target consumer with the proposition that it is perfectly fine to be afraid of certain things. The brand communication shows the protagonist in challenging situations like jumping off a cliff into the sea. In one of its latest campaigns, the ads show a group of boys engaged in a rescue mission training on high seas. As other boys happily take the plunge, one of the boys upon being asked to take a jump is shown to be seized by fear at the moment of action. The protagonist then gathers himself, overcomes his inhibitions and takes the plunge to emerge victorious. The brand story focuses that it is perfectly fine to fear and be afraid of certain situations but success lies in overcoming fear to face a challenge. The brand's essence is condensed core proposition which says that victory is next to fear.

The above and many other brands provide testimony to the emergence of a new psychological makeup especially amongst the younger generation. The ideas that are closer to reality seem to be building up acceptance. Unlike tried and tested success formula which dichotomized between the 'reality' and the 'projected' to cultivate resonance based on the unreal, the new format is given to realism with all its flaws and defects. Now it is 'cool' to 'being' rather than 'becoming'. The term 'cool' captures a condensed distilled identity of new emergent young consumer which militates against the accepted idealized identity stereotypes.

RATIONALE

'Cool' is irresistible property for a brand to have. Brands gain value if they are added with 'cool' property. But it is not easy to achieve because 'cool' is nebulous and hard to apprehend. This is the reason why phenomena of 'cool' merits attention.

'Cool' Adds Value

Notwithstanding the difficulty in understanding cool, being cool matters. It is associated with popularity and attractiveness which can endow a person considered to be cool with a wide range of benefits including personal, social and economic (Hamermesh, 2011; Zebrowitz & Rhodes, 2004). It is common understanding that some people and products are considered to be cool and being bestowed with this nebulous; hard to define property certainly gives them a position of advantage. For many people a brand like Apple is cool and this gives the brand in question a position of relative superiority in terms of brand desirability, price advantage and brand stickiness. Cool is especially important in children and youth segments. A lot of importance is attached to being cool. But consensus is difficult to reach as to what cool is and what it is comprised of and who personifies it the best (Adler & Adler, 1998; Pountain & Robins, 2000). Cool does not lend itself to easy articulation. It is elusive but sought after and it is something to be coveted (Dick & David, 2000). The property of cool holds tremendous marketing worth especially during the time of dawning of commoditization.

Coolness is an appropriated property. It is not manufactured or assembled in factories. With the dawning of pervasive parity across product categories, brands are forced to tap into the culture reservoir to create differentiation. Being a cool brand can reverse the effects of free market system and provide escape routes by abstracting a quantifiable value into a symbol of expression and identification. Grossman (2003) called Cool as the 'most precious natural resource' which can make an otherwise substitutable product 'fantastically valuable'. Cool in this contextualization assumes position like a marketing resource which is used to drive consumer behavior. The phenomenon of cool in branding context is capable of influencing consumer response. It is through this consumer influence brands can gain significantly. Cool is key favorability driver across wide products and services and it is important across all age groups (Sturgess, 2013). The new globalised economy leaves very little scope to set one product apart from the other. Consider a pair of jeans or drinks or sneakers or automobiles, none differ substantively. Many brands like Rolex, Ray Ban, Aston Martin and Absolut defy commodity gravity by the power of 'Cool'. Becoming Un-cool is a sure run down the lane of oblivion. In the post-industrial America, cool started as a counter culture and it went on to become a dominant aesthetic category, and a tool to conquer the global society (Kohlenberger, 2016).

'Cool' and Transformation

The arrival of Cool in branding context is interesting and intriguing. Brand gains power by developing a perception of being Cool, which implies that it appropriates myths and mythology from a non-branding context. A brand achieves value transformation by tapping into meaning reservoir housed in socio-cultural construction of the society. Cool refers to meanings which are shared between the members of a society (Sundar et al., 2014). These meanings have facets of attitude, behaviors and values (O'Donnell & Warlow, 2000).

Becoming cool is about dissolution of objective reality and subtly replacing it with a subjective and perceptual construction. It is about de-centering, shifting the product from the center of perceptual frame and filling the center with the imagined reality. Coolness is one of the important ways to make a brand stay hot. Among other things, coolness is essential ingredient of a youth brand (Bergh & Behrer, 2016).

'Cool' and Cultural Capitalism

The capitalist model relies upon consumption for its sustenance and growth. The cycle of consumption must spin for business enterprises for sales and surpluses. This requires customers to be pushed into endless cycle of consumption. The contemporary consumption is not about meeting needs rather living new lifestyles and experience of the novelty. The markets are no longer the places where products are exchanged rather they are suppliers of meanings. This transformation is what Rifkin (2001) calls cultural capitalism. In contrast to industrial capitalism which was about supply of products, the wealth generation process has subtly shifted to production of symbols and marketing of attitudes and lifestyles (Žižek, 2009). The consumption is no longer about satisfaction of needs. It has seeped into culture and now money is made by commodification of culture (Rifkin, 2001).

Cool in this regard offers great opportunity for the marketers. By inventing and discovering new meanings products are transformed into 'cool' attractive offerings. For consumers it is cool to wear Nike not for its shoe quality rather and attitude and spend time at Starbucks not as much for coffee but for an experience. Cool fuels markets and gives businesses opportunities to make money on the strength of cool symbolism. The cool is hunted by professional cool hunters from the sites where youth congregate and is used by brands to extend their meaning in to realm of contemporary youth culture. Cool is defining characteristic of contemporary capitalism (Ferguson, 2011). The media diffuses cool across continents and gives rise to new markets. What was once a North American phenomenon has reached to Greenland (Kjeldgaard & Askegaard, 2006). What is cool diffuses through consumption which generates money but at the same time it also renders it uncool creating the need for the discovery of new cool. Therefore, its quest is an endless one (Danesi, 1994) and fuels endless consumption (Bell, 1979).

'Cool' is Nebulous

What is meant by cool remains elusive and invites diverse set of explanations including edgy to hip to 'sweet'. Cool Brands Council develops a list of cool brands based on expert and expert opinions. In deciding which brand is cool, the rating begins with understanding the concept of 'cool'. The idea of cool remains subjective and personal. In spite of subjectivity surrounding the concept, the Council relies upon factors such as style, innovation, originality, authenticity, desirability and uniqueness to rate brands (Cool Brands Council, 2013). At the heart of the debate lies seeking answer to the most fundamental question, what is cool? Is it a particular way of behaving or a personality or a kind of sensibility or an ideology or an attitude (Poutain & Robins, 2000)?

'Cool' is an elusive aspect but adds definite value. Its inroad into marketing is visible in product design and promotion. A study of tweens identified three drivers of 'cool': design, innovation and uniqueness (Jones, 2012). Consumers, especially in younger age groups want 'cool' things. The importance of 'cool' is widely shared by people who market to children. DelVecchio (1997) found that peer group influences the perceptions of what is cool for children. 'Cool' aesthetics has become significant in youth culture as is visible in youth music and fashion (Botz-Bornstein, 2010). Whatever it is, its trans- formational effects are undeniable. Consider a brand like Apple which has long thrived on being perceived as cool while Levis has suffered because it is no longer considered to be cool a pair of jeans. Rival brand like Calvin Klein is cool and happening. Cool is transformational. The perception of cool can catapult a commodity into a higher value spectrum and thereby create liking and preference. Cool however remains a fuzzy concept and it defies description into finely crafted definition. What it stands for has been changing with time and differs across cultures. For instance, cool's early slang meaning dates back to early seventeen hundred which stood for large sums of money which went on to evolve into somebody 'calmly audacious' in early eighteen hundred. Again in a radical break from its earlier meaning 'cool' came to stand for 'fashionable' in the 1930s (Kiplinger, 2012).

Sign and Meaning Underneath

Any sign operates in language through a two-tiered meaning structure. The first order signification is about denotation or dictionary meaning. The second order signification stands for culturally adjusted meaning (Barthes, 1967). 'Cool' as a linguistic sign carries a literal or denotative meaning. In order of signification the word 'cool' literally stands for a particular level of temperature which cannot be designated hot or warm or is of comparatively low temperature or less hot. 'Cool' in this frame of meaning offers little marketable value capable of value transformation. However, a cool attribute or character can be of use for products whose functionality is tied to generating coolness. Acceptance of 'cool' in brand building probably has something to do with the hidden signified which conveys meaning at the denotation level. 'Cool' is a double entendre word, its meaning operates in marketing through connotative signified with its movement into the cultural realm; the sign assumes psycho-social dimensions at the expense of its literal meaning Considering the second level the word assumes a new meaning by an act of subversion of its literal signified. This opens up opportunities for strategists to fill their brands with a psychological substance relevant to consumers, especially the youth segment. But before 'cool' could be exploited as marketable value adding dimension, it is important to explore what it stands for.

THE STUDY

There is no denying the fact that phenomenon of 'cool' is both real and impalpable. Marketing strategists fully appreciate the value that it can add to an otherwise commoditized good or service or place. Consider what Judge Colin Briss presiding over an infringement suit between Apple and Samsung observed. In response to Apple's claim that Samsung's Galaxy tablet was very similar to iPad, the Judge disagreed on two counts. First Galaxy did not have the same simplicity as iPad's created by its design and secondly it was not as cool. Brands differ in their cool quotient. It is a powerful differentiator. It is a strange reality and some brands like Apple and Aston Martin become hot by being 'cool'. This apparent irony between hot brand and 'cool' characterization raises many questions. First of all how the 'cool' and 'hot' dichotomy assumes significance in popular consumer culture. How do brands gain consumer liking and preference by appropriating meaning? How does a brand achieve value transformation by becoming 'cool'? What symbolic qualities does a brand need to appropriate in order to be 'cool'? What does 'cool' have to do with the culture of consumption? The search for capturing the fluid and transitory phenomena of 'cool' has given rise to a whole new industry of cool hunters (Gladwell, 1997; O'Donnell & Wardlow, 2010). The race is on to capture construct of cool and develops cool products. This study was guided by the following objectives:

1. To explore the genesis of the concept of 'cool' and its meaning;
2. To understand how 'cool' has become a marketable property capable of value transformation;
3. To find what does the term 'cool' mean.

A step research design was adopted to pursue research objectives. A discovery of meaning of the concept of 'cool' necessitated a dig into literature; accordingly, literature published on the subject was collected, analyzed and synthesized. The purpose was to locate and explore the origins of the concept of cool and what consumer behavior processes were responsible for its translation in general domain of marketing and branding in particular. Armed with this background, an exploratory study was undertaken with a young group of respondents between age group of 18-24 years. A convenient and judgment sampling procedure was adopted in which college students were asked to express their ideas as to what does the term 'cool' signify to them. The respondents were asked to write down their ideas on a piece of paper in a free flowing manner. Often identity disclosure causes the participants to provide guarded responses. In this survey the participants were not asked to reveal their identity. The responses were collected from a sample of two hundred respondents. The data were also collected by asking the same question on a Facebook page which had over thirteen hundred friends from the target age group. We managed to get forty usable responses on the Face book page. The collected responses were content analyzed to explicate the meaning and dimensions present in the construct of 'cool'. Content analysis is a widely used tool in fields like social psychology, communication and advertising (Pollay, 1985) and image analysis (Zimmer & Golden, 1988). The data used for the study comprised of text and involved meaning based analysis (Polkinghorne, 1983). This process involved multiple steps from multiple iterative readings of text and development of codes and categories (Spiggle, 1994). The authors instead of using the external coders performed the task of coding themselves because the external coders may be efficient but may lack subject knowledge which is essential for effective interpretation of the data (Spencer et al., 2014).

FINDINGS AND DISCUSSION

The study led to the following findings which are divided into three sections. First the genesis of the 'cool' is reported followed by its migration into popular culture. Finally results from the survey are reported.

Genesis

Theological Connection

'Cool' in its latest avatar appears to have originated and found currency in marketing. But quite contrary to popular opinion, the concept of 'cool' is traced to religious discourses. It shares some connection with the concept of equanimity. Equanimity is one of the four fundamental tenets of Buddhist religious orientation. Buddhism is about cultivation of virtues of compassion, loving kindness, sympathetic joy and equanimity. Equanimity is English translation of 'pali', 'upekkha' and Sanskrit 'upeksha'. This stands for a mental stage of an individual who is not ruled by passion, desires, likes and dislikes. A top Buddhist monk and scholar described it as:" a state of inner equipoise that cannot be upset by gain and loss, honor and dishonor, praise and blame, pleasure and pain. Upekkha is freedom from all points of self-reference; it is indifference only to the demands of the ego- self with its craving for pleasure and position, not to the well-being of one's fellow human beings (O'Brian, 2013). The corresponding Sanskrit word 'upeksha' translates into a state of non-attachment, non-discrimination, even mindedness and letting go (ThichNhatHanh, 1998). It is about liberation from the influence of passion, desires, likes and dislikes.

According to Hindu philosophy, human life is all about a negotiation between opposites. It is characterized by duality of pleasure and pain, success and failure, happiness and misery. Breaking from these dualities is a movement toward realization of the true self. The suggestion is to be indifferent to the pair of opposites- *nirdvandvonisthyasattvastho* (Sargeant, 2009). Being affected by the dualities is not the true nature rather being unaffected and unattached is. One of the core principles of Bhagavad Gita is winning over senses and not be affected by opposites that surround life. Living in the world of pleasures and pains and yet being unaffected and unattached is core to reaching the real being. The key to liberation is action without getting attached to fruits of action. Yoga is all about development of indifference and evenness of mind (Minor, 1986). Yoga has been used in many ways in Gita but its use is special in the sense of cultivation (*Smatwam Yoga Uchachyate*) of indifference to sorrow or happiness (Karan, 1989). Being in a world full of pleasures and pains and yet keeping a practiced detachment and indifference is core to existence. It is about transcendence from attachment to joys and aversion to miseries. One should remain unaffected by these opposites of life.

The idea of tranquility and indifference to is also found in western religious beliefs. Found by Zeno, Stoicism focuses on virtue of 'Ataraxia' or painlessness. An indifference to the opposites of pleasure and pain, life and death, wealth and poverty is central to the Stoic idea. The pain and adversities should be endured as they are beyond control. And this endurance should be done without display of emotions. Conflicts and turbulence are givens to which a mindful indifference should be cultivated. Sufferings are unavoidable therefore a life of 'apatheia' or apathy is desirable. The idea is not to kill emotions, rather to achieve transformation in them by developing clear judgment and inner calm (Graver, 2009). Passions, desires, and impulses depend on us and therefore these must be mastered. Whatever comes your way must be accepted with equanimity rejoicing in victories or despairing in failures. Destructive emotions are caused by errors in judgment. Stoic calm is about achieving freedom from anger, envy, and jealousy (Russell, 1972).

'Cool' and Slavery

'Cool' as is presently known in marketing as an aesthetic concept is quite different from its origins. The 'cool' in its original context was invented as a survival strategy of slaves who were subjected to extreme and adverse human conditions. The people on the fringes of society like prisoners and slaves were subjected to brutal behavior and inhuman conditions. Disobedience or retaliation attracted severe punishment like burning, mutilation, and hanging. The identity of African men in pre-colonial, pre-slavery period was defined in terms of its connection to battles and tied to strength, bravery, capability to uphold 'manhood creed' established by strong community bonding and family relation- ships. When placed within the context of an enslaved existence, the physical abuse and soul wrenching conditions challenged this African notion of masculinity to fight back and revolt but only to invite accentuated degradation of existence (Mazrui, 1977). 'Cool' was invented as a mask to hide inner reality. It emerged as a counter strategy. To act out composed, unruffled, nonchalant, detached, un-perturbed and unexcited against the agitating body, mind and soul (Morris, 1969).

For African-Americans the society continues to be restrictive and oppressive in many ways. The social system and institutional organization denies them fair opportunities to join the mainstream culture. The black males living in inner cities of the US adopt behavioral patterns entailing unique postures, impressions, and carefully orchestrated performances de- fined under the rubric of 'cool pose' (Majors & Billson, 1992). All this is done to establish an identity defined by pride, strength and control. Four reasons that propel black males to adopt postures of coolness are as: a navigation strategy of their world; systems to establish their manhood; source of resilience and form of aggression, strength and power (Connor, 1995). 'Cool' is paradoxical. It is about manufacturing a calculated inconsistency between manifest and the manifested. Cool in this sense is ironic. It represents paradoxical fusion of submission and subversion (Bornstein, 2010). Cool is about curbing and concealing the internal agitation and presenting an unaffected façade outside. Cool is about disguise.

'Cool' and Popular Culture

A word in linguistic system is called a sign which stands for a signified or a concept. Words have the capacity to carry multiple meanings. This phenomenon is represented by the concept of 'polysemy' of signs. A text has potential to generate infinite range of meanings (Hebdige, 1979). Words evolve and change their meanings with time. Slang and popular culture contribute adding of multiple meanings to words.

A word like 'gay' has acquired a completely different meaning from what it used to mean. In popular culture it represents 'homosexual'. The dissimilarity of backgrounds between the producer of text and its reader can potentially alter the interpreted meaning. There are always several meanings attached to a sign (Gottdiener, 2001). The lack of boundaries surrounding the sign 'cool' has contributed to its usage in different contexts including popular culture. The word 'cool' is used as slang to mean several things.

The polysemic nature of 'cool' prompted one study to investigate what all is associated with it (Dar-Nimrod, Hansen, & Proulx, 2012). It found that the term is used in somewhat different sense in popular culture far removed from its original meaning and original context. Cool seeped into mainstream consciousness through music and musicians, especially Jazz. Jazz is certainly a form of music but it also stands for a camouflaged idea. It represented a confrontation between an imposed structure of prejudices, exclusion and restrictions and a fragment trying to challenge it. As a music form of blacks, it stood for defiance and confrontation (Shapiro, 1999). Jazz evolved in the 1920s and was perceived as something hostile to old values. For the people in establishment jazz represented anti-establishment attitudes, decadent values and culture of unencumbered living. Later, taking a further detour 'cool' went on to get linked with a category of people called 'hipsters'. The term hipster is a derivative of the word 'hip'. Hip is slang for a person who is aware of new fashion and stylish. Hip is opposite of traditional or square. 'Hipster' and 'hippie' represented a counter sub-culture. Their lifestyle was influenced by the Jazz music and musicians influencing their dressing sense, language, attitude, drugs usage and liberal sexual orientation. To Frank (1977) hipster is amoral, anarchistic, gentle, and over civilized to the point of decadence. Hipster subscribed to an alternative lifestyle that ran counter to the mainstream ideology, they subscribed to their own liberal sensibility. Jezer (1992) calls it as an amorphous movement without ideology, way of 'being' without reason 'why'.

Criticism notwithstanding, hipster way of 'being' did enjoy a magnetic pull over a lot of people including whites. Norman Mailer (1957) explored how the then emergent black counterculture attracted younger whites. The pull of lifestyle related to blacks was so strong among younger whites that many of them voluntarily adopted artifacts, language, dress, music and values associated it. Norman Mailer's book's title 'The White Negro' is oxymoronic, a White cannot be a Negro and vice versa. One excludes the other in physical sense but a White can be Negro in thinking. Rejection and rebellion are two core ideas governing hipster beliefs and attitude. The quest for new meaning of life through sensory gratification by use of stimulants and sex by the marginalized ended up building up its own followers. The values like self-expression, unrestricted hedonism and sensory orientation, adoption of the forbidden, liberal sexual orientation unique to a select group came to define 'cool' as sections of people from the adjacent. For marketers, these values presented powerful symbolic resource for brand building. Brands by incorporating 'cool' values make the 'cool' available off the shelf. But 'cool' is transient. It can be incorporated on after it has been identified.

What is 'Cool'?

The phenomenon of 'cool' has attracted numerous meanings and interpretations yet it remains elusive and difficult to capture. It is this ephemeral quality that makes cool desirable (Kohlenberger, 2016). The range goes on to include 'cool' as an attitude of individuality, series of movements and flash of red hot radiation (Rice, 2002). It is often said that the very act of finding 'cool' is to lose it simultaneously. The discovery of 'cool' is paradoxical; 'cool' if discovered is no longer 'cool' and it takes a flight. There is circularity in 'cool's discovery and its loss (Gladwell, 1997), the moment of its find is also the moment of its loss. The cycle of discovery and loss gave rise to a new set of marketing professionals given to hunting what is cool. They immerse themselves in the natural habitat to discover cool trends to be used by marketers. Discovering what is cool is an ongoing process. Brands can develop edge over competition by seizing and incorporating cool trends and thereby earn superior returns (Runyan et al., 2013). Cool endows brand with liking and preference over competition (Sturgess, 2013).

The word 'cool' is used in different con- texts and with different imports. Some of the expressions that are used by the young include 'cool money', 'cool looks' 'cool treatment', 'that's cool', 'cool song', 'cool person', 'cool down', 'cool it man', 'cool movie' and 'take it cool'. The wide range of its usage implies its amorphousness and flexibility. There is certain amount of attraction that 'cool' enjoys but the question arises what do young have in mind when they use the term 'cool'. Is it an attitude, a way of being, or behavior pattern? What is it that invests it with mythical power that it has become an important coin in the currency of youth communication? What sits at the heart of the concept of cool? Our attempt to apprehending the meaning of 'cool' led to understanding that it is a term characterized by multiplicity of meaning. The lack of a fixed meaning leaves it open to adaptation in different contexts. The range goes on to include different aspects of a person including moral character of person, way of behaving, external appearance, orientation towards life and personality. It is a human phenomenon. Our study led to identification of four aspects of 'cool': 'cool' as temperament; 'cool' as reconciliation of opposites; 'Cool' as an external façade and 'cool' as moral character.

- **Composure as 'Cool':** This perspective comes closest to its genesis during the time of slavery when people subjected to extreme inhuman treatment and harsh conditions cultivated an attitude to remain calm under stress. It also is in harmony with the religious or theological interpretation of 'cool' as equanimity or a sense of detachment. The 'cool' sits here as a survival device. The character of James Bond is 'cool' for he always remains unruffled and unaffected by the pressures of the situation. Mahatma Gandhi was 'cool' for his capacity to stick to his resolve irrespective of pressures. Not getting agitated or excited during situational provocations is the reverse of ordinariness. It is common to break down under stress. But it is 'cool' to have 'never break under pressure 'personality.

There is no dearth of examples when people give in under pressure but to remain calm and poised under stress is extraordinary. Tag Heuer, a well-known Swiss watch brand in one of its advertising campaigns used a tag- line that went as 'What are you made of?' The brand is endorsed by the likes of Shah Rukh Khan and Tiger Woods. Both of the endorsers epitomize extreme mental strength and remain calm in challenging situations. This was in continuation of its earlier 'Don't crack under pressure' campaign. The 'Inner Strength' ads ran in nineties and focused on the supremacy of mental strength. A recently launched brand by the name of Attitude by Amway taps into this 'cool' psychology to build customer pull. One of the ads of Attitude shows how a model walking down the ramp suddenly gets tipped over due to high heels. Instead of hiding her face in embarrassment and shame, she pulls herself together with surprising ease, smiles, and continues her walk.

- **Paradox as 'Cool':** Actor Shah Rukh Khan's character in Bollywood blockbuster 'Kuch Kuch Hota Hai' exemplifies 'cool' as juxtaposing apparently irreconcilable opposites. He is portrayed as a typical cool dude in his appearance and his demeanor is typical American student on the edge. He wears designer clothes, is hyper active, bursting with energy, spontaneous, and reckless to a point of appearing insensitive, rash and carefree. But what makes this character appealing is the juxtaposition of his external 'hip' and reckless façade with sensitive and value oriented hidden character ('A man bows his head only in front of three women – in front of his own mother, in front of Goddess Durga and.... ;('We live once, we die once, we marry once and love also happens only once'.) The 'cool' is about effortless negotiation of culturally imposed binaries. The culturally imposed binaries that have been imposed to regulate behaviors are now in state of flux. To be 'cool' is to challenge these apparently dichotomized gender roles, notion of time, hierarchies and values. The 'cool' is the concept of 'the complete man', a concept appropriated by brand 'Raymond'. Completeness is about transcending muscular notions of manliness and combining it with the soft and emotional side associated with women.

Consider developments in beauty and wellness space. Beauty has made inroads into masculine gender and muscularity is new womanhood. The cricketer Virat Kohli epitomizes 'cool' for the youth in our study. He juxtaposes his brash, aggressive, arrogant and outspoken side with extreme talent which makes a 'cool' idol. It is 'cool' to escape from binaries like love and hate, study and play, rich and poor, reverence and ridicule. The 'cool' is to be able to juxtapose contradictory emotions and beliefs like love with hate, and reverence with ridicule. This is in consonance with postmodern cultural reality (Foster, 1983; Hutcheon, 1988; Wilson, 1989). 'Cool' is reconciliation between apparently irreconcilable cultural opposites.

- **Good as 'Cool'**: What characterizes the Occupy Wall Street, Arab Spring, Protest against Delhi Rape/ violence against women, anti-corruption protests, and other movements of similar nature led by youth across the globe? What unites these protests amidst their different locations and issues is that they all are voices against the ills that plague our society. Over the years, society across different continents has witnessed tremendous depreciation in values. The value system acts like an extra judicial system that does not allow human behavior to violate established socio-cultural norms. But ethical and economic breaches that have rocked the conscience in the recent times have brought conscience back into both popular and serious academic discussion. The youth is frontrunner in this movement.

Quite contrary to popularly held notion that being rich and successful is cool, our study discovered that young find it 'cool' to be a man of conscience. It is 'cool' to bear a torch against degraded human character which is at the root of social, economic and environmental malaise. To protest is perfectly legitimate; to be able to hold on scruples is 'cool'. In the business space Narayana Murti and Azim Premji is cool, so is anti-corruption crusader Anna Hazare. Dr Devi Prasad Shetty credited for compassion in health care is 'cool'. In response to our question defining cool, we found that to be able to keep to your conscience and uphold ethical values against pervasive moral degradation is 'cool'. Beneath the apparent glitz and gloss of consumerist society, for some sections of youth, 'cool' is to resist moral derailment caused by 'money culture'. Consider the television advertisement of Mahindra's newly launched compact sedan Verito Vibe. The ad exhibits a young person driving the car happily with a group of friends. On the road another person riding a scooter with his son gets involved in a race like situation. Both race to beat each other, but eventually the car driver deliberately slows down to let the kid on the scooter win. The ad appeals to moral value of sacrifice for the sake of others. Similar theme is visible in currently run advertisement featuring actor Salman Khan for Dicxy Scott innerwear for men. The ad features an arm wrestling competition wherein the actor wrestles against a man. Upon discovering that his opponent's son is watching the game and he wants his father to win, the actor gives up only to make the child happy.

- **Cheeky as 'Cool'**: This concept of cool is epitomized by ad campaign for Sprite, a brand of cool drink signed off by the statement '*chalo apni chal* ('walk your own way'). In is one of TVC shows three boys are being taught violin by a beautiful teacher. One of the boys is smitten with his violin teacher. As the teacher arrives and wishes them good morning, all the boys try to impress her by their skills by playing violin. Then the protagonist is asked to perform. Instead of impressing her with skilled play of violin he starts to make a screeching noise. The teacher impressed with the performance of other boys, tells them to meet her next week and asks the protagonist to spend two hours with her every day. The campaign appeals to a sense of overcoming odds with tongue in cheek humor and ingenuity.

Our study discovered that tongue in cheek humor and ability to poke fun at yourself is 'cool'. The 'cool' in this strand challenges the idea of being overly rational and thought driven. This 'cool' thrives on unexpected juxtapositions that produce humor. The Dollar Club ad begins with Akshay Kumar pretending to be blind. He stumbles his way into a bank being robbed by armed men. He fights the robbers off in a high voltage action sequence. After having beaten all the robbers he comes back into his blind garb. He puts on dark glasses and begins to walk with a stick. One of two staffers witness to the sequence visually stupefied by his heroic feat observes 'sad, he is blind but what a fighter'. In a shocking tongue in cheek reply he reminds her that her button is unfastened ('but- ton khulahaiaapka'). 'Cool' in this perspective implies thinking out of box and acting smart. It is harmless cleverness to produce results. 'Good looking rascals' campaign of readymade apparel brand John Miller comes very close to this category of 'cool'. The brand is targeted at young men and seeks to develop connect based on difference between hard-worker and smart worker. The ads show how a young good looking executive smartly maneuvers a difficult office situation.

The four perspectives in Table 1 were found to be dominant in our investigation. Besides this conceptualization of 'cool', some other minor fragments were also found. These perspectives were not categorized separately for the reasons of low frequency of mentions. Some of the popular meanings that 'cool' rep- resented include 'great', 'excellent', 'clever', 'skilled', 'socially adept' and 'acceptable'. Two noticeable minor categories of 'cool' were found to be making money quickly and being well-dressed. It is 'cool' to rise up the wealth spectrum either through entrepreneurship or climbing corporate ladder. Being 'cool' is also found to be construed narrowly in terms of external appearance of a person. Being

Table 1. Different perspectives of 'cool'

Perspective	Meaning	Illustrative brands
Composure as 'Cool'	Clam under stress, un-crushable spirit	Tag Heuer, 7 UP, Attitude
Paradox 'Cool'	Juxtaposition of cultural binaries	Raymond, Bajaj, Wills
Good 'Cool'	Conscientious, moral and ethical	Dixcy Scott, Verito Vibe
Cheeky as 'Cool'	Smart ingenuity, out of box, harmless clever	John Miller, Sprite, Pepsi, Ell 18

in the know of contemporary fashion and keeping pace with it is also 'cool'. The phenomenon of 'cool' extends on to include attitudinal or dispositional, behavioral and aesthetic sphere of human existence.

Aspects of Cool Visa-s-Vis Uncool

In the Figure 1, these aspects of cool have been contrasted with uncool by locating their dictionary opposites.

Cool derives its value from its scarcity. For manufactured products the scarcity of inputs drives its value. Symbolically brands become cool when they appropriate meanings that are rare and uncommon. This initial exploration of cool revealed that cool is defined by four temperamental and behavioral aspects. What renders these aspects as cool bestowing ingredients? This is brought to surface when these semantic terms are contrasted with their dictionary opposites.

Meanings are decoded in the interpretation process by linking what is present with what is absent. And when cool categories are compared with their opposites and related with contemporary cultural categories and narratives a confrontation is revealed. And this is what sparks cool. The first aspect defines cool as composure. The linear response to the situation when things are unsettling and stressful is to get agitated. The predictability and ordinariness is un-cool. Further, the greed and selfishness have emerged as dominant defining ideals of individuals. Siding with this overarching ideal defining the contemporary existence is un-cool when it is compared with the aspect 'cool as good' discovered in this study. Next, 'Cool as paradox' defines that taking a challenging position is cool. The finding that cool is when opposites are juxtaposed suggests assuming an uncommon position. This implies it is un-cool to fit neatly into dominant categories. Acceptance of the given is un-cool. Cool is for brave few who seek to create new category by seeking to reconcile the apparent dichotomies. Finally, it is un-cool to abandon one's own identity and live a life of pretentions and appearance. Containing the inner drives and spontaneous action is un-cool for this is what everybody does. Cool resides in an ability to be true to oneself. To sum up, our study indicates that cool emerges from a confrontation with dominant and taken for granted narrative enshrined in cultural categories. Underneath cool resides a powerful force of challenge that confronts the dominant and dominance.

Figure 1. Aspects of cool visa-s-vis uncool

Composure • Calm - stressed • Equanimity - agitated • Detachment - affected	**Paradox** • Reconciliation of opposites - one-sided • Challenge dichotomies - accept dichotomies • Masculinity and softness combined - masculine or feminine
Good • Conscience- Non-conscientious • Being good - degraded • Ethical - unethical	**Cheeky** • Sense of Humor - Serious • Out of Box - In the Box • Spontaneous - Calculated

BRANDING IMPLICATIONS

Brands create value therefore the route to value creation sits at the center of branding process. The value path cannot be independent of consumer imposed structure of needs and wants.

Three facets of satisfaction are: functional, symbolic and experiential (Park, Jawarski, & MacInnis, 1986). Consumers look for self-expression and self-fulfillment from products and brands (Brownlie, Hewer, & Traynor, 2007). Marketers invest in their product with symbolic meanings designed to help customers achieve their identity related goals (Belk, 1988; Holt, 2005; Fournier, 1998). A study in Indian context found that cool is an invisible construct which could manifest through seven indicators like being self, sense of humor, calm and groomed (Duggal & Verma, 2019). These indicators accord opportunity for brands to add value by creating identity based on these aspects. A brand apart from functionality can also be instrument of identity expression and creation. Brands are consumed for their symbolic properties (Levy, 1959; Elliott, 1999). The embedded symbolism serves crucial need for meaning creation for consumers. The concept of cool can also be used in achieving greater effectiveness in social marketing programs (Mohiuddin et al., 2016). Social welfare is imbued with the vital quality of cool by the cool brands (Concha-Ferreira, 2014). A study by Warren and Campbell (2014) reported that cool is about creating a deviation from the norm in an appropriate but not radical manner. Cool in this regard is about divergence from the norm.

Marketed goods and services constitute raw material to create, foster and develop identity (Eliott & Wattanasuwan, 1998). Products are transformed into brands by sign making process to become 'live information system' (Douglas & Isherwood, 1996). The drive to form self- identity is likely to be more pronounced in early states of life. During this stage brands and consumption assume significance in identity creation reinforcement process (McCracken, 1986). Many brands follow identity-oriented strategy linking consumer, brand and identity intended to create self-conceptual links (Reed, 2002) to produce superior results. Consumers respond positively to brands whose image is synchronous with their identity. Only brands with meaning and purpose and a positive connect with the customers are able to stay cool in the fierce marketplace (Bergh & Behrer, 2016). Cool's relevance for marketers and brand managers stems from the fact that cool is about: possessing cool things, performing cool activities and being cool (Horton et al., 2012).

Functionality is essential but is not sufficient to create attachment. The brand-consumer bonding is critical to profitability and consumer lifetime value (Thomson, MacInnis, & Park, 2005). A sense of unity or oneness with brand translates into love and attachment which starts with building cognitive and emotional linkages between consumer and brand (Chaplin & John, 2005; Escalas, 2004). It happens when a brand ceases to be an entity extraneous to consumer and becomes the part of consumer's identity. At the heart of identification lies a feeling being in oneness with the brand, a sense of belongingness (Bhattacharya et al., 1995). By moving into the realm of perception, brands appropriate relevant consumer identity dimensions and strike a fit to create identification (Mael & Ashforth, 1992). 'Cool' in this background offers symbolic resource for brand building (see Figure 2). It is involved with products, designer labels and aesthetics and niche brands (Nancarrow et al., 2002). Cool is also used as a medium to enable the trading of nations by cultivating nation's social imaginary on the notions of "cool" (Valaskivi, 2016). 'Cool' is important for consumers for three reasons. First, brands like other cultural artifacts contain abstract cultural meaning (McCracken, 1986) and embedded 'coolness' implies a specific meaning. Consumers buy brands to extract meaning contained therein to use them in their lives (Fournier, 1998).

Figure 2. 'Cool' and brand building

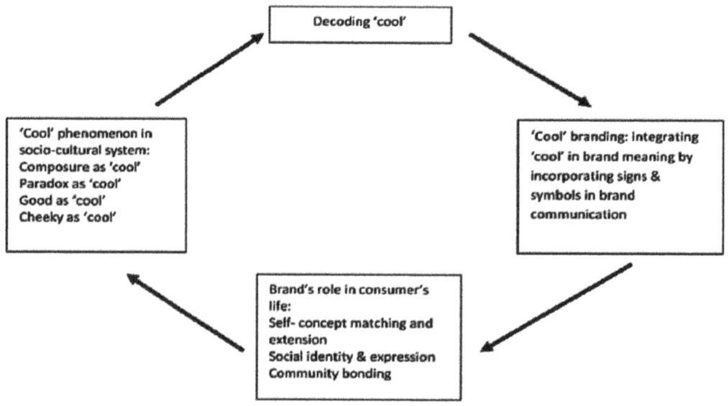

'Cool' as symbolic brand property permits consumers to realize self-concept related goals by identity matching and extension. Consumers' cool is initiated by a product symbolizing cool on the basis of consumer's peculiar actions (Wang & Dalton, 2016). Second, brands are also helpful in creation of social identity and negotiation of social roles. The desire fit-in and stand out is inherent to people. Brands help consumers express who they are and who they are not. For instance, by buying a Mercedes a person joins a select group of moneyed social class yet at the same time he also conveys that he is not a yuppie. 'Cool' brand allows consumers to leverage its symbolism for their purpose of expressing and constructing social identity. Lastly, consumers form communities based on their consumption or use of brands (Muniz & O'Guinn, 2001). Some of the known brand based communities are that of Apple, Beetle, Saturn and Harley Davidson. Commonness of a brand among people can pull them together and create communal bonds. 'Cool' here offers opportunity for brand community creation in which people could be bound together by a specific concept of 'cool' integrated in a brand. Danesi (1994) observes that people in teenage group with sensitivity to self- concept are likely to affiliate with others based on traits like 'coolness'. They are likely to both conform to group norms and also retain their individual identity.

Being 'cool' bestows power upon a brand but becoming un-coolness robs it off. Brands lose appeal when they become un-cool. A brand can become 'uncool' when it fails to sync itself with changing concept of 'cool'. Levis lost its ground in the youth market as it lost its 'cool' appeal. Behind the manifested sales statistics lies a reason that the quintessential blue pair of jeans is no longer perceived as 'cool' (Poutain & Robins, 2000). 'Cool' is not a static phenomenon. It evolves with time. It is through this evolutionary process, today's 'cool' can become 'un-cool' tomorrow. The blue double seamed riveted denim became 'cool' during fifties and sixties for associations and imagery with working class. Levis drew its 'cool' by becoming symbol of anti-establishment generation. It was a sartorial expression of a rebel identity. The anti-conformity, anti-establishment symbolism held sway till seventies and eighties. But as time passed rebel of then aged and came to signify the establishment. How can a pair of blue jeans worn by the parents be 'cool'? The new brands gained grounds taking 'against' position by building and evolving rebel phenomenon in contrast of Levis which by now had come to symbolize jeans of old generation. For the new youth and their rebel identity Levis was not the right pair. Brands such as Calvin Klein and Tommy Hilfiger filled in this space. Tommy Hilfiger built its brand by harnessing of ghetto cool into marketable commodity. The brand feeds off the alienation in American race relations. It sells fetishization of black style to white youth and fetishization of white wealth to black youth (Klein, 2000).

The Cool Brands Expert Council compiles a list of brands which are considered to be cool on the basis of expert and consumer opinions. Their evaluation criteria are based on the premise that the concept of 'cool' is very personal and subjective. Hence no objective definition forms the background in their selection process. However, there are factors that do guide selection process which include style, innovation, originality, authenticity, desirability and uniqueness (Cool Brands Council, 2013). The top cool brands of 2012-13 include names like Apple, YouTube, Aston Martin, Twitter, BBC, Virgin Atlantic, Bang & Olufsen, Sony, Bose, HaagenDazs, and Vogue. The technology brand Apple pushed high-price luxury automobile car Aston Martin from its previous top position which has consistently occupied the top slot for many years. This may imply that being expensive, super luxury and aspirational may not be all that which makes a brand cool. Affordability does not prevent a brand from being perceived as cool. Aspects like free spirit, enjoyment and pleasure have assumed significance in defining coolness in a brand. There are over 39 different explanations of 'cool', ranging from edgy to luxurious and from hip to 'sweet'.

CONCLUDING REMARKS

There are brands and there are cool brands. Cool brands manage to forge superior connections with their target customers. These connections enable brands to generate superior marketing outcomes. Brands are no longer bought for their product elements. The product parity renders them blunt and sterile. The forces of competition have created markets into fierce battlegrounds where brands fight with each other to make a way into customer's heart and mind. In this background, brand managers especially catering to youth segment struggle hard to find ways to achieve value transformation of their offerings. Cool in this context has emerged as a magical value transformation device. Cool factor can transform an object of need to an object of desire and thereby create several positive side effects. But adding a cool dimension to a brand is not easy because what cool stands for remains shrouded in mystery.

Cool is not a substance that can be manufactured in factories. It is an invisible layer which is added to brand to transform its meaning in the perception of target customers. Therefore, cool's meaning needed to be hunted, appropriated and conveyed through resonating brand narrative. The brand in this regard assumes position of a signifier of cool which is bought by target customers for its role in identity construction and signification. This study was undertaken to develop concrete understanding of cool in India. This was done on the premise that cool being a socio-cultural construction is likely to have different facets in comparison to western countries.

The study found that cool in Indian context has roots in theology especially Hinduism and Buddhism. If on the one hand it belongs to the popular consumption culture space, it is paradoxically also belonging to spiritual-religious area. The Bhagawat Gita and Buddhist preaching lay stress on the cultivation of a temperament of indifference or equanimity to the binaries of pain and pleasure. Cool, however emerged from the sites of slave exploitation as a means of survival. Calm is inextricably linked to the concept of cool which stands for an ability to maintain composure in pressure and pain. Our study found four facets of cool: cool as composure, essential goodness, being cheeky or harmlessly cleaver and combination of something inherently paradoxical. Unlike the US where cool has been predominantly a counterculture virtue or anti-establishment stance, cool in Indian context stands for very different facets. The findings suggest that if a brand seeks to create cool image in India, it must create narrative based on insights how cool is constructed by youth in India. This study provides a useful starting point in this direction.

REFERENCES

Adler, P. A., & Adler, P. (1998). *Peer power: Pre- adolescent culture and identity*. New Brunswick, NJ: Rutgers University Press.

Barthes, R. (1967). *Elements of semiology* (A. Lavers, & C. Smith, Trans.). London, UK: Jonathan Cape.

Belk, R. W. (1988). Possessions and the extended self. *The Journal of Consumer Research, 15*(2), 139–168. doi:10.1086/209154

Bell, D. (1979). *Cultural Contradictions of Capitalism*. London, UK: Heinemann.

Bergh, J. V. den, & Behrer, M. (2016). *How cool brands stay hot: Branding to generations Y and Z*. London, UK: Kogan Page.

Bhattachargya, C. B., & Sen, S. (2003). Consumer –company identification: A framework for understanding consumers' relationship with companies. *Journal of Marketing, 67*(2), 76–88. doi:10.1509/jmkg.67.2.76.18609

Bornstein, T. B. (2013). What does it mean to be cool? *Philosophy Now, 80*, 1–3.

Brownlie, D., Hewer, P., & Treanor, S. (2007). Social- ity in motion: Exploring logics of tribal consumption among cruisers. In B. Cova, R. V. Kozinets, & A. Shankar (Eds.), *Consumer Tribes* (pp. 109–128). Oxford, UK: Elsevier.

Chaplin, N., & John, R. E. (2005). The development of self-brand connections in children and adolescents. *Journal of Consumer Behaviour, 32*(1), 119–129. doi:10.1086/426622

Concha-Ferreira, I. (2014). Brands, welfare and 'welfare-cool'. *Ephemera: theory and politics in organization, 14*(1), 109-117.

Connor, M. K. (1995). *What is cool? Understanding black manhood in America*. New York, NY: Crown.

Cool Brands Council. (2013). *Cool brands-an insight into some of Britain's coolest brands*. Retrieved from http://s3.coolbrands.uk.com/files/2012/09/CB2012- 13-Selection-Process-mRb8d2.pdf

Danesi, M. (1994). *Cool: The signs and meanings of adolescence*. Toronto, Canada: University of Toronto. doi:10.3138/9781442673472

Dar-Nimrod, I., Hansen, I. G., Proulx, T., Lehman, D. R., Chapman, B. P., & Duberstein, P. R. (2012). Coolness: An empirical investigation. *Journal of Individual Differences, 33*(3), 175–185. doi:10.1027/1614-0001/a000088

DelVecchio, G. (1997). Creating ever cool: A marketer's guide to a kid's heart. Gretna, LA: Pelican.

Dick, P., & David, R. (2000). *Cool rules: Anatomy of an attitude*. London, UK: Reaktion Books.

Douglas, M., & Isherwood, B. (1996). *The world of good: Towards an anthropology of consumption*. London, UK: Routledge.

Duggal, E., & Verma, H. V. (2019). Cool perspectives, Indian cool and branding. *South Asian Journal of Business Studies, 2*(8), 130–145. doi:10.1108/SAJBS-07-2018-0083

Elliott, R. (1999). Symbolic meaning and post- mod- ern consumer culture. In D. Brownlie, M. Saren, R. Wensley, & R. Whittington (Eds.), *Rethinking marketing*. London, UK: Sage.

Elliott, R., & Wattansuwan, K. (1998). Brands as symbolic resources for the construction of identity. *International Advertising*, *17*(2), 135–145.

Escalas, J. E. (2004). Narrative processing: Building consumer connections to brands. *Journal of Consumer Psychology*, *12*(1-2), 168–179. doi:10.120715327663jcp1401&2_19

Ferguson, S. (2011). A global culture of cool? Generation Y and their perception coolness. *Young Consumers*, *12*(3), 265–275. doi:10.1108/17473611111163313

Foster, H. (1983). *Postmodernism: A preface. The anti-aesthetic: Essays on postmodern culture*. Port Townsend, WA: Bay Press.

Fournier, S. M. (1998). Consumer and their brands: Developing relationship theory in consumer research. *The Journal of Consumer Research*, *24*(4), 343–373. doi:10.1086/209515

Frank, T. (1977). *Jazz: A history*. New York, NY: W.W. Norton & Co.

Gladwell, M. (1997). Annals of style: The cool hunt. *New Yorker (New York, N.Y.)*, (March): 17.

Gottdiener, M. (2001). *The theming of America – American dreams, Media Fan*. Boulder, CO: Westview Press.

Graver, M. (2009). *Stoicism and emotion*. Chicago, IL: University of Chicago Press.

Grossman, L. (2003). The quest for cool. *Time*, *162*, 48.

Hamermesh, D. S. (2011). *Beauty pays: Why at- tractive people are more successful*. Princeton, NJ: Princeton University Press.

Holt, D. B. (2005). How societies desire brands: Using cultural theory to explain brand symbolism. In S. Ratneshwar, & D. G. Mick (Eds.), *Inside consumption* (pp. 273–291). London, UK: Routledge.

Horton, M., Read, J. C., Fitton, D., Little, L., & Toth, N. (2012). Too Cool at School - Understanding Cool Teenagers. *PsychNology Journal*, *10*(2), 73–91.

Hutcheon, L. (1988). *A poetics of postmodernism: History, theory, fiction*. New York, NY: Routledge. doi:10.4324/9780203358856

Jones, M. C. (2012). Cool for kids. *The Marketing Week*, July, 1.

Karan, K. K. (1989). *George Bernard Shaw and the concept of Superman*. New Delhi, India: Vani-Prakashan.

Kiplinger. (2012). *Cool its earliest slang meaning*. Retrieved from http://kiplinger.tumblr.com/post/25889435298/cool-its-earliest-slang-meaning-dates-to-1728

Kjeldgaard, D., & Askegaard, S. (2006). The globalization of youth culture: The global youth segment as structures of common difference. *The Journal of Consumer Research*, *33*(2), 231–247. doi:10.1086/506304

Klein, N. (2000). No logo. London, UK: Hammersmith.

Kohlenberger, J. (2016). The New Formula for Cool: Science, Technology and The Popular in the American Imagination. New York, NY: Columbia University Press.

Levy, S. (1959). Symbols for sale. *Harvard Business Review, 37*(4), 117–124.

Mael, F., & Ashforth, B. E. (1992). Alumni and alma mater: A partial test of the reformulated model of organizational identification. *Journal of Organizational Behavior, 12*(2), 103–123. doi:10.1002/job.4030130202

Majors, R., & Billson, J. M. (1992). *Cool pose: The dilemmas of black manhood in America*. New York, NY: Lexington Books.

Mazrui, A. (Ed.). (1977). *The warrior tradition in modern Africa*. Leiden, The Netherlands: Brill.

McCracken, G. (1986). Culture and consumption: A theoretical account of the structure and movement of cultural meaning of consumer goods. *The Journal of Consumer Research, 13*(1), 71–84. doi:10.1086/209048

Mikulincer, M., & Shaver, P. R. (2007). *Attachment to adulthood: Structure, dynamics and change*. New York, NY: The Guilford Press.

Mohiuddin, K. G. B., Gordon, R., Magee, C., & Lee, J. K. (2016). A conceptual framework of cool for social marketing. *Journal of Social Marketing, 6*(2), 121–143. doi:10.1108/JSOCM-07-2015-0046

Morris, W. (Ed.). (1969). *The American heritage dictionary*. New York, NY: American Heritage Publishing Co., and Houghton Mifflin.

Nancarrow, R., Nancarrow, P., & Page, J. (2002). An analysis of the concept of cool and its marketing implications. *Journal of Consumer Behaviour, 1*(4), 311–322. doi:10.1002/cb.77

Norman, M. (1957). *The white negro*. Dissent Publishing Associates.

O'Brian, B. (2013). *Buddhism and equanimity: Why equanimity is an essential Buddhist virtue*. Retrieved from http://buddhism.about.com/od/basicbuddhistteachings/a/Buddhism-And-Equa- nimity.htm

O'Donnell, K. A., & Wardlow, D. L. (2010). Atheory on the origins of coolness. *Advances in Consumer Research. Association for Consumer Research (U. S.), 27*(1), 5.

Park, C. W., Bernard, J. J., & Deborah, J. (1986). Strategic brand concept-image management. *Journal of Marketing, 50*(4), 135–145. doi:10.1177/002224298605000401

Polkinghorne, D. (1983). *Methodology for the Human Sciences*. Albany, NY: Human Science Press.

Pollay, R. W. (1985). The subsiding sizzle: A descriptive history of print advertising, 1900-1980. *Journal of Marketing, 49*, 24–37.

Pountain, D., & Robbins, D. (2000). *Cool rules*. London, UK: Reaktion Books.

Reed, A. (2002). Exploring the links between brand name and consumer identity. Knowledge@Wharton, Nov. 6 (pp. 1-4).

Rice, J. (2002). *What is cool? Notes on intellectualism, popular culture, and writing*. Retrieved http://www.ctheory.net/articles.aspx?id=338

Rifkin, J. (2001). *The Age of Access: The New Culture of Hypercapitalism Where All of Life is a Paid-for Experience*. New York, NY: TarcherPerigee/Putnam.

Runyan, R. C., Noh, M., & Mosier, J. (2013). What is cool? Operationalizing the construct in an apparel context. *Journal of Fashion Marketing and Management, 17*(3), 322–340. doi:10.1108/JFMM-01-2012-0001

Russell, B. (1972). *A history of Western philosophy*. New York, NY: Simon & Schuster, Inc.

Shapiro, H. (1999). *Waiting for men*. London, UK: Helter Skelter.

Spencer, L., Ritchie, J., Ormston, R., O'Connor, W., & Barnard, M. (2014). Analysis: principles and processes. In J. Ritchie, J. Lewis, C. McNaughton Nicholls, & R. Ormston (Eds.), Qualitative Research Practice (pp. 269–293). London, UK: Sage.

Spiggle, S. (1994). Analysis and interpretation of qualitative data in consumer research. *The Journal of Consumer Research, 21*(3), 491–503. doi:10.1086/209413

Sturgess, D. (2013). *A breakthrough in measuring cool*. Forbes. Available at www.forbes.com/sites/onmarketing/2013/03/07/a-breakthrough-in-measuring-cool

Sundar, S., Tamul, D., & Wu, M. (2014). Capturing cool: Measures for assessing coolness of technological products. *International Journal of Human-Computer Studies, 72*(2), 169–180. doi:10.1016/j.ijhcs.2013.09.008

ThichNhatHanh. (1988). *The heart of the Buddha's teaching*. Berkeley, CA: Broadway Books.

Thomson, M., MacInnis, D. J., & Park, C. W. (2005). The ties that bind: Measuring strength of consumers' attachment to brand. *Journal of Consumer Psychology, 15*(1), 77–91. doi:10.120715327663jcp1501_10

Valaskivi, K. (2016). *Cool Nations: Media and the Social Imaginary of the Branded Country*. London: Routledge. doi:10.4324/9781315794662

Wang, L., & Dalton, A. (2016). How and Why Wearing Sunglasses Makes For Cool Consumers. In P. Moreau, & S. Puntoni (Eds.), *NA - Advances in Consumer Research, 44* (pp. 663–664). Duluth, MN: Association for Consumer Research.

Warren, C., & Campbell, M. (2014). What makes things cool? How autonomy influences perceived coolness. *The Journal of Consumer Research, 41*(2), 543–563. doi:10.1086/676680

Wilson, E. (1989). *Hallucinations: Life in the post- modern city*. London, UK: Hutchinson Radius.

Zebrowitz, L., & Rhodes, G. (2004). Sensitivity to "bad genes" and the anomalous face over generalization effect: Cue validity, cue utilization, and accuracy in judging intelligence and health. *Journal of Nonverbal Behavior, 28*(3), 167–186. doi:10.1023/B:JONB.0000039648.30935.1b

Zezer, M. (1982). *The Dark Ages: Life in the United States 1945-1960*. Cambridge, MA: South End Press.

Zimmer, M. R., & Golden, L. L. (1988). Impression of retail store: A content analysis of consumer images. *Journal of Retailing, 64*(Fall), 265–294.

Žižek, S. (2009). *First as Tragedy, Then as Farce,* The RSA, November, 24. Retrieved from https://www.thersa.org/discover/videos/event-videos/2009/11/first-as-tragedy-then-as-farce

Chapter 4
Analysis of Social Value of TV Dramas Based on Audience Comments

Xinye Liu
Tianjin University, China

Xiaotong Zhang
Tianjin University, China

Tao Wang
Tianjin University, China

Kun Cheng
Tianjin University, China

Shangbing Jiao
Tianjin University, China

Diqing Liu
Tianjin University, China

Jia Su
Tianjin University, China

Xiaoqian Wang
Tianjin University, China

ABSTRACT

This chapter analyzes the social value of the TV drama Entrepreneurial Age through the mining of the audience's comments, so as to provide reference for the TV drama producers in topic selection, casting, and script design. Design/methodology/approach: The research is based on a three-step approach including data crawling, two-dimension data tags, and the random forest algorithm design. Findings: This chapter finds that there are three factors related to demand of TV drama:1) the appearance and acting skill of actors; 2) the closeness between TV plays and real life; 3) whether the topic of TV plays has high attention. Value: Based on the big data of audience comments, this chapter explores the factors that influence the number of TV plays. It provides an important reference for TV drama producers on how to design the plot of TV drama, how to choose actors, and how to create topics.

DOI: 10.4018/978-1-7998-0357-7.ch004

INTRODUCTION

Today, human society is entering the era of big data and it is affecting all fields of society. Especially, big data analysis is changing the TV dramas in essence.

In terms of the audience, with the popularity of the Internet, more and more people are watching on the mobile terminal instead of televisions. The audience's online watching behavior forms a huge data source. Every click, search, and comments of the users will leave traces on the Internet. Audience's comments have its social value. From the comments, the people's attitude towards some popular phenomena and their preference for artistic expression can be shown. By analyzing the data, the TV dramas can be produced more specifically.

In terms of the producers of TV drama, in recent years, many platforms have tried to produce and broadcast homemade dramas and customized dramas. The essential difference between the traditional media and new platforms is that the latter adopt a data-driven program to make production, price and distribution decision. For example, Nexfilx analyzed the watching behavior of nearly 60 million registered users. The results showed that users who like BBC dramas, the director David Fincher and the actor Kevin Spacey have something in common. Therefore, House of Cards, which meet the three elements, came into being. Big data analysis is more detailed, more comprehensive and more advanced than traditional ratings survey in the aspect of audience analysis. It also can be applied in TV theme selection, topic setting, casting formation, distribution selection, marketing strategies and rating forecasting.

Connecting the above two aspects, the authors raise the following questions: What social value can we dig from the audience's comments on TV drama? How can we use that information to guide the production process of TV drama?

To solve the questions, the authors carry out this study. Based on the audience' comments of the popular TV drama named Entrepreneurial Age, the authors expect to explore the social value behind it and provide the TV producer with reference to the subject selection, casting selection and script design. The authors plan to crawl the data of related websites and conduct deep data mining and text analysis. Then the authors will design an algorithm to analyze the users' watching or click habits and their emotional polarity, reducing the blindness in theme selection, actor matching, shooting production and broadcast arrangement in TV drama production. Finally, the implications for practitioners and academia are discussed and conclusions drawn.

The authors collaborate in the following ways: Liu Xinyun and Jiao Shangbing are responsible for designing algorithms. Wang Tao and Cheng Kun are responsible for data mining and processing. Zhang Xiaotong, Liu Diqing and Wang Xiaoqian are responsible for the remaining working.

LITERATURE REVIEW

The Application of Machine Learning in Text Analysis

Emotion classification technology is to divide human emotion into negative emotion and positive emotion. Currently, the most popular research methods are machine learning-based method (Mullen & Collier, 2004) and semantic-based method (Turney, 2002). The approach based on machine learning regards emotion analysis problem as a classification problem. The machine learning algorithm trains the marked training set to obtain the classification model, which can be used for future emotion classification (Liu & Liu, 2012).

At present, there are three main research aspects of machine learning algorithm for text classification. The first one is the study of the machine learning algorithm itself. Liu and Liu (2012) find that IG is a better feature selection method compared to SVM methods and TF-IDF methods. The combination of the three algorithms performs best in the emotional classification of Weibo. Chatterjee et al. (2019) combine both semantic and sentiment-based representations with a novel Deep Learning based approach to detect emotions. The second one is combining machine learning with other ways for text analysis. Jiang and Xia (2017) propose a sentiment analysis method via a combination of rule-based and machine learning methods. An effective integration feature set is obtained by adding various rule-based features to the basic feature set after expanding and converting them. Zablith and Osman (2019) propose a text mining and processing procedure that ①trains machine learning algorithms to predict comment dimensions; ②predicts emotions in other comments based on the training set; ③changes the extracted measures from the comments. The third one is the application of machine learning algorithm in text analysis. Amrit et al. (2017) identify and predict cases of child abuse using text mining and analysis.

The development of machine learning for text analysis has been comprehensive both in method itself and in application field. However, there are few researches that use machine learning algorithm to classify the evaluation of TV series, so as to provide references for the production of TV series.

The Application of TV Drama Comments

Practice has proved that audience comments analysis has multiple uses. To explore the television's influence on trends toward cultural homogenization in American society, Neuman (1982) collects viewers' comments in order to analyze the connection between the program and important issues of society and culture. Müller and Hermes (2010) conclude their findings by analyzing audience reactions to a Dutch multicultural reality show named *WestSide*, which was aimed to develop a better environment for intercultural understandings and tolerance. Another research tries to illuminate key elements of collaborative program of transnational media by working on fan and audience studies (Li, 2009).

Similar to the TV drama industry, audience's online comments of films are a useful information source to help film producers reduce potential risks (Koh et al., 2010). By applying methods of machine learning and sentiment analysis, Yao and Chen (2013) study the relationship between audience online comments for a film and the film's revenue performance. With the data from Douban, a famous online social network in China, an original Big Data processing framework was proposed to investigate a user-generated popular culture content (Yang & Yecies, 2016).

Investment in the television and film industry is huge and there exists too much uncertainty. So the producers usually take a high risk that they may not get the desired return. Fortunately, Dong (2015) says that the effectiveness of market forecasting can be improved with the help of web data mining and data analysis.

After sorting out and summarizing the above research, the authors find that there have been many studies on TV drama comments, but there is a lack of research aiming at the social significance behind it. Besides, data mining, data processing and algorithm are seldom used for accurate analysis in previous study, so our research can make up for the blank in this respect. Thus, the authors believe that their work would be very valuable and meaningful.

The Application of Machine Learning in Audience Comment Analysis

Nowadays, user comment plays a more and more important and useful role in people's information resource. Therefore, automatic user comment mining and summarization has become a hot research topic recently.

As for the mining and summarization of user comments, machine learning algorithms are widely applied. Lei Cen et al. (2014) collected a dataset of user comments from Google Play and those CSPI within it were described by a dimensional label system. As for detecting various kinds of CSPI described by this label system, a supervised multi-label learning method utilizing comment expansion was then adopted. Ashish Sureka (2011) utilized machine learning method to automatically detect comment spammer in YouTube forums based on user comments crawled from YouTube. Yang and Yecies (2016) proposed a modified Apriori algorithm based on MapReduce for analyzing Douban movie reviews. Wang et al. (2010) identified the main aspects of user comments by starting from few seed keywords which are fed into a bootstrapping-based algorithm. Zhuang et al. (2006) extracted a list of features about users' opinions based on a multi-knowledge approach incorporating movie knowledge, WordNet and statistical analysis together and created a comments summary. Ganesan and Zhai (2012) extracted entities from user comments for the purpose of recommending to users comments which appeals to them. Popescu and Etzioni (2007) proposed an unsupervised information-extraction system to obtain product features from user comments. Abbar et al. (2013) formed new article diversity measure incorporating entities and sentiment obtained from user comments, and then nearest neighbor algorithms is utilized to address the problem of online recommendation.

As a result, in the analysis of social value of TV dramas based on audience comments, the applicability and superiority of machine learning algorithms in audience comment analysis is fully verified. However, the application of machine learning in TV series' audience comment still remains to be explored.

DATA PROCESS

Data Collection

The Procedure of Data Collection

We collected users' comments on the TV series Entrepreneurial Era from Youku video website (http://youku.com/). There are three reasons for us to choose this series. Firstly, this is the latest new play, so the audience have more comments on the plot which we are interested in. Secondly, the leading actor and actress are so famous that many people have great interest in their performance and the play's broadcast volume is very impressive. Thirdly, the score of this play on Douban is pretty low, so it is necessary to analyze its pain points.

The software we use to collect comments data is Octopus. The operation we need is simple and the wizard mode is enough to finish the whole procedure of data collection.

The Results of Data Collection

A total of 29,841 valid pieces of data were collected, and each piece of data includes 7 dimensions, namely, username, comment time, comment content, approval quantity, objection quantity, comment quantity and episode number. The first six dimensions were collected by the Octopus itself, and the seventh dimension (episode number with a total of 54 episodes) was manually added.

Data Tagging

We sort the collected data in an ascending order according to the approval quantity, objection quantity and comment quantity, and then manually tag the first 10,000 pieces of 29,841 pieces of data. We consider every piece of data is characterized by the content attribute and content has three classes of different sentiments, positive emotion, negative emotion and neutral emotion.

Content Tag

The content comments can be divided into five categories: actor related, plot related, social phenomenon, post production related, and meaningless filling, as showed in Table 1.

1. **Actor Related:** It mainly consists of opinions on acting skills and actors. Such as, "When the heroine arrives at the emotional scene, her face is expressionless." and "The heroine doesn't look good so that I don't want to see it anymore."
2. **Plot Related:** Attitudes to character setting and plot development are all related to the plot. For example, "This kind of plot can come out?", "The hero's personality is not good shape. It is difficult to do things like this for a manic person."
3. **Social Phenomenon:** Every TV play series can trigger the audience's thinking about the reality, the TV play series we have chosen to analyze is no exception. Such as "Magic crystal is WeChat, Li Benteng is Ma Huateng, and Qilin is QQ." and "I am now in the same situation as Guo Xinnian in 2016. It is really hard to start a business."
4. **Post Production Related:** Post production is almost as important as the plot and acting skills. It greatly affects the viewing experience. Like filter, picture quality, advertisement, detail processing and so on, such as "Smart phones, HP ultra-thin books have been universal, however, voice messages are still being developed, which pays too little attention to details!"
5. **Meaningless Fillings:** Many comments don't reflect on any effective information. These comments are invalid data. We need to filter them out.

Note that some of the information may be hidden. Comments like "This noon, I watched totally 16 episodes." it doesn't directly express the opinions on the TV play series but implies that the plot is attractive. In addition, some comments include two or more categories information, such as "the theme and the story are good, but the acting is not good enough." The opinion related to the plot is discussed before, and the opinion related to the actor is discussed later. We weight the proportion of every category and then classify it into the category which occupies a higher proportion.

Table 1. Content categories

Label	Category	Description	Examples
1	Actor Related	Opinion on actors/ Acting skills	When the heroine arrives at the emotional scene, her face is expressionless.
2	Plot Related	Plot development /Character setting	The hero's personality is not good shape. It is difficult to do things like this for a manic person.
3	Social Phenomenon	Thinking about the reality	I am now in the same situation as Guo Xinnian in 2016. It is really hard to start a business.
4	Post Production	Filter/Picture quality	Smart phones, HP ultra-thin books have been universal, however, voice messages are still being developed, which pays too little attention to details!
5	Meaningless Filling	Invalid comments	Please give me 10 likes!

Sentiment Tag

As illustrated in Table 2, comments have three different sentiments, positive emotion is marked by "1", neutral emotion is marked by "0", and negative emotion is marked by "-1".

The contents of every category basically have three classes of sentiments, so comments are divided into 15 categories. Overall, if the positive emotion accounts for a larger proportion than negative emotion, we can roughly judge that the TV play series has a high quality.

Tag Result Statistics

Appling with semi-supervised learning, all review data is made statistical analysis through data perspective. It can be seen that both the number of comments in each episode and the fluctuations in the emotional proportion of each episode are very obvious. Most of the comments are located at the beginning, middle, and end. It indicates that the audience may only focus on the beginning, climax, and ending of the story. So the scriptwriter should pay more attention to make the story compact. Here we introduce the single-valued moving range diagram control graph in quality management to identify the anomaly.

Control Chart is a quality method based on hypothesis and it is a statistically designed diagram that monitors, records, and evaluates process quality characteristics to monitor whether the process is in a controlled state. There are three straight lines parallel to the horizontal axis: CL(Control Line),Upper Control Line(UCL),and Lower Control Line(LCL) .Also, there are a sequence of strokes with sample statistic values extracted in chronological order. UCL, CL, and LCL are collectively referred to as a

Table 2. Sentiment categories

Label	Category	Description	Examples
1	Positive emotion	Content / Praise	This is a particularly good TV play series, I feel what the heroine fells.
2	Neutral emotion	No emotions	The situation of the heroine is miserable.
3	Negative emotion	Discontent / Criticism	The setting of the plot is contrary to reality.

Control Line, and the control limit is usually set to a position of between positive three and negative three standard deviation. The centerline is the average of the controlled statistics, and the upper and lower control limits are several times the standard deviation from the centerline. If the traces in the control chart fall outside the UCL and LCL or the alignment between the UCL and the LCL is not random, the process is abnormal.

It can be seen in Figure 1 that in the control chart of the total number of comments 1, 2, 18, 19, 50, and 51 are abnormal values, and we respectively check the number of sets of abnormalities. For example, the first episode, has a hot spot because of the hot discussion, and the characters in the play are played by the current hot actors, so the heat is very high. And the main body of the play is relatively novel and the reality, so it received a very high degree of attention. The total number of comments in the first episode reached a peak state of the whole drama. Other episodes are not in the narrative.

Content Classification Statistics

As shown in Figure 2, the proportion of comment related to story is most, accounting to 63%. It demonstrates people pay more attention to plots. The scriptwriter should set some attractive story to obtain more audience.

From Figure 3, we can see that in the initial stage, audience discuss actors more. Gradually, the comment on the plot exceeds the comments of actors. It illustrates flow actors don't necessarily retain the audience to a certain extent.

We can observe that the proportion of comment about story plot stay high in Figure 4. Flow actors would attract a lot of fans in the early stage, but the story will mainly retain the audience in the later stage.

Figure 1. I-MR control chart of total comments

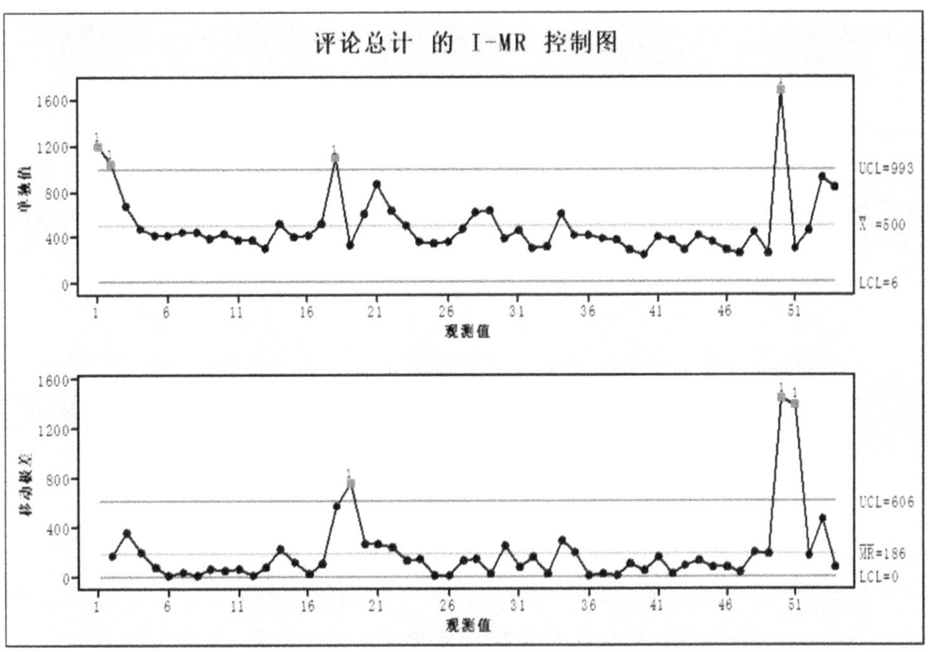

Figure 2. Total content classification

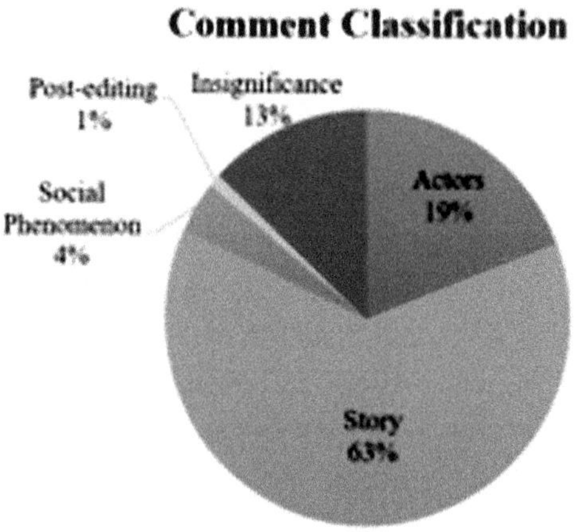

Emotion Classification Statistics

Figure 5 shows that audiences having negative attitude to teleplay is more.

As is shown, the negative emotions to each main character is more than positive emotions and neutral emotions. This shows that the audience is not very satisfied with the role, such as acting.

Through the analysis of more than 30,000 pieces of comment data, we get the main role discussion hot TOP4 of TV play series "Entrepreneurship". They are Na lan played by Yang Ying, Guo Xinnian played by Huangxuan, Wendi played by Songwei, and Luowei played by Zhou Yiwei. Among the five roles, netizens has the most positive attitude toward "Nalan", with a positive sentiment of 38% and the worst attitude towards "Wendi", with a negative sentiment of 59%. On the whole, there are many negative emotions. In the commentary area, there are more and more spit on the plot, characters, and plot settings. The main reason is the ungrounded character setting and the existence of many detailed bugs that are out of real life. This reflects to a certain extent that the audience's quality requirements for TV dramas are

Figure 3. Absolute value statistics of each type of content

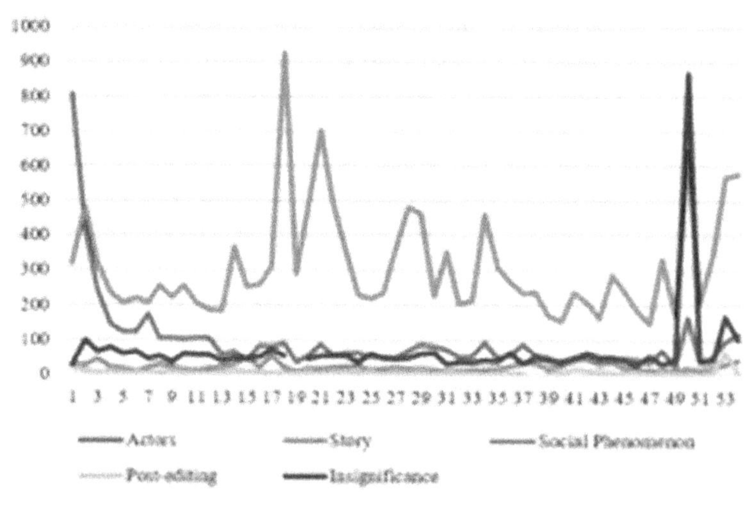

Figure 4. Relative value statistics of each type of content

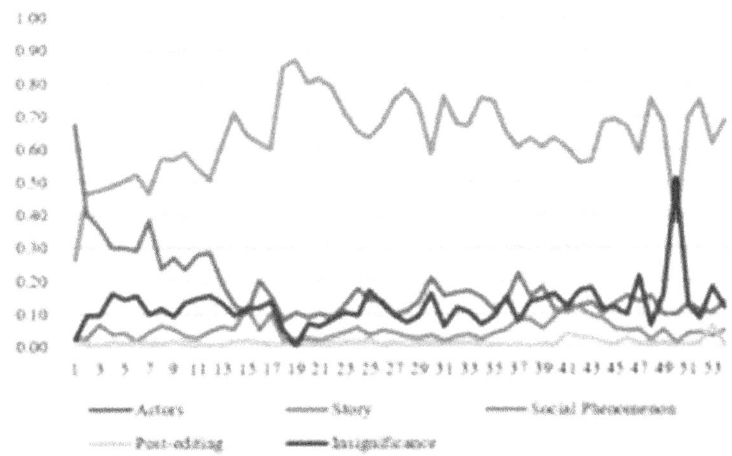

getting higher and higher. It illustrates flow niche, old trouper and the general production can't attract the audience, the simple plot's patching has been difficult to satisfy the audience. The strange plot and the protagonist's incompetent acting can't satisfy the audience completely. A lot of old troupers can't save the low evaluation. To attract the audience, the plot setting needs to be based on the real reality.

Figure 5. Emotion classification

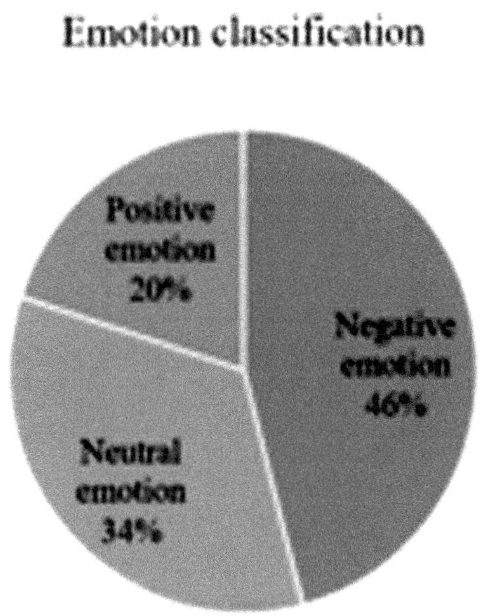

Figure 6. Audience's emotional attributes to Nalan and Guo Xinnian

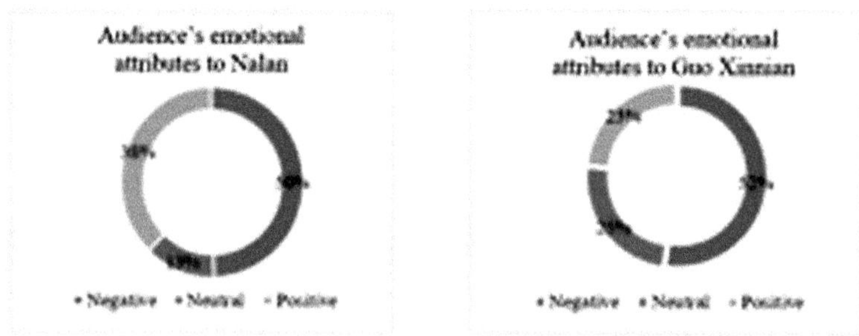

DATA ANALYSIS AND RESULTS

Proposed Method

In this article, Jieba is first used to implement word segmentation. Then, we use word2vec method to transform word text to word vector. Later, several machine learning method are used to predict the comments' label. The details are following.

Algorithm Implementation

Step 1: The Participle Part

Word segmentation is carried out through Jieba class library. Firstly, the data of text type is transformed into word array. On the one hand, we need to deal with the emoticons contained in the original comments. On the other hand, we need to create a stop word list to delete the stop words in comments.

Among them, stop words include some common symbols and Chinese modifiers.

Secondly, the stop words in each word array are deleted using the stop word list. Finally, the emoticons are removed using coding method.

Step 2: Word Frequency Statistics

In order to judge the importance of words and the audience's attention to TV plays, it is necessary to count the frequency of effective words in the results of word segmentation. The results show that there

Figure 7. Audience's emotional attributes to Wendi and Luowei

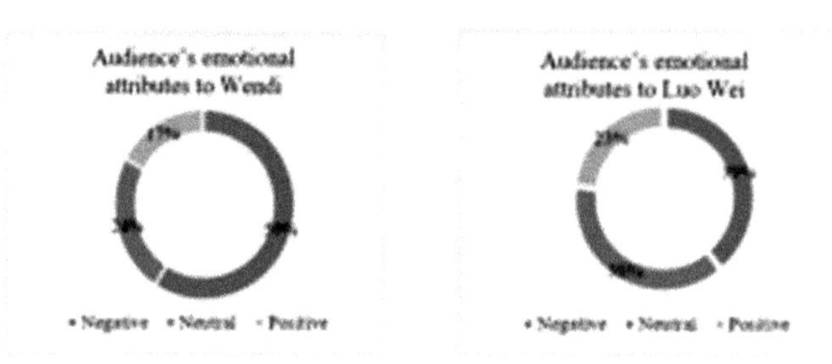

Table 3. Some of the original comments

	Comment
1	'杨颖明明演的很好,喜欢她,没有杨颖我就不看这部电视剧'
2	'唉,黄轩啊,跟谁合作不行'
…	…

Table 4. Word segmentation results

Index	Comment
1	'杨颖' , '明明' , '演' , '喜欢' , '杨颖' , '这部' ,' 电视剧'
2	'黄轩' , '合作' , '不行'
…	…

are more concerns about "acting skills", "plots" and "actors" in the reviews, which can be used for subsequent information mining through these words and their related words.

Step 3: Generating Word Vector

Based on Word2Vec class library, we transform Chinese word into word vector. To get the sentence vector, we sum all the word vectors in the sentence and compute the average value of them

Prediction Algorithm

With the rapid development of information technology, data mining method has been relatively mature, and now it can extract valuable information from data. As a basic data processing method, data classification method is widely used. Machine learning is a widely used data classification method. We used machine learning algorithm to train and predict comments data, which greatly improves the reliability and robustness of prediction.

Generally speaking, the data involved in machine learning algorithms are divided into two parts: training data and test data. The training data is used for model learning. After the model learns knowledge from the training data, the test part is used to evaluate the performance of the model. For each part, the data is divided into two parts: attribute and label. Attribute is the part describing data characteristics, and label is the part identifying data categories. In this paper, we first use Word2vec to vectorize the

Table 5. The results of word frequency statistics

Index	Word	Frequency
1	喜欢	819
2	演技	799
…	…	…

Table 6. Performance of algorithms

	Recall	Precision	Accuracy	Fmeasure
Tree	0.5940	0.6002	0.5940	0.5968
Svm	0.5561	0.4582	0.5561	0.3985
Bayes	0.2108	0.5140	0.2108	0.1695
Lr	0.6510	0.5948	0.6510	0.5723
Rf	0.7071	0.6864	0.7071	0.6752
Gbdt	0.6970	0.6642	0.6970	0.6575

content of the text. By learning the data after vectorization, we can continuously improve the ability to predict the types of comments and emotional tendencies.

The machine learning algorithms used in this paper include decision tree, support vector machine (svm), naive Bayesian algorithm (bayes), logical regression (lr), random forest (rf), gradient lifting tree (gbdt) to predict the tags of comments.

As shown in Table 3, the performance of six algorithms under four evaluation indicators is shown. The experimental data are 10,000 labeled data. Before introducing the experiment, we first introduce the meaning of each index. Recall is the recall rate, and it is a measure of coverage, Precision is the precision rate, meaning the proportion of correctly predicted samples of one certain class. Accuracy is the proportion of correct samples to all samples. Fmeasure is the harmonic mean of Recall and recision. The formula is as (3-2):

$$Fmeasure = \frac{2 \times Recall \times Precision}{Recall + Precision} \text{(3-2)}$$

The experimental results are the average results of 30 experiments on the test set. In each experiment, the data set is randomly divided into training set and test set according to 8:2 ratio, and the six algorithms use the same training set and test set in each experiment. Comprehensive comparison shows that the performance of random forest algorithm is better than other algorithms. The first reason is that random forest is an ensemble learning algorithm based on Bagging, which can fully guarantee the randomness of sample selection and feature selection. On the other hand, the sub-classifiers of random forest algorithm are diversing, which avoids over fitting. Therefore, we decided to use random forest algorithm to predict the remaining unlabeled data.

Result Analysis

Plot Analysis

In this section, the author analyzes the plot with outliers in the control chart. Mainly by watching the plot, the barrage and related discussions. The author personally views the content of each episode, and combines the control map to find out the cause of the abnormal value of the event in the total number of comments, content or emotion, and analyzes the final result in accordance with the classification of the control chart. It is divided into nine categories: total comment, Actor related, Plot related, Social Phenomenon, Post-editing, Insignificance, Negative Emotion, Neutral Emotion, Positive Emotion

Total Comment

The total commentary of the six episodes of the whole drama has abnormal values, which are 1, 2, 18, 19, 50 and 51 episodes.

The reason for the outliers in the first episode is, because it is the first episode, it has always been easy to become a hot spot because of the hot discussion, and the characters in the play are played by the current hot actors, so the heat is very high.

The episodes of the 19 episodes and the 18 episodes have fluctuated greatly. The hero and the sopitha established a relationship of love. This is in sharp contrast with the previous plot, and the female actor's actions made the audience feel very angry, this also led to the generation of negative emotions and an increase in the total number of comments.

In the 50 episodes and 51, the audience in order to consume time and earn energy. They fill in meaningless comments, which pushes up the total number of comments. So a lot of comments are similar to "Only take energy, don't care other things."

Actor Related

The total commentary of the 8 episodes of the whole drama has abnormal values, which are 1, 2, 3, 4, 5, 6, 7, and 8 episode.

The heroine is played by Yang Ying. Because she is not a professional actor and has less experience in her performance, Yang Ying's acting has not been recognized. It is even worse than other actors, so she was squandered by the majority of the audience in the first few episodes. On the contrary, other actors' performances have been recognized, such as the male lead Huang Xuan, as well as Zhou Yiwei, Song Yi, and so on. Many people think that Song Yi's acting as a sopitha is much better than Yang Ying. There are also many viewers who are fans of one of the actors who have expressed their support for their idols. It can be clearly found that the discussion of the actors is concentrated in the front. The author analyzes the reasons as follows: First, the audience gradually accepts the actors' acting skills in the process of watching the story, so the heat is declining. Second, The audience is attracted by the plot or other aspects, so the actors' acting skills are ignored, resulting in a drop in heat.

Plot Related

The plot related of the 5 episodes of the whole drama has abnormal values, which are 1, 18, 19, 50, and 51 episode.

Since innovation and entrepreneurship is a hot topic in the society in recent years, and has formed a trend, the content theme of the drama has also aroused the resonance of many audiences. Many users expressed that it is difficult to understand entrepreneurship, and the plot reflects the current situation of society. However, some viewers expressed doubts and doubted the logic of the plot and content. I think that the setting of the background of the times is vague and chaotic. In that era, there should have been voice software and mobile phone mailboxes. At the same time, some viewers expressed doubts about the technical level in the play.

The episodes of the 19 episodes and the 18 episodes have fluctuated greatly. The hero and the sopitha established a relationship of love. This is in sharp contrast with the previous plot, and the female actor's actions made the audience feel very anger, this also led to the generation of negative emotions.

The number of comments related to the plots in episodes 50 and 51 is below average. The overall reaction in the TV drama commentary is that the plot is too procrastinating, so the audience has very little discussion about the plot, and the dragged story affects the heat of discussion.

Social Phenomenon

The plot related of the 5 episodes of the whole drama has abnormal values, which are 15, 16, 18, 41, and 42 episode.

The plots of these 5 episodes are closely related to the reality of society. Guo Xinnian's various problems encountered on the entrepreneurial road triggered the netizens to think about the real society. They believed that reflecting the reality, entrepreneurship is not easy, and that the spirit of the role of the drama in the unremitting entrepreneurship is worth learning. In the play, Luo Wei and Guo Xinnian's products are also corresponding to the actual life of WeChat, Alipay and so on. The 16 episodes are closely related to commercial reality.

In the following plots, the protagonist's entrepreneurial road has obvious fluctuations, and the concept of speed and new technology in the Internet industry has a unique elaboration, and Guo Xinnian proposed that even if the magic crystal will be removed tomorrow, the software should be perfected today. This fluctuating plot clearly resonates with the audience.

Post-Editing

The plot related of the 3 episodes of the whole drama has abnormal values, which are 41, 53, and 54 episode.

These two episodes are the last two episodes of the entire TV series. In terms of plot, Guo Xinnian has experienced hardships again and again, and finally the new technology has been promoted. In the most difficult period of Guo Xinnian, he harvested the love of Na Lan. With the encouragement of his lover, he realized that the current difficulties are not the end, but a new beginning. In the end, Guo Xinnian and Na Lan joined hands and continued to go on the road to entrepreneurship. The audience is generally satisfied with the plot, so more attention is paid to the post-production of the TV series, and music, sequels and other aspects are discussed. It is worth noting that in terms of scheduling the broadcast time, it used to be broadcast two episodes a day. In the last two episodes, it was changed into a one-day update episode, which caused a big dissatisfaction among the audience. Many comments reflected "why not updated", etc. So there are obvious fluctuations in the post-production.

Insignificance

The plot related of the 2 episodes of the whole drama has abnormal values, which are 50 and 51 episode.

The number of comments in the 50 episodes and 51 concentrated irrigation is obviously increased. According to the playing time of Youku video software, the playing time of these two episodes is before the "Double Eleven". You can watch this drama for more than three minutes, obtaining the energy in the double eleven activities. Although the energy gathering activities can encourage more people to watch the video, it does not play the role of urging the audience to watch the drama seriously. Many viewers do not come for TV dramas, but In order to consume time and earn energy. They fill in meaningless comments, which pushes up the total number of comments. So a lot of comments are similar to "Only take energy, don't care other things."

Negative Emotion

The plot related of the 3 episodes of the whole drama has abnormal values, which are 1, 2, 18, 19, 20, and 21 episode.

The reason for the abnormal fluctuations in negative emotions in the first two episodes is mainly because of Yang Ying's unsatisfactory acting skills and some details in the plot are not reasonable.

In the episodes of the 18 and 19 episode. Guo Xinnian and Wendy established a relationship of love. This is in sharp contrast with the previous plot. Because Wendy deceived Guo Xinnian, making Guo Xinnian admit the wrong person. And the Wendy's actions made the audience feel very angry. The contents of the 20th and 21st episodes are a continuation of the previous plot, and many bad things have happened under the guidance of Wendy's mistakes. These all led to the generation of negative emotions.

Neutral Emotion

The plot related of the 6 episodes of the whole drama has abnormal values, which are 1, 19, and 50 episode.

From the analysis of negative emotions and positive emotions, it can be found that in episodes 1 and 19, these two emotions are very high, so the neutral mood is relatively low. In the insignificance analysis of Section 6, it can be seen that due to the activities of the 50th episode of energy collection, most viewers only pay attention to absorbing energy, not the content of TV dramas. The comments made are mostly meaningless or unrelated to the plot, so neutral emotions in this episode reach a peak.

Positive Emotion

The plot related of the 4 episodes of the whole drama has abnormal values, which are 6, 11, 18, and 19 episode

To the 6 episode, the audience expressed envy and appreciation for the friendship between Guo Xinnian, Yang Yangyang, and Luca. They felt that the friendship was very precious, and the Oolong that the three people made was also very interesting.

In the 11th episode, the Na Lan's intelligence improved and began to detect the greasy and investigated Wendy. The interaction between Luca and the Yang Yangyang was very interesting so the audience liked it. The audience was very satisfied with Bobo's assists, because of the assists Na Lan discovers the secrets of Luo Wei and poke the lie of Luo Wei. The audience also believes that Yang Ying's acting skills have improved. The pleasant cooperation between Guo Xinnian and angel investors is also popular with the audience.

The episodes of the 18 episodes and the 19 episodes have fluctuated greatly. Guo Xinnian and Wendy established a relationship of love. As the previous analysis shows, the negative sentiment of these two episodes is strong, resulting in lower positive emotions.

Statistical Analysis

This part is the analysis of relationship between TV series broadcast amount and different comments.

The leading cast of the TV we choose is strong. People who really have entrepreneurial experience, however, gave the TV series a bad rating after watching it. Indeed, the combination of the hot topic and the super star is good. But what causes douban score such low? Does this mean that hot topic and the super star can not contribute to the demand of TV series?

First of all, several factors that are significantly related to TV series broadcast amount are determined, including actor-related, plot-related, social phenomenon, post-production related and meaningless information. Through the analysis of the relationship between the number of reviews of different categories and the number of TV plays, we found that the number of reviews of the evaluation of actors is significantly positively correlated with the broadcast amount of TV plays. TV drama reviews are generated by users' active behavior, which reflects users' willingness to pay attention to TV dramas. We think the number of reviews in different categories reflects the degree to which users value different elements of the show.

The first we do is to probe the role of super stars and hot topic in TV drama on-demand. Using Stata to analyse the correlation between actor-related comment and TV series demand, we find that the comment amount of actors is significantly positively correlated with the broadcast amount. There is a phenomenon called herd effect (or group psychology) which may result in it. The herding theory tells us that people tend to do the same things as other. Thus, we hold that the more willing an audience is to discuss actors, the more likely they are to watch relevant episodes, leading to the increase of TV broadcast amount. In terms of commercial value of the TV series, of particular important is the attention and the flow. We can see that although many people don't think highly of the TV series, super stars and hot topic do result in a lot of attention for the TV series and increase the broadcast amount of the TV series.

In addition, we also find that the number of audience comments on actors is similar to the Ebbinghaus Forgetting Curve, indicating that the flow effect of super stars and hot topic actors is exponentially degenerated. For the TV drama manufacturers and video website, which takes the broadcast amount as the main indicator, we think that the producers of TV series should guide the audience's discussion on actors at regular intervals to slow down the decreasing herd effect so as to improve the TV series broadcast amount.

Secondly, the influence of plot on TV series broadcast amount is also analyzed. Through the previous analysis on the number of comments of various categories, we find that excluding extreme cases (the first episode and the second episode), the amount of comments on the plot is the largest, which indicates that the audience attache high importance to the plot. And the correlation analysis shows that the more the number of comments on the plot, the less the number of broadcast amount. Does this mean that the audience's attention to the plot inversely reduces the broadcast amount? Through the analysis of the emotional tendency of the audience in the plot, we find that the audience have more negative comments on the plot in this TV series, which actually suggests that the bad plot will affect the TV series broadcast amount, such as unreasonable plot and antisocial plot. The reasonableness of the plot has a great impact on the broadcast amount, and the fluctuation curve of the number of plot comments shows that the fluctuation of the plot will also affect the audience's psychology. Therefore, it is suggested that the TV series producers, before shooting the TV series, should do some investigation to make the plot more realistic. In addition, they should try to display the plot that is too controversial. And the plot that is likely to arouse people's antipathy should be reduced as much as possible. From the plot review curve, we can see that there are quite a lot of negative comments in episode 18 and 19. Meanwhile, we can know, from the correlation analysis, that these plots which make audience have negative comments will significantly reduce the broadcast amount.

Therefore, in terms of realizing the commercial value of TV series, it is necessary to focus on hot topics and actors. Meanwhile, it is also necessary to avoid plots that are likely to arouse people's negative emotions. Otherwise people are likely to abandon TV series, resulting in the decline of broadcast amount. At the same time, in terms of the realization of the artistic value of the drama, the rationality of the plot is necessary.

CONCLUSION AND SUGGESTIONS

Conclusion

With the development of society, TV drama plays a more and more important role in people's life. According to an authoritative research, every year, more than 1.28 billion Chinese watch TV dramas through various channels, such as website, App, TV and so on. Furthermore, TV drama not only exert a significant implications on people's entertainment, but also deeply influence people's sense of worth, world view and so forth. In this sense, it is necessary to create and present excellent TV dramas which is full of positive value. Therefore, in this paper, the researchers conduct social value analysis on the popular TV drama named *Entrepreneurial Age*, based on its audience commence. After that, the researchers will try to provide the TV producer with suggestions about the subject selection, casting selection and script design.

To explicitly tackle with the problem, the researchers divide the whole process into several procedures. Firstly, the researchers utilized Bazhuayu.com to crawl a considerable amount of data of audience comments from website. The researchers totally obtained 29,841 valid data from 54 episodes; in 7 dimensions. Secondly, the researchers tagged comments from two dimensions, which are content classification and emotional polarity classification and give comments tags respectively. Because the number of comments is too many, the researchers designed a prediction model to tag the comments. It needs us to select a part of comments as training set and label the training set. After comparisons among several prediction model, the researchers picked the random forest algorithm to predict the remaining unlabeled data due to its high accuracy and efficiency. Finally, the researchers did analytical interpretation of the results and draw the conclusions. The conclusion can be organized from four aspects.

Through the label summary, the result of emotional polarity classification shows that negative comments outweighs the positive comments in every episode, especially in the middle of the TV drama. And the result of content classification indicates that the most comments at the beginning of the TV drama is about actor, while most comments after a while is about plot.

Through the role emotion analysis, the researchers find that audiences' negative emotions take most place. And the proportion of positive emotion about the leading actress- Na Lan is most, which occupies 38%. In the meantime, the percent of positive emotion about the second actress- Wen Di is least, which occupies 17%. It is because that the role is far away from reality and there are many detailed bugs.

Through social influence analysis, in which the researchers use the control chart to monitor the proportion of social phenomenon labels of comments and identify the outliers. It is easy to observe that most kinds of comments exceed the control limit at around 18th episode and 51st episode. At around 18th episode, the development of TV drama reaches its climax. In other words, the story at this point reflects the real life and is close to the audience. In this way, it incredibly appeals to audiences and arouse their resonance.

Through analysis of relationship between TV series on demand and different comments, several factors that are significantly related to TV series on demand are taken into the linear regression model to analyze the correlation between actor-related comment and TV series demand. It can be found that they are significantly positive correlated with each other. Moreover, there was a significant negative correlation between the number of plot-related negative comments and the TV series demand, especially in 18th and 19th episodes, which have a lot of negative comments. So it is necessary to avoid plots that are likely to cause people's extreme negative emotions. Otherwise, people are likely to abandon the drama because of that, leading to a decline in TV series demand.

Suggestions

Based on the audience commentary of the recent era of a popular TV drama, the researchers look forward to exploring the value behind the TV series. Here, value can be divided into business value and social value, business value can be measured by the amount of play, and social meaning is the embodiment of total social value.

When it comes to the analysis of business value, the previous analysis has already drawn that the audience's emphasis on the plot is relatively high, the traffic effect brought by high-value and high-topic actors exists and the bad plot will affect the TV drama on-demand. The rationality of the plot has a great influence on the amount of on-demand, and there are fluctuation curves of the number of plot comments. The researchers can see that the fluctuating plot will also affect the psychology of the audience. Therefore, it is recommended that the TV drama producers, before the start of the TV series, it is best to train the main creative staff of the TV series to make the plot closer to reality. Moreover, it is necessary to avoid controversial episodes, and the plots that are likely to cause people to resent should also be minimized. Therefore, in terms of realizing the commercial value of TV dramas, hot topics and actors are still necessary, while avoiding easy to cause people. The plot of negative emotions and as much as possible to ensure a reasonable plot.

When it comes to the analysis of social value, the researchers try to use the comments to analyze the psychology of the audience, classify the comments according to the content and motivation, and divide the comments into the motivations of the drama, the details of the comments and the language, the understanding of the content of the story, and the comments and answers. Barrage and other content, analyze what kind of spiritual needs the audience is, and under what circumstances it is easier to express their inner feelings and express what they write. Secondly, the online data and various indexes are summarized and analyzed, and relevant statistics on user data can be used to construct crowd portraits from the geographic, age, gender, and interest dimensions. For example, searching for "Entrepreneurial Times" on Xiaohongshu, you can find that the protagonist's makeup, professional wear and the same jewelry have become topics of discussion. Accurate marketing after accurate user portraits, accurately predicting their psychological status, and making public opinion monitoring are an important manifestation of social value.

Big data has played a huge role in TV drama theme selection, topic setting, team formation, delivery methods, channel selection, marketing strategy, narrative style, and viewing prediction. At the beginning of the production of TV dramas, TV stations can make full use of big data technology, take market appeal as the core, and take the audience as the principle, and strive to achieve the synergy of economic and cultural values.

In general, the contribution of big data in speculative topics, targeting audiences, precision marketing, getting feedback, making explosive spots, etc. can help to find relative certainty in the uncertain market of the film industry and ensure its economic benefits. It also maximizes social value.

ACKNOWLEDGMENT

This research was supported by Professor Zhang Xi, Hong Yalan, Zhu Wenwen, Wang Xinyu and You Wenbo from College of Management and Economics of Tianjin University.

REFERENCES

Abbar, S., Amer-Yahia, S., Indyk, P., & Mahabadi, S. (2013). Real-time recommendation of diverse related articles. In *Proceedings of the 22nd international conference on World Wide Web* (pp. 1-12). ACM.

Amrit, C., Paauw, T., Aly, R., & Lavric, M. (2017). Identifying child abuse through text mining and machine learning. *Expert Systems with Applications*, *88*, 402–418. doi:10.1016/j.eswa.2017.06.035

Cen, L., Si, L., Li, N., & Jin, H. (2014). User Comment Analysis for Android apps and CSPI Detection with Comment Expansion. In *PIR* (pp. 25–30). SIGIR.

Chatterjee, A., Gupta, U., Chinnakotla, M. K., Srikanth, R., Galley, M., & Agrawal, P. (2019). Understanding Emotions in Text Using Deep Learning and Big Data. *Computers in Human Behavior*, *93*, 309–317. doi:10.1016/j.chb.2018.12.029

Dong, Y. (2015). On the Data Management and Data Mining of Film and TV Drama Played On-Line. *In Conference on Engineering And Technology Management* (p. 69). GCETM.

Ganesan, K., & Zhai, C. (2012). Opinion-based entity ranking. *Information Retrieval*, *15*(2), 116–150. doi:10.100710791-011-9174-8

Jiang, J., & Xia, R. (2016). Microblog sentiment classification via combining rule-based and machine learning methods.

Koh, N. S., Hu, N., & Clemons, E. K. (2010). Do online reviews reflect a product's true perceived quality? An investigation of online movie reviews across cultures. *Electronic Commerce Research and Applications*, *9*(5), 374–385. doi:10.1016/j.elerap.2010.04.001

Li, X. (2009). Dis/locating audience: transnational media flows and the online circulation of East Asian television drama (*Doctoral dissertation, Massachusetts Institute of Technology*).

Liu, Z., & Liu, L. (2012). Empirical study of sentiment classification for Chinese microblog based on machine learning. Jisuanji Gongcheng yu Yingyong(*Computer Engineering and Applications*), *48*(1), 1-4.

Mullen, T., & Collier, N. (2004). Sentiment analysis using support vector machines with diverse information sources. *In Proceedings of the 2004 conference on empirical methods in natural language processing*.

Müller, F., & Hermes, J. (2010). The performance of cultural citizenship: Audiences and the politics of multicultural television drama. *Critical Studies in Media Communication*, *27*(2), 193–208. doi:10.1080/15295030903550993

Neuman, W. R. (1982). Television and American culture: The mass medium and the pluralist audience. *Public Opinion Quarterly*, *46*(4), 471–487. doi:10.1086/268745

Popescu, A. M., & Etzioni, O. (2007). Extracting product features and opinions from reviews. In *Natural language processing and text mining* (pp. 9–28). London, UK: Springer. doi:10.1007/978-1-84628-754-1_2

Sureka, A. (2011). Mining user comment activity for detecting forum spammers in youtube. *arXiv preprint arXiv*:1103.5044.

Turney, P. D. (2002, July). Thumbs up or thumbs down?: Semantic orientation applied to unsupervised classification of reviews. *In Proceedings of the 40th annual meeting on association for computational linguistics* (pp. 417-424). Association for Computational Linguistics.

Wang, H., Lu, Y., & Zhai, C. (2010). Latent aspect rating analysis on review text data: a rating regression approach. *In Proceedings of the 16th ACM SIGKDD international conference on Knowledge discovery and data mining* (pp. 783-792). ACM. 10.1145/1835804.1835903

Yang, J., & Yecies, B. (2016). Mining Chinese social media UGC: A big-data framework for analyzing Douban movie reviews. *Journal of Big Data*, *3*(1), 3. doi:10.118640537-015-0037-9

Yao, R., & Chen, J. (2013, December). Predicting movie sales revenue using online reviews. *In 2013 IEEE International Conference on Granular Computing (GrC)* (pp. 396-401). IEEE.

Zablith, F., & Osman, I. H. (2019). ReviewModus: Text Classification and Sentiment Prediction of Unstructured Reviews using a Hybrid Combination of Machine Learning and Evaluation Models. *Applied Mathematical Modelling*, *71*, 569–583. doi:10.1016/j.apm.2019.02.032

Zhuang, L., Jing, F., & Zhu, X. Y. (2006). Movie review mining and summarization. In *Proceedings of the 15th ACM international conference on Information and knowledge management* (pp. 43-50). ACM.

Chapter 5
Disruption in HR Through Inclusive Emotional Culture:
Key to Sustainable Growth in Indian Businesses

Nitu Ghosh
https://orcid.org/0000-0002-5744-1670
REVA University, India

ABSTRACT

Diversity in workforce has demanded the need to develop inclusive culture that suffices the needs, interests, and emotions of all equally. High-performing organizations have incorporated emotions in their mission, creating an emotional culture that accentuates emotional intelligence of their talent force rather than suppression of emotions. The chapter explains how and why organizational cultures are transforming into inclusive emotional cultures. Through an analysis of cases of Indian multinationals, the antecedents and outcomes of Inclusive emotional culture is described through Inclusive Emotional Culture Framework model. Views of HR managers are expressed through content analysis qualitative technique. The chapter provides a key to sustainable development of Indian businesses amidst the doldrums of global business environment through inclusive emotional culture integrating values of long-term sustenance and development.

INTRODUCTION

Sustainability through inclusiveness, emotional intelligence capabilities and values- driven business processes are the main agendas of sustainable development in the twenty-first century. The present era has seen a revolution in the science of emotions and sustainability through emphasis on emotional culture and emotionalytics. Although there have been fast advancements in technology and digitization, yet organizations shifted focus from perfect rationality organisational culture to bounded rationality-oriented culture. Massive disruption in HR practices and composition of organisational culture has been characterising

DOI: 10.4018/978-1-7998-0357-7.ch005

organisations under the influence of global integration over the last two decades. Building an inclusive emotional culture has been the trend in the 21st century in order to achieve the goals of sustainability. It is not just the cognitive culture (shared intellectual values, norms, artefacts, and assumptions that serve as a guide for the group to think and behave), but importantly the emotional culture (the shared affective values, norms, artefacts, and assumptions indicating which emotions people have and display at work and which ones are been suppressed) that governs how innovative, team-oriented, customer-focused, or competent the employees are supposed to be (Barsade & O'Neil, 2016). Robust transitions in structure, culture and climate of business organisations has brought about shifts in demographic and psychographic trends of the workforce including their motives, emotions, expectations and work engagement levels.

Workplace dynamics constituting diversity in terms of gender, age, culture, emotions, innovative and motivational drives have called for a re-emphasise on aspects related to human capital management practices and crafting of an inclusive emotional culture, which caters to the needs of the highly diversified workforce. Inclusive culture that recognises, addresses and appreciates the emotions of every employee on an everyday basis can be instrumental in building engagement, leadership, ownership, innovation and high performance at every level of the organisation. High performing organizations such as Google, Apple, Coca Cola, PepsiCo, Southwest Airlines, HCL Technologies, Whole Foods Market, The Container Store, Zappos, Happiest Minds, Mahindra, Sap Labs India, Intuit India, Adobe India and many others have incorporated emotions in their mission statement. Some others like Cisco Finance, Ubiquity, C&S Wholesale Grocers, Camden Property Trust, Vail Resorts and many start-ups, emphasize the importance of fun to innovate and succeed. By doing so, they create an emotional culture that not only appreciate their emotions and feelings but enhances the Emotional intelligence of their talent force rather than suppressing.

A lot of disruption in HR practices has been happening in the new millennium in most organisations. To sustain, grow and remain competitive, it is crucial to retain talent force. However, the question arises, what exactly is the sustainable solution for talent acquisition, retention, high engagement and performance? Despite the hefty investments into HR strategies for accentuating employee satisfaction, commitment and productivity, engagement levels across the United States' workforce have remained stagnant and those worldwide considerably low. Survey reports state that the United States witnesses $450 billion to $550 billion of lost productivity and labour turnover due to low satisfaction among employees every year (Head, ACC, & Freedman, 2014). Hence the question arises- Why are employees detached from their jobs, their bosses and each other? How can organizations re-design their HR practices so as to build high performance work teams and a culture of commitment? How is it that the Indian businesses are soaring high despite the global slowdown? What is the secret behind the sustainable growth of Indian businesses?

Shaver P. et al. (1987) found that people can reliably distinguish among 135 emotions. But understanding the most basic ones like-joy, love, anger, fear, sadness is the initial step for any leader to manage his team and build a positive emotional culture. Thus, the present chapter presents a detailed literature review on the concept of emotional culture, dimensions of emotional culture and how to build positive emotional culture. The constructs of emotional intelligence and emotional labour have been thoroughly dealt. Through case based and content analysis qualitative study, practices building a sustainable emotional culture in Indian organisations have been discussed.

India is the fastest growing economy with a GDP rate of 8.4 percent in 2018. According to the World Bank, India overtook China to become the fastest-growing major economy in the world as of 2015 (ET, 2015). The robust growth in India's economy, businesses, literacy rate, technology, research, socio-cultural factors and the demographic dividend has placed this country in an envious position on the world map.

India's rise to power and speculations of 'soon to become superpower', has raised expectations about how it is going to change the global order. As per a report by Ernst and Young, India's workforce population is going to rise to an estimated 900 million level by 2020, while American population (third largest in the world) is just 160 million people (Chris Matthews, Oct 30, 2017). Hence, hopes and speculations are increasingly placing a lot of importance on the fast growth and sustainability of Indian businesses, industries, economy, and society. Researchers have started focussing on understanding the organisational practices, culture and ethos in Indian businesses that have enabled them to sustain in economic recession and growing throughout. This study plays a significant role in understanding the organisational culture and people management practices that focus on accentuating the workforce strength and making these businesses stronger over the long term.

WHY SUSTAINABILITY?

In 1987, United Nation's Brundtland Commission's report stated sustainable development to be "development that meets the needs of the present without compromising the ability of future generations to meet their own needs" (Brundtland Commission, 1987). Sustainability is a universal approach that is based on the three pillars of ecological, social and economic sustenance and growth, realizing the need to drive all the three together for lasting prosperity. In 2010, the Office of Sustainability at the University of Alberta proposed a more comprehensive view of sustainability that included the existence and role of people. The report defined Sustainability as "the process of living within the limits of available physical, natural and social resources in ways that allow the living systems in which humans are embedded to thrive in perpetuity" (Sustainability Summary Report and Recommendations, 2011). Sustainability is rather a value that is commonly shared by many individuals (of different types) and organizations who demonstrate this value in their policies, systems and culture.

Sustainable development of people, planet, prosperity, peace and partnership are the SDGs identified for universal development in the United Nations General Assembly with the 2030 Agenda for Sustainable Development (Wilkins, Hugh, 2008). Sustainability of economy, businesses and society revolves around effective diversity management, which has to be built on principles, embodied as pillars of an inclusive culture. Building on the principle of "leaving no one behind", the agenda puts impetus on a holistic approach to achieving sustainable development for all. Thus, sustainability in business also calls for equal opportunity, equal treatment, accessibility and development of people of all categories. In the polarised business world, sustainability and retaining talent force involves focus on inclusion and not diversity, which implies catering to the emotions, feelings, needs and interests of all types of individuals belonging to different genders, ethnicity, disabled, age, colour, special categories like LGBT (Lesbians Gay Bisexual, and Transgenders) etc. in the work groups. This definitely mandates sustainable HRM (Stefanie App, Marion Büttgen, 2016) and an inclusive organisational culture. To achieve the goals of sustainable development, transformations in individual behaviour patterns and decision-making processes is necessary. Cognition and emotions play decisive roles on sustainable behaviour (Kals & Maes, 2002).

The psychology of sustainability and sustainable development has recently become an upcoming field of research. It mainly emphasizes on the sustainability of interpersonal and intrapersonal competencies of individuals, groups and communities, including aspects of reflexivity, values, purpose, and progress for the sustainability of projects harmonizing the different perspectives (Di Fabio, & Kenny, 2018). The United Nations stated in its recent conference titled "Unlocking Your Emotions" in May 2019 that emotional intelligence is a powerful tool to support the implementation of sustainable development goals. Daniel Goleman quoted while deliberating, "emotional intelligence skills are key to the jobs of the future and a better predictor of workplace success than IQ and other measures of intelligence" (Academic Impact UN Report, April 18, 2019). It was concluded that EI, mindfulness and nonviolent communication are effective tools that could be implemented to help achieve sustainable development and a more peaceful and prosperous world.

Sustainability in the business context describes the ability of the entity (brand, company, or management) to deliver on its promises consistently to all its stakeholders (customers, employees, government, public, etc.) at all levels over time for better employee outcomes (Backhaus, 2004). It can be perceived through different approaches as per various researches. Substance oriented approach towards sustainability emphasizes on the balance between the consumption and regeneration of its human resources for the survival of an organisation (Müller-Christ & Remer, 1999). For such a balance they focus on creating value for their potential and existing employees by investing on their long-term retention and utilisation (through effective health and safety benefits, socialisation, managing aged or special category workers, female employee friendly policies, competency development, culture of lifelong learning). Sustainable HRM facilitates organisations to achieve its goals of fulfilling the expectations of its employees and also to attract, acquire, retain high quality and satisfied workforce (Ehnert, 2009; Jabbour & Santos, 2008). Building and promoting the culture of sustainable HRM and being perceived as a psychological benefit by its users, is very crucial for an organisation to successfully operate, leveraging maximum potential of its committed and competent workforce. This is possible only when a working environment is nurtured that replicates a corresponding image of 'sustainability' in the minds of its employees (Stefanie App, Marion Büttgen, 2016).

IS INCLUSIVE CULTURE REQUIRED?

Inclusivity is the degree to which an employee (with individual differences) is accepted as an insider by the organisation and accepted in all work systems (Pelled, Ledford, & Mohrman, 1999). It is the sense of belongingness, ownership and equal participation in all work processes irrespective of differences (Lirio et al., 2008). The philosophy of 'inclusion' directs the group effort towards the term 'all' equally (April & Blass, 2010). The twenty first century has seen the transition of the discourse on diversity to the concept of inclusion (Nair & Vohra, 2015) with a purpose to create a culture that perceives, appreciates, manages and utilises each person's emotions, values, beliefs, cognitions in a manner that seems equitable to all. Inclusion is considered to be a process of change as well as an affective outcome or feeling of belongingness, respect, fairness etc. (Dobusch, 2014). Culture of Inclusion are mainly driven by perceptions of fairness, respect, value, equity and belonging (Nair & Vohra, 2015). Inclusion is required wherever diversity or differences exist. The polarised economy thus calls for inclusion in every aspect of social life, education and business processes.

Building a winning and sustainable culture is always a common vision and fruitful outcome (Smircich, 1983; Schein, 1985). A culture of inclusion depends on the level of individual differences perceived in the group and extent of mutual recognition demanded. 'Inclusion' is a word more important for sustainability than diversity. Organisations such as AT &T, Deloitte, HCL Technologies, Verizon and many other Fortune companies emphasized on hiring people with diversified backgrounds- skillsets, gender, race, ethnicity, minority, disadvantaged and ages. Focus of leadership, management policies and HR practices should be to distribute the diversified workforce evenly and equitably across the company's divisions to create conducive and inclusive working culture, creating opportunities for all (Morais et al., 2014; Lundrigan et al., 2012; SHRM, 2009). Promoting an inclusive culture implies that respect is paid to the plurality of subcultures, and none are exempted from this ethical discourse. Each subculture is provided an equal opportunity to participate in shaping the cultural backbone in the organisation, which includes values, norms, policies etc (Pless & Maak, 2004). Thus, an inclusive culture is empowering, engaging and high performing as each one is heard and involved equitably in decision-making, problem-solving, innovation and moulding the organisational culture towards one that sustainably adds value to the company's performance (Pless & Maak, 2004). A culture of inclusion is said to exist when strategies, processes, systems, values and norms are based on diversified knowledge and perspectives of the overall group (Holvino, Ferdman, & Merrill-Sands, 2004). Open communication channels, diversity-based recruitment, training, promotion, compensation and career growth and planning practices that establishes an aura of 'working climate for one and all' constitutes the building blocks of an inclusive culture (Daya, 2014). The feeling of one-ness, respect, fairness, belongingness, tolerance, empathy and commitment results in creating psychological contract and engagement, which eventually leads to sustainable development of people, practices and organisations. No doubt, Inclusivity is essential to achieve the sustainability goals and also to create sustainable values and HRM practices in organisations.

As per a Deloitte report (2012), a workplace where just 10% more employees feel 'inclusive', the company would be able to increase work attendance by almost one day per year per employee. In addition to this, the business performance is higher in terms of people innovation, responsiveness, customer service and team collaboration. As per the monthly Labour review of the United States Bureau of Labour statistics (Dec. 2015), the projected workforce is expected to reach 163.8 million in 2024, would include more aged workforce, more women with a share of women in the workforce projected to be 47.2% in 2024, more educated employees (with technical qualifications), more Asians and Africans, with good representation from the minority LGBT group. The millennial workspaces of present and future, is thus characterised with more diversity, dynamism and soaring expectations, which can be sufficed through inclusive work culture with high emotional intelligence and practices that nurture positive emotions conducive for accentuated creativity and team performance.

Surveys have revealed that employee perceptions about inclusion in Indian workplaces accounted for 43 percent of team citizenship behaviour, which implies that focus on inclusion is essential to harness the potential of diversity (Prime & Salib, 2014). India being a highly diversified country with multicultural teams in its workplaces has been experimenting with a plethora of HR practices right from diversity-based recruitment to mentoring practices that bring in higher inclusivity in organisations. This has helped them to develop sustainable HRM practices and loyal workforce.

Inclusive work culture in Indian organisations have focussed on the core values of mutual respect, tolerance, spirit of integrity and harmony, collaboration, open communication, belongingness, trust and engaging. Emotions rather than cognition has been guiding most of the decisions regarding inclusive people practices in these organisations. Emotions are fundamentally important for sound moral development and healthy interpersonal relationship, which is invariably crucial in building a culture that fosters inclusion and recognizes differences. This has helped Indian industries to function more smoothly that its political environment and also to sustain the global economic recession. Godrej group is one such FMCG company that has brought about a turnaround in its business outcomes through its Diversity and Inclusivity (D&I) initiatives. With a strong belief that passionate, rounded individuals with diverse interests are better performers, they have institutionalised a culture and HR practices that encourage employees to 'bring out their whole self and show their full potential at work'. Their health insurance benefits allow the members to add their 'partners' as a beneficiary not just their spouse, thus bringing in an open outlook in work culture too. Their regular diversity - sensitisation workshops across different geographical locations facilitates employees to share the common belief that integrating openness and inclusiveness begins with people and workplace.

EMOTIONAL CULTURE: A SUSTAINABLE HR SOLUTION

Emotions and feelings are a critical component of an inclusive culture and when that is met, much of the goals of inclusiveness is attained. Studies on Pennsylvania Railroad System employees (Hersey, 1932) and factory employees (Roethlisberger & Dickson, 1939) stated the importance of caring, affection, compassion, and tenderness, and the negative effects in the absence of these emotions, particularly in interpersonal relationships with foremen. As mutual recognition is usually facilitated through emotional recognition and solidarity (Honneth, 1994; Maak, 2006), thus emotional capabilities are critical in any interpersonal transaction. A study on an analysis of 25 common competencies stated that demand for social and emotional skills will grow across all industries by 26% in the United States and 22% in Europe, due to automation, between 2016 and 2030. Thus, companies should gear up to focus on developing emotional and social skills among their employees to help withstand changes, challenges, and progress (Jacques Bughin et al., 2018).

Emotions are crucial in social analysis as they play an impactful role in the process of socialization. Emotional culture is a regime or way of life that favours the perceiving, managing, controlling and utilising emotions of each and every member in an effective manner so as to achieve the overall purpose. It places importance on positive emotional expression, emotional management and satisfaction of desires. Organisational climate that implements HR practices nurturing emotions like joy, satisfaction, respect, trust, appreciation, achievement, compassion, pride etc is always high performing, creative and sustaining, rather than one with anger, frustration, fear and mistrust. Positive emotional culture constituting affection, pride, compassion, empathy, joy and gratitude foster supportive employee behaviours and organisational citizenship among employees. This invariably enhances employee commitment, motivation, engagement and retention. Symmetrical communication and responsive leadership builds a positive emotional culture in organisation (Linjuan Rita Men & Cen April Yue, 2019), therefore promoting sustainable HRM and people development. Thus, it is important to institutionalise an organisational culture that promotes the expression and sharing of positive emotions and does not coercively suppress negative emotions. Moods affects and discrete emotions (satisfaction, trust, joy, pride, anger, fear) have their own appraisal antecedents, subjective experiences, and action tendencies that direct people to respond to their existing situation. These emotions also have intrapersonal impact on the person experiencing them in terms of attention, motivation, creativity, information processing and judgment, and well-being, which drive impulsive or mature expressions as per the ability of the individuals to control them (Cynthia Fisher, 2019).

Alice Isen (1999) posited that a conducive emotional culture in the organization cultivates creativity, mental efficacy, making people efficient at comprehending and analysing information, adopting decision rules in judgmental problems, and in adopting flexible decision-making styles. As Daniel Goleman (2000) in his bestseller 'Working with Emotional Intelligence' explains,

If you see a star performing team, you're seeing a very high group IQ. But what predicts the actual productivity or effectiveness of a team is not the potential—that is, the best talents of every person—it is how people are valued on that team. It is how people feel there's harmony, that we get along, that we surface simmering issues, that we take time to celebrate, that we know each person's strengths and that we step aside when it is time for this person to come forward. In other words, that we are a team that has a high emotional intelligence.

Every job has its affective component and if that is not re-enforced due to weak emotional culture, the effectiveness on job performance declines. For example, healthcare jobs demand compassion and love over impatience and anger while security and investment jobs involve fearlessness over recklessness. Ripple effects of unexpected emotions expressed in workplace are immense and eventually lead to poor interpersonal relations between colleagues and boss, ineffective communication, role conflicts, underperformance, frustration, psychological contract breach, burnout and even high labour turnover. Organisational culture is usually a set of norms, beliefs, values and emotions integrative to the mission statement; and is also often enacted in the 'micro moments' of daily organizational life (Barsade & O'Neil, 2016). These constitute small gestures like acts of kindness, support, enjoyment or sense of achievement rather than bold declarations of feelings. Cultures that are weak in companionate love, expressions of affection, caring, or trust among employees, show indifference and callousness towards each other, due to low intensity and crystallization (Jackson, 1966; Barsade & O'Neil, 2014).

Figure 1 illustrates the variable of emotional culture of compassion, main mechanism involved in creating the emotional culture (through 'feeling the emotion' equivalent to emotional intelligence and emotional labour and through normative enactment) and the outcomes (in terms of employee outcomes and stakeholders' outcomes and better quality of work life and engagement of the employee) (Barsade & O'Neil, 2014).

Various HR practices from hire to exit cycle, emotional modelling (managers consciously enacting expected emotions to display a 'feel good' factor), company interiors, workspace designing, posters, furniture, emotions tracking Apps, etc also makes up the organisational culture, reminding employees and management to adhere to such emotional requirements and help integrate inclusivity in work culture.

Figure 1. Theoretical model of emotional culture of companionate love
Source: Barsade, Sigal, & A O ' Neill, Olivia. (2014).

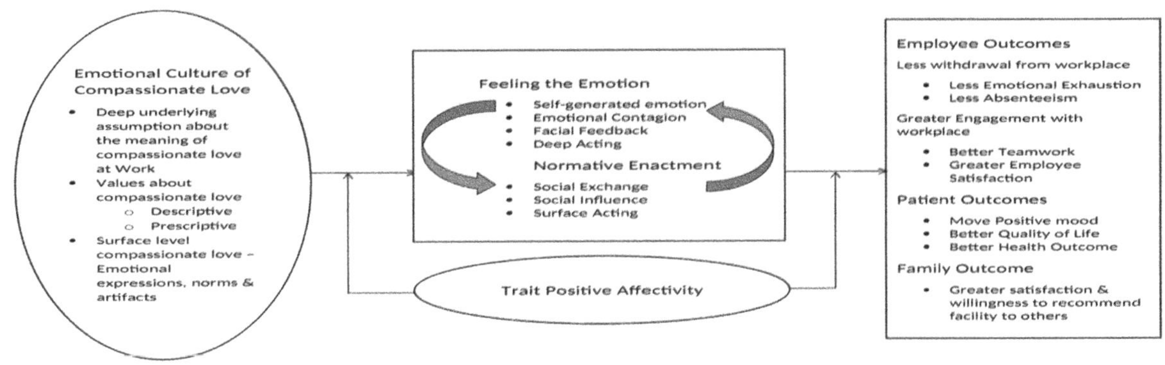

For example, the Indian company, Religare Enterprises has recruitment and selection practices that assess candidates based on the emotional intelligence and other emotions demanded by their jobs, EI based training and 720-degree appraisal system that appraises employees based on emotions as an important criterion. Jubilant Organosys keeps EI of its employees on the priority list of competency development and routinely checks on the EI levels of its employees. HCL's excite-enthuse-engage HR policy is world famous in turning around the company and building an employee-first culture.

Thus, it is significant to note that every organization has an emotional culture, even if it's one of suppression (Barsade & O'Neil, 2016). Organisations are rapidly moving towards a culture that nurtures its employees' emotions, directs them towards positive emotions, beliefs, values, common vision and utilises those emotions in driving their job motivation, psychological contract and workplace behaviour. Global companies are adopting HR practices to continually develop the emotional competencies of their employees and keep them emotionally connected through an integrated inclusive emotional culture that caters to the needs and interests of diversified workforce. Researches posit that emotional culture influences employee satisfaction, burnout, teamwork, and even hard measures such as financial performance and absenteeism (Barsade, & O'Neill, 2016).

Disruption is happening in the HR space in building an emotionally sustainable culture. Some have included emotions in their mission statement, management principles and organization culture, eg- Coca Cola, PepsiCo, Tata Group, Whole Foods Market, and Zappos have listed 'love' or 'caring' among their corporate values. Emotional artificial intelligence technologies, EI tracking apps and emotionalytics are adopted in work culture to track and measure the moods and emotions of employees on a frequent basis. Training programs and performance appraisals integrating EI and engagement into their system is a regular practice in most organizations. Infosys, Indian Railways, Entropik tech, Tata sky and many millennial organizations are now moving ahead to develop a much stronger emotional culture by fostering the EI of their employees, providing an engaging experience for higher talent retention and performance outcomes.

DIMENSIONS OF EMOTIONAL CULTURE

Emotional Intelligence

Emotions play a pivotal role in almost all actions. People who are able to manage emotions are the best performers and leaders. About seventy five percent of an organization's success can be attributed to the emotional intelligence of its employees. Emotions are the drivers of motivation and attention; identify and facilitate change management, and influence behaviour and actions (Fredrickson, 2001). Emotions have been found to be determining factor in developing the ability of resilience and capacitate people to cope with challenges proactively, build new vision, enhance one's competencies and connect with others for effective teamwork (Fredrickson, 2001; Fredrickson & Branigan, 2005). Emotional Intelligence is an individual's capability to perceive, understand, control and manage emotions of oneself and others leading to better personal effectiveness, interpersonal relationships and social behaviour (Salovey and Mayer, 1990). Emotional culture emphasizes on practising of certain emotions by ensuring the development of high EI (or EQ) and expressing expected emotional labour display acts. EI is defined as an array of noncognitive capabilities, competencies, and skills like intrapersonal skills, interpersonal skills, adaptability, stress management skills and general moods that influence one's ability to succeed in coping with environmental demands and pressures (Bar-On, 1997, p. 14). It is a decisive factor in success of those individuals possessing higher levels of EQ (Bar-On, 1997). The importance of EI at work cannot be understated, as it has been found to influence leadership abilities by facilitating an effective handling of emotions, better ways of handling the needs of colleagues, and motivating them, thus making them feel contended and accomplished at work (Goleman, 1998).

EI has been stated to be a wide constellation of emotion-related self-perceptions and dispositions about self and others, assessed through self-report (Petrides & Furnham, 2003, p. 40). Studies conducted on managers' performance behaviours under two dimensions: "what they accomplish" and "how they accomplish it" through MSCEIT clearly illustrated that EI plays a more significant role in how managers do their work (leadership abilities) rather than in what they accomplish (Rosete, 2005, 2009). Emotions stimulate interest, focus attention, indicate the need for transformations, and direct people to appropriate actions (Fredrickson, 2001). They influence the way people show resilience and cope with challenges, define new goals and means to reach the ends, learn new competencies, build relationships and yield higher outcomes with the support from others (Fredrickson, 2001; Fredrickson & Branigan, 2005). Occupational self-efficacy and commitment is also found to be higher in individuals with increased EI (Rathi & Rastogi, 2009). Extant researches assert that emotional intelligence involves the capacity to reason accurately with emotions and emotional information, and about one's and others' emotions to enhance thoughts, decisions and actions (Mayer, Salovey, Caruso, & Cherkasskiy, 2011).

Therefore, building an emotional culture, where people are encouraged to sustainably develop their emotional capabilities and EI, while also contagiously pursuing positive emotional (beliefs/values) norms, becomes a more fulfilling, enthusing and engaging place to work. Much of the objectives of sustainable HRM and sustainable development of organisations are accomplished through building a high EI inclusive emotional culture. As the 21st century witnesses the transformation of business enterprises to social enterprises with the fast-changing expectations of the millennial workforce, the emotional competencies of organizational leaders are the crucial decisive factors in accomplishing organizational mission and goals, besides building an engaged and valued workplace for all (Barthwal & Som, 2012). Indian companies like ITC, Indian Railways, Whirlpool, GE, Gillette, Jubilant Organosys, Infosys, Coco Cola India, Hay Group India, Pepsico are few of the Indian companies that have realized the need for ongoing EI assessment and training to reap higher ROI in the longer term. Thus, emotional culture is not only defined by the emotions expressed and followed by all, but also the EI of its employees.

Emotional Labour

Impulse is the medium of emotions, seed of all impulse is a feeling bursting to express (Daniel Goleman, 1998). Excessive expression of positive or negative emotions (such as excitement, joy, anger, frustration, etc.) causes an adverse effect on the interpersonal environment, which should be invariably controlled for better performance. Hochschild (1983) introduced a breakthrough concept in organisational psychology called 'emotional labour' in 'The Managed Heart- Commercialization of Human Feeling'. Emotional labour (EL) has been defined as the "the management of feelings to create a publicly observable facial and bodily display" (Hochschild, 2003, p. 7).

Most of the jobs today involve interpersonal interactions for business development especially the service sector jobs that have certain expected norms and standards for display of emotions to their stakeholders. EL is crucial in this context, when most jobs call for a capability to deal with people rather than with machines, for more interpersonal skills and behavioural transactions and fewer mechanical skills (Hochschild, 2003, p. 9). Service employees like educators (Meier et al., 2006), restaurant employees

(Hallett, 2003) etc. are required to manage their own emotions in order to manage the emotions of customers (Leidner, 1999) for better work environment, goodwill and high performance. Hence, in order to perform better and fulfil organizational expectations, employees in service organizations frequently regulate and manage their emotions, a process generally called Emotional Labour (Hochschild, 1983; Grandey, 2000). Employees usually resort to four types of emotional labour displays – deep acting, surface acting, genuinely expressed emotions and suppression. During this process of emotional labour and stress, they often develop some negative emotions that contradicts with the organizationally desired emotional display, leading to dissonance, suppression, intrapersonal as well as interpersonal conflict and performance deterioration. This may lead to behavioural and decision-making issues in the employee and may lead to adverse results in terms of employee and organizational performance.

Gosserand and Diefendorff (2005) defined Emotional labor (EL) as the process of regulating one's emotional displays in response to work display rules so that goals can be achieved. The main objective of having pre-determined display rules is to dictate the emotions that employees are supposed to express in order to facilitate the attainment of work goals. (Zapf et. al, 1999) prescribed a high amount of empathy and emotional involvement in order to avoid treating other people like objects in difficult interpersonal situations (specially with clients, customers, subordinates and children). For example – service employees in hospitality sectors like in Café Coffee day, Pizza Hut etc are trained and reminded through smileys, posters and their own work display rules that despite any challenges they always need to smile and be polite. This owes to a great extent in determining the inclusive emotional culture of the organisation.

EL facilitates task effectiveness and self-expressions and is often the major cause for customer dissatisfactions and performance issues at workplace (Ashforth & Humphrey, 1993). Thus, managing EL is an important strategy for building an emotional culture and manage inclusivity. There are usually two perspectives of emotional labour as predictors of burnout and employee ineffectiveness: job-focused emotional labour and employee- focused emotional labour (Brotheridge & Grandey, 2002). Through such EL strategies employees can manage their emotional displays effectively while interacting with colleagues, bosses and customers of diversified backgrounds and struggling with their work pressures, ending up with satisfying transactions. Emotional labour and attributional ambiguity theories have the potential to provide prescriptive understanding about how the communication of emotions and evaluation contribute to the issues like marginalization and emotional costs faced by minority employees in a diversified workforce at the personal and organizational level (Wong, Kathleen, 2007). Indian services and BPO sector have seen a robust boom over the past two decades. Much of the demand for Indian workforce in these sectors is mainly because of the high-grade soft competencies that Indian employees possess in terms of emotional adaptability and emotional labour that suffices the demand for high quality customer service. Emotional culture with right emotional labour displays is responsible for promoting happiness and engagement leading to sustainable development of people and services.

Indian Business Cases of Disruptive HR Through Inclusive Emotional Culture

Companies in majority of Indian industries have integrated disruptive HR practices into their culture to build a sustainable emotional culture. The importance of EI, appraisal of certain emotions and eliminating rather than suppressing certain other negative emotions in workplace has been recognised and applied in Indian companies. Emotions have been aligned in the mission and corporate philosophy, EI assessments and training is a usual activity. Emotions are tracked adopting various types of gamification and emotioanlytics software applications (artificial intelligence software to track and measure the EI).

Indian railways, the country's national railway system operating under the Indian Central government and the fourth largest railway network in the world (by size of 42,988 miles route distance as of April 2019), has adopted a milestone intervention to bring about a turnaround in its decision-making body. In order to bring down the number of trade union strikes and protests, accidents, derailments, train collisions; and improve the job-related behaviour of train drivers, officers, and managers emotional intelligence assessment and training process has been undertaken on a large scale. Around 1800 Railway officers (250 in the rank of additional secretaries, general managers, principal HODs; 150 in the rank of joint secretaries, chief workshop managers and additional DRMs; and 1400 officers of rank of deputy secretary, director, junior administrative) have been identified to undergo the EI testing and EI training workshop to enhance their decision making capabilities and job performance. The benefits that are expected to be accrued from this process which costs 423 USD to 595 USD per employee has a high perceptual value, since the initiative of building a culture of emotional intelligence among its officers is unprecedented in the Indian government. As quoted by Railway Board chief Ashwani Lohani, "the focus is on 'satyanishtha' - a mission for inculcating ethics and integrity among the railway employees. EI is an input in that direction" (Sharma, Dec. 2018). It is believed by the higher authorities of Indian Railways that if emotional intelligence and spiritual intelligence is enhanced, rapport with trade unions would improve resulting in lesser strikes and better IR (industrial relations) as well as customer service. As a result of this initiative focussed on building a positive emotional culture, the world's fourth largest railway network has aimed to improve the quality of its services to customers and accentuate employee engagement.

Happiest Minds Technologies, a next generation digital transformation, infrastructure security and product engineering services company located in the IT capital of India, Bengaluru has adopted positive emotions in its name, mission, vision and values to build a work culture where team spirit, joy, mindfulness, integrity and social responsibility connect employees, leaders, work, clients and customers together. As the name suggests, their mission says- "Happiest People, Happiest Customers". Their vision speaks about being the Happiness Evangelists for each other, customers and society, be recognised for thought leadership, be leader in social responsibility initiatives. Thus, as the name of the company suggests, their ultimate aim is to create happy customers through nurturing happy employees who are highly creative, innovative and empowered to offer the best service. The acronym 'SMILES' describes the company's ethos, and stands for the six core values that every employee proudly imbibe and follow- sharing, mindful, integrity, learning, excellence and teamwork.

The organisational culture has clearly mentioned certain predictive factors that are antecedents of its success and happiness irrespective of the nature of the business. They are organizational factors- fairness, transparency and joy in workplace, and individual oriented factors such as wellness, enriched personal life and giving back to society. The Happiest People Framework is built on the 7Cs of Culture, Credibility, Collaboration, Contribution, Communication, Community and Choice. The organisation has well-crafted wellness program as one of its disruptive HR practices that adds extra value to its happy employees. The Happiest Minds' Wellness Program constitutes the 7Ws of Physical Wellness, Spiritual Wellness, Intellectual Wellness, Professional Wellness, Social Wellness, Emotional Wellness and Environmental Wellness. These are nurtured by aligning activities, logistics, facilities and expertise of the organization through an array of Wellness schemes.

Emotional Wellness programs in Happiest Minds is cultivated through a wide range of activities such as Good Samaritan Network, Buddy programs, Stress Management lessons, Emotional Intelligence training and assessments, Emotions gamification and Mentoring. Programs such as Group activities,

Celebrations, Get-togethers, Blood donation camps, Circle of Happiness and Diversity initiatives such as Aura, the Women's council promote social skills, team spirit, trust and joy to work inclusively among all. Competency Building, 360-degree feedback-based Performance Enhancement Processes and rigorous Talent Reviews facilitate in improving professional effectiveness among the employees. Social Responsibility is a core value of Happiest Minds. Such disruptive practices make this company culture the most unique one to foster the feeling of wellness, happiness and sustainability in mankind.

Pidilite Industries Limited, Indian-based adhesives manufacturing company focus on building a culture based on creativity, relationship-building and people friendly HR policies that respect the emotions of all. Pidilite Industries President (HR), Rahul Sinha stated-

At around 8pm every day, we take a round of the office to see if anybody is working late. If we find someone repeatedly working till late, we have a chat with his/her boss and ask them to review the employee's workload. Similarly, if somebody wants to work on a Saturday, they need to take special permission from HR.

Pidilite offers to its employees one day per week as "work from home". These initiatives have emerged from their "listening" culture. The organisational culture emphasizes on trust, empathy, transparency and maintaining high ethical standards as the most critical values to attain sustainability and high performance. Thus, the company boasts of not only being a pioneer in adhesives that create the ever-lasting bonds, but also nurturing ever-lasting relationships with employees.

TCS, the Indian IT giant works with a partner company called 'Courageous Success' to maintain its emotional culture centred around EI building of its talent force. As quoted by TCS manager of HR,

We work with coaches to focus on our own emotions, and also to understand the emotions, feelings and triggers of our closest colleagues. This can be hugely helpful in knowing what pushes an emotional button for someone.........We are also piloting a meditation week to help team members work through their emotions at work, how to control stress, anxiety and difficult relationships with colleagues. (Kati Chitrakorn, March 31, 2016)

Organisations are increasingly adopting apps to help individual employees and teams log their emotional reactions to various activities and make the connection between their moods and productivity. Through emotionalytics, the trends in emotions experienced and recorded by every employee is statistically understood and strategies taken to appropriately cater to positive or negative trends in employee emotions. Entropik Tech is India's only Emotions - AI start-up offering a full suite of emotion recognition technology options to its clients for assessing emotions of its customers and other stakeholders. The designers of this Emotionalytics start-up have been nurturing their creativity within a culture of fun, wellness, humour, flexibility, innovation and social responsiveness.

INCLUSIVE EMOTIONAL CULTURE IN INDIAN ORGANISATIONS

Cultivating the right emotional culture, mandates people to feel the emotions valued by the organization or team for achieving their common vision. This may involve emotional modelling, enhancing emotional intelligence to actually internalise those emotional norms or adopting any emotional labour strategies (deep/surface acting, genuine expression or suppression). Through harnessing what people already feel, modelling and imitating the emotions that are required and motivating employees to fake the emotions till they genuinely feel it, a positive emotional culture can be created and sustained (Barsade & O'Neil, 2016).

Recently a survey and focus group discussions of HR experts from three multi-national organizations located in the Silicon Valley of India was conducted. Opinion about existence and structuring of emotional culture has been solicited and analysed by the author. HR experts were identified based on the suitability and position in the company for the study to gather perspectives on certain aspects like-

- Can emotional culture suffice the requirement of inclusivity in workplace?
- How can emotional culture be built?
- What ways do you propose to cultivate an emotional culture in your workplace?
- How can emotional culture actually help the organization to sustain?

Semi-structured interviews and focus group interviews of HR managers were conducted for thirty minutes each and responses recorded. The HR managers belonged to three companies coded A, B and C (actual identity kept confidential) from different industrial sectors. Each HR manager was administered through two sets of interviews – first individual interview and the other, focus group interview. Consistency of the opinions was checked and validated. Thereafter the primary data was analysed and interpreted. Company A belonged to healthcare sector, privately owned, headquartered in Bengaluru, India, with employee strength of 19,000. Company B belonged to IT sector, headquartered in India, with public ownership and employee strength of 225,501 employees, operating in 50 countries and revenue of 11.8 billion USD. Company C belonged to retail sector, privately owned, headquartered in Mumbai, India, employee strength of 55,000 and revenue of US$6 billion. Based on thorough analysis of information recorded from the focus group interviews the following facts about emotional culture and HR disruptive practices were identified. Table 1 shows the analysed interpretations from the interview conducted for HR managers of Company A (healthcare sector), Company B (IT sector) and Company C (Retail sector).

The content analysis of the opinion of HR managers have shown direction to conceptualising a model for structuring an Inclusive Emotional Culture within the organisation. Figure 2 illustrates the Inclusive Emotional Culture Framework (IECF). Based on the opinions sought from the HR managers, the IECF framework has been conceptualised. This framework shows clear direction as to how an inclusive emotional culture can be built to attain sustainability over a longer term.

The IECF framework shows that the first step in the process of determining inclusive emotional culture is understanding its need through the recognition of the extent of workforce diversity existing in the organisation. Based on degree of differences and workforce diversity aspects, goals of inclusion are determined and analysed. Goals of Inclusion become the determinant for building the emotional culture in the organisation. Emotional intelligence of each and every individual influences their emotional labour displays and strategies. EI and EL influence the determination of emotional culture that aligns with the goals of inclusion of the organisation as well as the emotional set up of each groups of employees. Conducive emotional culture can be built by clearly stating positive emotions like fun, trust, creativity, affection, support etc. in the mission statement of the organisation. Values to be followed by each employee should be clearly mentioned in the culture statement descriptively and prescriptively. Goals based on emotions fulfilment to be framed and based on these goals HR practices have to be designed. For example- framing the goal of 'becoming the happiest and emotionally intelligent company within two years". This goal definitely emphasizes on fostering a positive emotional culture with fun, happiness and

Table 1. Emotional culture in Indian organisations

Company A (Health care)	Company B (Information Technology)	Company C (Retail)
By identifying and expressing emotions as per organisational requirement, we have been able to create a workplace for all without differences.	Respecting emotions of employees and aligning them with organisational vision, we have been able to innovate and work in teams in an inclusive manner	Assessing emotions of employees have enabled in developing a focussed workforce and eliminating them, who could not align their emotions with others and with organisational purpose.
By developing emotional goals and integrating emotions in mission, vision statements	Identifying and defining values to be practiced by every member of the organisation	Creating emotions and value oriented organisational culture. HR policies to be based on emotions pre-stated in the company philosophy.
EI training, Emotions tracking, emotional appraisal, gamification to orient employees about emotions that needs expression/ elimination.	EI training and assessment, Emotionalytics and tracking emotions through apps	HR practices like recruitment, selection, training, appraisal, career growth based on job related emotions/values to be followed in the organisation. Artefacts, interiors of workspace, posters, emoticons, EI tracking apps can help nurture positive emotional culture
Making sustainability development the emotional goal of every employee	Integrating sustainable development and innovation into organisational culture through emotions employees	Sustainability is a goal, that can be accomplished by driving people's emotions in that direction

emotional intelligence at its forefront. Eventually its HR practices are aligned and executed to achieve these goals and the organisational workforce becomes highly productive, contended and engaged. The outcomes of emotional culture are high levels of inclusion, employee satisfaction, happiness quotient (HQ), deeper employee engagement, less absenteeism and turnover, competency development, commitment and higher productivity among employees.

Practical Implications and Future Research Directions

The IECF model provides systematic flowline to HR managers for cultivating positive emotional culture that fosters sustenance of multiple cultures and people together. Diversity and inclusiveness are the precursor of any changes needed. Several studies on organisational psychology have identified the impact of cognitive, emotional and social processes on diversity- inclusion within groups (Kearney et al., 2009; Stockdale & Crosby, 2004; Hobman et al., 2003). In early anthropological studies of group dynamics, strategic emotional expression was found to facilitate group cohesion by dominating individual feelings and synchronizing interpersonal behaviour. Social identity theory illustrated that identifying and managing differences at intrapersonal and interpersonal level when working in groups leads to better performance outcomes (McKay et al., 2009).

Studies on emotional contagion shows that people in groups "catch" feelings from others through behavioural observation and imitation, and subsequent changes in brain function leads to a ripple effect of the same emotions throughout (Barsade, 2002). This can be practically applied in all teams and groups with no investment of money, but with conscious investment of positive emotions, such as a smile or positive greeting. Cognitive re-appraisal, listening and eliminating negative emotions through consulting, artefacts and posters exhibiting positive emotions like joy, fun, excitement, curiosity, respect, trust can spread the required emotions in workplace. Happy pictures of moments with family and team members at workplace develops a culture of affection, care and trust. Emotional modelling by manag-

Figure 2. Inclusive emotional culture framework (IECF)

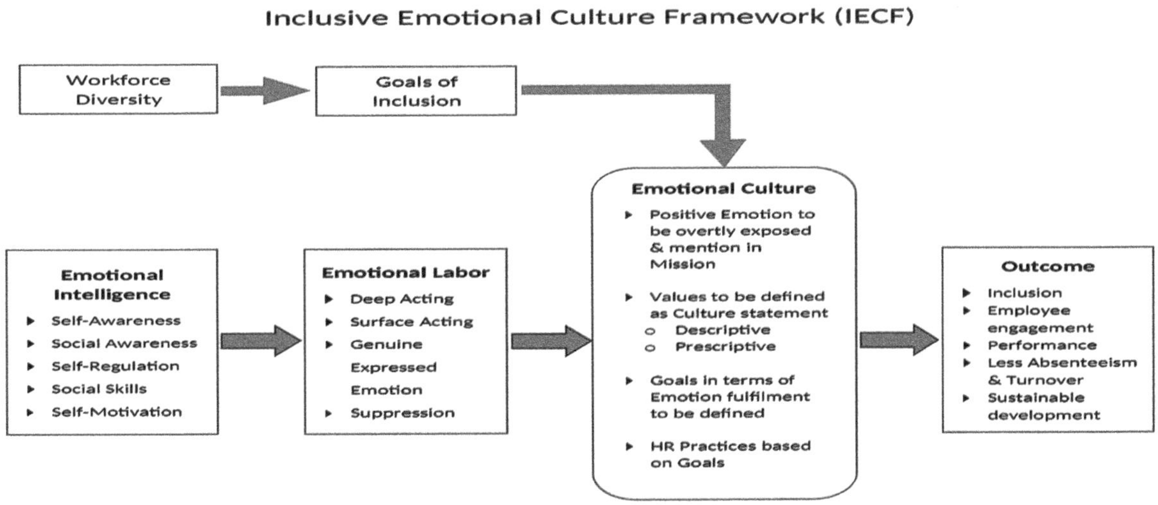

ers gives examples to subordinates that this is how they should behave and work. Integrating emotions in company's mission, vision and values makes the culture emotionally positive and highly creative. Emotional intelligence-based recruitment, training and appraisals practically integrates emotional capabilities in the company's formal people acquisition and management policies. Investing on AI based apps to track emotions and re-enforce positive emotions, gamification etc are additional measures that companies can take to transit towards an inclusive emotional culture. In order to succeed, it is important that the entire team works together as a single entity. This is possible only when the emotions of all are identified, appreciated, fulfilled and re-enforced.

Globalised economy demands high mobility and sharing of resources. Human resources mobility, skilled migrant workers and increasing demographic as well as psychographic dividend is characterising today's labour, which should be taken into consideration in crafting HR policies. Emotions and feelings cannot be isolated from humans. Hence, it is wise to shift from mechanistic structures to decentralized and knowledge-based operational models (Adler, 2001; Cho & McLean, 2009) governed by positive emotions. It is crucial to recognise and appreciate each entity's emotions and feelings to harness their contributions to the maximum. Consequently, the role of talent forces has become critical with an increased drive towards participation in organizational decision making, striking involvement in strategic business operations, higher ownership quotient with strong psychological contract and demand for challenging work context that pays them with a sense of self-actualization and esteem (Burke & Ng, 2006). Another emerging challenge is virtual workforce and virtual organisations. How an emotional culture would cater to the leadership and inclusivity needs of a virtual organisation is still to be answered. This chapter is significant in realizing the importance of emotional culture, its components and its outcomes. Since many organisations are progressively recognising the importance of emotions and emotional culture in organisational productivity and ROI in longer term, the practices should be popularised and further researches should be conducted to know more.

CONCLUSION

Sustainability is an issue in the present context as globalisation has brought about intense competition and hardly any loyalty. Workforce demographics and psychographics are changing with the speed of advancement of technology. Organisational growth and development are constraint due to rising employee turnover and issues of employee commitment, psychological contract, competency development and performance. Sustainability of business depends upon business values, organisational culture and people management practices. Thus, meeting the emotional requirements of people of diverse backgrounds is critical to fostering the feeling of 'oneness' in the workplace. This chapter provides an exhaustive review on the available literature related to sustainability, inclusiveness, emotional culture, emotional intelligence (EI) and emotional labour (EL). Various management practices of how Indian organisations have been able to develop a culture of sustainability through positive emotional culture and inclusiveness has been discussed. Disruption in HR practices have been happening in many ways, some through the integration of technology, AI and bots and other through the integration of the science of emotions in managing people. The cases help in understanding how fostering inclusive emotional culture have enabled Indian organisations to sustainably develop and grow.

Clear communication channels, primal leadership, HR practices that recruit, train, allocate, utilise and retain the employees based on EI (emotional capabilities as per job and organisational requirements), and aligning emotions into the mission and vision are the ways through which positivity can be ensured in the working culture and systems. Business cases of Indian organisations have been described to explain how they brought about disruption in HR practices to manage a positive emotional culture and employee wellbeing. Well managed happy employees eventually yield higher returns on investment through the emotional connect, psychological contract and engagement they develop with the organisation. Based on focus interviews of HR managers of Indian multinational companies in three different industrial sectors, a model has been proposed called as Inclusive Emotional Culture Framework (IECF) that offers systematic guidelines to HR leaders for cultivating a powerful positive emotional culture for 'All' in a highly diversified polarised organisation. The flowline states the identification of inclusion goals, perceiving the emotions of employees, aligning required emotions with the organisational mission, vision, values (descriptively and prescriptively), defining clear emotional goals meant for all, communicating these goals to all and designing HR practices based on the emotion's goals. A continuous process of follow up and re-enforcing the emotions is significant for sustainable HRM in the organisation.

REFERENCES

Academic Impact UN Report. (2019, April 18). *UNAI to Host Conference on Unlocking Your Emotions to Achieve the SDGs*. Retrieved from https://academicimpact.un.org/content/unai-host-conference-unlocking-your-emotions-achieve-sdgs

App, S., & Büttgen, M. (2016). Lasting footprints of the employer brand: can sustainable HRM lead to brand commitment?, *Employee Relations*, *38*(5), 703-723.

April, K., & Blass, E. (2010). Measuring Diversity Practice and Developing Inclusion. *Dimensions*, *1*(1), 59–66.

Ashforth, B. E., & Humphrey, R. H. (1993). Emotional Labour in service roles: The influence of identity. *Academy of Management Review*, *18*(1), 88–115. doi:10.5465/amr.1993.3997508

Backhaus, K. (2004). An exploration of corporate recruitment descriptions on monster.com. *Journal of Business Communication*, *41*(2), 115–120. doi:10.1177/0021943603259585

Bar-On, R. (1997). *The Emotional Quotient Inventory (EQ-i): Technical manual*. Toronto, Canada: Multi-Health Systems.

Barsade, S., & O'Neil, O. A. (2016, January-February). Managing your emotional culture. *Harvard Business Review*, 58–66.

Barsade, S., & O'Neill, O. A. (2014). What's Love got to do with it? A longitudinal study of the culture of compassionate love and employee and client outcomes in the long-term care setting. *Administrative Science Quarterly*, *59*(4), 551–598. doi:10.1177/0001839214538636

Barsade, S. G. (2002). The Ripple Effect: Emotional Contagion and its Influence on Group Behavior. *Administrative Science Quarterly*, *47*(4), 644–675. doi:10.2307/3094912

Barthwal and Som. (2012). Emotional Intelligence as a measure of an employee's overall effectiveness, *Drishtikon: A Management Journal, 3*(2).

Brotheridge, C., & Lee, R. T. (2002). Testing a conservation of resources model of the dynamics of emotional labour. *Journal of Occupational Health Psychology*, *7*(1), 57–67. doi:10.1037/1076-8998.7.1.57 PMID:11827234

Brundtland Commission. (1987). *Our common future*. Oxford, UK: Oxford University Press.

Bughin, J., Hazan, E., Lund, S., Dahlström, P., Wiesinger, A., & Subramaniam, A. (2018, May). *Skill shift: Automation and the future of the workforce*. McKinsey Global Institute. McKinsey & Company.

Chitrakorn, K. (2016, March 31). *You feel me? Why Emotional Culture matters at work*. Retrieved from https://www.businessoffashion.com/articles/careers/you-feel-me-why-emotional-culture-matters-at-work

Daya, P. (2014). Diversity and inclusion in an emerging market context, *Equality. Diversity & Inclusion*, *33*(3), 293–308. doi:10.1108/EDI-10-2012-0087

Deloitte. (2012). *Waiter, is that inclusion in my soup? A new recipe to improve business performance*, Deloitte Research Report, Australia.

Di Fabio, A., & Kenny, M. E. (2018). Connectedness to Nature, Personality Traits and Empathy from a Sustainability Perspective. *Current Psychology (New Brunswick, N.J.)*. doi:10.100712144-018-0031-4

Dobusch, L. (2014). How exclusive are inclusive organisations?. *Equality, Diversity and Inclusion: An International Journal, 33*(3), 220-234.

Ehnert, I. (2009). *Sustainable Human Resource Management: A Conceptual and Exploratory Analysis from a Paradox Perspective*. Heidelberg, Germany: Physica-Verlag. doi:10.1007/978-3-7908-2188-8

Fisher, C. (2019). *Emotions in Organisations, Oxford Research Encyclopaedia of Business and Management*. USA: Oxford University Press.

Fredrickson, B. L. (2001). The role of positive emotions in positive psychology. *The American Psychologist, 56*(3), 218–226. doi:10.1037/0003-066X.56.3.218 PMID:11315248

Fredrickson, B. L., & Branigan, C. (2005). Positive emotions broaden the scope of attention and thought-action repertoires. *Cognition and Emotion, 19*(3), 313–333. doi:10.1080/02699930441000238 PMID:21852891

Gallup. (2006). Gallup study: Engaged employees inspire company innovation: national survey finds that passionate workers are most likely to drive organisations forward. *The Gallup Management Journal.*

Gallup state of the American workplace: Employee engagement insights for U.S. business leaders. (2013). Retrieved from http://www.michaeljbeck.com/documents/State

Goleman, D. (1998). *Working with emotional intelligence.* New York, NY: Bantam.

Gosserand, R. H., & Diefendorff, J. M. (2005). Emotional Display Rules and Emotional Labor: The Moderating Role of Commitment. *The Journal of Applied Psychology, 90*(6), 1256–1264. doi:10.1037/0021-9010.90.6.1256 PMID:16316278

Grandey, A. A. (2000). Emotion Regulation in the Workplace: A New Way to Conceptualize Emotional Labor. *Journal of Occupational Health Psychology, 5*(1), 95–110. doi:10.1037/1076-8998.5.1.95 PMID:10658889

Hays-Thomas, R., Bowen, A., & Boudreaux, M. (2012). Skills for diversity and inclusion in organizations: A review and preliminary investigation. *The Psychologist Manager Journal, 15*(2), 128–141. doi:10.1080/10887156.2012.676861

Head, J. H. ACC, & Freedman, J. (2014, Jan. 2). *Inspiring Employee Engagement through Emotional Intelligence.* Retrieved from https://www.6seconds.org/2014/01/02/employee-engagement-emotional-intelligence/

Hersey, R. B. (1932). *Workers' Emotions in Shop and Home: A Study of Individual Workers from the Psychological and Physiological Standpoint.* Philadelphia, PA: University of Pennsylvania Press.

Hobman, E. V., Bordia, P., & Gallois, C. (2003). Consequences of feeling dissimilar from others in a work teams. *Journal of Business and Psychology, 17*(3), 301–304. doi:10.1023/A:1022837207241

Hochschild, A. (1983). *The managed heart.* Berkeley, CA: University of California Press.

Hochschild, A. R. (2003). *The Managed Heart: Commercialization of Human Feeling.* Los Angeles, CA: University of California Press.

Holvino, E., Ferdman, B. M., & Merrill-Sands, D. (2004). Creating and sustaining diversity and inclusion in organizations: Strategies and approaches. In M. S. Stockdale, & F. J. Crosby (Eds.), *The psychology and management of workplace diversity* (pp. 245–276). Malden, MA: Blackwell Publishing.

Honneth. (1994). The social dynamics of disrespect: on the location of critical theory today, *Constellations* 1(1), 255-269.

India's growth rate set to surpass China this year: World Bank. The Economic Times. Retrieved September 11, 2015.

Isen, A. M. (1999). Positive affect. In T. Dalgleish, & M. J. Powers (Eds.), *Handbook of Cognition and Emotion* (pp. 25–521). Hoboken, NJ: Wiley.

Jabbour, C. J. C., & Santos, F. C. A. (2008). The central role of human resource management in the search for sustainable organizations. *International Journal of Human Resource Management*, *19*(12), 2133–2154. doi:10.1080/09585190802479389

Jackson, J. (1966). A conceptual and measurement model for norms and roles. *Pacific Sociological Review*, *9*(1), 35–47. doi:10.2307/1388305

Kals, E., & Maes, J. (2002). *Sustainable Development and Emotions*. doi:. doi:10.1007/978-1-4615-0995-0_6

Kearney, E., Gebert, D., & Voelpel, S. C. (2009). When and how diversity benefits teams: The importance of team members' need for cognition. *Academy of Management Journal*, *52*(3), 581–598. doi:10.5465/amj.2009.41331431

Leidner, R. (1999). Emotional Labor in service work, *THE. The Annals of the American Academy of Political and Social Science*, *561*(1), 81–95. doi:10.1177/000271629956100106

Lirio, P., Lee, M. D., Williams, M. L., Haugen, L. K., & Kossek, E. E. (2008). The inclusion challenge with reduced load professionals: The role of the manager. *Human Resource Management*, *47*(3), 443–461. doi:10.1002/hrm.20226

Lirio, P., Lee, M. D., Williams, M. L., Haugen, L. K., & Kossek, E. E. (2008). The inclusion challenge with reduced-load professionals: The role of the manager. *Human Resource Management*, *47*(3), 443–461. doi:10.1002/hrm.20226

Lundrigan, M., Tangsuvanich, V. L., Wu, S., & Mujtaba, B. (2012). Coaching a diverse workforce: The impact of changing demographics for modern leaders. *International Journal of Humanities and Social Science*, *2*(3), 40–48.

Maak, Th., & Pless, N. M. (2006). Responsible Leadership: A Relational Approach. In Th. Maak, & N. M. Pless (Eds.), *Responsible Leadership*. London, UK: Routledge. doi:10.4324/9780203002247

Matthews, C. (2017, Oct. 30). *Here's why India is going to become the World's most important economy.* Retrieved from http://fortune.com/2016/04/08/india-economy/

Mayer, J., Salovey, P., Caruso, D., & Cherkasskiy, L. (2011). Emotional Intelligence. In R. Sternberg, & S. Kaufman (Eds.), The Cambridge Handbook of Intelligence (Cambridge Handbooks in Psychology, pp. 528-549). Cambridge, UK: Cambridge University Press. doi:10.1017/CBO9780511977244.027

McKay, P., Avery, D. R., & Morris, M. (2009). A tale of two climates: Diversity climate from subordinates' and managers' perspectives and their role in store unit sales. *Personnel Psychology*, *62*(4), 767–791. doi:10.1111/j.1744-6570.2009.01157.x

Men, L. R., & Yue, C. A. (2019). Creating a positive emotional culture: Effect of internal communication and impact on employee supportive behaviours. *Public Relations Review*, *45*(3). doi:10.1016/j.pubrev.2019.03.001

Morais, U. P., Jacqueline, P., Kevin, S., Lucien, S., Roiner, R., & Yesenia Rivera, M. B. (2014). Managing Diverse Employees at Starbucks: Focusing on Ethics and Inclusion. *International Journal of Learning & Development, 4*(3), 35–50. doi:10.5296/ijld.v4i3.5994

Nair, N., & Vohra, N. (2015). *Diversity and Inclusion at the Workplace: A Review of Research and Perspectives*, IIMA Working Papers WP2015-03-34, Indian Institute of Management Ahmedabad, Research and Publication Department.

Panel of Experts Discuss Using Emotional Intelligence to Achieve the SDGS. (2019, May). Retrieved from https://academicimpact.un.org/content/panel-experts-discuss-using-emotional-intelligence-achieve-sdgs

Pelled, L. H., Ledford, G. E., & Mohrman, S. A. (1999). Demographic dissimilarity and workplace inclusion. *Journal of Management Studies, 36*(7), 1013–1031. doi:10.1111/1467-6486.00168

Pelled, L. H., Ledford, G. E., & Mohrman, S. A. (1999). Demographic dissimilarity and workplace inclusion. *Journal of Management Studies, 36*(7), 1013–1031. doi:10.1111/1467-6486.00168

Petrides, K. V., & Furnham, A. (2003). Trait emotional intelligence: Behavioural validation in two studies of emotion recognition and reactivity to mood induction. *European Journal of Personality, 17*(1), 39–57. doi:10.1002/per.466

Pless, N., & Maak, T. (2004). Building an Inclusive Diversity Culture: Principles, Processes and Practice. *Journal of Business Ethics, 54*(2), 129–147. doi:10.100710551-004-9465-8

Prime, J., & Salib, E. R. (2014). *Inclusive leadership: The view from six countries*. New York, NY: Catalyst.

Rathi, N., & Rastogi, R. (2009). Assessing the Relationship between Emotional Intelligence, Occupational Self-Efficacy and Organizational Commitment. *Journal of the Indian Academy of Applied Psychology, 35*, 93–102.

Road map towards the implementation of the United Nations Millennium Declaration: Report of the Secretary General. (2001). UN GAOR, 56th Sess., Annex, Agenda Item 40, UN Doc. A/56/326 55.

Roethlisberger, F. J., & Dickson, W. J. (1939). *Management and the Worker*. Cambridge, MA: Harvard University Press.

Rosete, D. (2009). *A leader's edge – what attributes make an effective leader?* Manuscript in preparation.

Rosete, D. (2005, June). *A leader's edge – what attributes make an effective leader?* Paper presented at the Fifth Annual Emotional Intelligence Conference, The Netherlands.

Salovey, P., & Mayer, J. D. (1990). Emotional intelligence. *Imagination, Cognition and Personality, 9*(3), 185–211. doi:10.2190/DUGG-P24E-52WK-6CDG

Schein, E. (1985). *Organizational Culture and Leadership*. San Francisco, CA: Jossey-Bass.

Sharma, S. N. (2018, Dec.). *Why Indian Railways is testing emotional intelligence of its officers*, Economic times, Retrieved from https://economictimes.indiatimes.com/industry/transportation/railways/can-emotional-intelligence-training-for-staff-make-indian-railways-safer/article

SHRM. (2009). *Global Diversity and Inclusion-Perceptions, Practices and Attitudes*. Society for Human Resource Management.

Smircich, L. (1983). Concepts of Culture and Organizational Analysis. *Administrative Science Quarterly, 28*(3), 339. doi:10.2307/2392246

Stockdale, M. S., & Crosby, F. (2004). *The Psychology and Management of Workplace Diversity*. Boston, MA: Blackwell.

University of Alberta Students' Union. (2011). *Sustainability Summary Report and Recommendations, University of Alberta*. Retrieved from https://www.su.ualberta.ca/media/uploads/538/2011assessment.pdf

Wilkins, H. (2008). The Integration of the Pillars of Sustainable Development: A Work in Progress. *McGill International Journal of Sustainable Development Law & Policy, 4*(2). Available at https://ssrn.com/abstract=2623221

Wong, K. (2007). *Emotional Labour of diversity work: Women of colour faculty in predominantly White Institutions*. doi:. doi:10.13140/RG.2.1.2880.6562

World Commission on Environment and Development. (1987). *Our common future*. Oxford, UK: Oxford University Press.

Zapf, D., Vogt, C., Seifert, C., Mertini, H., & Isic, A. (1999). Emotion Work as a Source of Stress: The Concept and Development of an Instrument. *European Journal of Work and Organizational Psychology, 8*(3), 371–400. doi:10.1080/135943299398230

KEY TERMS AND DEFINITION

Cognitive: It is a psychological process occurring in the brain that involves acquisition and understanding of knowledge, experience and senses from the stimuli received.

Culture: It is a set of shared values, beliefs, and behaviours that are practiced, learned, and shared by a group of people.

Emotion: It is a mental state associated with the nervous system brought on by neuro-chemical changes, associated with thoughts, feelings, behavioural responses, and a certain level of pleasure or displeasure.

Emotional Culture: It is a set of shared affective values, feelings, emotions, artefacts, beliefs that are set as a norm and supposed to be followed by every person of a group.

Emotional Intelligence: It is the ability of a person to perceive, understand, regulate and manage one's own and others' emotions.

Inclusive: The degree to which each different individual is considered to be one entity of a group and included within a group.

Performance: It includes the actual output or results of an entity as measured against its intended outputs.

Retention: The process of retaining employees within the organisation and preventing them from leaving due to some or the other reasons.

Chapter 6
Employee Experience Design: An Innovation for Sustainable Human Capital Management Practices

Nitu Ghosh
https://orcid.org/0000-0002-5744-1670
REVA University, India

Urmila Itam
REVA University, India

ABSTRACT

The philosophy of employee experience has been advocated as an effective tool to achieve the highest level of employee engagement in a business environment characterized by increased dynamism in workforce demography, psychography, competencies, and expectations. Employee experience encapsulates the overall journey of an employee in an organization from hire to exit, encompassing their experiences, what they feel, do, and achieve from the company. It is a comprehensive view of the relationship shared and experienced between the individual and the organization right from the application stage to stage they join the alumni group after their exit. The chapter explores this new philosophy in current human capital management practices that aims at revamping the HR practices and policies so as to create an experience that provides the highest level of engagement. Based on content analysis qualitative survey of various HR managers, opinions and new age thoughts in HRM practices has been highlighted in this chapter, giving an innovative dimension to HRM.

INTRODUCTION

Radical innovations have been witnessed in human capital practices due to changing workforce and workplace dynamics in the Twenty-first century. Present millennium is characterised by Learning organisations with a millennial workforce that typically demands individualistic attention, care and engaging experience from their employer. The human capital trends of 2017 boast about millennial workforce with changing demography and psychograph. Not only diversity in the workforce has intensified but also the

DOI: 10.4018/978-1-7998-0357-7.ch006

emotions, feelings, beliefs, work attitude, expectations and needs. Millennial employees are looking for a productive, engaging and enjoyable work experience from the beginning of their employment journey to their exit (Bersin, 2016). According to a 2016 Millennial Survey report by Deloitte, Millennials are found to associate value to career opportunities based on good work-life balance as the most important factor, opportunities of progression and leadership opportunities, work flexibility and meaningful nature of work. Organisations believe that happy, committed and engaged employees add higher dividends to the company's financial performance rather than any other resources. That's why competent and committed employees are referred to as human capital, that is an investment with higher returns over a longer term.

Present business organisations emphasize on experiences they provide to their stakeholders for business development, expansion, branding (including employer branding) and overall growth of the organisation. Design thinking has been applied successfully for enhancing customer experience and user experience in numerous services and products, but in case of human resources management focuses on enhancing employee experience is just the beginning of the endeavour towards sustainable development (Plaskoff, 2017). Human capital management in this millennium is going through a paradigm shift in the same direction; through the designing of policies and practices that provide an enriching and enthusing experience to employees throughout their employment journey in the organisation rather than just increasing financial incentives and employee recreation.

Employee experience is the synchronised perceptual outcome of all interactions occurring between employees and the organization (Jacob & Goldsmith, 2017). These interactions are influenced by three factors, which includes the 'physical space' (workspace/work environment) that employee uses every day, the 'culture' (norms, beliefs, values, policies) of organization and the 'technological environment' provided by the employer (Morgan, Dec 15, 2015). When organisations focus on enhancing the quality of these three factors by adopting digitization, artificial intelligence, inclusive culture, charismatic leadership, challenging and meaningful nature of jobs, conducive work environment etc, employees tend to enjoy the overall interactions they have with the organisation, work and other organisational members. This provides them with satisfying, delighting and motivating experiences that drive their motivation towards organisational excellence.

In this era of virtualisation and automation, employee experience is essential for companies to compete (Dhingra, Emmett, & Samadani, 2018). Employee experience has recently gained popularity in many organizational setups, especially in the services sector in order to build employee loyalty and engagement by holistic designs and innovations at the workplace. Over the last thirty years, organisations have been struggling to provide best practices for employee work-life balance and effective diversity management. But present millennial work climate transcends work-life balance to provide an experience to 'each and every employee' that gives them a 'feel good' factor to cherish and be self-driven towards maximisation of their contributions for goals accomplishment. A lot of effort has to go into understanding each employee deeply and co-creating experiences with them that bear high perceptual value and display care (Plaskoff, 2017). Dramatic changes in the work climate have often blended with a host of issues focused on seeing employees as "whole persons" (Harrington & Ladge, 2009). Inclusivity and sustainability are the new HR goals for any organisation, leading to innovative HR practices providing enriching experiences and perceptions about interactions with work, manager and work environment.

Organizational challenges like downsizing, outsourcing, increased performance and efficiency issues, longer work hours, work-life balance conflicts, stress etc. impel employees to seek out organizations that seem to have alignment with their mind and heart (App, Merk, & Buttgen, 2012). This challenging psychographic trend of the workforce creates constraints on organizations in sourcing and managing competent, committed and motivated workforce to achieve their economic goals in the longer term (Ehnert, 2009). Sustainability in the human resource management (HRM) context is the current demand of business environment (Clarke, 2010; Ehnert, 2009; Jabbour & Santos, 2008) mandating innovative HR practices to enhance employee experience, thereby yielding high engagement, retention and employer branding.

As organizations move towards a more collaborative, open, information-loaded structure, the new millennium HR managers can improve employee engagement, leadership, decision making and performance through a focus on employee experiences. IBM Globoforce survey report (2018) has revealed that firms that score in the top twenty-five percent on EEI claim to have nearly three times the return on assets and double the return on sales. This chapter critically aims at describing the disruptive philosophy of employee experience. It discusses Human capital management practices that help in fostering an ideal employee experience design in today's workplace. The key drivers of positive employee experience have been identified and explained. Best practices of Accenture and IBM have been discussed that foster an ideal EE design to achieve performance outcomes. Content analysis qualitative study has helped in illustrating how the drivers facilitate in acquiring positive employee experience at the workplace. The chapter also discusses the innovative ways through which the HR managers have been able to build positive employee experience.

WHY DO WE NEED TO FOCUS ON EMPLOYEE EXPERIENCE?

Decades ago, the employee relationship with the organization and the employer was absolutely transactional and unidirectional. The era of engagement changed the notion of hard work into smart work and focused more on how and why employees work for the organization. Disruptive innovations in science, technology, and demographics have changed the nature of work and transferred the bargaining power on employees and consumers (Nelson, 2017). Organisations that transform and provide the best employee experience and customer experience are said to have sustainable development in terms of market growth, agility and innovation. Practitioners and researchers have contributed to the concept of employee engagement and confirmed that engaged employees are more productive, committed to the organization and stay happy and healthy. However, employee experience is stated to develop a culture of extreme engagement and foster undeterred commitment and sustainable competency development. Highly engaged employees constantly put forth discretionary effort to actuate the business forward.

For decades, employee engagement had been the main thrust for all businesses, investing heavily into programs designed to make people happier and more satisfied at work, but unfortunately, employee engagement levels across the U.S. have remained tenaciously static and those worldwide surprisingly low. Survey reports state that the United States witnesses $450 billion to $550 billion of lost productivity due to low employee engagement every year (Head, ACC, & Freedman, 2014). Majority of engagement strategies in these organizations focussed on external mechanisms to accentuate employee motivation, job satisfaction, and performance, which failed to have a long-lasting impact. Ironically, HRM practices like financial incentives, fun-filled organisational culture and climate and various other initiatives failed to suffice the higher order motivation needs of employees, thus creating a gap in employee motivation. True engagement is born out of the intrinsic motivation and enduring experiences of the employee in a diverse context. It emerges from the employee's relationship or psychological contract with the employer and with the work itself. Experiences encountered by the employee at the workspace and in their operational context determine their future intent and motivation towards their work, profession and organisation. Employee experience involves holistic approach encapsulating intangible aspects of organizational life and converting them to tangible outcomes, building solutions organically (Plaskoff, 2017).

Multinational companies like Accenture, Deloitte, Southwest Airlines, Coca Cola and GE have adopted strategies like 'Employee First approach' and 'translating customer experience (CX) principles to talent management' to empower workers to give their all, to build superior employee experiences that drive employee engagement, retention and productivity. High performing organisations are steadily realizing the importance of employee experience as the new battleground for competitive advantage (Liley, Feliciano, & Laurs, 2017). Companies are aiming to develop into exponential organisations through disruptive HR transformations that foster employee experience. Studies reveal that companies with employee experience outperform the Standard & Poors (S&P) 500 companies by 122 percent (Chamberlain, 2015). Companies with highly engaged workers are found to be twenty-one percent more profitable than those with poor engagement levels as per a Gallup survey report (Gallup Q12 Meta-analysis report, 2016). Expectations of millennial workforce has been changing dramatically, demanding relevant, convenient and engaging experiences they have various extraneous factors replicated on the job. They want the opportunity to shape their workplace experiences on their own terms through emphasis on hyper-personalization. Employee experience is about giving employees a range of options that will help them actualise their full potential and drive business value. Cultivating an inclusive work culture facilitates personalised opportunities of self-actualisation possible for every employee giving them an enduring experience (Liley, Feliciano, & Laurs, 2017)

Various global organisations have adopted practices that track into employees' emotions and cater to their feelings and emotional needs. This creates an enduring feeling of affection and psychological contract between the employee and organisation. Experiences of employees strongly influence their attitudes and motivations, which in turn determines their behaviour towards expected outcomes. To meet sustainable goals and achieve organisational vision, such employees are required who develop psychological contract with the organisation and extend their contributions beyond expectations. A study on more than 5,000 U.S. employees, revealed that smaller companies tend to have higher engagement levels. About 60 percent of employees working in the smallest companies we looked at, with 100 employees or less, are moderately or highly engaged, compared with only 49 percent of employees at the largest companies, with more than 10,000 employees. The research also shows that engaged employees tend to have more customer contact, higher positions within their organization, earn more money, and have higher levels of education (Temkin, 2016). An example of a strategy leading to positive employee experience is 'Employee Hackathon program' organised by Cisco and LinkedIn, where the purpose is to break down particular practices within the company and rebuild them for problem solving at grass-root level and achieving higher efficiency. Involving employees in problem solving, innovation and initiating organisational change process through such gamified strategies gives them a self-actualizing and empowering experience. Studies have identified various reasons cited by global organisations for investing in employee experience. Table 1 describes the reasons why employee experience (EX) needs focus in present context.

Employee experience is not about just collecting employee feedback or assessments and analysing it, but to listen, understand and act with precision and conviction, so as to make employees feel valued and contribute further (Albert, 2019). This results in nurturing a positive work culture, ownership and positive business outcomes.

Table 1. Reasons for investing on employee experience

Employee oriented factors	Organisational goal-oriented factors
• Scaling Culture for growth (Airbnb invested on EX to scale culture)	• Raise productivity (General Electric)
• Personalization at scale (IBM's data analytics and personas guide managers to make better decisions for employees).	• Restructure under new business model (ABN Amro focus on EX and digital HR to win)
• Integrate two distinct cultures (Start-up LearnVest utilized EX in its acquisition - integration strategy with Northwestern Mutual's cultures)	• Regain market share and competitive advantage (Maxis market share outperformed competitors after restructuring around EX)
• Start with 'Why?' (World Food Program realized volunteers' motivation through EX to maximise global impact)	• Digital Strategy (ING extended their DS for employees)

Source: Compiled by authors from published report (Nelson, 2017).

EMPLOYEE EXPERIENCE DESIGN IN MILLENNIAL ORGANISATIONS

The concept of employee experience has been considered to be a much progressive step to employee engagement in the organisations. The probable failure of engagement practices in attaining long term goals of the organisation as stated by some research studies highlights the importance of employee experience to achieve the same goals. Employee experience leads towards a purposive direction to achieve the highest level of employee engagement through an ultimate experience provided by the organisation to its employees throughout their 'hire – exit – alumni' journey. Concept of employee experience (EX) was proposed based on customer experience (CX) management, which begins at the heart of an organization (Harris, 2007; Liley, Feliciano & Laurs, 2017) and continues much beyond through the 'loyalty' built between the two entities. Similar to the evolution of customer satisfaction to overall customer experience, organizations have started restructuring their HR strategies and practices to evolve from the philosophy of employee satisfaction to holistic employee experience.

'Experience' as a delivery model was first operationalized in the service industry and outcomes accounted led to a rise in sales and financial performance. This has led to the transformation of the service economy into 'attention economy', emphasizing on creating customer experience that grabbed their attention through creative marketing strategies (Davenport & Beck, 2002), 'entertainment economy' focusing on creating unique experience through entertainment (Wolf, 1999), 'emotion economy' that aimed at creating an emotional connect with target customers to create an unique experience for them (Gob´e & Zyman, 2001), and ultimately an 'experience economy' (Pine & Gilmore, 1999; Schmitt, 1999). The same philosophy has been adopted in the field of human capital management, by tapping the attention, emotions and social dimensions of employees though contemporary practices, work environment and technology(Liley, Feliciano, & Laurs, 2017).

Hyper-personalization is considered to be the hallmark of experiences, inherently personal and outcome of one's feelings, belief, cognitive perception and emotions about external stimuli that gain selective attention by the individual (Belk, 1975; Gardner, 1985; Hirschman & Holbrook, 1982; Zuckerman, 1971). Experiences that are targeted towards the actualization of specific needs or emotions result in long-lasting loyalty and stronger bonding (Davenport & Beck, 2002; Gob´e & Zyman, 2001; Pine & Gilmore, 1998, 1999; Reichheld, 1996; Schmitt, 1999). This is not only true for the customers but also for employees. Companies are delivering targeted experiences for individual employees by segmenting the workforce into clusters that share common characteristics(Liley, Feliciano, & Laurs, 2017). By delving deeper into psychographics of employees (emotions, attitude, beliefs, expectations and perceptions) rather than demographics alone, adopting techniques like sophisticated analytics tools or social listening (to understand individuals' needs, preferences, attitudes, intentions and motivations), the organization can understand and design specific type of experiences that will boost their engagement and productivity. The experiences are built through changes in work environment, job structure and nature, interpersonal relationships, HR practices and other motivational cues that induces positivity, competitiveness, joy and creativity within the employee. Experiences need to be individual-specific or segment- specific as each individual employee is different.

Social exchange theory can be considered to be the roots of employee experience, eventually leading to stronger psychological contract through positive experiences (Cropanzano & Mitchell, 2005; Bhattacharya, Trehan & Kaur, 2018) and instilling the values of trust, commitment and loyalty. Studies assert that whenever there is higher fulfilment of employee demands and expectations by the organization, the level of engagement tends to be higher (Kahn,1990; Saks,2006; Turnley, Bolino, Lester & Bloodgood, 2003), as the employees feel psychologically more obliged and contribute better into their jobs (Coyle Shapiro & Kessler, 2002). The perception about psychological contract depends upon the tenure of an employee in the organization and their experiences (good or bad) throughout their interactions in the work journey (Bal, De Cooman & Mol, 2013; Bhattacharya, Trehan & Kaur, 2018). The experiences that an organization provides to their newly recruited employees is very crucial as it has a veritable effect on their overall attitude towards the job and the organization, molding the employee's future intentions and performance, and resulting into high engagement and sense of obligation (which are eventually the outcomes of employee experience).

Positive employee experience has become a sort of new contract between employer and employee leading to a sense of loyalty, commitment and trust (Deloitte Global Human Capital Trends, 2017). Various studies have defined employee experience by focusing on employee perceptions and value attributed to the circumstances and interactions they undergo throughout their employment journey. It is defined as a set of perceptions that employees have about their experiences at work in response to their interactions with the organization (IBM and Globoforce, 2016 pp.3).Various disruptive HR strategies are been implemented in millennial organisations to fulfil the expectations of millennial workforce and delight them, like providing enriching and 'funified' workspaces, team-based projects, meaningful nature of job design, challenging opportunities, intrapreneurship opportunities, employee benefits, career planning and development, rewards and reinforcements and other employee engagement strategies besides financial incentives. These facilitate in providing enduring experiences to the employees, resulting in emotional bonding and stronger psychological contract formed between them. The outcomes are higher motivation, job satisfaction, long term engagement, performance and retention. Employee experience can be defined as – the feelings, perceptions and emotions that an employee undergoes through their involvement in work and within the organisational environment, which provides them positive reinforcement to develop, contribute, engage and sustain within the organization for good.

Reports suggest limited improvement in the areas of employee personal and professional work-life demands, alignment between employee and personal goals with corporate purpose, programs for different age groups, and using design thinking as part of employee experience (Deloitte Global Human Capital Trends, 2017; Foresee, 2014; IBM and Globoforce, 2016). Need for disruptive practices in HRM that are sustainable is been highlighted in these reports. A survey of 7000 respondents from 130 different countries asserted the need for radical system redesign in HR practices (Deloitte Global Human Capital Trends, 2017 pp.53). About 89% of Indian companies responded that goals for achieving high employee experience is very important for sustainable performance and documented many challenges in implementing the philosophy. Companies are now revamping their HR practices by integrating technology, automation and AI to integrate performance management, workforce diversity, inclusion management, engagement, leadership, employee well-being, workplace design, and goal setting as a unique framework. Globally 79% of companies in the list of Fortune-500 companies have ranked employee experience as the most desired. Organizational culture and engagement practices are the core determinants of employee experience throughout an employee's hire-to-exit-to-alumni cycle.

Companies are progressively adopting predictive analytics, artificial intelligence, digitization and Big data analytics in developing strategies that keep analyzing the employee journeys, needs, expectations, emotional trends, motivators, and overall employee experience using net promoter score, thereby re-designing jobs, workplace, well-being, and work productivity systems for the present digitally-augmented workforce. HR accounting is now a popular practice in organisations. The dramatic transition in human capital management practices is now evident, as every interaction between the employee and the workplace is accounted, analyzed, quantified, predicted and workplace re-invented to yield higher employee satisfaction and performance. High-performance organizations globally are scuttling to catch up with the expectations and motives of the millennial workforce in order to transform them into positive and productive employee experiences. However, it is difficult measuring the employee experience in a way that provides actionable insights for the development of successful employee initiatives with quantifiable outcomes. Employee experience is about building connection and relationship with employees and delegating them a type of ownership to shape the organization they are working for. Employee experience gets input on what matters to employees from various sources like interviews, focus groups, hackathons, instant feedback apps and predictive analytics. This provides information about what employees feel most strongly about and their emotions about what they are going through (Morgan, 2017), providing a clear comprehensive picture of employee experience on a one to one basis.

Paul Davies, Head of EX at General Electric (GE) defines employee experience as "Enabling our people to do the best work of their lives through moments that matter". GE usually leverages employee experience practices like personas, design thinking, storyboards, emotional listening etc. The frequent practice of these HR initiatives enables them to connect to the people's heart and minds, which is significant in achieving positive employee experience. Paul quotes, "We don't go an hour without using one of these. We ask our employees, our people leaders and our candidates what matters most to them, we listen to their stories, listening especially for emotions" (Nelson, 2017).

Business Cases of Employee Experience Practices

The Fortune Global 500 company, Accenture is a global leader in the consulting business. Accenture's strategy for achieving employee experience (EX) lies in its customer experience (CX) framework. While its CX framework generates customer loyalty, sales and revenues, its EX model contributes to its employer branding, attracting talent force, boosts engagement, retention and performance (Accenture Strategy, 2017). Accenture's 3-pronged strategy aims to create individualized employee experience comparable to CX in the next couple of years. Its Three-pronged strategy focuses on –

- Hyper-personalization at work

- Technology as an enabler of employee experience and collaborator
- 'Moments that matter' for each one individually

Accenture's EX design process encompasses the strategy of workforce segmentation based on deeper demographic analysis, sophisticated analytics and social listening for understanding individual needs, expectations and intentions. By accessing social sites like Glassdoor and Kununu, HR gets a deeper insight into its employee perceptions. Thus, customised individualised experiences are provided through investing in technologies like internal collaboration tools, predictive analytics-based employee services tools and self-service apps that create consumer-like experiences for employees. Social media apps like 'Know me' tools, artificial intelligence tools to boost personal productivity, emotions-tracker tools and apps are been progressively used to track individual employee's state of mind and offer them experiences accordingly. HR strategies are designed to identify individualistic personal moments (based on employee needs and emotions) and experiences offered on three dimensions- physical, human and digital.

Gen Z workforce demands excitement, challenges, an enthusing work environment with career advancements while also managing work-life balance. Co-creating experiences through a collaborative culture is the key strategy adopted towards achieving success in employee experience in the global consultancy firm. Accenture's HR aims at achieving EX at different segments of its workforce through identifying values or business outcomes it aims to achieve incrementally and designing physical, human or technological experiences (for example- introducing a new ERP system or app for use of sales force to boost enthusiasm among them) accordingly. In order to empower employees towards higher productivity, HR focuses on hyper-personalizing experiences by identifying 'moments that matter' through frequently collected survey data, analytics and social listening. Design thinking principle is been adopted at all levels and employees are engaged in order to co-create the most relevant and valuable experiences. This becomes the basis for HR decisions like training, rewards, workplace environment, behavioural requirements etc. New experiences at all three dimensions (physical, technological and social) are pilot tested on smaller groups of critical workforce and impact of new experiences assessed on employee performance and business value.

The American multinational company, IBM has designed the Employee Experience Index (EEI) as a new global measure for a human workplace based on a set of leadership and organizational practices to create more positive employee outcomes. The IBM Smarter Workforce Institute and Globoforce's Work Human Research Institute collaborated to conduct extensive research studies and surveys on positive employee experience. Through a detailed analysis, IBM developed a 5-dimension and 10-item index to define and explore the core facets of employee experience. The five dimensions were identified as – belongingness, purpose, achievement, happiness and vigour. Assessment of employees through the EEI index enabled IBM to categorise employees based on their index scores and analyse the consequences of higher scores (linking them to better work performance, higher discretionary effort and retention). Their strategies emphasized on enhancing leadership abilities to drive employee interests towards organisational purpose, HR practices and building an environment where practices of organizational trust, co-worker relationships, meaningful work, recognition, feedback, empowerment, growth and work-life balance drive positive employee experience. Based on their survey reports, IBM implemented strategies of diagnosis and acting. By listening to employee voices through forums such as census, pulse surveys, social listening etc, and building culturally relevant collaborative workforce practices, the HR has successfully enabled employees to understand the deeper meaning of their work and its contribution to wider organisational goals. Therefore, IBM has effectively designed practices linking continuous performance to social recognition and foster positive employee experience through its extensive researches.

KEY DRIVERS OF EMPLOYEE EXPERIENCE

Employee experience is a holistic approach which helps organizations to identify better workplace practices in order to scale new pathways to the performance that is radically different. It has often been observed that organizations fail to offer outstanding employee experience to their internal stakeholders due to lack of clarity and unfamiliarity in identifying the employee's experiential needs and desires (Schmitt, 2003, pp. 219). A study stated that Employee experience is different from HRM and proposed six guiding principles on which organizations can build better employee experience – first, deeply understand employees and their needs; embrace expansive and holistic thinking; make the intangibles tangible; insist on radical participation; iterate and experiment; and finally trust and appreciate the process (Plaskoff, 2017, pp. 138-140). EX represents the integration of several emerging elements, like- digitization, analytics, design thinking and behavioural science.

The core facets of employee experience have been identified as belongingness, purpose, achievement, happiness, and vigour. According to Morgan J. (2018), employee experience comprises of seventeen attributes that are abbreviated under three categories – "ACE technology, COOL physical spaces, and CELEBRATED culture"(pp.28). Based on a research survey of top-level executives from more than 150 global organizations, seventeen employee experience variables have been identified – availability to everyone, consumer-grade technology, employee needs versus business requirements, option to bring in friends or visitors, offers flexibility, organizations values are reflected, leverage multiple work options, dedicated to employee health and wellness, executives and managers are coaches and mentors, the ability to learn new things, referrals come from employees, believes in diversity and inclusion, employees feel like they are part of the team, legitimate sense of purpose, everyone feels valued, company is viewed positively (pp.29). Therefore, organizations that truly believe in designing positive employee experience must be empowered by technology and invest in employee expectations, needs and emotions, while also keeping the required tools in place. Through innovative HR interventions such as integrated employee self-service tools, design thinking tools to promote employee creativity, employee journey maps for planned performance-centric career growth, wellness and fitness apps to act as stress-busters and innovation enhancers for employees, collaborative and team-operational model, HR departments are strategically devising ways to holistically improve the employee experience at work, so as to result in better performance and financial outcomes. Leaders and managers play a vital role in establishing a positive work environment and set the right tone and direction with a high level of clarity (IBM and Globoforce, 2016).

There are two primary factors that help organizations to drive towards building a positive employee experience. The first factor is workplace practices integrated with meaningful work, supportive management, positive work environment, growth opportunity, and technology; whereas the second factor is leadership and management behaviours and actions.

HOW DO ORGANISATIONS IMPLEMENT EMPLOYEE EXPERIENCE?

"The employee journey has many milestones and interactions, and the quality of employee experience has a direct influence on employee satisfaction, engagement, commitment and, in the end, performance" (Plaskoff, J. 2017 pp. 141). The experimental design thinking procedure has five steps that are followed in employee experience. Firstly, it allows employees to learn by observing, listening and engaging with others. Secondly, synthesis of their learning to define opportunities is initiated; thirdly, generating ideas for addressing those opportunities and developing experimental versions and implementation. Next step involves enhancing the opportunities and collaborating it with agility (Plaskoff, J. 2017 pp. 140). By involving employees in the process, they feel that they can voice together, build a relationship, develop understanding, and create ownership (pp. 141).

When the employees are provided with a conducive work climate, organization culture of trust, respect, and empowerment - employees experience true engagement with the employer and everlasting loyalty. These are the HR values that extend a competitive advantage to the organization. Studies on employee experience listed intent to stay, advocacy, initiative, recommend a brand, recommend employer to job seekers, and customer orientation as the outcomes of positive employee experience (Foresee, 2014). It is believed that positive employee experience is linked with better workforce outcomes such as individual and job performance, intention to stay (loyalty) and discretionary effort.

As per the Deloitte Global Human Capital Trends report (2017), the capabilities required by an organisation to achieve higher employee experience include understanding and adopting design thinking as part of employee experience. Adopting design thinking in developing HR and talent programs in the organisation can contribute to the achievement of higher employee experience. Another strategy in enhancing employee experience is providing programs for the younger, older and multi-generational workforce. This brings in a high level of diversity and therefore employees tend to get positive experience working in diversified teams. Considering diverse employee preferences in designing work settings, work roles and workspaces is an added competency in achieving employee experience fostering inclusivity at the workplace. By building a strong and differentiated employee experience brand and by aligning employees and personal goals with common corporate purpose, high employee experience can be achieved. Studies have also shown that by facilitating employees balance their personal and professional life expectations and by integrating social, community and corporate programs into HR practices, positive and enduring experiences can be provided to employees. These strategies can play a crucial role in enriching employee experience in the workplace. ForeSee's Employee Experience Measurement survey (2014) reveals scores for the key elements that drive the employee experience. It clearly illustrates that the key drivers for higher employee engagement, employee satisfaction, performance outcomes and retention are – job, workload, work environment, manager, teamwork, compensation, career advancements, leadership, work support through training, corporate culture etc. Hence HR strategies that fulfil these needs and expectations of employees can lead to positive experiences undergone, leading to sustainable human capital management and business sustainability.

METHODOLOGY

The value for human experiences (CX or EX) in businesses has a significant impact on its branding. Companies still prefer to have face-to-face interactions during decision making, in order to give value to experiences shared in such interactions of cognition or intuition. Researches revealed that 39% of the graduates prefer direct interactions with their colleagues over web-based tools/online interactions; 59% people love to work in cultures characterised by fun, positive social atmospheres even at lesser remuneration status (Micheal, Particia, & Alex, 2017). The new paradigm of human resource management addresses the accounting of both professional and personal moments of employees.

In this study, we also examine the perceptions of HR managers related to the philosophy of employee experience and how ideal EX framework can be used as a starting point to bridge the gap in our understanding and to add to the body of theory in this area. A qualitative research method was adopted to collect first-hand data from the HR managers where researchers felt that it was beneficial and appropriate to gain practitioners perceptions for the advancement of the EX theory. Three indepth interviews were carried out for the study and purposive sampling was used to establish the sampling frame (Given in Table 2). From the reviews, key drivers and specialist practices were identified and the sampling frame was constructed to reflect the key drivers namely leadership and management practices and HR values.

Interviews were semi-structured in nature where each HR respondents were asked about the terms 'successful' and 'unsuccessful' applied to employee experience. They were asked to list out the practices fall under the control of top management and HR managers. Finally, respondents were asked about their take on employee experience and its impact on employee performance. Finally, the respondents were asked to define employee experience with reference to their workplace. Verbatim transcripts of the interviews were coded as per the qualitative research guidelines and data has been categorised un-

Table 2. Profile of the company, participants and the interview type

	Profile	Participant profile	Interview
Company A	• Ownership: Member firms have different legal structures (USA and UK: Limited Liability Partnership) • Headquarters: London UK • Industry: Financial services and consultancy • Area served: worldwide • Revenue: US$34.8 billion (2018) • No of employees: 270,000 (2018) • Service lines: Assurance, Advisory, Tax, Transaction Advisory Services • Country Headquarters: Gurgaon National Capital Region (NCR) • Number of people: More than 37000 people in 32 offices across 17 cities	Designation: HR Experience: 8yrs Location: Bangalore office India	Face to face interviews
Company B	• Ownership: Private sector, • Industry: Healthcare • Headquarter: Bangalore • Area served: National • Revenue: US$ 5 billion (2018) • Total employees: 23000	Designation: HR Experience: 11yrs Location: Bangalore office India	Focus group interviews
Company C	• Ownership: Private sector, • Industry: Retail • Headquarter: Mumbai • Area served: Global • Revenue: US$ 9.5 billion (2018) • Total employees: 40000	Designation: HR Experience: 10yrs Location: Bangalore office India	Face to face interviews

Source: Website of the respective company and personal interview with respondents

der identified themes as part of the interpretive analysis. The identified themes were grouped to create conceptual consistency with the identified framework for the study.

Interpretation

The respondents of the study were remarkably consistent in sharing the views on employee experience and their perpetual consistency facilitated the analysis process. From the three interviews, HR managers felt that it is important to think in terms of the employee experience as something concrete and academic, and not purely as a task of management. The point is articulated below -

By creating a culture that respects diversity and inclusion, attracts talent and retains highly engaged employees, having an efficient and transparent feedback system, the participation of the employee in decision making and problem-solving at every level and a feeling that employees' ideas and suggestions matter to the organisation in its growth journey...we build positive employee experience (Company A).

Three HR managers felt that firms exhibited varying degree of success when it came to engaging in, and managing, employee experience.

Realistic and consistent information communicated to the employees from the beginning with the recruitment process and throughout the journey while on the job and organization. The experience is believed to be positive when the information delivered is believed to be genuine and true by the employees. This transparent communication system creates positivity in employees through loyalty, positive word of mouth, improved productivity, and fewer turnover. When employees agree that their job utilises their skills and interests to the maximum and provides the scope for development and exploration, employee experience is developed. We start with people... Each moment that matters in an individual's relation with the group, from recruitment all the way to the exit dialogue is designed and accounted around our people – employees, managers and HR professionals (Company B).

Top management in association with HR teams plays a vital role in setting the overall tone and stage for positive employee experience. Providing more clarity about the organizations' goals and directions to employees is important. Senior leadership positions hold this responsibility and they should align employee and their behaviours with the organizations' goals.

Emphasizing on building a platform for personal development and creative self-expression, by providing opportunities to explore, initiate, lead and develop ownership, to be responsible in achieving organisational goals, to listen to the feelings and emotions of our people at different levels, is the best way through which employee experience can be achieved. It's not about investing huge capital on our employees; it's about investing time and emotions with them. Employee experience is the way we relate, connect, identify and interact with the organization leading to optimal engagement, collaboration, free flow of energy and ideas, wellbeing and a sense of purpose (Company C).

Employees believe that personalized support and commitment from middle management plays a crucial role in team building. Leadership practices set the momentum towards a positive employee experience whereas the best HR practices will lead the momentum forward. Working environment, job and the person are the three criteria that the HR team need to pay attention while thinking about positive employee experience. Studies suggested that organizational integrity and co-worker support matters at the workplace; meaningful work, feedback, recognition and growth has a positive association with employee experience; and similarly, employee empowerment and voice, work-life balance link with better experiences (IBM and Globoforce, 2016; Deloitte Global Human Capital Trends- Rewriting the rules for the digital age, 2017; Jacob & Goldsmith, 2017).

- Company-A believes that practices like- providing the positive organisational culture with a completely new mindset and way of doing things, shifting to the outside-in thinking process, inclusion centric and mirroring what is being done for customers, enhance employee experiences.
- Company-B asserts that communication with the employee is significant in providing an enriching experience. Focus on employee journeys in the organisation and giving due concern to 'moments that matter to employees' makes the difference, besides other practices.
- Company-C stated that importance is given to social life/personal life experiences at work and humane HR policies enhances the experience. Multiple channels listening, identifying holistic journeys, listen to the emotions, create personas, design experiences, create and launch digital HRM strategies facilitates in fostering positive employee experiences.

Long term sustenance and strive for excellence is not so easy for any organisation in this global business environment unless the approach of design thinking at scale is adopted. Design thinking approach can be one of the models for implementing EX framework in the organisation, although different organisations have resorted to different disruptive practices in doing so. For example, at Ford, global listening practice is implemented to understand employee emotions, feelings, grievances and joys. This eventually leads to feeling of satisfaction, ownership and mutual trust among employees endowing them with cherishing experiences. Listening practice has become an effective employee experience strategy in many other organisations (Deloitte Global Human Capital Trends, 2017 pp.57). Similarly, Facebook, Google, LinkedIn, Gensler, Leesman, and many others designed their work climate and organisational culture integrating values of innovation, personalized recreation, and collaboration in a novel way, offering unique experience to their employees (pp. 58). Hence, employee experience is an intersection of expectations, needs and wants of employees with the organization design (Jacob & Goldsmith, 2017). From the organization's perspective, the concept of employee experience is what they designed and created for their employees; but for the employees, the experience is simply the reality of what it's like to work. Experience is subjective in nature because people have emotions and their perceptions, attitudes and behaviours, which are dynamic and contextual. To balance those differences organizations have invested time and effort in designing the unique experience spaces for employees and they make sure that the practice is implemented properly. Being in an experience economy currently, organisations are focussing on driving business results through workplace experiences by the ability to combine experience data (X-data) with operational data (O-data)(Albert, 2019). Most organisations have abundance of operational HR data in terms of who is leaving, who is been hired, who has accepted or rejected offer and how much time is required to get new hires for full productivity. However, most organisations lack in compiling the X data which infers why such occurrences have taken place. The newly launched SAP Qualtrics Employee Engagement solution enables HR leaders and managers to assess every experience that is significant to employees, which facilitates managers to automatically predict where to focus and what actions to be taken to avoid employee underperformance and turnover through analysis of key experience, engagement, and productivity drivers (Albert, 2019).

PRACTICAL IMPLICATIONS AND CHALLENGES

Various studies assert that employee experience and engagement go hand in hand to reflect employee state of work. This association helps organizations to understand the desired state of employees with the job and organization (Macey & Schneider, 2008). Engaged employees take pride in talking about the company they work with and its practices. However, employee experience gives a holistic view of the employees' feelings, emotions, attitude and drive towards excellence at the individual, group and organizational level. Employees who experience a sense of belongingness, purpose, achievement, vigour and happiness are more likely to perform at higher levels and contribute beyond expectations (IBM and Globoforce, 2016). Companies that invest heavily on employee experience were listed as the best place to work, earned four times more than the average profit and two times the average revenue, also the highest stock values (Denise 2018). Companies need to be alert with the holistic change and allow HR managers to lead the momentum of employee experience forward.

Companies worldwide have been experimenting with various HR practices to achieve high employee experience for sustainable growth. However, identification and implementation of the best suitable practices could be the biggest challenge to the heads. Challenges identified in executing EE practices were – choosing the right workplace solutions, creating a true employee-centric organization, focusing on right motivators, building a smart and digital workplace, collecting data and making it actionable (Dukes E, 2019). To overcome the mentioned challenges, management needs to try new technology by developing user-friendly mobile apps for the employees to access their day-to-day activities quickly and efficiently. A new idea termed as 'Prospective Employee experience' is been proposed. Prospective Employee Experience (PEE) could be a significant practice in the recruitment process, indicating that the company is dedicated to ensuring a positive and enduring experience to prospective candidates and the new hire's success. PEE can attribute to employer branding, employee engagement and retention to a great extent. Encouraging autonomy, participative decision making, providing clear career path and advancement, flexibility, appreciation and offering the best are the strategies that can be considered in promoting the positive connection between employees, job and organization (Deloitte Global Human Capital Trends, 2017 pp.57).

Usually, feedback methods like surveys, face to face interviews, focus groups, feedback apps and hackathons are the methods that most organisations adopt conveniently and effectively for assessing employee experience. Various leading companies have integrated new strategies to gather employee feedback like design thinking methodology, value stream mappings, Kaizen sessions, Yammer and other online communication tools, Glassdoor and custom made online tools. Some of these methods still depend upon human and digital interaction, while some others on a continuous feedback system. However, the efficiency of these measures is still to be assessed. Another key issue still to be resolved is how to measure employee experience accurately (Deloitte Global Human Capital Trends, 2017 pp.58).

CONCLUSION

Employees appreciate being empowered and heard while operating in a rewarding environment. Leadership must aim at strategically managing their human capital for an optimal return and sustainability. Employee experience is the ultimate level of fulfilment that an employee can feel and derive from their interactions with work and work environment in the organisation. It would invariably lead to the highest level of engagement, empowerment and commitment that an organisation can ever think about leading to employee delight and customer delight. At this point, it can be stated that sustainable HRM is practised in the organisation to achieve sustainability goals. However, understanding the nature and effects that the employee experience has on the organization can be difficult to quantify and manage.

Several suggestions emerged in the process of designing the employee experience model for cultivating better work experience and achieving positive outcomes. An exploration of digital and technology-enabled HR systems like gamification, artificial intelligence enables learning management systems (training and development) and digital performance management systems at the workplace. These developments can definitely infuse higher employee experience at the workplace. The productivity and collaboration apps, engagement survey apps, performance management system apps, well-being and welfare apps, and many other technological tools add value to the existing HR practices. Infact, these integrated tools bring independent HR systems with technology managers to create end-to-end employee experience strategy and programs. Frequent listening to the voice of the workforce and continuous diagnosis of the cultural issues help organizations to understand their workforce deeper and better.

The purpose of this chapter has been to explore the concept of EX and its various drivers which help in enhancing Employee experience at the workplace. Emphasis is laid on leaders to strategically tone their future directions of work with supporting manager' behaviours and actions with employees. When an organization tries to elevate positive employee experience and consider it as a priority, it will invariably lay the path to high-performance groups who are believed to be loyal and work beyond expectations. The purpose of employee experience is to create and retain such a workforce that delivers beyond limits and set expectations. This is the next step to employee engagement and can be seen as the solution to designing enduring and most admired employer work systems.

REFERENCES

Albert, L. (2019, May 7). *Delivering Exceptional Employee Experiences Requires Identifying and Closing Experience Gaps*. Academic Press.

App, S., Merk, J., & Buttgen, M. (2012). Employer Branding: Sustainable HRM as a Competitive Advantage in the Market for High-Quality Employees. *Management Review*, *23*(2), 262–278.

Bal, P. M., DeCooman, R., & Mol, S. T. (2013). Dynamics of psychological contracts with work engagementand turnover intention: The influence of organizational tenure. *European Journal of Work and Organizational Psychology*, *22*(1), 107–122. doi:10.1080/1359432X.2011.626198

Belk, R. (1975). Situational variables and consumer behavior. *Journal of Consumer Behaviour*, *2*(3), 157–164. doi:10.1086/208627

Bersin, J. (2016). *Predictions for 2017: Everything Is Becoming Digital*. Deloitte Development LLC.

Bhattacharya, S., Trehan, G., & Kaur, K. (2018). Factors Determining Psychological Contract of IT Employees in India. *International Journal of Human Capital and Information Technology*, *9*(1), 37–52.

Chamberlain, A. (2015). *Why Is Hiring Taking Longer? New Insights from Glassdoor Data*. Mill Valley, CA: Glassdoor.

Clarke, J. (2010). Revitalizing Entrepreneurship: How Visual Symbols are Used in Entrepreneurial Performances. *Journal of Management Studies*, *48*(6), 1365–1391. doi:10.1111/j.1467-6486.2010.01002.x

Coyle Shapiro, J., & Kessler, I. (1998). The pshycological contract in the UK public sector. In S. J. Havlovic (Ed.), Employee and employer obligations and contract fulfillment (pp. 1-7). NPS.

Cropanzano, R., & Mitchell, M. (2005). Social Exchange Theory:An Interdisciplinary Review. *Southern Management Association*, *31*(6), 874–900.

Davenport, T., & Beck, J. (2002). *The attention economy: Understanding the new currency of business*. Boston: Harvard Business School Press.

Deloitte Global Human Capital Trends- Rewriting the rules for the digital age. (2017). *The employee experience - Culture, engagement, and beyond*. Deloitte University Press.

Denise, L. Y. (2018, June 2). *2018 Will Be the Year of Employee Experience*. Retrieved 06 03, 2019, from Forbes: https://www.forbes.com/sites/deniselyohn/2018/01/02/2018-will-be-the-year-of-employee-experience/#1af202ae1c8f

Dhingra, N., Emmett, J., & Samadani, M. (2018, March 12). *Employee experience: Essential to compete*. McKinsey & Company.

Dukes, E. (2019). *5 Big Employee Experience Challenges to Overcome in 2019*. Houston, TX: Office.

Ehnert, I. (2009). *Sustainable Human Resource Management- A conceptual and exploratory analysis from a paradox perspective* (1st ed.). Physica-Verlag Heidelberg.

Foresee. (2014). *Measuring employee experience to drive positive employee engagement*. A Foresee White Paper.

Gardner, M. (1985). Mood states and consumer behavior:A critical review. *The Journal of Consumer Research*, *12*(3), 281–300. doi:10.1086/208516

Gob´e, M., & Zyman. (2001). *Emotional branding: The new paradigm for connecting brands to people*. New York: Allworth Press.

Harrington, B., & Ladge, J. (2009). Work–Life Integration:Present Dynamics andFuture Directions for Organizations. *Organizational Dynamics*, *38*(2), 148–157. doi:10.1016/j.orgdyn.2009.02.003

Harris, P. (2007). We the people: The importance of employees in the process of building customer experience. *Brand Management*, *15*(2), 102–114. doi:10.1057/palgrave.bm.2550123

Head, J. H., & Freedman, J. (2014, Jan 2). *Inspiring Employee Engagement through Emotional Intelligence*. Retrieved June 01, 2018, from EQ Business: https://www.6seconds.org/2014/01/02/employee-engagement-emotional-intelligence/

Hirschman, E. C., & Holbrook, M. B. (1982). Hedonic consumption: Emerging concepts, methods, and propositions. *Journal of Marketing*, *46*(3), 92–101. doi:10.1177/002224298204600314

IBM & Globoforce. (2016). *The Employee Experience Index- A new global measure of a human workplace and its impact*. Software Group.

Jabbour, C., & Santos, F. (2008). Relationships between human resource dimensions and environmental management in companies: Proposal of a model. *Journal of Cleaner Production*, *16*(1), 51–58. doi:10.1016/j.jclepro.2006.07.025

Jacob, M., & Goldsmith, M. (2017). *The Employee Experience Advantage: How to Win the War for Talent by Giving Employees the Workspaces they Want, the Tools they Need, and a Culture They Can Celebrate*. New York: John Wiley & Sons.

Kahn, W. A. (1990). Psychological Conditions of Personal Engagement and Disengagement at Work. *Academy of Management Journal*, *33*(4), 692–724.

Liley, M., Feliciano, P., & Laurs, A. (2017). *Employee Experience Reimagined*. Accenture.

Macey, W. H., & Schneider, B. (2008). The meaning of employee engagement. *Industrial and Organizational Psychology: Perspectives on Science and Practice*, *1*(1), 3–30. doi:10.1111/j.1754-9434.2007.0002.x

Micheal, L., Particia, F., & Alex, L. (2017). *Employee Experience Reimagined*. Accenture Startegy.

Morgan, J. (2018). The technological environment in employee experience. *Leader to Leader*, *2018*(87), 28–35. doi:10.1002/ltl.20340

Nelson, E. (2017). *Employee Experience-How to build an EX-centric organization*. Kennedy Fitch.

Pine, B., & Gilmore, J. (1999). *The experience economy*. Boston: Harvard Business.

Plaskoff, J. (2017). Employee Experience- The new human resource management approach. *Strategic HR Review*, *16*(3), 136–141. doi:10.1108/SHR-12-2016-0108

Pullman & Gross. (2004). Ability of Experience Design Elements to Elicit Emotions and Loyalty Behaviors. *Decision Sciences, 35*(3), 551-578.

Reichheld, F. (1996). *The loyalty effect: The hidden forces behind growth, profits,and lasting value*. Boston: Harvard Business School Press.

Saks, A. (2006). Antecedents and Consequences of Employee Engagement. *Journal of Managerial Psychology*, *21*(7), 600–619. doi:10.1108/02683940610690169

Schmitt, B. (1999). *Experiential marketing*. New York: The Free Press.

Schmitt, B. H. (2003). *Customer Experience Management*. Hoboken, NJ: Wiley.

Schwartz, B. (2015, August 28). Rethinking Work. *The New Times*. Retrieved from http://www.nytimes.com

Temkin, B. (2016, January 11). *Temkin Group's 2015 Research: Insights For CX Success*. Academic Press.

Turnley, H., Bolino, M., Lester, S., & Bloodgood, J. (2003). The Impact of Psychological Contract Fulfill-ment onthe Performance of In-Role and OrganizationalCitizenship Behaviors. *Journal of Management*, *29*(2), 187–206. doi:10.1177/014920630302900204

Wolf, M. J. (1999). *The entertainment economy—How mega-media forces are transforming our lives.* New York: Times Books, Random House.

Zuckerman, M. (1971). Dimensions of sensation seeking. *Journal of Consulting Psychology*, *36*(1), 45–52. doi:10.1037/h0030478

Chapter 7
From Entertainment Device to IoT Terminal:
Smart TV Helps Define the Future Living in Smart Home

Bu Zhong
Pennsylvania State University, USA

Fan Yang
University at Albany (SUNY), USA

ABSTRACT

Following the trend of home appliances and devices being growingly interconnected into the Internet of Things (IoT) system, smart TV helps define the future of human living as part of the smart home IoT system. This research explores viewers' perceptions toward four emerging interactivity functions of smart TV: the interaction between viewers and TV functions, between viewers and friends, between viewers and programming, and viewers' interaction with products appearing in TV shows. The technology has been mature to provide the four interactivity functions, though they are yet to be adopted by all smart TV sets so far. Our research shows that the viewer's residence was a significant moderator in the preference for the interactivity functions. Viewers from developed regions enjoyed the four functions more than those living in underdeveloped regions. Social media habits and the power usage of information and communication technology are positively associated with the preference, while need for cognition, age, and gender had little effect.

INTRODUCTION

Smart TV, functioning as part of the Internet of Things (IoT) system in a smart home, helps define the future of human living in the IoT era. Watching TV has remained as an indispensable part of human communication since color TV was introduced to the world in the 1950s. TV viewing has had an enormous impact on viewers' social life, whose influence spills over onto a spectrum of social activities, in

DOI: 10.4018/978-1-7998-0357-7.ch007

particular, the perception of social reality (Robinson, 2011). Indeed, new functions and services were added to modern TV sets like remote control, voice control, video streaming and TV apps. But watching TV itself did not experience lots of significant changes until the pervasive use of information and communication technology (ICT) (e.g., smartphone, social media) and the rise of IoT. People used to sit in a couch and wait for TV programs aired on a fixed schedule, during which little interactivity happened with the content. A new trend of TV viewing emerges when social TV appears, during which people may choose to co-view TV content and social media information simultaneously (Doughty, Rowland, & Lawson, 2011). While watching TV, viewers today tend to multitask on smartphones, tablets, or laptop computers so to stay social with friends by sending tweets, sharing photos, chatting online, and post comments (Cohen & Lancaster, 2014; Doughty et al., 2011). A survey shows that 87% of viewers using a second screen device when watching TV (Flomenbaum, 2015). The phenomenon indicates that audiences seek to be an active user of information, rather than being a "couch potato" who received information in a passive way.

Guided by the literature (Cohen & Lancaster, 2014; Hunt, 2014; Shah, Hanna, Bucy, Wells, & Quevedo, 2015; Shin, 2016), this chapter aims to access four proposed interactivity functions and explores viewers' disposition and preferences toward them in the context of the (ICT power usage, personality traits and demographics. The interactivity functions under study represent the future development of TV industry as they facilitate viewers' interaction with 1) smart TV, 2) online friends, 3) TV content, and 4) shopping for products appearing on TV. Some current smart TV sets may have one or two of the above functions, but none had all the four features so far. Research on the interactivity functions should shed light on our evolving TV viewing habit and a better understanding of the smart TV role in the smart home IoT system.

The Rise of Smart TV

Smart TV is defined as "technological convergence between computers, television sets, and set-top boxes integrated with online connectivity", providing "internet service, online interactive media and on-demand streaming media" (Wang & Chen, 2018, p. 87). In reality, smart TV, also known as social TV, offers a mediated TV viewing experience incorporating watching TV and staying connected online at the same time, which is "a rewarding social experience" due to the enhanced social interaction (Cohen & Lancaster, 2014, p. 512). Prior studies mostly examined smart TV by treating it as a co-viewing practice of TV programming and social media information, during which a second screen like a smartphone, a tablet, or a laptop computer is involved (Cohen & Lancaster, 2014; Shin, 2009; Shin et al., 2013). In this scenario, viewers are multitasking on a TV screen and ancillary devices. Few studies, however, had examined the phenomenon by studying how viewers interact with TV content and friends on the same TV screen partly due to the fact of limited interactivity on smart TV.

To attract social media users, TV manufacturers have worked hard to make their products stay relevant in the living room as an entertainment and information center (Shin et al., 2013). This requires their TV sets to be growingly connected with the Internet and integrated with the IoT system. Smart TV thus represents the trend of technological convergence that combines computers, set-top boxes and TV sets to bring viewers an interactive viewing experience. Albeit researchers recorded a slight drop in TV watching among young people, the Internet has been found enhancing TV viewing due to the interactivity functions on smart TV (MarketingChart, 2014). The new trend endorses the importance of critically studying viewers' perceptions of interactivity functions on smart TV.

Going beyond functioning as an entertainment device, smart TV has provided viewers a growing range of services that enrich the viewing experience and smart living. Meanwhile, smart TV could play a central role in the smart home IoT system. The IoT global market was expected to worth $11 trillion by 2025 (Shuhaiber & Mashal, 2019), connecting billions of smart objects like smart TV, heartbeat monitors, smart lighting, power meters, smart locks, temperature sensors, robotic vacuum cleaners, and digital assistant devices. A smart home is thus defined as "a residence equipped with smart technologies aimed at providing tailored services for users" (Marikyan, Papagiannidis, & Alamanos, 2019, p. 139). The word "smart" is used as an umbrella term for innovative and intelligent technology, which possesses some degree of artificial intelligence (Marikyan, et al., 2019; Yang, Lee, W., & Lee, H., 2018). These AI-powered devices share a key attribute – the ability to acquire information from the surrounding environment and react accordingly (Marikyan et al., 2019). The smart devices and applications are growingly interconnected through the IoT system to turn a regular home into a smart home, in which smart TV may play a central role of displaying and coordinating the connected devices. To fulfill the new assignment, smart TV itself needs to be more interactive in a smart home.

Perceived Interactivity on Smart TV

Smart TV can stream media and run TV apps for accessing online services, such as video on demand, Internet surfing, social networking and online banking (Darnell, 2015). In the era of IoT, smart TV would play a bigger role in smart homes as an IoT terminal, enabling innovative analytics on IoT-captured data from smart homes (Yassine, Singh, Hossain, & Muhammad, 2019). To meet the changing needs, TV manufacturers had invested heavily in developing smart TV with enhanced interactivity functions. The current smart TV, however, has a major setback: limited interactivity. When viewers need to post comments on TV programming, for instance, they have to switch to another device, as the TV does not display the video content and social media posts on the same screen. Smart TV, therefore, is not that smart as its screen is not truly interactive (Shin, 2016).

To understand the interactivity on smart TV, this chapter employs the theoretical frameworks of the theory of reasoned action (TRA) and the theory of planned behavior (TPB). Based on the premise that overt behavior is a function of users' behavioral intention, the TRA posits that the best predictor of behavior (e.g., using certain smart TV functions) is behavioral intention, and behavioral intention is, in turn, shaped by the attitude toward the behavior and subjective norm (Ajzen, 1985; Fishbein & Ajzen, 1975). Ajzen (1985) extended the TRA and proposed the theory of planned behavior by adding a new component of perceived behavioral control, i.e., the degree to which users believe they can control the behavior. The TPB is widely used to explain and predict behavior based on behavioral intention, defined as the attitude about the likelihood that the behavior will lead to the expected outcome and the subjective evaluation of the risks and benefits of such outcome. The TRA and TPB are well suited to explain viewers' perceptions and attitudes towards smart TV (Shin et al., 2013), which was used to explain the intention in using IPTV (Shin, 2009).

Interactivity is one of the five qualities of communication on the Internet, and the other four are multimedia, hypertextuality, packet switching and synchronicity (Newhagen & Rafaeli, 1996). Perceived interactivity can be defined as the perceptions of engagement with responsive actual or virtual objects or people, and perceived interactivity mediates the effect of actual interactivity on attitude during the interaction process (Wu, 2005). Hence, this chapter aims to explore the following four interactivity functions: 1) the viewer-TV interaction, e.g., voice control or facial recognition; 2) the viewer-friend interaction, or the interaction between viewers and their online friends, e.g., real-time chatting or sharing comments on TV shows; 3) the viewer-content interaction, e.g., video on demand or programming recommendation; and 4) online shopping, viewers' interaction with commercial products appearing in TV programs. With these functions in place, smart TV is revolutionizing traditional TV viewing experience by offering new services never available before. On smart TV, all these activities can be done on the same screen and viewers can choose to add picture-in-picture-style windows on the TV screen or expand one of them into a full-screen mode (see Figure 1 for the imitated smart TV interface).

Media interactivity is defined as "a measure of a media's potential ability to let the user exert an influence on the content and/or form of the mediated communication" (Jensen, 2008, p. 129). Jensen (2008) divided the concept of interactivity into four dimensions, which are called, respectively, 1) transmissional interactivity, referring to media's ability to let the user choose content from a continuous stream of information in a linear way (e.g., multichannel TV); 2) conversational interactivity, referring to the media's ability to let the user generate content and share it with others; 3) consultational interactivity, referring to the media's ability to let the user choose, by request, from an existing selection of pre-produced information; and 4) registrational interactivity, referring to the media's ability to register information from and thereby adapt or respond to users' needs and actions (e.g., online shopping). Among the four dimensions, little difference exists between smart and traditional TV regarding transmissional interactivity. This chapter will thus focus on examining four interactivity functions inspired by the other three dimensions of interactivity.

The Viewer-TV Interaction

This function belongs to consultational interactivity, a major feature that sets smart TV apart from the traditional TV. Some smart TV brands have supported the viewer-TV interaction through voice control, face recognition, or motion detection. The voice recognition technologies allow viewers to do a verbal search on TV. Besides, smart TV can authenticate viewers through identifying the characteristics of viewer voices or faces (Kinnunen, Tomi, & Li, 2010). Motion detection is to detect any changes in the

Figure 1. The imitated interface of a futuristic smart TV

position of an object relative to its surroundings and vice versa (Singh & Kaur, 2012). With motion sensing detectors, smart TV can capture viewers' body movements to provide a new interactive experience.

The Viewer-Content Interaction

This personalization function belongs to both conversational and consultational interactivity, offering personalized services based on viewers' interaction with TV content. For example, while watching TV, one's viewing habits (e.g., favored genres, using options such as pause, fast forward, etc.) were documented and stored. By tracking such behaviors, smart TV could generate a detailed model of the personal viewing experience, according to which specific TV programs may be recommended (Yang et al., 2014).

The Viewer-Friend Interaction

This function belongs to consultational interactivity. Users can enjoy both real-time and asynchronous communication with friends through smart TV, like using a smartphone. Given the popularity of social media, having access to and staying connected on social media has become an increasingly important component of TV viewing experience (Harboe et al., 2008). Nielsen (2014) found that 15% of viewers reported watching TV more enjoyable when social media were involved.

The Viewers' Interaction with Products in TV Content

This TV shopping function belongs to transmissional interactivity. Exclusive shopping channels have been incorporated into TV since the 1970s. Because of its interactive functions, shopping on smart TV may be enticing as it can deliver highly personalized shopping service (Yang & Zeng, 2014). Unlike traditional shopping channels that present product information linearly, smart TV offers customers opportunities to view products according to their needs through shopping apps. For example, when viewers see a purse an actress uses on TV, they can start to search and buy it in a picture-in-picture window without interrupting watching TV. Shopping apps can also display the stores nearby based on the viewers' geo-locations. Since little had been done to investigate viewers' perceptions towards these interactivity functions, RQ1 was raised:

RQ1: Among the four interactivity functions, which one is most favored by viewers?

Viewer Demographics and TV Viewing

A series of user demographics have been identified for their considerable influences on media use. For TV viewing, viewers tend to spend more time watching TV as they age (Harwood, 2007; Nielsen, 2009), due to reasons like more leisure time in retirement or a rise in information needs (Vandebosch & Eggermont, 2002). Gender also plays a role as studies constantly note differences between males and females in TV viewing, including the finding that women often watch more TV than men (Comstock, 1989). Meanwhile, male TV viewers tend to be more purposeful in TV content selection to meet personal needs (e.g., acquiring information, managing mood) (Morley, 1986). Female viewers, compared to the instrumental approach of males, are more expression-oriented (Nathanson, Perse, & Ferguson, 1997). To women, watching TV often becomes a way of social interaction (e.g., spending time with friends) to fulfill their interpersonal and emotional needs (Morley, 1988). Viewers in various geo-locations tend to watch TV differently as well (Nielsen, 2015). Compared to those in big cities, viewers in small towns often have distinct TV watching habits partly due to economic and social differences (Morley, 1988).

Some psychological traits also play a notable role in TV viewing. Henning and Vorderer (2001) found that how much people enjoyed effortful thinking – which can be measured by the scale of need for cognition (NFC) (Petty, Cacioppo, & Kao, 1984) – was positively associated with how much TV they watched. Individuals with lower NFC spent more time watching TV as a way of killing time and escaping the reality, while those with higher NFC watched it less as they took TV as a distraction from thinking (Henning & Vorderer, 2001). NFC is also associated with the motivation of information processing (Petty & Cacioppo, 1984), which can be further moderated by situational factors (Cacioppo, Petty, Feinstein, & Jarvis, 1996).

The above patterns of TV viewing, however, were mostly discovered in the age of traditional TV. New changes could occur as smart TV converges ICT applications. As smart TV begins to shift the paradigm of TV viewing, some previous findings, including elder people watching more TV than younger people, might no longer hold true. In the past, traditional TV often turned away viewers with high NFC because of the lack of "pause and think" button (Henning & Vorderer, 2001), smart TV, nevertheless, supplements such functions by enabling them to pause and think as much as they like. To investigate the TV viewing experience on smart TV, the following research questions are formulated:

RQ2: What are the gender differences concerning viewers' preferences for the interactivity functions of smart TV, if any?

RQ3: What are the age differences concerning viewers' preferences for the interactivity functions of smart TV, if any?

RQ4: What are the differences, if any, between people living in cities that are more economically developed and those in less developed areas concerning their preferences for the interactivity functions of smart TV?

RQ5: Does viewers' need for cognition have an impact on the preference for the interactivity functions of smart TV?

Smart TV as IoT Terminal

Smart TV represents a convergence of TV viewing, social networking and other online activities (Harboe et al., 2008). On a smart TV screen, viewers may see statics of athletes in sports competition, comments from friends, utility bills, and data visualization from healthcare applications at home. Moreover, users may interact with their healthcare providers through smart TV, who in turn may monitor the health status of patients living inside a smart home, tracking their usage of appliances and detect their routine or abnormal activities that could indicate signs of health problems (Yassine et al., 2019). Meanwhile, privacy can be a concern, which is beyond the scope of this chapter.

When the above online activities happen on a TV screen, viewers' social media use may influence their preferences for the interactivity functions of smart TV. Prior research discovers that elder people often report more negative attitudes toward ICT-powered devices than young people due to the elder's low familiarity with technologies (Teo, 2001). How well viewers use the interactivity functions may depend on their ICT knowledge and skills (Marathe, Sundar, Nije Bijvank, Van Vugt, & Veldhuis, 2007). Zhong (2013) proposed the concept of ICT power users, referring to those who use ICT applications more innovatively, efficiently and thoroughly than ordinary users. In many cases, ICT power users tend to be innovators or early adopters of new technologies (see Rogers, 1962). Hence ICT power users are more likely to favor the interactivity functions under study as they could harvest more benefits out of the new features. Finally, two hypotheses are articulated:

H1. Heavy social media users may report more appreciation for the four interactivity functions on smart TV than light social media users.

H2. Those with a high level of ICT power usage may report more appreciation for the four interactivity functions on smart TV than those low in ICT power usage.

METHOD

Sample and Procedure

By using the snowball-sampling technique, this study conducted an online survey by recruiting 283 college students, their parents, family members and friends living in Hong Kong, several cities and rural areas in Mainland China from March 11th to April 1st, 2014. The participants, with an average age of 27.2 (SD = 1.17), were then divided into three age groups: 19-25 (n = 121, 74.9%), 26-45 (n = 40, 14.1%), and 46-58 (n = 31, 11%) years old. The sample was divided into the three age groups in consideration of each group's TV viewing habits and needs. In the first group (19-25), most participants were college students or young adults who freshly graduated from college. Most in the second group (26-45) were young or middle-aged participants who lived a family setting, and those in the third group (46-58) were the oldest in the sample who may have more free time and higher social statuses than the other two groups. Research shows that age-related changes and generational differences strongly alter TV viewing habits among viewers in Europe and the United States, and older adults face more barriers to enjoy the benefits of interactive TV (Rice & Springett, 2015).

Most participants (70%, or 199) were women. Over half of them (65%, or 184) self-reported that they had a middle-level household income in their cities or regions, while 11%, or 30 had a household income lower than the middle level. Only a few came from either high-income (n = 8, 2.8%) or low-income families (n = 2, 0.7%). Most were well educated, with 263 (91.2%) having a baccalaureate or above – Bachelor's degree (n = 198, 70%), Master's degree (n = 60, 21.2%), or Ph.D. (n = 5, 1.8%). Only 12 (4.2%) had an associate degree, 6 (2.1%) with a high school diploma, and 2 (0.7%) had a junior school diploma or below. A handful of participants (n = 121, 74.9%) lived in centrally administered municipalities (e.g., Beijing, Shanghai and so on), 77 (27.2%) in province-level cities, 115 (40.6%) in prefecture-level and county-level cities, 2 (0.7%) in village-level regions, 19 (6.7%) in special administrative region/SAR (Shenzhen), and 6 (2.1%) lived overseas.

A password-protected Qualtrics link was sent out to recruit participants through email. After participants indicated their consent to ensure the voluntary nature of the study, they were instructed to answer questions about the demographics, need for cognition, ICT power usage, and social media habits. Afterward, one imitated interface of smart TV (see Figure 1) was shown to the participants and each of the four proposed interactivity functions was explained in detail. The whole procedure was conducted in the same order for all participants. Participants were asked to indicate their preferences for each of the four functions. Upon completion of the survey, they were thanked for participation, but no incentive was provided.

Independent Variables

Viewer Demographics

Viewer demographics were measured by asking participants to indicate their gender ("male" and "female"), age and residence. There were three age groups of participants: "19-25", "26-45", "46-58" years old. For residence, participants were asked "where do you live?" and they selected one out of the following options: "centrally administered municipality", "province-level city", "prefecture-/ county-level city", "village-level region", "special administrative region/SAR" or "overseas." Centrally administered municipalities like Beijing and Shanghai and special administrative region/SAR (i.e., Shenzhen) are cities with the highest levels of economic development, followed by province-level cities in Mainland China such as Chengdu. The prefecture-/county-/village level cities, on the other hand, are usually the least developed economically. Therefore, residence was re-coded into "developed city" ("centrally administered municipality" and "special administrative region/SAR"), "developing city" ("province-level city"), and "under-developed city" ("prefecture-/county-level city" and "village-level region"). "Overseas" was discarded due to too few cases ($n = 6$).

Need for Cognition

Need for cognition was measured by an 18-item short version of the need for cognition (NFC) scale (Petty, Cacioppo, & Kao, 1984). Participants rated their responses on a Likert scale of 1 = "strongly disagree", and 5 = "strongly agree". Sample items include "The notion of thinking abstractly is appealing to me", and "I would prefer complex to simple problems" (Cronbach's $\alpha = .85$, $M = 3.19$, $SD = .52$).

Social Media Habit

An eight-item scale was developed for this study based on previous literature on peoples' social media use to measure socialization while watching TV (Wohn & Na, 2011). Participants rated their responses on a Likert-type scale of 1 = "never", and 5 = "always" on items including "Discussing with friends who are watching the same program via mobile phone or the Internet", and "Posing comments about TV programs on social media (Weibo, online forums and WeChat, etc.)" (Cronbach's $\alpha = .80$, $M = 2.52$, $SD = .59$).

ICT Power Usage

ICT power usage was measured by a six-item scale adapted from Agarwal and Prasad (1998) and Zhong (2013). Participants rated their responses on a Likert scale of 1 = "strongly disagree", and 5 = "strongly agree" on items including "I am good at using searching engine to find answer that I want" and "People around me think I am an expert in using mobile media device" (Cronbach's $\alpha = .78$, $M = 3.50$, $SD = .71$)

Dependent Variables

Preference for the Viewer-Friend Interaction

Preference for the viewer-friend interaction on smart TV was measured by a five-item developed for the current study. Participants rated their responses on a Likert-type scale of 1 = "not at all interested", and 5 = "very interested" on items like "I could make video calls with friends simultaneously", and "I could share television programs to friends" (Cronbach's $\alpha = .81$, $M = 3.49$, $SD = .83$).

Preference for the Viewer-TV Interaction

Preference for the viewer-TV interaction on smart TV was measured by a four-item scale based on the functions of voice-control and facial recognition. Participants rated their responses on a Likert-type scale of 1 = "not at all interested", and 5 = "very interested" on items including "I could input text comment by speech recognition", and "I could switch between different channels by gestures replacing the remote controller (motion sensor)" (Cronbach's $\alpha = .75$, $M = 3.71$, $SD = .84$).

Preference for the Viewer-Content Interaction

Preference for personalization function of smart TV was measured by a 6-item scale developed for this study. Participants rated their responses on a Likert-type scale of 1 = "not at all interested", and 5 = "very interested" on items including "It would recommend me similar TV programs based on what I watched", and "I could view the TV program ranking by audience rating and other indicators and could thus click to view" (Cronbach's $\alpha = .79$, $M = 3.60$, $SD = .73$).

Preference for Online Shopping

Preference for online shopping on smart TV was measured by a 2-item scale. Participants rated their responses on a Likert-type scale of 1 = "not at all interested", and 5 = "very interested", and the items are "I could click on the items that appear on the program, in order to obtain purchase information", and "When items appear in the program, the relevant shopping information will pop out at the corner of the screen if clicked" (Cronbach's $\alpha = .85$, $M = 3.12$, $SD = 1.19$).

RESULTS

A confirmatory factor analysis was first conducted, and the results show the four functions were nicely loaded together, which confirmed each function as a distinct factor (see Table 1). To answer RQ1 – the most popular function among the four interaction functions under study, a one-way repeated measures ANOVA employing a multivariate approach was conducted to examine the differences between participants' preferences for the four features. The analysis reveals significant differences for the functions, Wilks' $\Lambda = .79$, $F(3, 280) = 25.43$, $p < .001$, partial $\eta^2 = .21$. Bonferroni sequential post-hoc comparisons indicate that the viewer-TV interaction – features such as voice recognition, facial or motion detection – was rated as the most popular among the four proposed functions ($M = 3.71$, $SE = .05$), followed by the

Table 1. Confirmatory factor analysis of the preference for four interactivity functions

	Preference for interactivity functions			
	1	**2**	**3**	**4**
Factor 1: Preference for socialization				
I could view my friends' comments on the film or program	**0.76**	-0.09	0.02	-0.06
I could make video calls with friends simultaneously	**0.54**	0.21	-0.11	0.06
I could share television programs to friends	**0.68**	0.14	-0.04	0.05
I could check what my friends are watching	**0.76**	-0.17	0.14	0.00
It could link my micro-blogging and other social media accounts for me to comment, share, etc.	**0.57**	0.02	0.05	0.02
Factor 2: Preference for the viewer-TV interaction				
I could input text comment by speech recognition	0.20	**0.64**	-0.04	-0.08
I could search programs by speech recognition	-0.08	**0.81**	0.01	-0.09
It would recommend me a program list when I tell it what I feel at this moment, such as "I just broke up with my girlfriend and need comfort."	-0.10	**0.58**	0.17	0.08
I could switch between different channels by gestures replacing the remote controller (motion sensor)	-0.02	**0.59**	-0.10	0.10
Factor 3: Preference for the viewer-content interaction				
I could view the TV program ranking by audience rating and other indicators and could thus click to view	0.10	0.03	**0.59**	-0.07
It would display the filtered videos with only the most popular fragments	0.00	-0.07	**0.67**	-0.01
I could browse programs by short trailers	0.08	0.04	**0.56**	-0.02
I could do something like leaving messages or voting in support of players /directors /actors / screenwriters	0.20	0.15	**0.41**	-0.01
It would recommend me similar TV programs based on what I watched	-0.04	0.08	**0.60**	0.09
I could view real-time audience rating of a certain TV program	-0.05	-0.09	**0.70**	0.02
Factor 4: Preference for TV shopping				
I could click on the items that appear on the program, in order to obtain purchase information	0.08	-0.06	0.00	**0.84**
When items appear in the program, the relevant benefit information will pop out at the corner of the screen	-0.06	0.06	0.02	**0.87**

functions of viewer-content ($M = 3.60$, $SE = .04$) and viewer-friend interaction ($M = 3.49$, $SE = .05$). Online shopping was reported as the least favored function ($M = 3.12$, $SE = .07$).

RQ2, RQ3 and RQ4 are interested in the impacts of demographics (i.e., gender, age, and residence) on the preferences for the four functions. A mixed model factorial ANOVA analysis was conducted to explore the participants' preferences, with the functions of viewer-TV interaction, viewer-friend interaction, viewer-content interaction and online shopping as the within-subject variables, and gender, age, and residence as the between-subject variables. As shown in Table 2, a significant main effect was obtained for function, $F(3, 256) = 16.85$, $p < .001$, partial $\eta^2 = .17$. No significant main effects were found for gender, $F(1, 258) = 1.25$, $p = .27$, or age, $F(2, 258) = 1.96$, $p = .15$, or residence, $F(2, 258) = .47$, $p = .63$.

The main effects of function, however, need to be interpreted in light of a significant interaction of function and residence, $F(6, 512) = 2.43$, $p < .05$. As shown in Figure 2, preferences for smart TV

Table 2. Preferences for the four proposed functions of the futuristic smart TV

TV Interaction	*Viewer-friend*	*Viewer-TV*	*Viewer-content*	*Online Shopping*
	3.42(.09)$_b$	3.71(.09)$_a$	3.60(.08)$_a$	2.84(.13)$_c$
Gender	*Male*	*Female*		
	3.31(.10)$_a$	3.47(.10)$_a$		
Age	*19-25*	*26-45*	*46-58*	
	3.49(.06)$_a$	3.32(.11)$_a$	3.36(.17)$_a$	
Residence	*Developed City*	*Developing City*	*Underdeveloped City*	
	3.32(.09)$_a$	3.43(.17)$_a$	3.43(.08)$_a$	

Note: Cell entries are means. Standard deviations are reported in the parentheses. Means with different subscripts differ at $p < .05$ using Holm's Sequential Bonferroni post hoc comparisons.

did not vary much as a result of TV functions to people living in cities that were fairly economically developed. For viewers living in these central- or special-administered cities, they favored the four interactivity functions in a similar fashion. However, for audiences living in the developing cities, that is, province-level cities, they reported a drastic dislike toward the online-shopping function as opposed to the other three functions. A similar pattern was found for residents in small cities or rural areas, who had an unfavorable attitude toward the online shopping function on smart TV.

RQ5 asks whether there is an association between preferences for the four interactivity functions and need for cognition (NFC). Series of simple linear regressions were conducted and suggested no significant associations between NFC and preferences for the viewer-friend interaction ($\beta = .09, p = .13$), viewer-content interaction ($\beta = .09, p = .12$), and online shopping function ($\beta = -.01, p = .84$). Nevertheless, a positive association between NFC and viewer-TV interaction emerged with marginal significance, or $p < .10$ (see American Psychological Association, 2010). Higher level of NFC led to more favorable attitudes towards viewer-TV interaction function of smart TV ($\beta = .10, p < .10$).

H1 and H2 predict that viewers who are more used to interacting with others on social media while watching TV and those with higher levels of ICT power usage are more likely to favor the four interactivity

Figure 2. The interaction of function and residence on preference for smart TV

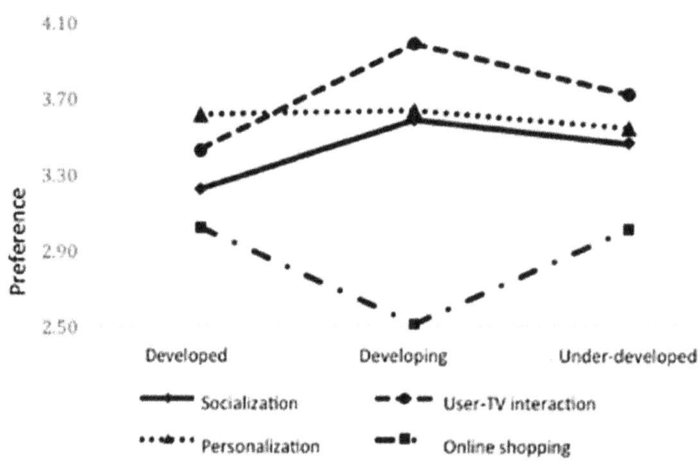

Table 3. The impact of social media habits and power usage on the preference for the four interactivity functions of smart TV

	Preference for Smart TV Functions			
	Viewer-friend interaction	*Viewer-TV Interaction*	*Viewer-content interaction*	*Online shopping*
Social Media Habits	.25***	.17**	.12*	.13*
Power Usage	.10†	.11†	.09	-.01
	$F(2, 280) = 10.82$, Adjusted $R^2 = .27$, $p < .001$	$F(2, 280) = 5.58$, Adjusted $R^2 = .20$, $p < .01$	$F(2, 280) = 3.19$, Adjusted $R^2 = .15$, $p < .05$	$F(2, 280) = 2.58$, Adjusted $R^2 = .14$, $p < .10$

Note: † $p < .10$, * $p < .05$, ** $p < .01$, *** $p < .001$.

functions. Series of multiple regressions were conducted to test these hypotheses. Table 3 reports statistics associated with the analyses and revealed that, together, they accounted for a significant proportion of variance in preference for the functions of viewer-friend interaction ($R^2 = .27$), viewer-TV interaction ($R^2 = .20$), viewer-content interaction ($R^2 = .15$), and online shopping ($R^2 = .14$).

More social media interaction was positively associated with a higher preference for the viewer-friend interaction, $\beta = .25$, $p < .001$, viewer-TV interaction, $\beta = .17$, $p < .01$, viewer-content interaction, $\beta = .12$, $p < .05$, and online shopping functions, $\beta = .13$, $p < .05$. Power usage, on the other hand, was positively associates with preference for the viewer-friend interaction, $\beta = .10$, $p < .10$, and viewer-TV interaction, $\beta = .11$, $p < .10$ with marginal significance. However, no significant relationship was found between power usage and preference for viewer-content interaction, $\beta = .09$, $p = .11$, or between power usage and online shopping, $\beta = -.01$, $p = .88$.

DISCUSSION

Going beyond the current smart TV functions, this chapter proposes four interactivity functions on smart TV: the viewer-TV interaction, viewer-friend interaction, viewer-content interaction, and viewers' interaction with products appearing in TV content (online shopping). So far no smart TV had provided an aggregate display of the four interactivity functions on the same screen by the time of writing this chapter, although, technically, they could be easily incorporated into smart TV. As the emerging ICT devices continue to reshape the media landscape, the four functions represent the future advances of smart TV industry (Harboe et al., 2008; Robinson, 2011; Samsung Tomorrow, 2015; Shin et al., 2013), especially in the IoT era.

The results show that the viewers rated the function of viewer-TV interaction most favored. They preferred the idea of interacting with smart TV by employing voice control, facial recognition and other features, suggesting that most viewers had a positive attitude toward interacting with their TV sets in a creative fashion. Such interest in a more interactive TV may be traced back to human beings' natural tendency to apply social rules to non-human items, including computers and machines (see Sundar & Nass, 2001). Given that television may be one of the most widely used ICT devices around the world, it is possible that viewers may treat television, especially, smart TV, as a social actor in their day-to-day lives (see Nass, Steuer, & Tauber, 1994), rather than just an entertainment device. It is clear that today's viewers prefer to actively engage in the viewing experience. With its viewer-TV interaction function, smart TV provides a strong sense of scalability by enabling more interactions among viewers and between viewers and online activities (Shin, 2016).

This research also discovers that some demographics and personality constructs play significant roles in influencing viewers' perceptions of smart TV features. Unlike what's been suggested on traditional TV viewing, this study found no significant gender or age differences in terms of the preference for the interactivity functions under study. The finding indicates that these functions of smart TV might have enriched the meaning of "watching TV" and reshaped the traditional TV viewing experience for all viewers, regardless of their age or gender. For instance, compared to females, males were previously found being more apt to instrumental TV viewing, i.e., viewing traditional TV as a way to fulfill personal needs rather than to involve in social interaction. Smart TV, incorporating social media like Facebook, Twitter, and Instagram, might have blurred such distinction like gender or age difference because smart TV is no longer a passive, disconnected and linear tube, but an active, responsive and interactive platform. Smart TV viewers enjoyed staying connected and other online activities while watching TV. This may also explain why the previous phenomenon of elder people watching more TV than young people no longer holds true for smart TV viewers.

This research reveals an important demographic factor – viewers' residence – in moderating the preference for the smart TV functions. The findings demonstrate that viewers living in developed cities favored the four interactivity functions indiscriminately. This is probably due to the fact that they live in more economic-developed regions, and consequently, are more likely to be exposed to the latest smart TV technologies than those in less developed areas. Hence, they are more likely to accept the four functions as a whole package of interactivity on smart TV. This study also found that online shopping was least favored by residents from less developed areas, which might be partially due to the poor or unreliable delivery systems in those areas.

The positive association between social media habits and preferences for the interactivity functions provides empirical support that smart TV could offer much more than what traditional TV did. To viewers, smart TV was an intelligent and interactive media device that could fulfill both information and sociality needs in daily life. While power usage positively associates with the preference for viewer-friend and viewer-TV interaction functions with marginal significance, it is interesting to note the non-significant relationships between power usage and the preference for viewer-content interaction and online shopping confirms previous findings that viewer-content interaction and online instant-gratification offers (e.g., coupons) do not necessarily appeal to power viewers (Zhang, Wu, Kang, Go & Sundar, 2014).

Theoretical Implications

Two major theoretical implications stem from this study. Despite the recent popularity of smart TV, little has been known regarding the public's attitudes toward and preferences for smart TV. This study, therefore, expands previous literature on traditional TV viewing by investigating viewers' disposition toward four interactivity functions, which stand at the frontier of smart TV development. By pinpointing the key demographic (i.e., residence), psychological (i.e., need for cognition), social (i.e., social media habits), and technological (i.e., ICT power usage) variables, the findings shed light on how people watch TV in the age full of ICT devices. The associations between these key variables and viewer preferences for smart TV functions portray a nuanced picture of smart TV viewing different from traditional TV.

Furthermore, the result that both gender and age had little impact on the preferences for smart TV contradicts what we knew about traditional TV viewing. This and other findings call for more academic attention to a different viewing experience on smart TV.

Practical Implications

This study should benefit manufacturing future smart TV. First, the viewer-TV interaction function was found to be the most popular smart TV function, indicating that viewers are passionate about interacting with TV sets in diverse ways. Smart TV manufacturers, thus, need to focus on developing more creative and viewer-friendly options, for instance, eye scrolling. The current study found that viewers from different geographic areas favored different interactivity functions. Thus, it is important to take viewers' residence into consideration when designing smart TV. For TV manufacturers, customizing various models of smart TV by considering customers' local economic development should better serve their needs and eventually help promote adoption of smart TV. The strong positive association between social media habits and preference for interactivity functions suggests the value of positioning smart TV as a social platform rather than another household appliance.

Limitation and Future Studies

This study has several limitations, upon which future studies can be built. First, it employed a convenience sample that was female-dominated. Due to the snowball-sampling technique used for the current study, the findings might not be generalized to other populations in other regions. Results of this study should also be interpreted in considering possible sample biases. In addition, since the exploratory survey was conducted among the participants from Hong Kong and Mainland China who were mostly well educated and had a fairly higher income than the average locals. Future studies may expand the sample by including more diverse participants.

Another limitation is that the study mainly centered on viewers' disposition toward interactivity functions, which cannot be taken as the only contributor to the dynamic viewing experience on smart TV. TV programs in different formats might also significantly impact the viewing experience but was not considered in this study. More insights should emerge when both functions and content on viewers' disposition toward smart TV functions are investigated. Future studies on cross-cultural comparisons of East-Western viewers' responses should contribute to the understanding of the evolving TV viewing experience around the world. As smart TV plays an increasingly important role in the IoT system of a smart home, more research is called on to investigate data flows from smart TV and users' privacy concerns in the changing market realities.

CONCLUSION

Like other human activities, TV viewing is an evolving and complex experience that can be influenced by a long list of external and internal variables. This paper reveals that viewers favor the viewer-TV interaction more than the viewer-friend interaction, viewer-content interaction, and their interaction with products appearing in TV programs. The finding indicates that viewers were excited about interacting with smart TV, instead of treating it as a mere TV tube. Where viewers live also has a significant impact on the attitude and preference for interactivity functions. Viewers living in more economically advanced areas tend to welcome the four proposed interactive functions as a whole package, while those from less developed regions indicate a strong reluctance to the online shopping function on smart TV. Furthermore, viewers' social media habits positively predict their preferences for all of the four functions. Power usage, on the other hand, was found to be associated with a higher preference for the viewer-friend and viewer-TV interaction function. As the world is entering the IoT age, the findings in this research should provide insights into the future development of smart TV. Without these interactivity functions, smart TV could be readily replaced by other smart home appliance. With enhanced interactivity, it could significantly enrich people's life in a smart home. In this way, smart TV has the potential of defining the future of human living as it becomes a terminal connecting other smart home appliances in an IoT system.

REFERENCES

Agarwal, R., & Prasad, J. (1998). A conceptual and operational of personal innovativeness in the domain of information technology. *Information Systems Research*, *9*(2), 204–215. doi:10.1287/isre.9.2.204

Ajzen, I. (1985). From intentions to actions: A theory of planned behavior. In J. Kuhl, & J. Beckmann (Eds.), *Action Control: From cognition to behavior* (pp. 11–39). Berlin, Germany: Springer-Verlag; doi:10.1007/978-3-642-69746-3_2

Ajzen, I. (1991). The theory of planned behavior. *Organizational Behavior and Human Decision Processes*, *50*(2), 179–211. doi:10.1016/0749-5978(91)90020-T

American Psychological Association. (2010). *Publication manual of the American Psychological Association* (6th ed.). Washington, DC: American Psychological Association.

Arons, L., May, M. A., & Brodbeck, A. J. (1963). *Television and human behavior: Tomorrow's research in mass communication*. New York, NY: Appleton-Century-Crofts.

Cacioppo, J. T., Petty, R. E., Feinstein, J. A., & Jarvis, W. B. G. (1996). Dispositional differences in cognitive motivation: The life and times of individuals varying in need for cognition. *Psychological Bulletin*, *119*(2), 197–253. doi:10.1037/0033-2909.119.2.197

Cohen, E. L., & Lancaster, A. L. (2014). Individual differences in in-person and social media television coviewing: The role of emotional contagion, need to belong, and coviewing orientation. *Cyberpsychology & Behavior*, *17*(8). PMID:24950260

Comstock, G. (1989). *The evolution of American television*. Sage.

Darnell, M. J. (2015). The interactive TV experience: Where we came from and where we are going. In R. Nakatsu, M. Rauterberg, & P. Ciancarini (Eds.), *Handbook of digital games and entertainment technologies*. New York, NY: Springer. doi:10.1007/978-981-4560-52-8_52-1

Doughty, M., Rowland, D., & Lawson, S. (2011). Co-viewing live TV with digital backchannel streams. In *Proceedings of the 9th International Interactive Conference on Interactive Television* (pp. 141-144). New York, NY: Association for Computing Machinery. 10.1145/2000119.2000147

Fishbein, M., & Ajzen, I. (1975). *Belief, attitude, intention, and behavior: An introduction to theory and research*. Reading, MA: Addison-Wesley.

Flomenbaum, A. (2015). Accenture Report: 87% of consumers use second screen device while watching TV. *Adweek*. Retrieved from http://www.adweek.com/lostremote/accenture-report-87-of-consumers-use-second-screen-device-while-watching-tv/51698

Harboe, G., Metchalf, C. J., Bentley, F., & Tullio, J., Massey, N., & Romano, G. (2008). Ambient social TV: Drawing people into a shared experience. In M. Czerwinski, & A. Lund (Eds.), *Proceedings of the SIGCHI Conference on Human Factors in Computing Systems* (pp. 1-10). New York, NY: ACM. 10.1145/1357054.1357056

Harwood, J. (2007). *Understanding communication and aging: Developing knowledge and awareness*. Thousand Oaks, CA: Sage.

Henning, B., & Vorderer, P. (2001). Psychological escapism: Predicting the amount of television viewing by need for cognition. *Journal of Communication*, *51*(1), 100–120. doi:10.1111/j.1460-2466.2001.tb02874.x

Hunt, T. (2014). Has social media changed television? The Internet and social media have enhanced TV viewing. *Social media Explorer*. Retrieved from http://www.socialmediaexplorer.com/social-media-marketing/has-social-media-changed-television/

Jensen, J. F. (2008). The concept of interactivity - revisited: Four new typologies for a new media landscape. In *UXTV '08 Proceedings of the 1st International Conference on Designing Interactive User Experiences for TV and Video* (pp. 129-132).

Kinnunen, T., & Li, H. (2010). An overview of text-independent speaker recognition: From features to supervectors. *Speech Communication*, *52*(1), 12–40. doi:10.1016/j.specom.2009.08.009

Marathe, S., Sundar, S. S., Nije Bijvank, M., Van Vugt, H., & Veldhuis, J. (2007). Who are these power viewers anyway? Building a psychological profile. *Paper presented at the International Communication Association*, San Francisco, CA.

Marikyan, D., Papagiannidis, S., & Alamanos, E. (2019). A systematic review of the smart home literature: A user perspective. *Technological Forecasting and Social Change*, *138*, 139–154. doi:10.1016/j.techfore.2018.08.015

MarketingCharts. (2014). Are young people watching less TV? (Updated – Q4 2013 Data). Retrieved from http://www.marketingcharts.com/wp/television/are-young-people-watching-less-tv-24817/

Morley, D. (1986). *Family television: Cultural power and domestic leisure*. London, UK: Comedia.

Morley, D. (1988). Domestic relations: The framework of family viewing in Great Britain. In J. Lull (Ed.), *World families watch television* (pp. 22–48). Newbury Park, CA: Sage.

Nass, C., Steuer, J., & Tauber, E. R. (1994). Computers are social actors. In *Proceedings of the SIGCHI conference on Human factors in computing systems* (pp. 72-78). ACM.

Nathanson, A. I., Perse, E. M., & Ferguson, D. A. (1997). Gender differences in television use: An exploration of the instrumental-expressive dichotomy. *Communication Research Reports*, *14*(2), 176–188. doi:10.1080/08824099709388659

Newhagen, J. E., & Rafaeli, S. (1996). Why communication researchers should study the internet: A dialogue. *Journal of Communication*, *46*(1), 4–13. doi:10.1111/j.1460-2466.1996.tb01458.x

Nielsen. (2009). Online engagement deepens as social media and video sites reshape the Internet. Retrieved from http://blog.nielsen.com/nielsenwire/wp-content/uploads/2009/04/nielsen-online-global-_pr.pdf

Nielsen. (2014). Living social: How second screens are helping TV make fans. Retrieved from http://www.nielsen.com/us/en/insights/news/2014/living-social-how-second-screens-are-helping-tv-make-fans.html

Nielsen. (2015). Local watch: Where you live and its impact on your choice. Retrieved from http://go.brandlift.com/rs/vizu/images/q1-2015-local-watch-report.pdf

Petty, R. E., & Cacioppo, J. T. (1984). The effects of involvement on response to argument quantity and quality: Central and peripheral routes to persuasion. *Journal of Personality and Social Psychology*, *46*(1), 69–81. doi:10.1037/0022-3514.46.1.69

Petty, R. E., Cacioppo, J. T., & Kao, C. F. (1984). The efficient assessment of need for cognition. *Journal of Personality Assessment*, *48*(3), 306–307. doi:10.120715327752jpa4803_13 PMID:16367530

Rice, M., & Springett, M. (2015). Digital interactive television and the older generation. In R. Nakatsu, M. Rauterberg, & P. Ciancarini (Eds.), *Handbook of digital games and entertainment technologies* (pp. 1–28). Singapore: Springer. doi:10.1007/978-981-4560-52-8_38-2

Robinson, J. P. (2011). IT, TV and time displacement: What Alexander Szalai anticipated but couldn't know. *Social Indicators Research*, *101*(2), 193–206. doi:10.100711205-010-9653-0 PMID:21475389

Rogers, E. M. (1962). *Diffusion of innovations*. New York, NY: Free Press.

Saad, L. (2013). TV Is Americans' main source of news. Retrieved from http://www.gallup.com/poll/163412/americans-main-source-news.aspx

Samsung Tomorrow. (2015). History of Samsung smart TV. Retrieved from http://global.samsungtomorrow.com/infographic-history-of-samsung-smart-tv/

Shah, D. V., Hanna, A., Bucy, E. P., Wells, C., & Quevedo, V. (2015). The power of television images in a social media age: Linking biobehavioral and computational approaches via the second screen. *The Annals of the American Academy of Political and Social Science*, *659*(1), 225–245. doi:10.1177/0002716215569220

Shin, D.-H. (2009). An empirical investigation of a modified technology acceptance model of IPTV. *Behaviour & Information Technology*, *28*(4), 361–372. doi:10.1080/01449290701814232

Shin, D.-H. (2016). Do users experience real sociability through social TV? Analyzing parasocial behavior in relation to social TV. *Journal of Broadcasting & Electronic Media*, *60*(1), 140–159. doi:10.1080/08838151.2015.1127247

Shin, D.-H., Hwang, Y., & Choo, H. (2013). Smart TV: Are they really smart in interacting with people? Understanding the interactivity of Korean Smart TV. *Behaviour & Information Technology*, *32*(2), 156–172. doi:10.1080/0144929X.2011.603360

Shuhaiber, A., & Mashal, I. (2019). (in press). Understanding users' acceptance of smart homes. *Technology in Society*, 1–9. doi:10.1016/j.techsoc.2019.01.003

Singh, A. I., & Kaur, G. (2012). Motion detection method to compensate camera flicker using an algorithm. *International Journal of Computational Engineering Research*, *2*(3), 919–926.

Sundar, S. S., & Nass, C. (2001). Conceptualizing sources in online news. *Journal of Communication*, *51*(1), 52–72. doi:10.1111/j.1460-2466.2001.tb02872.x

T. R. A. C. Media. (1988). Demographics. Retrieved from http://www.tracmedia.com/Library/Concepts/Demographics/Default.aspx

Teo, T. S. (2001). Demographic and motivation variables associated with Internet usage activities. *Internet Research*, *11*(2), 125–137. doi:10.1108/10662240110695089

Terra, D. (2013). How the Internet works. Retrieved from http://www.giantbomb.com/forums/off-topic-31/how-the-internet-works-kinda-1432819/

U.S. Department of Labor. (2012). American time use survey – 2012 results. Retrieved from http://www.bls.gov/news.release/pdf/atus.pdf

Vandebosch, H., & Eggermont, S. (2002). Elderly people's media use: At the crossroads of personal and societal developments. *Communications*, *27*(4), 437–455. doi:10.1515/comm.2002.002

Wang, C.-H., & Chen, T.-M. (2018). Incorporating data analytics into design science to predict user intentions to adopt smart TV with consideration of product features. *Computer Standards & Interfaces*, *59*, 87–95. doi:10.1016/j.csi.2018.02.006

Wohn, D. Y., & Na, E. K. (2011). Tweeting about TV: Sharing television viewing experiences via social media message streams. *First Monday*, *16*(3). doi:10.5210/fm.v16i3.3368

Wu, G. (2005). The mediating role of perceived interactivity in the effect of actual interactivity on attitude toward the website. *Journal of Interactive Advertising*, *5*(2), 29–39. doi:10.1080/15252019.2005.10722099

Yang, H., Lee, W., & Lee, H. (2018). IoT smart home adoption: The importance of proper level automation. *Journal of Sensors*, *2018*. doi:10.1155/2018/6464036

Yang, Y., Liu, C., Li, C., Hu, Y., Niu, Y., & Li, L. (2014, July). The recommendation systems for smart TV. In *Proceedings of the 2014 International Conference on Computing, Communication and Networking Technologies (ICCCNT)* (pp. 1-6). doi:10.1109/ICCCNT.2014.6963095

Yassine, A., Singh, S., Hossain, M. S., & Muhammad, G. (2019). IoT big data analytics for smart homes with fog and cloud computing. *Future Generation Computer Systems*, *91*, 563–573. doi:10.1016/j.future.2018.08.040

Zhang, B., Wu, M., Kang, H., Go, E., & Sundar, S. S. (2014). Effects of security warnings and instant gratification cues on attitudes toward mobile websites. In *Proceedings of the 2014 Annual Conference on Human Factors in Computing Systems (CHI'14)* (pp. 111-114). 10.1145/2556288.2557347

Zhong, B. (2013). From smartphones to iPad: Power viewers' disposition toward mobile media devices. *Computers in Human Behavior, 29*(4), 1742–1748. doi:10.1016/j.chb.2013.02.016

Chapter 8
Is Construction of Timber Focused Exchange Traded Fund Plausible in SAARC Countries With Reference to India?

Nivedita Mandal
https://orcid.org/0000-0002-6877-9343
Adamas University, India

Rituparna Das
https://orcid.org/0000-0002-6877-9343
Adamas University, India

ABSTRACT

Given the fact that timber is neither traded as commodity nor is any timber-focused exchange-traded fund (ETF) available in the SAARC countries, the objectives of this chapter are to examine the plausibility of making use of the wood resources of SAARC countries in the form of ETFs, e.g. timber-focused ETF, the nature of the legal and policy frames in these countries as implemented in timber business with reference to their suitability for opening ways to construction of timber-focused ETF and the theoretical plausibility of constructing a timber-focused ETF with reference to India. It is found that in comparison with other SAARC countries, India has developed timber-based industries whose stocks are traded in the bourses and amenable to analysis for ETF construction purposes. While Bangladesh, Pakistan, Sri Lanka, and Nepal have certain legal and institutional artifacts for preservation and maintenance of forests toward gradual but by-and-large unplanned industrialization of the timber industry, Maldives and Bhutan have hardly any.

DOI: 10.4018/978-1-7998-0357-7.ch008

INTRODUCTION

Increasing trade in timber (timber, wood, lumber, log etc. are used synonymously each being conceived to be a member of generic class 'wood') and forest products support economic growth. The rate of growth of investment in timber can be one of the highest rates of growth owing to the continuous compounding. There are meager correlations between timber investments and assets. The price-expectation for timber may depend on the derivative contracts in lieu of econometric forecasting and hence may be used as a component in the investment portfolio in order to cut the portfolio risk. Further, timber investments may facilitate managing the risk of inflation. Overall timber investments are expected to moderate depreciation and volatility of the portfolio. The cash flow from timber investment is akin to a long-term discount bond. Within the timber component of the investment portfolio there may be further diversification in terms of alternative products like lumber and pulp wood. This however needs a vibrant, if not deep, timber-focused asset market.

TIMBER BASED FINANCIAL PRODUCTS

An exchange-traded fund is a hedging instrument against the risk of one specific stock, rather the return from it generates out of the concerned sector as a whole. As a result, not only institutional investors but also retail investors are benefited by it. Since investment in timber is immune to inflation, pays higher returns than do stocks, investment in land yields positive return even during recession and finally it is possible to invest in timber without physical possession of timber via investing in timber-focused exchange-traded funds (ETFs), the timber investment is increasingly receiving importance from the global investing communities. It can be proved that investments in the above products can enable an investor in portfolio diversification such as to achieve an equilibrium level of utility but the success of the investment depends on the financial and legal artifacts of the specific country. It is popular in the developed western countries. For example, in USA two popular timber funds are S&P Global Forestry & Timber Indexed Fund (symbol WOOD) and Timber ETF (symbol CUT). The proof is given below.

Suppose x is the weighted portfolio return in R. The investor's utility is $f(x)$. D is the set of constraints, for the time being, without any timber-focused ETF. If we now assume that timber-focused ETF is launched in the economy and the investor chooses to include it in her portfolio for diversification purpose, i.e. the ETF return moves in the opposite direction to all other returns, or it props up $f(x)$ during slump and pulls down during boom thereby pushing the investor towards a steady state equilibrium $f(x) = f(z)$

Proposition

Irrespective of the boom or slump in the economy the timber-focused ETF enables an investor to achieve the steady state equilibrium position.

Axiom

The portfolio returns are real numbers.

Proof

We assume that the timber-focused ETF is launched in the economy and the investor chooses to include in her portfolio for diversification purpose, i.e. the weighted return in the portfolio not containing the ETF and the ETF return are complementary to each other.

If there are two alternative outcomes where all the return is following a normal distribution:

1. The economy is going up wherein the investor's utility is approaching the maximum at some set of weighted portfolio return x such that $f(x) \geq f(y)$ for all $y \subset$ D,
2. The economy is going down wherein the investor's utility is approaching the minimum at some set of return z such that $f(z) \leq f(y)$ for all $z \subset$ D.

The ETF return moves in the opposite direction to all other returns, or it props up $f(x)$ during slump and pulls down during boom thereby pushing the investor towards a steady state equilibrium $f(x) = f(z)$ as depicted in Figure 1.

This is plausible when the weighted portfolio return in R, i.e. the set 'x' contains real numbers varying from negative infinity - ∞ to positive infinity + ∞ through density property, which share the same properties. Hence in pursuit of the duality principle as per De Morgan's Law the proposition is proved. At the same time the rate of return on timber-focused security should be more than the investor's minimum expected return. Let us consider an investor's lifetime utility maximization activity. If the Hamiltonian of Chiang and Wainwright (2005) is

$$H = U(C(t)e^{-\delta t}) + \lambda(t)[rK(t) - C(t)]$$

with the first order derivatives

$$\frac{\delta H}{\delta C} = U'(C)e^{-\delta t} - \lambda = 0$$
$$K' = rK(t) - C(t)$$

and

Figure 1. Steady State Equilibrium

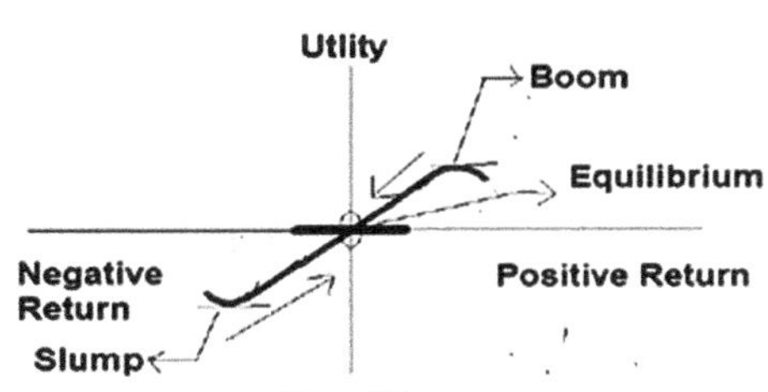

$$\lambda' = -\frac{\delta H}{\delta K} = -r\lambda$$

Chiang and Wainwright (2005) derived the following from above

$$\frac{-U''((C'(t))}{U'(C(t))}C'(t) = r - \delta$$

where *C(t)* is consumption as a function of time, *U(C(t))* is utility as a concave function of consumption, *K(t)* is wealth as a function of time, *r* is the market rate of interest, δ is the investor's personal rate of time preference in terms of Chiang and Wainwright (2005), i.e. the reservation rate or the minimum expected rate of return. Because of concave utility function $U'(t) \geq 0$ and $U''(t) \leq 0$, the sign of $C'(t)$ is positive or negative according to $r - \delta > 0$ or $r - \delta < 0$. If r is considered to be probabilistic, one may examine the daily return of a stock like Kitply over a year from 1 October 2015 to 20 September 2016. It follows Cauchy's distribution with mean 0 and variance 0.00989 in Figure 2.

The daily return

$$r = r_0 + \gamma\sqrt{(\frac{1}{p\pi y} - 1)}$$

where r_0 is location parameter. γ is scale parameter and $p = f(r)$ is the probability density function. The value of the scale parameter γ is too small due to a too narrow variance as less as below 1%. For the expression inside the square root sign to be positive, *p* need to be sufficiently small, i.e. for any unfavorable movement of the return on some day the probability is miniscule. This indicates the artifact of the stock market, the legal and policy frames of the country and the state of implementation of the

Figure 2. Cauchy distribution

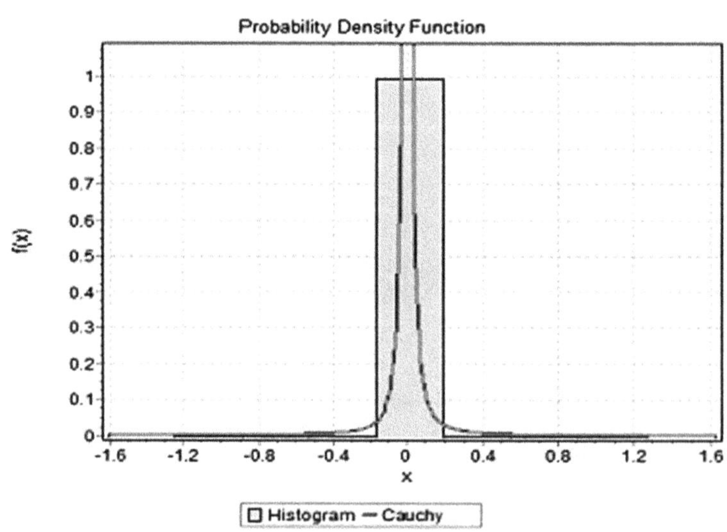

same. In India the special economic zone (SEZ) policy and export processing zone policy seem to be not implemented in timber business.

LITERATURE REVIEW

Ramage et al. (2017) presented a holistic picture ranging in scale from the science of the cell wall to the engineering and global policies that could maximise forestry and timber construction as a boon to both people and the planet. There are other following works that are noteworthy among those on economic perspectives of businesses and investments in timber, forestry and agriculture. Fisher (1930) is considered to be a textbook to start with on varying types of income streams and returns from farming, forestry etc and investments opportunity in the raising of a crop. There are several works afterwards on economics of forestry and natural resources but a detailed treatment of the time paths connected to investment horizon, interest cost, return and profit from decision-making view-point at different stages of development like feudalism and capitalism pertaining to timber business are available in Helmedag (2008). Before it, a work tracing back to the literature of political economy and then landing on the neoclassical paradigm of optimization was authored by Samuelson (1976). A detailed overview of commodity-based products for investment like exchange-traded funds is given in Engenka and Yuen (2008). A work on timber production and timber market catering around the optimum age of a tree for harvesting is available in Johansson and Lofgren (1985) as reflects in the review by Norbury (1986).

BACKGROUND

Performance of Timber-Focused ETF/Natural Resource-Focused ETF in Developed Countries

WOOD-iShares S&P Global Timber and Forestry Index ETF

The iShares Global Timber & Forestry ETF (WOOD) seeks to track the investment results of an index composed of global equities in or related to the timber and forestry industry. It is composed of 27 companies dealing with forestry products related to timber-focused ETF. During the first quarter of the financial year 2018-19, its net asset value (NAV) at NASDAQ varied from 77.28 to 78.28. The return chart is in Figure 3.

These companies are the largest publicly traded companies engaged in the ownership, management or upstream supply chain of forests and timberlands. These include forest product companies, timber real estate investment trusts ("REITs"), paper-product companies, paper packaging companies and agricultural product companies of the countries like: Brazil, Canada, Finland, Ireland, Japan, South Africa, Sweden, the United Kingdom (the "U.K.") and the USA (ishares, 2018).

CUT-Claymore/Beacon Global Timber Index ETF

This ETF provided by Beacon Indexes LLC gives investors indirect exposure to the price of timber through a basket of fifteen global equities which own or lease forested land and harvest the timber for

Figure 3. History chart of WOOD

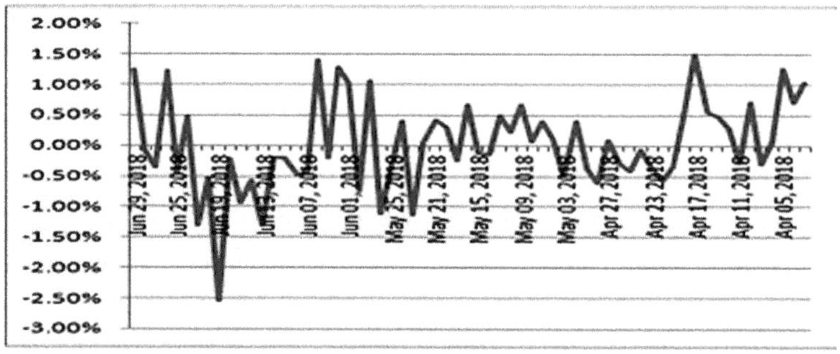

commercial use and sale of wood-based products. The profitability of these companies is linked to the prevailing market price of timber, and they often trade as a leveraged play on spot prices. CUT's holdings are internationally diversified in developed countries, and increased investment in infrastructure and a recovery in the housing market will further benefit the timber industry (ETFdb, 2018). During the first quarter of the financial year 2018-19, its closing price at NYSE varied from 32.08 to 32.43. The return chart is in Figure 4.

Timber Industry in SAARC Countries: An Unexplored Horizon

An ETF based on underlying timber, wood, etc, like gold ETF, cannot be constructed because wood or timber is not traded in any commodity exchange in India. In India, none of the four commodity exchanges: Indian Commodity Exchange (ICEX, 2019), The Multi Commodity Exchange (MCX, 2019), and National Commodity and Derivatives Exchange Ltd (NCDEX, 2019) deal with wood-based products.

In the Afganistan Global Commodity Exchange, the only bourse in Afghanistan no wood-based product is traded.

In Bangladesh, the only commodity bourse The Bangladesh Jute and Commodity Exchange does not deal with any wood-based product (BDCOMEX, 2018).

In Bhutan and Maldives, there is no commodity bourse, but there is the Commodity Trading Company (commoditiestradingcompany, 2019) which does not deal with any wood-based product.

Pakistan's only commodity bourse the Pakistan Mercantile Exchange does not deal with wood-based products (PMEX, 2019).

Figure 4. Historical chart of CUT

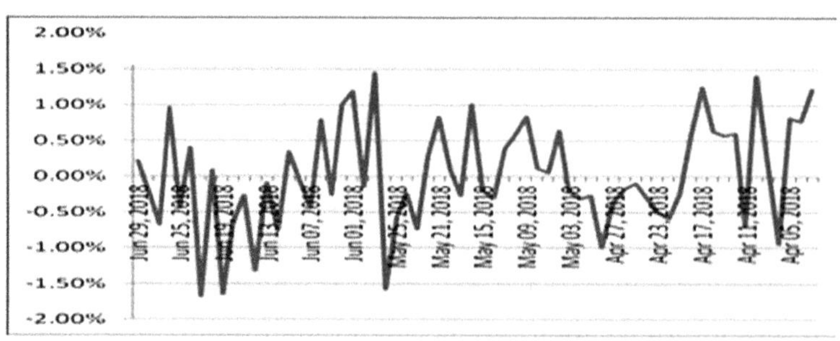

Sri Lanka does not have currently any commodity exchange.

The commodity bourses of Nepal, Mercantile Exchange Nepal Limited (Mercantile Exchange of Nepal Limited, 2019), Commodities Futures Exchange Limited (CFX, 2019) and Derivative and Commodity Exchange Nepal Limited (DCXNepal, 2019) do not deal with any wood-based product.

Business Environment of Timber Industry in India

The global hike in timber cost affected the timber industry (Sarda Plywood, 2017). It was expected that growth in the housing and infrastructure sector will not gain momentum in the days ahead. The same work observed timber to be a sensitive product as far as environmental policy is concerned world over, the availability of which depends on the licenses given by local authorities so much so that any delay or problem in issuing licenses by the governments of the countries from where timber is imported may affect the availability of raw material. Timber business is reported to be affected adversely by poor off-take from the country's real estate sector, postponement of purchase by consumers because of demonetization, pricing by informal sector producers 40% lower than Centuryply's product price (Centuryply, 2017). There are problems of interrupted power supply and hikes in the prices of the raw materials, fire woods and resins facing Mangalam Timber in Odisha. Due to the exposure of the Indians to global-ergonomically designed solutions in the residential and work places the increasing demand for space-saving multi-functional ready-to-use modular furniture solution is making timber and laminar products more and more unviable (Mangalam Timber, 2017). The risks of economic slowdown, decline in product demand, threat from unorganized players, increasing competition, quality contents, environmental issues, unavailability of raw material, inadequate funds in timber business are mentioned in Greenply (2018). There is still a considerable chunk of plywood business operated in the unorganized sector. Uniply Industries (2017) reports exposure to the risks posed by economic fluctuations, credit risk, labour risk, growing competition, exchange rate fluctuations and socio-political changes. However annual reports of all firms in general expressed happiness caused by goods and services tax (GST), which is expected to mitigate competition of the unorganized sector.

Globally billions of dollars in timber investments are lost each year because of downward pressure on timber prices caused by illegal logging. A further investigation to the factors behind the above kind of loss reveals that poor communities who are completely dependent on forests lose out to powerful interests, logging companies and migrant workers who reap most of the benefits. There is however little evidence that increased demand for forest products has brought some financial benefits for poor people living near forests. This is one of the reasons why investment in legal timber is not taking off in the emerging economies including India though increased demand for forest products has brought some financial benefits for poor people living near forests. The second reason is inefficiently small, lacking economies of scale and usage of outdated technology.

OBJECTIVES OF THE WORK

A survey of the mutual funds traded in India and the products traded in the commodity exchanges of India reveal that the timbers and timber derivative products are not present in India as well as in any other SAARC countries. So the present work is an attempt (a) to explore the plausibility of investment in natural resources in the form of ETFs, e.g. Timber-focused ETF in the SAARC countries and (b) to examine the nature of the timber industry stocks, the legal and policy frames and the state of implementation of the same and whether the special economic zone policy, export processing zone policy and similar policies in these countries are implemented in timber business and (c) to check the theoretical plausibility of constructing a timber linked ETF with reference to India.

INVESTMENT OPPORTUNITIES IN TIMBER-FOCUSED ETFS IN THE SAARC COUNTRIES

The SAARC region is expected to remain a net importer for at least the next 15 years because domestic demand is estimated to grow faster than domestic production. The gap between recorded supply and consumption is reportedly expected to widen further in the future, increasing reliance on imports. Forest law enforcement remains reportedly challenging in SAARC countries because of weak regulatory frameworks to control timber transport and verify the supply chain, a lack of resources and corruption in forest departments (EUFLEGT, 2017).

There are 15 timber and lumber companies in Bangladesh (Suppliers Directory, 2019a). The scope for woodworking machinery, furniture hardware, fitting & fixtures, panels & plywood, power tools, particle boards, wood composites and coatings is indicated by large scale presence of international companies. Reportedly the country's furniture sector is expanding both locally and globally and in the 90's the sector moved to mechanized-production oriented industries. Since then furniture business is reported to grow with modern machinery, innovative designs and use of diverse materials. Furniture industry of Bangladesh is reportedly shifting its focuses from dependency on teak for raw materials to processed wood. Around 60 per cent of raw materials, including timber, hardware, wood coating materials, hardware & accessories and fabrics, are imported. Exports have jumped by 253 per cent during the last five years. The demand has been reportedly rising in the international markets, especially in the US and the EU. At present about 41,560 enterprises and nearly two lakh skilled and semi-skilled people are reportedly engaged in this sector. Bangladesh Furniture Export Association (BFEA) and Bangladesh Furniture Industries Owners Association (BFIOA) are reported to be actively working on materializing the potential of this sector. Bangladesh Furniture Export Association (BFEA) has 12 members. The government of Bangladesh has reportedly declared the furniture industry as a 'Thrust Sector'. Apart from the large export-oriented enterprises, there is a number of furniture clusters reportedly in Dhaka at Badda, Sutrapur and Mirpur areas. A number of large companies reportedly have factories in Savar and Gazipur areas (BangladeshWood, 2019). There are 106 timber and lumber companies in Pakistan (Suppliers Directory, 2019b). Pakistan Stock Exchange is yet to launch ETF (The Express Tribune, 2018). There are 9 companies in Afghanistan dealing with timber and lumber (Suppliers Directory, 2019c). In Afghanistan businesses are disturbed by insurgents and illegal trade. In Maldives there are six business concerns (Timberweb, 2019a). In Bhutan there are two businesses (Timberweb, 2019b). In Sri Lanka there are 28 timber and lumber companies *Suppliers Directory, 2019d). In Nepal the timber industry is in nascent stage and in the category of cottage industry. The firms are mostly small-scale. The factories work with old machines (FAO, 2019a). Inferences about similar conditions can be made in Maldives, Bhutan and Sri Lanka.

The SAARC countries together have 18.2% forest area out of total land areas. It is reportedly a reservoir of great biodiversity, in and outside forests, and has untapped potential to develop the use of trees outside the forest. The subregion supports about 22 percent of the global population. India has reportedly the largest area of plantations in the sub-region for the production of industrial raw material and fuelwood, and Bhutan has the lowest plantation area. The current level of private planting reportedly exceeds public planting, which focuses on satisfying social (conservation and environmental) rather than commercial needs. This has reportedly changed the landscape picture across the sub-region over the last two decades. The most preferred plantation species in India, Bangladesh and Sri Lanka have been teak and eucalyptus while in Pakistan and Nepal it has been sheesham. The ability of natural forests to meet domestic timber and fuelwood requirements is reported to be continuously declining. The unsatisfied requirements are often reportedly met from private plantations or from illegal ad hoc harvesting in natural forests. Uncontrolled access and excessive use of forest resources in many places is leading to forest degradation, fragmentation and deforestation (FAO, 2019b). For creation of timber-focused ETF, there need to be a formal timber industry where the firms are listed in the stock exchanges, because the performances of the funds track the performances of the underlying stocks. The absence of interest in any of the SAARC countries in launching ETF despite the existence of reasonably sufficient forest resources could be traced to the stages of growth in these countries in line with W W Rostow, where only India is reported to have reached the last stage (Merinews, 2019). Costa et al. (2016) postulated the new theory of stages of growth – Stage 0. Malthusian trap; Stage 1. taking off into growth, Stage 2. catching up to the economic leader UK and Stage 3. joining the Economic Lead UK.[1] The aforesaid two ETFs, WOOD and CUT are traded in USA perhaps because USA is in the last stage of economic development (Sachs, 2004).

NATURE OF THE TIMBER INDUSTRY: LEGAL & POLICY FRAMEWORK

Availability of timber-focused ETF requires existence of vibrant timber business which in turn requires conducive policy and incentives from the government. In terms of the availability of forest land in the country the positions of the SAARC members in 2000 are furnished in Table 1.

India holds third position, but nevertheless India does not have any policy to encourage timber plantation although India has a National Policy on Biofuel (Press Information Bureau, 2018) that requires rise in sugar cane production for the purpose of ethanol generation. In India the farmers are not willing for timber plantation. They prefer to produce wheat and paddy due to absence of any support price for timber vis-à-vis other crops (Chaba & Jagga, 2016). In Afghanistan there are five listed timber and lumber companies. Bader et al. (2013) observed that the comingling influences of culture, tradition, and economic necessity, have made active and willing participants of local communities in the illegal commercial timber trade and natural forest ecosystems in Afghanistan are experiencing accelerated disturbance from timber harvest since insurgents began taking over the illegal timber trade. Bangladesh is known as net importer of wood. The policy to promote wood cultivation of timber business is not much known here but forestry and environmental education now receiving high priority from the government makes people aware of the benefits of timber and wood businesses (Choudhury & Hossain, 2011). Pakistan's Forestry Sector Master Plan aims at afforestation and reforestation (ADB, 2010). In Sri Lanka timber trade liberalization reduced deforestation (Weerahewa & Gunatilake, 2010). Maldives does not have any national forest programme. Forest management for timber production reportedly requires a long planning horizon. Long-term commercial lease on uninhabited islands is reportedly expected to be in the Maldivian circumstances the most suitable land tenure instrument, provided that the lease period is long enough to motivate for investments in tree planting and in improvement of the existing forest

Table 1. Share of forests in total land

Country/Area	Land Area	Forest Area 2000				
		Natural Forest	Forest Plantation	Total Forest		
	000 ha	*000 ha*	*000 ha*	*000 ha*	*%*	*ha per capita*
Bangladesh	13017	709	625	1334	10.2	n.s.
Bhutan	4701	2995	21	3016	64.2	1.5
India	297319	31535	32578	64113	21.6	0.1
Maldives	30	1	-	1	3.3	n.s.
Nepal	14300	3767	133	3900	27.3	0.2
Pakistan	77087	1381	980	2361	3.1	n.s.
Sri Lanka	6463	1625	316	1940	30	0.1
Total South Asia	**412917**	**42013**	**34652**	**76665**	**18.6**	**0.1**

Source: FAO, http://www.fao.org/docrep/004/y1997e/y1997e0s.htm

stands (Forestry Department, 2010). In Nepal in the aim to mitigate climate change by curbing carbon dioxide emissions stemming from the destruction of forests, the reducing emissions from deforestation and forest degradation, and enhancing forest carbon stocks in developing countries' (REDD+) – a proposed forest-carbon offsetting mechanism having emerged through ongoing United Nations Framework Convention on Climate Change (UNFCCC) negotiations is being implemented (ICIMOD, 2016). Bhutan has systematically codified rules on forests and uses of forestry but there is no specific incentive to promote forestry based industry (Royal Government of Bhutan, 2006)

CONSTRUCTION OF TIMBER LINKED ETF IN INDIA

Theoretical Framework

Sharpe Ratio

It is the ratio of the difference between the average return of the security and the risk-free return to the standard deviation of the security (Brealey, Myers, & Allen, 2011). Standard deviation is the measurement of risk of the stock. It denotes how much excess return over the risk-free return is expected by an investor against every unit of risk she bears. In the case where the above difference is positive, higher the quotient of the ratio, the riskier is the asset and vice versa. Sharpe ratio is a measure of risk in a mutual fund (Srivastava, 2011)

Efficient Frontier

An efficient frontier is the locus of the combinations of risk and return of a portfolio, where the return is the highest given the risk. In a portfolio of tradable or traded financial securities, the portfolio return is measured as the weighted average of the returns of the underlying securities and the risk is measured as the weighted standard deviation of the individual standard deviations of the same. Given a level of risk, the portfolio return that exceeds the risk-free return is considered an efficient portfolio. The higher the difference over the risk-free rate, the more efficient is the portfolio. Investments in stocks are considered to be for long run. The 10-year sovereign yield is generally considered to be the long-run, risk-free rate.

Timber-Focused ETF

If tradable financial securities are classified in terms of risk-return profile, mutual funds and ETFs are assets that are less risky than an ordinary share but are expected to earn a higher return than fixed income securities. A timber-focused ETF may be based on the prices of shares of companies selling products of timber, wood and plywood.

Listed and Traded Timber Stocks in India

In India there are following companies dealing with the products of wood, timber and plywood, listed in Bombay Stock Exchange (BSE):

1. Mangalam Timber
2. Woodsvilla
3. Dhabriya Polywood
4. National Plywood
5. Sterling Greenwoods
6. Sarda Plywoods

There following six timber and wood companies listed in the National Stock Exchange (NSE):

1. Greenply industries
2. Uniply industries
3. Kitply
4. Archid Ply
5. Century Plyboards
6. Mangalam Timber

METHODOLOGY

The risk of the scrips are tested in terms of ascertaining probability distributions and unit root tests using @risk and Eviews softwares and portfolio construction using 'solver'.

SAMPLE SELECTION

The above list is exhaustive. Out of the above, the daily trading prices are available on Archid Ply, Century Plyboards, Mangalam Timber, Greenply and Uniply in the NSE. There is no data available on Kitply.

The rest five are not traded. Mangalam Timber is traded in both the bourses - BSE and NSE. National plywood was traded only Aug. 10, 2018 last year (Yahoo finance, 2018a). Similarly, VR WoodArt was traded only on Dec. 4, 2017 in the last one year (Yahoo finance, 2018b). As per BSE (2018) in last one year Sterling Greenwoods was traded on six days, Woodsvilla was traded only in July 2018 and National Plywood has been traded in July and August in last one year and Dhabriya Polywood was very thinly traded in May 2018 and not at all traded in June 2018.

Therefore, based on frequency of trading, regularity of trading, and availability of data, the following scrips (or stocks or shares) are considered for further analysis:

1. Archidply (Archid)
2. Centuryply (Century)
3. Greenply (Green)
4. Mangalamtimber (Mangalam)
5. Uniply
6. Sardaply (Sarda)

DATA COLLECTION

The price of a security on a day means on a day when the security has been traded. The probability distribution of the daily data round the year is examined above for a scrip like Kitply. The risky character of the data, as is found below, continues to be similar even for samples of reasonable size like the sample of a quarter. The daily closing prices of the aforesaid six stocks in the above list for the quarter from April 1, 2018 to June 30, 2018 are collected from the sources like NSE, BSE and yahoo finance[2].

There are few issues in data filtration:

1. Mangalam Timber is traded in both NSE and BSE. So the quantity-weighted closing prices have been considered.
2. On May 20, 2018, trading did not take place both in the government bond and in the stock market. Hence the non-trading day is omitted from the sample.
3. Again on April 30, 2018, the government bond was not traded, but stocks were traded. It has been assumed that the government security's traded return does not change till the next day.

DATA ANALYSIS AND RESULTS

Stage I: Preliminary Analysis

From the daily closing prices of the stocks, the daily return (r) is calculated. For the period from April 1, 2018 to June 30, 2018, the daily yield, i.e. the risk-free return r_f of the 10-year sovereign bond is collected from NSE. During the above quarter this was the most popular risk-free security in terms of the maximum traded volume in 2018 (Investing.com, 2018).

The time series data in the selected sample scrips do not exhibit any trend. Rather it appears to be white noise in Figure 5.

Figure 5. Historical chart of returns

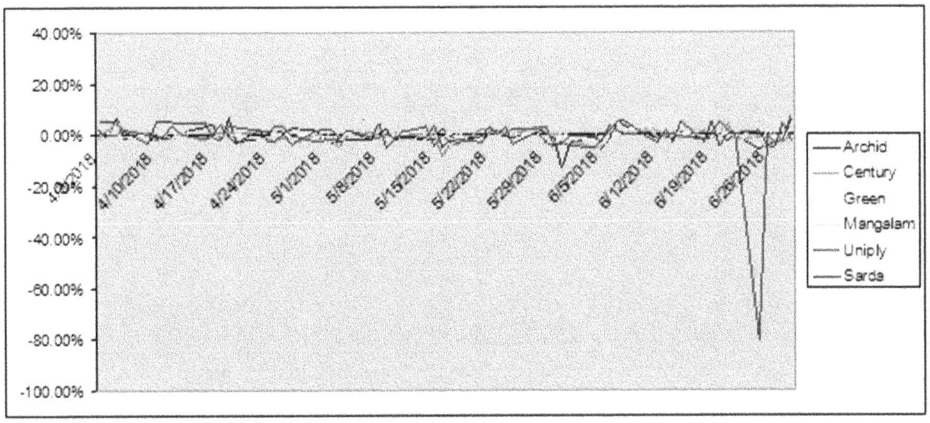

This means an investor cannot expect any considerable gain or loss. The average return of all the selected scrips is close to zero. There are two outliers - one with Mangalam and another with Archidply.

Now the mean (μ) and the standard deviation (σ) of the return series of each of the six stocks are calculated for the purpose of calculating Sharpe ratio.

It is found that the risk-free return exceeds the stock return for all the six stocks and the mean returns are negative except for Mangalam and Sarda as shown in Table 2.

Stage II: Efficient Portfolio for ETF

For construction of an ETF, it is necessary to construct an efficient portfolio of the stocks. In this case to construct an ETF linked with timber, a portfolio of 6 selected timber stocks in India are considered.

Following Docherty (2016) in line with Cvitani'c and Fernando (2004), an attempt has been made to construct an efficient portfolio with these stocks using 'solver' tool in MS Excel in Figure 6.

At first, random weights are assigned to each stock, the total weight being unity. Then simulation is run using 'solver' in order to find that combination of weights that maximizes the portfolio return.

For most of the trading days of the sample data, the result suggests to choose only a single stock, rather than a portfolio to maximize the return. The solver results over last two business weeks of the sample 18-29 June 2018 are given in Table 3.

From this, it can be seen that for each day the portfolio return is equal to the daily return of any one particular stock which has the highest positive return on that day compared to other stocks considered here. This proves the implausibility of a valid and efficient ETF linked to timber stocks in India.

Table 2. Mean return of the stocks

Risk Free	Archid	Century	Green	Mangalam	Uniply	Sarda
7.59%	-0.24%	-0.35%	-0.40%	0.43%	0.25%	0.23%

Figure 6. Solver result on the last day of the sample

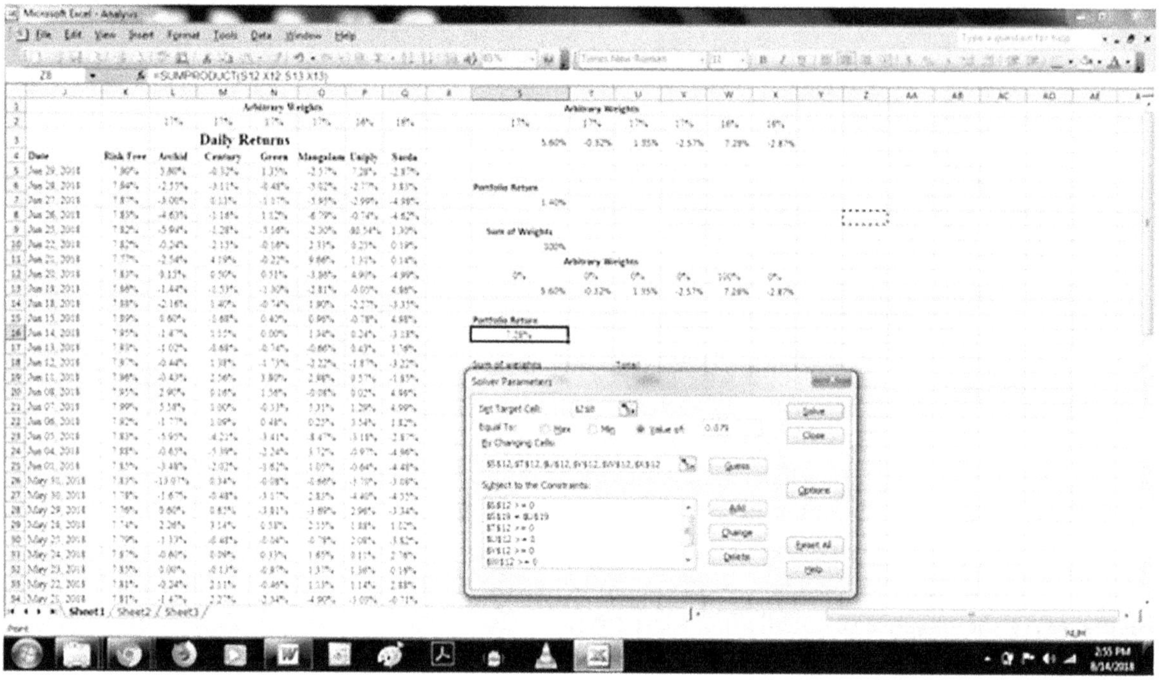

The reason might be that on most of the trading days, almost all the stocks exhibit negative returns, except a few with positive returns, albeit much lesser than the risk-free return. Hence the portfolio return is very likely to boil down to the individual stock return, which in turn is similar to invest in any individual stock. Moreover in India the timber stocks are the poor players in the overall market.

Table 3. Solver results on last two weeks of the sample

Date	Risk Free	Archid	Century	Green	Mangalam	Uniply	Sarda	Solver Results
Jun 29, 2018	7.90%	5.60%	-0.32%	1.35%	-2.57%	7.28%	-2.87%	7.28%
Jun 28, 2018	7.94%	-2.55%	-3.11%	-0.48%	-5.02%	-2.77%	3.83%	3.83%
Jun 27, 2018	7.87%	-3.00%	0.13%	-1.17%	-5.95%	-2.99%	-4.98%	0.13%
Jun 26, 2018	7.83%	-4.63%	-1.16%	1.12%	-6.79%	-0.74%	-4.62%	1.12%
Jun 25, 2018	7.82%	-5.94%	-1.28%	-5.16%	-2.30%	######	1.30%	1.30%
Jun 22, 2018	7.82%	-0.24%	-2.15%	-0.16%	2.33%	0.25%	0.19%	2.33%
Jun 21, 2018	7.77%	-2.54%	4.19%	-0.22%	9.66%	1.31%	0.14%	9.66%
Jun 20, 2018	7.83%	0.15%	0.50%	0.51%	-3.86%	4.90%	-4.99%	4.90%
Jun 19, 2018	7.86%	-1.44%	-1.53%	-1.30%	-2.81%	-0.05%	4.86%	4.86%
Jun 18, 2018	7.88%	-2.16%	1.40%	-0.74%	1.90%	-2.27%	-3.35%	1.90%

Stage III: Calculation of Sharpe Ratio

The Sharpe ratio is the ratio of the difference between the security return and the risk free return to the standard deviation of the former

$$S = \frac{r - r_f}{\sigma}$$

In this work, it has been found that the numerator value of the Sharpe Ratio for most of the stocks is negative, except Sarda and Mangalam with negligible small values (0.05%, and 0.13% respectively).

The mean returns of all the stocks are much less than the risk free rate of return. Only Uniply has 7.28% on the last day of the sample June 29, 2018, which is the closest to the risk free return compared to other stocks.

So the results from Sharpe ratio prove that the investors do not have enough incentive to invest on those stocks.

Stage IV: Limitation of Efficient Portfolio Construction using Solver

We construct 95% level confidence intervals the closing returns of the selected scrips in the sample considered here. We find that the return of Mangalam can rise up to 19.27% and that of Uniply can rise up to 10.87%. We also run unit root tests of these scrips and the NSE Infra index because the major demand from timber and wood products arise from infrastructure industry as a whole not only from construction or realty industry. The covarinaces of returns individual scrips with NSE infra index return are stronger than with NSE realty index return. The unit root tests reject all hypotheses of existence of unit roots. Then we calculate betas 'β' of these scrips with respect to NSE Infra index. The results go against the results of application of solver software in portfolio optimization in Table 4.

First of all solver does not recommend trading of Mangalam and Uniply everyday during the sample period, but the confidence interval suggests acquiring at least Mangalam everyday and may be Uniply also. But if the NSE infra index is in the portfolio, acquiring Uniply is at all not recommended because of its highest beta with the former.

DISCUSSION

This work uses time series data on scrips. There are several doctrines and models available to apply for analysis and processing of above data in the context of efficient portfolio construction. An investor is making rational expectation before investing means she is trying to insulate her decision against systemic shocks. Her decision may be vulnerable against shocks arising out of the factors specific to her. In this case the knowledge on unit root of the time series is as imperative as is the knowledge on whether the present values of the series carry the memory of the past values, i.e. lagged values. If the rational expectation comes to be true there should be long term relationship between the past values and the future realizations. The private investor here learns from past lessons and updates the forecasting techniques with new models and parameters. In this work the sample size is 90 days of a quarter. If this is considered

Table 4. Confidence interval and beta

Date	Archid	Century	Green	Mangalam	Uniply	Sarda	NS Infra	NSE Realty
29-Jun-18	5.60%	-0.32%	1.35%	-2.57%	7.28%	-2.87%	2.42%	2.43%
28-Jun-18	-2.55%	-3.11%	-0.48%	-5.02%	-2.77%	3.83%	-0.27%	-2.21%
27-Jun-18	-3.00%	0.13%	-1.17%	-5.95%	-2.99%	-4.98%	-2.04%	-1.79%
26-Jun-18	-4.63%	-1.16%	1.12%	-6.79%	-0.74%	-4.62%	0.22%	-0.72%
25-Jun-18	-5.94%	-1.28%	-5.16%	-2.30%	-80.54%	1.30%	-1.52%	-1.49%
22-Jun-18	-0.24%	-2.15%	-0.16%	2.33%	0.25%	0.19%	0.84%	0.16%
21-Jun-18	-2.54%	4.19%	-0.22%	9.66%	1.31%	0.14%	-1.00%	-0.79%
20-Jun-18	0.15%	0.50%	0.51%	-3.86%	4.90%	-4.99%	0.04%	1.10%
19-Jun-18	-1.44%	-1.53%	-1.30%	-2.81%	-0.05%	4.86%	-0.84%	-2.01%
18-Jun-18	-2.16%	1.40%	-0.74%	1.90%	-2.27%	-3.35%	-0.38%	-0.38%
15-Jun-18	0.60%	-1.68%	0.40%	0.96%	-0.78%	4.98%	-0.95%	-1.27%
14-Jun-18	-1.47%	1.15%	0.00%	1.34%	0.24%	-3.18%	-0.78%	0.00%
13-Jun-18	-1.02%	-0.68%	-0.74%	-0.66%	0.43%	1.76%	-0.70%	-0.05%
12-Jun-18	-0.44%	1.38%	-1.73%	-2.22%	-1.87%	-3.22%	0.61%	0.41%
11-Jun-18	-0.43%	2.56%	3.80%	2.98%	0.57%	-1.85%	0.21%	-0.62%
8-Jun-18	2.90%	0.16%	1.56%	-0.08%	0.02%	4.98%	-0.33%	0.36%
7-Jun-18	5.58%	1.00%	-0.33%	5.31%	1.29%	4.99%	0.57%	2.97%
6-Jun-18	-1.77%	1.09%	0.48%	0.25%	3.54%	1.82%	1.42%	1.51%
5-Jun-18	-5.95%	-4.21%	-3.41%	-8.47%	-3.18%	-2.87%	-1.57%	-1.05%
4-Jun-18	-0.65%	-5.39%	-2.24%	3.72%	-0.97%	-4.96%	-1.69%	-3.33%
1-Jun-18	-3.48%	-2.02%	-1.62%	1.05%	-0.64%	-4.48%	-0.65%	-1.25%
31-May-18	-13.07%	0.34%	-0.08%	-0.66%	-3.70%	-3.08%	0.07%	-0.71%
30-May-18	-1.67%	-0.48%	-3.17%	2.83%	-4.40%	-4.55%	-0.53%	-0.02%
29-May-18	0.60%	0.65%	-3.81%	-3.69%	2.96%	-3.34%	-0.09%	-0.98%
28-May-18	2.26%	3.14%	0.58%	2.15%	1.88%	1.02%	1.22%	1.54%
25-May-18	-1.33%	-0.48%	-0.04%	-0.78%	2.08%	-3.82%	1.05%	1.27%
24-May-18	-0.60%	0.09%	0.33%	1.65%	0.11%	2.76%	0.64%	-0.70%
23-May-18	0.00%	-0.13%	-0.97%	1.37%	1.36%	0.16%	0.05%	-0.86%
22-May-18	-0.24%	2.11%	-0.46%	1.13%	1.14%	2.88%	0.28%	1.23%
21-May-18	-1.47%	2.27%	-2.34%	-4.90%	-3.03%	-0.71%	-0.60%	-3.29%
18-May-18	-1.34%	-3.57%	-1.14%	0.51%	-1.93%	-3.34%	-2.08%	-1.50%
17-May-18	-0.52%	-3.89%	0.28%	-4.55%	-3.34%	-2.03%	-0.48%	0.10%
16-May-18	2.37%	-8.48%	-1.28%	10.41%	-0.09%	-4.97%	-0.54%	2.06%
15-May-18	-1.75%	-0.10%	-0.39%	-1.35%	-4.40%	3.63%	-0.29%	-2.00%
14-May-18	-1.55%	0.00%	-1.84%	0.13%	3.04%	-1.28%	-0.40%	-0.70%
11-May-18	-1.13%	-0.33%	0.00%	-3.68%	1.20%	-0.18%	-0.23%	-0.50%
10-May-18	-2.38%	0.85%	0.00%	0.72%	-1.30%	-1.65%	-1.03%	-1.97%
9-May-18	1.92%	-1.17%	-2.12%	-1.01%	0.14%	-4.95%	-0.01%	-0.09%
8-May-18	0.40%	-0.18%	-0.71%	2.76%	-1.13%	4.75%	-0.31%	1.18%
7-May-18	-2.21%	0.54%	0.31%	-0.31%	-0.80%	-1.72%	0.73%	1.11%
4-May-18	-1.58%	-0.21%	0.46%	0.24%	1.59%	1.46%	-0.59%	-0.14%

continued on following page

Table 4. Continued

Date	Archid	Century	Green	Mangalam	Uniply	Sarda	NS Infra	NSE Realty
3-May-18	-2.45%	-5.02%	-0.58%	1.70%	-1.52%	-1.37%	-1.01%	-1.66%
2-May-18	-3.14%	-0.61%	-0.37%	-6.55%	-1.88%	1.93%	-0.67%	-2.00%
30-Apr-18	1.46%	0.37%	-0.34%	24.71%	-2.72%	1.45%	0.92%	1.58%
27-Apr-18	2.46%	-3.39%	0.85%	-6.81%	-0.44%	-3.51%	0.89%	0.35%
26-Apr-18	0.92%	-0.52%	-0.24%	-12.26%	3.13%	0.14%	-0.95%	-0.69%
25-Apr-18	2.21%	-2.69%	0.15%	2.85%	3.46%	-2.03%	-0.23%	0.60%
24-Apr-18	0.56%	-1.75%	-0.60%	-1.31%	1.00%	-2.81%	0.42%	0.40%
23-Apr-18	-1.26%	1.04%	-1.50%	-1.13%	1.60%	2.36%	-0.11%	1.61%
20-Apr-18	-2.77%	0.45%	0.22%	0.46%	3.28%	-3.15%	-0.81%	-0.96%
19-Apr-18	6.53%	1.28%	-0.21%	-1.25%	3.45%	-0.13%	1.01%	0.64%
18-Apr-18	-2.17%	0.85%	0.18%	-1.51%	-2.16%	3.99%	0.12%	0.57%
17-Apr-18	4.47%	0.77%	-0.54%	1.95%	-0.89%	-0.07%	0.32%	1.19%
16-Apr-18	3.17%	-2.41%	-1.14%	4.73%	-1.10%	4.99%	0.65%	1.63%
13-Apr-18	1.21%	0.45%	0.06%	1.21%	0.06%	4.57%	0.00%	0.45%
12-Apr-18	-0.78%	-1.69%	-1.01%	-0.62%	3.58%	5.00%	-0.15%	-1.92%
11-Apr-18	-1.01%	0.40%	0.88%	-12.67%	-0.42%	5.00%	-0.01%	-0.39%
10-Apr-18	-1.00%	-2.14%	-0.13%	7.42%	-0.05%	4.98%	0.88%	1.09%
9-Apr-18	1.37%	0.82%	0.19%	10.70%	0.09%	-3.09%	0.28%	-0.03%
6-Apr-18	-1.41%	1.49%	-0.70%	4.05%	2.30%	1.42%	-0.35%	0.56%
5-Apr-18	7.06%	4.63%	0.74%	4.86%	0.33%	4.96%	1.55%	2.62%
4-Apr-18	-1.31%	2.25%	1.27%	-1.44%	0.29%	4.97%	-1.66%	-1.31%
3-Apr-18	2.16%	-1.45%	0.32%	-2.36%	1.00%	4.97%	0.86%	0.51%
	-0.47%	-0.41%	-0.46%	0.13%	-1.16%	0.05%	-0.12%	-0.13%
	Mean	Mean	Mean	Mean	Mean	Mean	Mean	
	3.09%	2.25%	1.36%	5.48%	10.42%	3.46%	0.89%	1.38%
	SD	SD	SD	SD	SD	SD	SD	
	S	S	S	S	S	S	S	
	0.01%	0.01%	0.00%	0.01%	0.03%	0.00%		
	Cov	Cov	Cov	Cov	Cov	Cov	with NSE infra	
	1.49%	0.61%	0.53%	1.06%	2.86%	0.55%		
	β	β	β	β	β	β	with NSE infra	
	UB	UB	UB	UB	UB	UB	UB: Upper bound of confidence interval	
	5.59%	4.01%	2.20%	10.87%	19.27%	6.82%		
	LB	LB	LB	LB	LB	LB	LB: Lower bound of confidence interval	
	-6.54%	-4.83%	-3.12%	-10.60%	-21.59%	-6.73%		
	0.00%	0.01%	0.00%	-0.02%	0.00%	0.00%		
	Cov	Cov	Cov	Cov	Cov	Cov	with NSE realty	
	0.20%	0.44%	-0.24%	-1.50%	-0.26%	0.15%		
	β	β	β	β	β	β	with NSE realty	

Table 5. Unit root test results

Dickey-Fuller test (ADF(stationary) / k: 10 / -3.63636363636364E-02):	
Tau (Observed value)	-11.276
Tau (Critical value)	-3.392
p-value (one-tailed)	< 0.0001
alpha	0.05
Test interpretation:	
H0: There is a unit root for the series.	
Ha: There is no unit root for the series. The series is stationary.	
As the computed p-value is lower than the significance level alpha=0.05, one should reject the null hypothesis H0, and accept the alternative hypothesis Ha.	
Phillips-Perron test (PP(no intercept) / Lag: Short / -3.63636363636364E-02):	
Tau (Observed value)	-29.671
Tau (Critical value)	-1.941
p-value (one-tailed)	< 0.0001
alpha	0.05
Test interpretation:	
H0: There is a unit root for the series.	
Ha: There is no unit root for the series. The series is stationary.	
As the computed p-value is lower than the significance level alpha=0.05, one should reject the null hypothesis H0, and accept the alternative hypothesis Ha.	

Table 6.

KPSS test (Level / Lag Short / -3.63636363636364E-02):	
Eta (Observed value)	0.143
Eta (Critical value)	0.475
p-value (one-tailed)	0.424
alpha	0.05
Test interpretation:	
H0: The series is stationary.	
Ha: The series is not stationary.	
As the computed p-value is greater than the significance level alpha=0.05, one cannot reject the null hypothesis H0.	

to be the length of short run and compared with the length of 1000 days or more, i.e. 16 quarters or 4 years considered to be the long run, the nature of (non-)stationarity should be different between the two. Using Eviews software this has been tested for Mangalam Timber Daily Closing Prices of 1022 days starting from Aug. 22, 2014 to Aug. 31, 2018 and still the data is found stationary in Table 5.

The stationary nature of data in short as well as long runs may lead to inference that the Mangalam scrip is left out of systemic developments and hence immune to systemic shocks. The same examination can be performed on other scrips and as a whole some conclusion can be arrived for the timber industry as a whole.

CONCLUSION

This work examined the environment of timber businesses in the SAARC countries with reference to India from both local and global perspectives. None of the SAARC members seems to have conscious official initiative to promote timber business such as to enable the entry of timber and lumber in the financial markets as commodities or underlying commodities. Despite having over 18% of total lands as forests, the SAARC region has a whole appearance to be in the early stages of economic growth and development whereby a strong timber and lumber sector could be conceived to be composed of many companies listed in stock exchanges. The legal and policy frameworks as to promotion of timber are strong in those countries having codified forest rules and policies like Bhutan, India and Pakistan. Deforestation and illegal trade are working as retarding factors in the way of development of timber industries in other countries. Most of the India's listed timber companies are found to ail in terms of stock performance, far to think of ETF construction. It is not found to be plausible to construct an efficient portfolio of timber stocks of India. The statistical exercise using solver tool suggested investment in a single stock only for the purpose of a rational investor's attaining steady state equilibrium in either boom or slump but time series exercise of calculating beta of susceptibility of individual scrip toward the sectoral index suggests a portfolio of more than one scrip because of covariance reasons. However in absence of appropriate policy framework and implementation drive, it is not recommended to formulate or to invest in any timber-focused portfolio or ETF in India, rather one may invest in some chosen single timber scrip out of the whole sector which could play the role of hedging where the portfolio as a whole is expected to contain scrips of other sectors also. An appropriate policy framework should provide for higher investment rates, application of improved technologies and programmes for building multi-dimensional awareness of people such as to unlock forest finance, create sustainable landscapes and develop forests of the SAARC countries as channels of alternative finance. This work adds value to research on private investments in environment, which is otherwise rare because relatively long gestation gap because of which only governments spend on environmental conservation and throws light on the risk of investors in the timber stocks where for several days the return may be nil or negative except for one or maximum two scrips in an industry that is stagnating because of lack of government supports.

REFERENCES

ADB. (2010). Pakistan Forestry Sector Project. Retrieved from https://www.adb.org/sites/default/files/evaluation-document/35571/files/in343-10.pdf

AGCXL. (2019). Product. Retrieved from http://www.agcxl.com

Bader, H., Hanna, C., Douglas, C., & Fox, J. D. (2013). Illegal Timber Exploitation and Counterinsurgency Operations in Kunar Province of Afghanistan: A Case Study Describing the Nexus Among Insurgents, Criminal Cartels, and Communities Within the Forest Sector. *Journal of Sustainable Forestry*, *32*(4), 329–353. doi:10.1080/10549811.2013.767913

BDCOMEX. (2018). Introducing the First Commodity Exchange in Bangladesh. Retrieved from http://bdcomex.blogspot.com/p/about-us-bangladesh-jute-commodity.html</eref>

Brealey, R., Stewart, M., & Allen, F. (2011). Principles of Corporate Finance, New York, NY: McGraw-Hill Irwin, 191

BSE. (2018) Stock Prices. Retrieved from https://www.bseindia.com/markets/equity/EQReports/StockPrcHistori.aspx?scripcode=512289&flag=sp&Submit=G

Centuryply. (2018) Annual Report 2017. Retrieved from https://www.centuryply.com/files/download/43a5eb8b870d3cc

CFX. (2019). Products. Retrieved from www.cfxnepal.com

Chaba, A. A., & Jagga, R. (2016). Agro-forestry: Poplar's popularity dip, no takers for PM Modi's timber farming call. Retrieved from http://indianexpress.com/article/india/india-news-india/agro-forestry-poplars-popularity-dip-no-takers-for-pm-modis-timber-farming-call-2819456/

Chakroborty, S., & Digal, S. K. (2013). Analysis of Investment Pattern of Mutual Funds Investors – An Empirical Study in Orissa, *GITAM. Journal of Management*, *11*(2), 192–207.

Chiang, A. C., & Wainwright, K. (2005). *Fundamental Methods of Mathematical Economics* (pp. 645–646). Boston, MA: McGraw-Hill.

Choudhury, J. K., & Hossain, A. A. (2011). Bangladesh Forestry Outlook Study, *APFSOS Working Paper*. Retrieved from http://www.fao.org/3/a-am628e.pdf

Commodities Trading Company. (2019). Commodities Trading in Bhutan. Retrieved from http://www.commoditiestradingcompany.com/commodity_trading_in_bhutan.html

Costa, D., Kehoe, T. J., & Ravindranathan, G. (2016). The Stages of Economic Growth Revisited, *Economic Policy Paper* 16-5. *Federal Reserve Bank of Minneapolis*. Retrieved from https://www.minneapolisfed.org/~/media/files/pubs/eppapers/16-5/epp-16-5-stages-of-economic-growth-revisted-part1

Cvitani'c, J., & Fernando, Z. (2004). *Economics and Mathematics of Financial Markets* (pp. 17–159). Cambridge, MA: MIT Press.

DCXNepal. (2019). Products. Retrieved from www.dcxnepal.com

Docherty, P. (2016). Excel to derive the efficient frontier. Retrieved from https://www.youtube.com/watch?v=qV1lDqbUwPo

Dubey, P. (2009). Prospects and Challenges for the Emerging Timber Import Market in India. *Journal of Forestry*, *107*(1), 23–28.

Engenke, L., & Yuen, J. C. (2008). Types of Commodity Investments. In F. Fabbozi, R. Fuss, & D. G. Kaiser (Eds.), *The Handbook of Commodity Investing* (pp. 549–569). Hoboken, NJ: John Wiley and Sons.

<eref>BangladeshWood. (2019). Bangladesh - Rapidly Expanding Market. Retrieved from http://www.bangladeshwood.com/#

ETF.com. (2016). CUT Fund Report. Retrieved from http://www.etf.com/pdf-version/CUT

ETFdb. (2018). Invesco MSCI Global Timber ETF. Retrieved from http://etfdb.com/etf/CUT/

European Foreign Institute. (2017). The Vietnam-EU Voluntary Partnership Agreement. Retrieved from http://www.euflegt.efi.int/publications/the-vietnam-eu-voluntary-partnership-agreement

Facility, E. U. F. L. E. G. T. (2017). What is FLEGT. Retrieved from http://www.euflegt.efi.int/publications/cross-border-timber-trade-in-the-saarc-area

FAO. (2019a). The State of Forestry in Napal. Retrieved from http://www.fao.org/docrep/w7719e/w7719e04.htm

FAO(2019b). Chapter 23: South Asia. Retrieved from http://www.fao.org/docrep/004/y1997e/y1997e0s.htm

Fisher, I. (1930). *The Theory of Interest as Determined by Impatience to Spend Income and Opportunity to Invest it*. London, UK: Macmillan.

Forestry Department. (2010). Global Forest Resources Assessment Country Reports Maldives. Retrieved from http://www.fao.org/docrep/013/al559E/al559E.pdf

Global Canopy Programme. (2009). The Little Climate Finance Book. Retrieved from https://www.ecbi.org/sites/default/files/Little_Climate_Finance_Book_-_English-libre_pdf

Greenply. (2018). Annual Report 2018. Retrieved from http://www.greenply.com/images/pdf/Annual-Report-2017-18.pdf

Helmedag, F. (2008). The Optimal Rotation Period of Renewable Resources: Theoretical Evidence from the Timber Sector. In F. Fabbozi, R. Fuss, & D. G. Kaiser (Eds.), *The Handbook of Commodity Investing* (pp. 145–166). Hoboken, NJ: John Wiley and Sons. doi:10.1002/9781118267004.ch6

ICEX. (2019). Products. Retrieved from https://www.icexindia.com/static/products.aspx

ICIMOD. (2016) Building Timber Value Chains for REDD+ in Nepal. *ICIMOD Working Paper 2016/9*. Retrieved from http://lib.icimod.org/record/32385/files/WP%202016_9_Value%20chain.pdf

Investing.com. (2018). India 10-year Bond Yield. Retrieved from https://in.investing.com/rates-bonds/india-10-year-bond-yield-historical-data

Investment Company Institute. (2015). 2015 Investment Company Fact Book. Retrieved from http://www.icifactbook.org/fb_ch3.html

Investopedia. (2016). Understanding Sharpe Ratio. Retrieved from http://www.investopedia.com/articles/07/sharpe_ratio.asp

Ishares. (2018). Prospectus. Retrieved from https://www.ishares.com/us/library/stream-document?stream=reg&product=I-TIMBR&shareClass=NA&documentId=925986~925792~926067~925559~925569&iframeUrlOverride=%2Fus%2Fliterature%2Fprospectus%2Fp-ishares-global-timber-and-forestry-etf-3-31.pdf

ITTO. (2009). Encouraging Industrial Forest Plantations in the Tropics. Retrieved from http://www.itto.int/direct/topics/topics_pdf_download/topics_id=2165&no=0&disp=inline

Mangalam Timber. (2018). Annual Report 2017. Retrieved from http://www.mangalamtimber.com/images/investors_desk_2016-17.pdf

MCX. (2019). Market Activity. Retrieved from https://www.mcxindia.com/home

Menon, R. (2013). Taking to the Trees. Retrieved from http://indiatoday.intoday.in/story/gujarat-farmers-switch-to-highly-profitable-cultivation-of-eucalyptus-trees/1/391831.html

Mercantile Exchange of Nepal Limited. (2019). Products. Retrieved from http://www.mexnepal.com/webpages/products_mex.html

Merinews. (2019). Stages of Economic Growth: An Analysis. Retrieved from http://www.merinews.com/article/stages-of-economic-growth-an-analysis/15811108.shtml

NCDEX. (2019). Products. Retrieved from https://www.ncdex.com/index.aspx

Negi, M. (2016). Indian Forestry: 10 Main Problems Faced by the Indian Forestry. Retrieved from http://www.yourarticlelibrary.com/environment/forest/indian-forestry-10-main-problems-faced-by-the-indian-forestry/13861/

Norbury, F., Johansson, P.-O., & Lofgren, K.-G. (1986). The Economics of Forestry and Natural Resources. *American Journal of Agricultural Economics*, *68*(3), 760. doi:10.2307/1241574

Patel, B. H. (2012). Export Market Challenges for Indian Timber Products. Retrieved from https://www.chathamhouse.org/sites/files/chathamhouse/public/Research/Energy,%20Environment%20and%20Development/10022012Patel1.pdf

PMEX. (2019). Products. Retrieved from https://www.pmex.com.pk/

Press Information Bureau. (2018). Cabinet approves national policy on Biofuels. Retrieved from http://pib.nic.in/newsite/PrintRelease.aspx?relid=179313

Ramage, M., Burridge, H., Busse-Wicher, M., Fereday, G., Reynolds, T., Shah, D. U., ... Scherman, O. (2017). The wood from the trees: The use of timber in construction. *Renewable & Sustainable Energy Reviews*, *68*(Part 1), 333–359. doi:10.1016/j.rser.2016.09.107

Royal Government of Bhutan. (2006). Forest and Nature Conservation Rules of Bhutan. Retrieved from www.moaf.gov.bt/download/.../Foest-and-Nature-Conservation-Rules-2006.pdf

Sachs, J. D. (2004). Stages of Economic Development. *Speech at the Chinese Academy of Arts and Sciences*. Retrieved from http://www.earth.columbia.edu/sitefiles/file/about/director/documents/china_speech061904.pdf

Samuelson, P. (1976). Economics of Forestry in an Evolving Society. *Economic Enquiry*, *14*(4), 466–492. doi:10.1111/j.1465-7295.1976.tb00437.x

Sarda Plywood. (2017). Annual Report 2017. Retrieved from https://www.sardaplywood.in/investors.php

Saunders, A., & Cornett, M. M. (2008). *Financial Institutions Management*. New York, NY: McGraw-Hill Irwin, 214-220

S&P Global. (2016). *S&P Global Timber and Forestry Index*. Retrieved from http://us.spindices.com/indices/equity/sp-global-timber-and-forestry-index

Srivastava, A. (2011). How to Calculate Mutual Fund Risk. Retrieved from http://karvy.com/articles/calculatemfrisk.htm

Suppliers Directory. (2019a). Timber and Lumber Companies from Bangladesh. Retrieved from https://www.lesprom.com/en/members/countries/Bangladesh/

Suppliers Directory. (2019b). Timber and Lumber Companies from Pakistan. Retrieved from https://www.lesprom.com/en/members/countries/Pakistan/

Suppliers Directory. (2019c). Timber and Lumber Companies from Afghanistan. Retrieved from https://www.lesprom.com/en/members/countries/Afghanistan/

Suppliers Directory. (2019d). Timber and Lumber Companies from Sri Lanka. Retrieved from https://www.lesprom.com/en/members/countries/Sri_Lanka/

The Express Tribune. (2018). Pakistan Stock Exchange to launch exchange traded fund this year. Retrieved from https://tribune.com.pk/story/1655461/2-pakistan-stock-exchange-launch-exchange-traded-fund-year/

The World Bank. (2019). GDP per capita growth. Retrieved from https://data.worldbank.org/indicator/NY.GDP.PCAP.KD.ZG

Timberweb(2019b). The Global Timber and Lumber eMarket. Retrieved from http://www.timberweb.com/Members/CountryResults/24/1.html

Timberweb. (2019a). The Global Timber and Lumber eMarket. Retrieved from https://www.timberweb.com/Contacts/CountryResults/210/AA/1.html

Uniply Industries. (2018). Annual Report 2017. Retrieved from https://www.uniply.in/pdf-excel/Uniply-Annual-Report-2017.pdf

US Department of Commerce. (2002). Environmental Technologies Export Market Plan. Retrieved from http://www.fao.org/3/a-am628e.pdf

Yahoo finance(2018b). V. R. WoodArt Limited. Retrieved from https://in.finance.yahoo.com/quote/VRWODAR.BO/history?p=VRWODAR.BO

Weerahewa, J., & Gunatilake, H. (2010). Timber Market Liberalization in Sri Lanka: Implications for Forest Conservation. *Sri Lankan Journal of Agricultural Economics*, *8*(0), 1–20. doi:10.4038jae.v8i0.1826

Yadav, M., & Basera, K. (2013). Status of Forest Products Production and Trade *Centre for Sustainable Forest Management and Forest Certification Working Paper*. Retrieved from http://iifm.ac.in/sites/default/files/working/IIFMWP-13-10-01.pdf

Yahoo finance. (2018a). National Plywood Industries. Retrieved from https://in.finance.yahoo.com/quote/516062.BO/history?p=516062.BO

ENDNOTES

1 A probable implication of the World Bank data on per capita GDP growth in light of Costa *et al* (2016) is that Afghanistan, Maldives and Sri Lanka are in the Malthusian trap having per capita average GDP growth over at least last 25 years to be less than 1%, while other SAARC countries entered the take-off stage only as per The World Bank (2019).

2 Only for Sarda Plywood the prices are taken from Bombay Stock Exchange. Mangalam Timber prices are taken from both NSE and BSE.

Chapter 9
Learning Path Recommendation Method Based on Knowledge Map

Xiaodong Zhou
Tianjin University, China

Yi Li
Tianjin University, China

Liping Yuan
Tianjin University, China

Gaofeng Ma
Tianjin University, China

Xinyun Tan
Tianjin University, China

Kaihua Zhang
Tianjin University, China

Lijiao Gong
Tianjin University, China

Boxiang Jia
Tianjin University, China

ABSTRACT

With the development of society, many industries and professions are more comprehensive and intersecting. Different industries have their own requirements for students with comprehensive backgrounds. For graduates, they may not know the skills required for various occupations, or what kind of jobs and occupations they can take based on their existing knowledge and skills, even how to acquire these abilities after they know the requirements of the jobs they want. In this chapter, authors combine the existing method to predict hot jobs with the analysis of knowledge map, aiming to achieve accurate recommendation of learning path for those who want to find a job. This chapter will help job hunters gradually master skills, and ultimately achieve the goal of optimizing resource allocation and saving social resources.

DOI: 10.4018/978-1-7998-0357-7.ch009

INTRODUCTION

The employment of college graduates has always been concerned by the society. However, facing with the increasingly heavy employment pressure, more and more college graduates have to choose the jobs that enterprises need and give up the jobs they like. This leads to the fact that the jobs they are looking for have little or no relationship with their major. As a result, more and more college graduates realize that they did not choose the right major when they entered the university.

This article finds that there are three main reasons that make it difficult for graduates to find suitable jobs. Firstly, they don't know which job is best for them based on their existing knowledge and abilities. Secondly, they don't know what knowledge and abilities they need for the jobs they want. Lastly, they don't know how to acquire these knowledge and abilities. If these three problems can be effectively solved, it will help individuals to improve their competitiveness and society to solve the contradiction between supply and demand in the human resources market.

In the field of Information Science, knowledge map is a semantic tool, which can describe knowledge model effectively at the level of knowledge management. The organic and multi-dimensional structure of knowledge can be revealed by knowledge map. Therefore, this article intends to use knowledge map to reveal the knowledge in job-hunting information to help graduates find jobs more easily. The basic ideas of this article are as follows. Firstly, this article takes the existing personal knowledge as the initial nodes. And then, a knowledge map is built with keywords extracted from related papers, which contain the knowledge and skills that graduates may need. Then the job information is analyzed, and the skills required by the job are regarded as the target nodes of the knowledge map. The last step is to find the path from the initial nodes to the target nodes. In this way, this article can establish an effective knowledge management tool, which can help college students to acquire professional knowledge according to their favorite jobs.

BACKGROUND

According to the "China Employment Report 2017", 56.7% of the fresh graduates indicated that their jobs were totally different from their majors. Only 25% of graduates had jobs directly related to what they had learned, while the rest graduates' jobs were partially linked to their majors. In addition, the latest statistics in 2018 saw an increase in the proportion of graduates whose jobs were related to their majors, but still less than 50%. And relevant research showed that salary (59.1%), promotion space (53.5%), and work prospect (34.6%) were the three most important factors affecting the employment of fresh graduates, which proved that most of those people did not consider their abilities and knowledge first, but their needs when looking for jobs. It's also worth noting that one of the most important abilities companies pay attention to when recruiting is work experience. For those graduates who have no working experience, they have to find ways to meet other requirements of the company.

In view of the problems reflected in the data, this article intends to recommend learning paths to graduates by combining text mining and knowledge map, so as to help them acquire the necessary knowledge better. The main technologies used include text mining, knowledge map, random walk method, etc. And this article reviews the literature on these aspects, which provides a basis for the proposed method.

Text Mining

In 1995, Feldman applied data mining technology to unstructured data, after which the concept of text mining was put forward. Text mining is a technology to discover potential and possible data patterns from a large number of unstructured text information, as well as the inherent relationship and development law between data (Sorensen, 2009). In recent years, with the rapid development of computer technology, text mining technology has made unprecedented progress and become a mainstream methodology.

The main processes of text mining can be divided into: data extraction, text preprocessing, text analysis and pattern evaluation. Data extraction refers to the capture of data from text databases or web pages. The commonly used technology is Spider, a web crawler. Text preprocessing transforms unstructured data into structured data which can be processed by computer. Generally speaking, text preprocessing includes word segmentation, removing stop words, part-of-speech tagging and feature selection. In feature selection, the most commonly used technology is "Term Frequency-Inverse Document Frequency" or "TF-IDF". The TF-IDF value increases proportionally with the number of words in the document, but is offset by the frequency of words in the corpus to balance the popularity of certain words (Yoshida et al., 2007). In addition, converting text into vector representation is also an important step in preprocessing. At the end of 2013, Google released Word2vec, which is considered as an excellent tool for word vectorization.

Text analysis refers to the use of statistical analysis, data mining or in-depth learning to analyze the structured data after processing to mine more information. The commonly used methods are classification, clustering, association analysis and so on. Text classification is a mining method that classifies each text into a certain category (Yin et al., 2007), and classification algorithms include KNN (Chang & Poon, 2009), support vector machine, decision tree (Forman & Kirshenbaum, 2008), neural network (Jo, 2010) and genetic algorithm (Khalessizadeh et al., 2006). The clustering algorithms of text mining include hierarchical clustering (Kavitha & Punithavalli, 2010), k-means clustering (Jain, 2010) and fuzzy clustering.

Model evaluation refers to the evaluation of the text mining model. Common indicators include accuracy, recall rate, error rate and so on.

Knowledge Map

Knowledge map has received increasing attention as an important sub-field of Knowledge Management in recent years. It helps to describe how and where to find useful knowledge within an organization, and several objectives can be acquired when utilizing knowledge map (Eppler & Simon, 2008). Some organizations view it as an activity that contributes to their strategic planning, while others may use it as a basis for knowledge transferring (Hellström & Husted, 2004).

In other cases knowledge map has been used to represent the participants' viewpoints and their relationship with other viewpoints, as well as illustrate the dependencies of learning paths and serve as the basis for the implementation of knowledge management programs (Dang, Zhang, Chen, & Larson, 2011; Einsfeld, Ebert, Kerren, & Deller, 2009; Ivanov & Cyr, 2006; Kim, Suh, & Hwang, 2003). Mahmoud et al. (2018) proposed Process Re-engineering Ontology-based knowledge Map Methodology (PROM) to solve BPR (business process re-engineering) problems and reduce the failure ratio.

In the field of Internet, knowledge map plays an important role in promoting organizational learning. After introducing the knowledge map-based Web platform, Marie-Hélène (2015) discussed the results of its use in the academic environment.

From the knowledge map, the users can recognize the important concepts and the relationships between them. After analyzing the number of relations required to identify the important ideas in the text, Lee (2012) compared knowledge map learning to document learning and found that knowledge map could identify more important ideas. Using knowledge map, Pyo (2005) proposed different tourism destination planning schemes, which provided a reference for the public sector to establish tourism destination knowledge base.

Learning Path Recommendation

In recent years, with the rapid development of information and communication technology, learning path recommendation has been widely concerned. Durand et al. (2013) demonstrated the method to address the problem of learning design recommendation in large repositories: reducing the problem space and using a greedy algorithm. Chen et al. (2017) proposed a novel coordinate system which was built to draw teaching or learning units to transform the unmeasurable concept map and information in syllabus into measurable data.

With the emergence of learning path requirement in e-learning domain, Dwivedi et al. (2017) proposed a learning path recommender system which was designed by employing Variable Length Genetic Algorithm (VLGA) to recommend optimal learning paths for learners by considering learners' requirements and preferences. Under the background of Interpretive Structural Model (ISM), Chungho S (2017) proposed an adaptive learning path recommendation system.

Random Walk Method

Random walk method is a mathematical and statistical model, which is widely used in the analysis of complex networks such as finance, physics and social media. The basic idea of Random walk model in graph application is to start with one or a group of nodes and access each node in graph by iterating randomly. Every time a node moves, the current node moves to its neighbor node with a certain probability. Each node in the graph will get a computed probability distribution of the current node walking to it. Every random walk takes place within a certain period of time. Finally, the probability will remain stable, and the probability of starting node to each node can be obtained. The simple random walk formula is as follows:

$$P(v_i)^{t+1} = W^t P(v_i)^t, \tag{1}$$

where v_i is the current node, W is the normalized matrix of adjacency matrix A corresponding to graph G after processing. The value of w_{ij} in matrix W represents the strength of the relationship between v_i and v_j in graph G. $P(v_i)^t$ is the probability of other nodes moving to node v_i through t-step in graph G. After a certain number of steps, the final probability will converge. Simple random walk is a reversible Markov chain with transfer matrix.

PROPOSED METHOD

Framework of Proposed Method

The proposed learning path recommendation method consists three processes.

Step 1: Knowledge Map Construction. The authors construct a knowledge map of Computer Science field by using papers crawled from CNKI.net and wanfang.net.

Step 2: Recruitment Information Analysis. The authors analyze the recruitment information crawled from Zhaopin.net to get the target knowledge.

Step 3: Learning Path Recommendation. A method which can automatically recommend learning path according to the initial knowledge nodes selected by user and target knowledge nodes according to user's ideal job is proposed.

Knowledge Map Construction

The authors crawl papers related to Computer Science from CNKI.net and wanfang.net and use the keywords of these papers to construct the knowledge map. Some keywords are too general to be a knowledge such as "application", "problem", so the authors eliminate these keywords manually. And the co-occurrence relationship of the keywords is regarded as the links in the knowledge map. If two keywords appear in a same paper, it means they have certain relationship in this field. The more they co-occurrence, the stronger the relationship is. The authors define the weight of the link as the frequency that two keywords co-occurrence. The weight calculation method is shown in Formula 2.

$$\mathrm{W}\left(n_i, n_j\right) = \frac{\mathrm{t}\left(n_i \cap n_j\right)}{\mathrm{t}\left(n_i \cup n_j\right)}, \tag{2}$$

where n_i, n_j are two nodes in the knowledge map, $\mathrm{t}\left(n_i \cap n_j\right)$ is the number of times that n_i, n_j co-occurrence, $\mathrm{t}\left(n_i \cup n_j\right)$ is the number of times that n_i, n_j occur independently.

Knowledge map is constructed based on these following steps:

Step 1: Calculating the occurrence frequency of each keyword. If it is bigger than a certain threshold, the keyword is used as a node in the knowledge map.

Step 2: For each paper, using co-occurrence relation of two keywords as a link in the knowledge map.

Step 3: Calculating the weight of each link. If it is larger than a certain threshold, the weight of link and link will be shown in the knowledge map.

Recruitment Information Analysis

Taking the "Intermediate Front-end Development Engineer" position as an example, which has five recruitment information. Table 1 shows the requirements for "Intermediate Front-end Development Engineer" after integrating the information.

Then the authors extract the features of the recruitment information, and extract some words that could represent the position from the recruitment information. Because words with higher frequency are more suitable to express the meaning of the text, this article adopts word frequency method to extract job

Table 1. Requirements of "intermediate front-end development engineer"

	Requirements
1	More than 2-year experience in front-end JAVA development
2	Familiar with Web front-end technologies such as JavaScript, jQuery, HTML, CSS
3	Knowledge of various front-end frameworks and relevant experience are preferred
4	Good learning ability, communication ability, teamwork ability, problem analysis and problem-solving ability, strong sense of responsibility
5	Experience in front-end optimization of B/S architecture is preferred
6	Experience in Internet industry or finance and taxation industry is preferred
7	Familiar with HTML/CSS/JavaScript and other front-end technologies, skilled in ES6 grammar
8	Master at least one JS framework, master its principle, and be able to independently develop common components
9	Some practical experience is preferred for ReactJS and RN
10	More than 3-year experience in program J2EE development
11	A solid JAVA programming foundation and good coding specifications

features. As shown in Table 2, this article uses the words "Front-end", "Development", "Frame", "JavaScript", "HTML" and "ES6" to represent the "Intermediate Front-end Development Engineer" position.

The authors compare these words with the knowledge map to see if there are same words in the knowledge map. If so, the word is regarded as a target node. If not, using the word similarity to find the most similar words as the target node. At present, there are two common methods for calculating word similarity. One is to calculate word similarity based on dictionary or some classification system. The commonly used dictionaries are HowNet, WordNet and Thesaurus. The other is Context-based Vector Space Model. Word2vec Vectorization Tool Based on Context Vector Space Model is the most famous tool. The authors use Word2vec to find the closest word to a given word in the knowledge map.

Learning Path Recommendation

When the users select current knowledge nodes and target knowledge nodes (by analyzing the requirements of user's target position), this system will recommend the learning path to the users to achieve their goal. Noticing that users may have many current and target nodes, so multiple learning paths may be recommended.

Recommending learning path is equal to recommend the path that can reach target nodes within a certain probability from current nodes. Random walk method is a widely used method in network analysis which calculates the probability of transferring from one node to another. So random walk method is used to recommend learning path to users in this article. It starts with the initial node in the network, and

Table 2. Word frequencies

Words	Front-end	Development	Frame	JavaScript	HTML	ES6	Demand
Frequencies	12	11	8	3	3	3	3

sets the value of the next nodes as the product of the initial node and the weight of the link. Candidate paths are updated after each iteration. Iteration terminates when the iteration time passes the maximum distance set. Only when the value of the target node is greater than the threshold will the path be chosen as a recommend learning path.

The process of learning path recommendation is as follows:

Step 1: Initializing the node. Set the value of the user's current knowledge node to 1 and the value of the remaining nodes to 0.

Step 2: Random walk. Starting from the initial node, each random walk updates the value of the adjacent node to the value of the current node multiplied by the weight of the link, and the updated node is added to the candidate path.

Step 3: Termination conditions. If it reaches the target node within the maximum distance set and the value of the target node is greater than the threshold, this path is considered to be an effective learning path.

SOLUTIONS AND RECOMMENDATIONS

Data Acquisition Process

Nowadays, China is developing rapidly in the field of economic and technological, and the demand for high level personnel is becoming more and more urgent, especially in the field of computer. Facing with this situation, this article selected Beijing, Shanghai, Guangzhou, Shenzhen and Hangzhou as examples, which are developing rapidly in China, and crawled the information about computer positions released by www.Zhaopin.com.

After the word segmentation and removing stop words of recruitment requirements, the TF-IDF method was used to extract the keywords with higher occurrences. In order to ensure the universality and validity of the data, about 2,500 job information were crawled from each of the above cities.

Finally, the authors crawled about 50,000 CNKI documents, including authors, titles, and keywords.

Data Analysis and Results

Recruitment Information Analysis Process

A total of 10 987 recruitment information was collected. Because the conditions of the same position were comparatively similar, the requirements of different companies for the same position were integrated. Finally, the position was taken as the basic unit for analysis.

Literature Data Analysis Process

This article crawled 54,330 papers in the field of Computer Science from CNKI.net and wanfang.net and constructed a knowledge map based on keywords of these papers.

In order to display knowledge map vividly to users, this article used an open source drawing software named Gephi to visualize the knowledge map.

Then the authors added nodes, links and weight into Gephi, after which the knowledge map of Computer Science was obtained, as shown in Figure 1. In Figure 1, the radius of nodes with higher occurrences

Figure 1. Knowledge map of computer science field

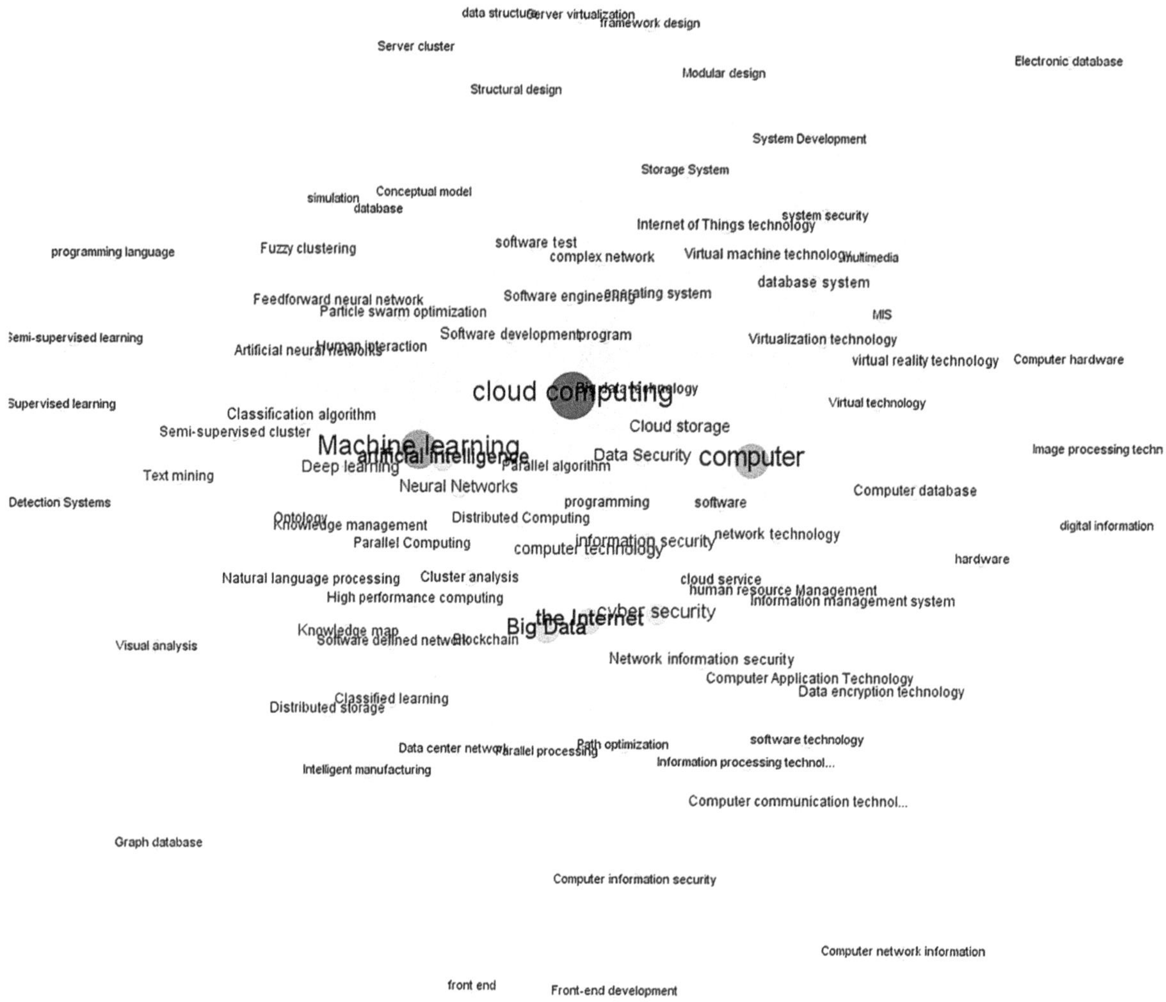

is larger and the color is darker. Meanwhile, those nodes are more likely to be located in the center of the graph. It can be seen that "computer", "cloud computing", "machine learning" and "big data" with larger radius and darker color, and they are located in the center of the graph, which indicates that these keywords appear more frequently and are more important in the field of Computer Science. And the thicker the links, the stronger the relationship between two nodes. And strongly related nodes are more likely to concentrate in a certain area of the graph. For example, "machine learning", "data mining", "neural networks", "support vector machines" and "genetic algorithms" have strong relationships.

Learning Path Recommendation Method

The authors selected the top 10 popular jobs as target positions and used this analysis method to get the learning path. Meanwhile, the authors selected "C language" as the current knowledge node. Then the authors used this learning path recommendation method to recommend learning path according to the current and target knowledge nodes. The recommendation results are given in Table 3. The first column involves the current knowledge nodes, the second column involves the target position, and the third column involves the target knowledge nodes. The authors set the maximum length to 4 and the threshold of target nodes' value to 0.001. The leaning path recommendation results are shown in the fourth column. This result may provide learning help to students who have studied C language and want to get those positions.

As can be seen from the results, since the target node is not unique, the recommended path is not unique. At the same time, it can be found that some target knowledge does not appear in the path because of the limited number of steps, or because the knowledge contributes too little to the position. Take the position of Hardware QA Engineer as an example. This position eventually includes two learning paths.

Table 3. Learning path recommendation results

Initial Knowledge	Target Job	Target Knowledge	Results of routine recommend
C language	Web Front-end Development Engineer	ES6, HTML, JavaScript, ReactJS, WEB, CSS, database, react	1. C language-software-B/S-HTML 2. C language-program design-internet-HTML 3. C language-program design-HTML-CSS 4. C language-program design-internet-JavaScript
C language	Cloud Computing Big Data Engineer	Cloud computing, GJB, ES6, database	1. C language-computer-information safety-cloud computing 2. C language-computer-bigdata-cloud computing 3. C language-computer-information system-database
C language	Hardware QA Engineer	Hardware, front-end, operating system, GJB, electromechanical, framework design	1. C language-embedded system-computer-hardware 2. C language-computer-web design-front end
C language	Image Processing Engineer	Cloud computing, picture processing, server, operating system, pattern recognition, network equipment, GJB, C	1. C language-computer-web design-image processing 2. C language-embedded system-machine vision-image processing
C language	Algorithm Engineer	Algorithm, data, operating system, GJB, C++, C, Linux, database	1. C language-software development-embedded system-Linux 2. C language-embedded system-system-Linux 3. C language-computer-information system-database
C language	Artificial Intelligence	AI, deep learning, GJB, image processing, Linux, C++, GB, python	1. C language-software development-embedded system-Linux 2. C language-embedded system-machine vision-image processing 3. C language-data mining-AI 4. C language-embedded system-convolutional neural network-deep learning
C language	Android Development Engineer	AI, android, GJB, C++, Linux, Java	1. C language-data mining-AI 2. C language-C++-Java
C language	PHP Senior Engineer	PHP, SQL, JQuery, JS, CSS	1. C language-program design-HTML-CSS
C language	Structural Engineer	Structural design, AutoCAD, Pro, BOM, product design	1. C language-information design-demand analysis-product design
C language	SEO Optimization Engineer	SEO, search engine, data analysis, keyword deployment, flow	1. C language-computer-big data-data analysis 2. C language-computer-software development-search engine

For a job seeker who masters C language, he can choose one of them to study, or he can study along two paths separately. Ultimately, he will systematically master the hardware and front-end knowledge, which is very helpful for applying for Hardware QA Engineer.

FUTURE RESEARCH DIRECTIONS

This article still has many limitations. Firstly, the scope of data collection is not wide enough, which only involves the field of computer. In fact, the current hot employment field may also refer to the finance, management, accounting, artificial intelligence and so on. If learning path of these areas are also constructed, the results of this article will be more representative. Secondly, based on the characteristics of the research results, this article needs to input the existing knowledge to get additional path recommendations on the knowledge map. It lacks a platform or interface to show the relationship between knowledge input and output more intuitively. According to the current researches, there is no good visualization method to express the application of knowledge maps more clearly.

In the future, this article will consider more factors with personal characteristics in the process of recommending learning paths. For example, suitable positions for job seekers can be recommended. In addition, this article plans to adopt more methods to match knowledge with recruitment information. Based on this article, more practical applications, such as job evaluation platform and learning recommendation platform, can be studied. On this kind of platforms, job seekers can access the learning resources related to skills and knowledge they need directly, which will be an important breakthrough in practice.

CONCLUSION

Nowadays the technology-driven world is not only full of opportunities, but also full of challenges. Some jobs which are performed by human-beings will be substituted by computer, while new jobs like programmer are created to respond the changes in the world. Unlike traditional jobs, these various kinds of jobs may confuse people on what they can do to obtain the skills the jobs need. This situation is aggravated by the rising unemployment rate and the emergence of new jobs.

Based on the literature data of CNKI.net, this article constructs a knowledge map about the field of computer. At the same time, this article captures a lot of recruitment information and replaces these positions with the most representative words. Then, on the basis of the existing work, this article combines the representative words of recruitment information with the nodes on the knowledge map. In the end, this article recommends learning paths precisely to users, so that they can learn the necessary knowledge with a special plan and get an ideal job. This method can draw people closer to their ideal future positions and recommend the most effective and convenient learning path. Those who follow the recommendations of this article will improve their skills by learning a series of related topics and improve their understanding of a particular field.

This article has certain significance in methodology and management. For methodology, this article makes full use of the advantages of knowledge map and learning path, and applies this method in the field of learning and employment. On the one hand, it broadens the application fields of these two methods, and on the other hand, it plays a good leading role for the similar studies in this field. In the sense of management, this article establishes the relationship between basic knowledge and job requirements, alleviates the contradiction between supply and demand in human resources market, and realizes the optimal allocation of resources. In addition, educational institutions could be encouraged to make use of this method to provide personalized educational resources and create more effective business models to meet the needs of students, which is conducive to the development and innovation of educational institutions.

ACKNOWLEDGMENT

This research was guided by Professor Zhang Xi. And this research received help from Xia Yuxi, Zhao Jinfan, Li Han and You Zhiwei, graduate students of Tianjin University.

REFERENCES

Abdellatif, M., Farhan, M. S., & Shehata, N. S. (2017). Overcoming business process reengineering obstacles using ontology-based knowledge map methodology. *Future Computing and Informatics Journal.*

Abel, M.-H. (2015). Knowledge map-based web platform to facilitate organizational learning return of experiences. *Computers in Human Behavior, 51,* 960-966.

Chang, M., & Poon, C. K. (2009). Using phrases as features in e-mail classification. *Journal of Systems and Software*, *82*(6), 1036–1945. doi:10.1016/j.jss.2009.01.013

Chen, M., Tong, M., & Liu, C. (2017). *Recommendation of Learning Path Using an Improved ACO Based on Novel Coordinate System.* Paper presented at the Iiai International Congress on Advanced Applied Informatics. 10.1109/IIAI-AAI.2017.90

Chungho, S. (2017). Designing and Developing a Novel Hybrid Adaptive Learning Path Recommendation System (ALPRS) for Gamification Mathematics Geometry Course. *Eurasia Journal of Mathematics, Science and Technology Education*, *13*(6), 2275–2298.

Dang, Y., Zhang, Y., Chen, H., & Larson, C. A. (2011). Knowledge Mapping for Bioterrorism-Related Literature. In C. Castillo-Chavez, H. Chen, W. Lober, M. Thurmond, & D. Zeng (Eds.), *Infectious Disease Informatics and Biosurveillance. Integrated Series in Information Systems* , 27, 311–338. Boston, MA: Springer.

Durand, G., Belacel, N., & Laplante, F. (2013). Graph theory based model for learning path recommendation. *Information Sciences*, *251*(4), 10–21. doi:10.1016/j.ins.2013.04.017

Dwivedi, P., Kant, V., & Bharadwaj, K. K. (2017). Learning path recommendation based on modified variable length genetic algorithm. *Education and Information Technologies.*

Einsfeld, K., Ebert, A., Kerren, A., & Deller, M. (2009). Knowledge generation through human-centered information visualization. *Information Visualization*, *8*(3), 180–196. doi:10.1057/ivs.2009.15

Eppler, M. J., & Simon, H. A. (2008). A process-based classification of knowledge maps and application examples. *Knowledge and Process Management*, *15*(1), 59–71. doi:10.1002/kpm.299

Feldman, R., & Dagan, I. (1995). *KDT-Knowledge Discovery in Texts.* Paper presented at the First International Conference on Knowledge Discovery (KDD).

Forman, G., & Kirshenbaum, E. (2008). *Extremely fast text feature extraction for classification and indexing*. Paper presented at the 17th ACM Conference on Information and Knowledge Management, California, USA. 10.1145/1458082.1458243

Hellström, T., & Husted, K. (2004). Mapping knowledge and intellectual capital in academic environments: A focus group study. *Journal of Intellectual Capital*, *5*(1), 165–180. doi:10.1108/14691930410512987

Ivanov, A., & Cyr, D. (2006). The concept plot: A concept mapping visualization tool for asynchronous web-based brainstorming sessions. *Information Visualization*, *5*(3), 185–191. doi:10.1057/palgrave.ivs.9500130

Jain, A. K. (2010). Data clustering: 50 years beyond k-means. *Pattern Recognition*, *31*(8), 651–666. doi:10.1016/j.patrec.2009.09.011

Jo, T. (2010). NTC (Neural Text Categorizer): Neural network for text categorization. *International Journal of Information Science*, *2*(2), 83–96.

Kavitha, V., & Punithavalli, M. (2010). Clustering time series data stream - A literature survey. *International Journal of Computer Science and Information Security*, *8*(1), 289–294.

Khalessizadeh, S. M., Zaefarian, R., Nasseri, S. H., & Ardil, E. (2006). Genetic mining: Using genetic algorithm for topic based on concept distribution. *Journal of Word Academy of Science. Engineering and Technology*, *13*(2), 144–147.

Kim, S., Suh, E., & Hwang, H. (2003). Building the knowledge map: An industrial case study. *Journal of Knowledge Management*, *7*(2), 34–45. doi:10.1108/13673270310477270

Lee, J. H., & Segev, A. (2012). Knowledge maps for e-learning. *Computers & Education*, *59*(2), 0-364.

Pyo, S. (2005). Knowledge map for tourist destinations—needs and implications. *Tourism Management*, *26*(4), 583-594.

Sorensen, L. (2009). *User managed trust in social networking comparing facebook, myspace and linkdin.* Paper presented at the 1st International Conference on Wireless Communication, Vehicular Technology, Information Theory and Aerospace & Electronic System Technology, (Wireless VITAE 09), Denmark.

Yin, S., Wang, G., Qiu, Y., & Zhang, W. (2007). *Research and implement of classification algorithm on web text mining*. Paper presented at the 3rd International Conference on Semantics, Knowledge and Grid, China.

Yoshida, K., Tsuruoka, Y., Miyao, Y., & Tsujii, J. (2007). *Ambiguous part-of-speech tagging for improving accuracy and domain portability of syntactic parsers.* Paper presented at the 20th International Conference on Artificial Intelligence, China.

ADDITIONAL READING

Balaid, A., Abd Rozan, M. Z., Hikmi, S. N., & Memon, J. (2016). Knowledge maps: A systematic literature review and directions for future research. *International Journal of Information Management, 36*(3), 451–475. doi:10.1016/j.ijinfomgt.2016.02.005

Driessen, S., Huijsen, W.-O., & Grootveld, M. (2007). A framework for evaluating knowledge-mapping tools. *Journal of Knowledge Management, 11*(2), 109–117. doi:10.1108/13673270710738960

Kenthapadi, K., Le, B., & Venkataraman, G. (2017). *Personalized Job Recommendation System at LinkedIn: Practical Challenges and Lessons Learned.* Paper presented at the Eleventh ACM Conference. 10.1145/3109859.3109921

Kretschmer, H. (2002). Similarities and Dissimilarities in Coauthorship Networks: Gestalt Theory as Explanation for Well-Ordered Collaboration Structures and Production of Scientific Literature. *Library Trends, 50*(3), 474–497.

Taabish Khan. (2016). *Learning Path: Data Science with R. Location.* Packt Publishing.

Yang, M. S., Joo, W. K., Choi, K. S., Kim, Y. K., & Kim, Y. J. (2017). Development of platform-based knowledge map service to get data insights of r&d institution on user-interested subjects. *Wireless Personal Communications, 98*(8), 1–21.

Zhao, W., Wu, R., & Liu, H. (2016). Paper recommendation based on the knowledge gap between a researcher's background knowledge and research target. *Information Processing & Management, 52*(5), 976–988. doi:10.1016/j.ipm.2016.04.004

Zhu, Y., Wang, P., Fan, Y., & Chen, Y. (2017). *Research of Learning path recommendation algorithm based on Knowledge Graph.* Paper presented at the International Conference on Information Engineering. 10.1145/3078564.3078567

KEY TERMS AND DEFINITIONS

Big Data: A field that treats ways to analyze, systematically extract information from, or otherwise deal with data sets that are too large or complex to be dealt with by traditional data-processing application software.

CNKI: CNKI (China National Knowledge Infrastructure) is a key national information construction project under the lead of Tsinghua University and has built a comprehensive China Integrated Knowledge Resources System, including journals, doctoral dissertations, masters' theses, proceedings, newspapers, yearbooks, statistical yearbooks, ebooks, patents, standards and so on.

Knowledge Map: A knowledge map is a navigation system for knowledge (both explicit and coded knowledge and implicit knowledge), and shows important dynamic links between different knowledge stores.

Learning Path: An effective way to provide the necessary knowledge and skills for job seekers by applying random walk method.

Random Walk Method: A method of random sampling in which the number of paces between sample points is determined by random numbers, usually drawn from random-number tables, and from each sample point a right-angle turn determines the direction of the next point, a coin being tossed to decide whether to turn left or right.

Recommendation Method: A method which can automatically recommend relevant items that user would prefer.

Text Mining: The process of exploring and analyzing large amounts of unstructured text data aided by software that can identify concepts, patterns, topics, keywords and other attributes in the data.

Chapter 10
Reverse Mentoring the Editing Edge in Management 4.0

Swati Sisodia
G. D. Goenka University, India

Neetima Agarwal
Amity University, India

ABSTRACT

Industry 4.0 is based on the implementation of a cyber-physical system, which includes sensors, networks, computers, offering digital enhancement and well-coordinated activities. This would create a great pool of all the workforce generations, having diverse experience, agility, and different modes of working. Millennials would add more of machine learning and Generation X and Y would be the richest source of tacit and operational knowledge. Together, they would develop solutions for catering complex and networked production and aggressive logistic management, meeting the challenges of the Industry 4.0. However, the benefits of digitization and automation can be achieved, if the different generations of workforce collaborate, cooperate, and postulate together in all the business processes. Reverse mentoring is a pristine concept and ingenious method to empower learning and encourage cross-generational connections. This chapter would elaborate on the advantage of reverse mentoring in crafting Industry 4.0 more acrobatic and quick-moving.

INTRODUCTION

The competitive index of some developed Asian countries such as China, Japan, Taiwan, and Singapore are higher as compared to other Asian countries like India, South Arabia, Malaysia, Sweden and others. The South Asian countries such as Bangladesh, Vietnam and others are criticized due to the lack of production, female leadership, technological application, working skills and factory infrastructure, all of which are identified as the major challenges for the growth of these countries. Most of the developing countries are lacking behind due to absence of latest technology, robust management system, optimized solutions, flexible environment, critical gaps in capabilities and socially responsible mission. Manage-

DOI: 10.4018/978-1-7998-0357-7.ch010

ment is critical to both the developing and advance countries. It is achievable with effective research & development, innovation and technological acceleration. Smart management would provide opportunities for career growth, continuous learning culture, and skill enhancement.

The advent of Industry 4.0 is converting traditional manufacturing into smart manufacturing, by using networked sensing devices and big data analytics. The production capacity can be enhanced, with technological integration. Industry 4.0 is noteworthy as it automates and exchange data in manufacturing and service sector. Moreover, Industry 4.0 would witness inverting of hierarchy and allowing millennials to act as mentors to Gen X and Gen Y. This process of feeding investments into millennials to create learning organization, would be called as "reverse mentoring". Reverse mentorship would breed the inclusivity by bridging the gap between populations, diversified in terms of age, exceptionalities, ethnicity and gender.

BACKGROUND

Millennials are the fastest growing generation at the work place and estimated to grow manifolds till 2020. Researchers in the various fields of psychology, political science, economic, humanities examined generational difference based upon social, educational, historical, experiences like second world war, global recession etc. Researcher states that the historical and social events shapes the value, behaviour, attitude and expectations of the generations. Generational difference is visible in commitments, ethics, work values, sportsmanship and work life balance.

The purpose of this chapter is to understand the difference between the generations and acceptance of reverse mentoring process in Industry 4.0. Gen X comprises of workforce born between 1965-1979. The core values of Gen X comprises of higher education, independent, balanced, high job expectation and self-reliance. Gen Y comprises of workforce born between 1980-1994. The core values of Gen Y includes high tolerance power, civic duty, stable families, great recession, climate change. Millennials are the young generation and were born between 1995-2015. Workforce of this generation has high confident level, extremely techno-savvy, street smarts, go getter, highly comparative, requires diversity and avid consumers.

The generational differences in the workforce makes the study significant. As now-a-days Asian industries are witnessing paradigm shift in the working culture. The workers are more tech-savvy as the industries are highly automatic and the process are censored controlled. In order to demonstrate in such sophisticated working environment, it becomes vital to upgrade the generations with the latest technical crafts and innovations. Thus, organizations are focusing on the economically optimal solution of Reverse mentoring.

INDUSTRY 4.0

Industry 4.0 is referred to as a fourth industrial revolution where computers are connected and communicated with each other to provide solutions with human entanglement. The ongoing debate on the digital revolution and disruptive competition advantage have led to the erection of a new concept to the business as Industry 4.0 (Glas & Kleemann, 2016). They forge ahead in industrialization and informalization techniques that prodded gigantic advancement in building up the ameliorate technology for next generation workforce. Germany is leading a transformation towards, fourth generational industry revolution. In 2013, among one of the ten 'Future Projects' distinguished by the German government as a major aspect of its High-Tech strategy. By 2020, Industry 4.0 venture is viewed as a noteworthy undertaking by Germany to build-up itself as a pioneer leader of integrated industry (Xu et al., 2018).

The amalgamation of the Internet of Things (IoT), Cyber Physical System (CPS), cloud computing, industrial integration and business process management forge industry 4.0 viable and the smart factory a reality. Moreover, with the aid and support of smart machines, it becomes easier as they gain admittance to more data, the factories will turn out to be increasingly effective, beneficial, productive and less inefficient. Eventually, it's the system of these machines that are digitally associated with each other. They invent and share information that upshot on the genuine intensity of industry 4.0. On the escarpment of Industry 4.0, the requirement for such synergetic and supple, teams is essential.

Due to the technological advancement the concept of industry 4.0 has been deemed as a tool and a strategy to enhance product quality and building operational processes more efficient. Industry 4.0 will make it conceivable to accumulate and dissect information crosswise over machines, faster, increasingly adaptable and progressively to create higher-quality products at the lesser expenses. This in turn will build fabricating profitability, economic shift, cultivate mechanical development and alter the profile of the workforce. This will ultimately change the intensity of companies. Thus, the amalgamation of reverse mentoring and industry 4.0 aided the organizations in building revamp work culture, velvety of operational activities and elevate productivity.

REVERSE MENTORING

Reverse mentoring is an exact form of classical mentoring which is divergent from traditional mentoring. The concept of mentoring began in 1980's (Chen, 2013). Kathy Kram's mentioned as phases 'Mentor Relationship' which had an influential impact in this field. Mentoring is a time worn process, followed and long established in the Industries and other Institutions (Biss & DuFrene, 2006). Over the years, mentoring turned into a built-up strategy for personnel development, improvement and revolution. Horde meanings of mentoring were created, reflecting in the critical number of contexts where this concept is utilized (Kaše et al., 2019). Therefore, a standard meaning of the term mentoring is difficult to set-up. Rather, it is conceivable to break-down essential components portraying the process of mentoring.

The concept of reverse mentoring has been around us for over a decade, but it is highlighted more eminently because of expeditious build-out of technological advancement (Industry 4.0) and innovation (Greengard, 2002). Reverse mentoring generally termed as reversing the role of mentoring wherein millennials act as a mentor for Gen X and Gen Y (Thomas, 2019). This is characterized by the shape of apprenticeship were the relationship between mentor and mentee is strengthened. Appropriate management practices are important to develop these dynamic capabilities and to stimulate learning and innovation in organizations. This will invigorate to individuals of Gen X and Gen Y to sustain in the changing world. The ken and respect will burgeon in a cross-generational relationship only if, both the parties are satisfied with the roles assigned to them.

A recent report of Morgan Stanley (2018) states that by 2020 most of the Asian Industries would witness the inclusion of more than 28% Millennials in their workforce, and India would become the world's youngest country with 64% millennials. Hence, the concept of reverse mentoring would certainly become the reality.

In fact the phrase 'Reverse Mentoring' is mainly being incorporated in the industries such as Information Communication and Technology (ICT), Manufacturing, FMCG and Education, etc. Wherein youth is more digital native & adaptive hence bringing new technological knowledge in the workplace. By incorporating this methodology there is a portion of advantage for mentor and protégé, such as engaged learning, cost efficiency, diverse learning and personal satisfaction (Carr, 2002; Cotugna & Vickery, 1998). The field of Human Resource Management (HRM) has progressively embraced inter-generational learning, which means at a single workplace different generations, including Gen X, Gen Y and Millennials are working together resulting in individual enhancement of knowledge, skills and values (Gerpott et al., 2017; Ropes, 2013).

Companies such as Tech Mahindra, Google, Cisco, Microsoft and Sodexo are encouraging the millennials to share their technological edge and insight with people of Gen X and Gen Y. Vodafone India, is striving to become a complete digital workplace, has posit a reverse mentoring programme known as "Digital Ninja". As a part of which, millennials are selected to reverse mentor the senior leaders, including the executive committee and business heads on exclusive digital tools.

Microsoft offers a reverse mentoring programme called as "Elevate". This leverages the generational diversity among teams. Microsoft Austria introduced their first reverse mentoring program in 2014 (Microsoft, 2015, July 23). They incite their manpower to ruminate and distinctively from the old tradition of hierarchical and traditional learning to inter-generational & cross-functional learning. According to, 'Magnus Svorstol lie' flipping the concept of mentoring relationship aid individuals of all the generations. Individuals are engaged in exchange learning methodology which is providing a better collaboration and leverage each other (Sodexo, 2017).

Pershing a financial services company has recently implemented a reverse mentoring program country-wide. The program has involved 77 millennials. Perishing has experienced 96% retention rate post implementation of the program. According, to the study conducted by Ernst and Young, millennials will comprise three-quarters of the global workforce by 2025. This ineluctable trend to threat the current employers of the organization. This pattern moves the businesses of today to discover new arrangements in such a manner that they stay alluring and future-proof in the term of employability. Engrafting reverse mentoring in organizational culture has a double benefit. First, the senior leaders are able to stay on the pulse of trends. Secondly, millennials perceive themselves more connected and perfused, as they are contributing to the improvement of the company.

For Millennials, born with gadgets in hands, eyes and ears; the mode of education, upbringing and co-curricular activities are completely technology-driven. For them, Maslow Need Hierarchy have shifted its base to wifi and chargers. Millennials are highly ambitious and talented quotient of the organization. They are constantly looking for the opportunities for their career advancement. Youth view technology as an extension of themselves. So, treating and competing with this generation needs a reverse mentoring.

METHODS OF IMPLEMENTING REVERSE MENTORING

Creating Acceptance in Between Generations

Changes are important in all the workplaces. At times changes are brought by new generations. As Millennials have high acumen on usability of headway technologies, they can bring advanced learning atmosphere in the organization. As Gen X and Gen Y are still working using conventional methods of work, which are relatively more time consuming. With the inclusion of smart technologies, the work can be performed in lesser time, more accurately and effectively.

Hence, it is overriding for the leaders to acknowledge the millennial which can accord profitable and valuable inclusion of technology with the conventional methods of work. This would encourage Gen X and Gen Y to adapt the new technology in optimal time.

Mixing it Rightly

There is a need of enriching mutual relationship in the process of mentoring. Most of the researchers examine the issues in mentoring relationship on behalf of protégé perspective. They reported issues such as mentees neglection to work, personality mismatch, lack of technical expertise and meticulous aptitude. However, management should analyze their behavioral pattern to avoid the ego issues (Eby & Allen, 2002; Eby et al., 2004). Usually, these disputes occur due to age-differences, work-style and cognitive processing.

It becomes the responsibility of the organization to conduct the appropriate match of mentor with mentee. This would provide comprehensive and balanced relationship for staunch buildout of mentoring programs. For, the effectiveness of the program there is a need to build a team which are best suited to work with each other. Mentors should not be selected just because they're young and mentees should be vetted for their candidness to learning new technology. The meetings between mentors and mentees should be designed in a structured manner, without violating the rules and parameters.

Coaching the Mentor for Greater Effectiveness

A mentor-mentee relationship can be effective if mentors are dedicated and understand the needs of mentees. Before training the mentees, it is important to train the mentors, on being polite to seniors, respect their thought process and providing weightage to their viewpoints. For developing effective mentor-mentee relationship, the harmony between the age groups is vital.

Organization should put efforts for training the mentors, before starting the reverse mentoring process. It becomes the responsibility of the organization to develop the platform, where all the generations are exultant to learn and un-learn together.

Building the Bridges

The knowledge gap is filled through reverse mentoring with respect to computerized change occurring invert coaching act as a bridge to reduce the technological gap between millennials and other generations. It offers a profitable chance to crack down the generational conflicts. This bridge aids, both the learners and the mentor as they acquire new skills and embellish the roles they perform. Figure 1 represents the effect of reverse mentoring on both mentor and mentee.

By counter parting leaders with employees from various backgrounds, both develop empathic vista and condenses unconscious biasedness. Moreover, when contemplate for organizational development strategy, these initiatives can activate a diverse pipeline of talents that will yield a needed percentage, intensifying in the diversity of top-level management across industries and sectors.

Figure 1.

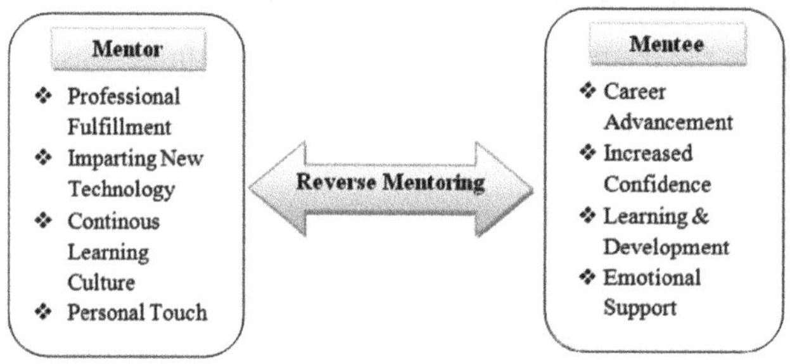

Making Them Feel Important

By encouraging reverse mentoring at the workplace, elder employees will get a chance to acquire knowledge and skills on the other hand, mentors are less experienced employees. Millennial acquires self-confidence to lead a project single handily. If, the concept of reverse mentoring is applied by the organization in its workplace. This confidence aids them in the development of their leadership and teamwork skills. When employers provide such gratifying opportunities, millennials become confident that their investment in the organization would be matched by the organization's investment in their future.

REASONS FOR RESISTANCE TO ACCEPTANCE

Now a day's reverse mentoring is a paramount part of an organizational learning and development culture. However, stiffness is evident in the process of reverse mentoring:

Conflicting Relationship

Gen X and Gen Y employees have a fear of losing their reputation. Employees of Gen X and Gen Y are predominantly the task masters as they are the most experienced workforce of the organization. Since, they are associated with the organization for a long period of time, therefore they are able to build a comfortable boundary around themselves. Task-masters are more stubborn to the change.

Lack of Leadership and Management Styles at the Top

If the organization's prevailing leadership style is that of commanding and instructing its individual's, that is how things ought to be done in a top-down way and concentrating just on the transient efficiency gain, at this point the training is probably not going to flourish. Thus, the organization should change its top-down culture. The organization, administration should be supportive, reassuring and rousing with a solid commitment for long-term reasonable development and sustainable growth. An organization generally practice autocratic or democratic management style. But, to motivate the involvement of the employee, there is a need to adopt a participative style of management.

Insufficient Reward and Recognition

The most commonly used phrase in human resource is 'what get measured gets done'. However, behind this dictum there is a lot of reality. In order to adopt a reverse mentoring culture, behavior and outcomes are requisite for performance management system. Mentors get measured and rewarded on the bases of the effectiveness of the result they achieve and reinforce successes. With the advancement in technology from 1.0 to 4.0 there is a need to train the Gen X and Gen Y. Hence, the organization's belief the urge of reverse mentoring and utilizing it as an imperative part of their learning and development mix.

Technical Illiteracy and Lack of Time Management

In the organization, employees are handling multiple assignments at the same time, this creates a bottleneck in bringing any change in the organization. In most of the cases, Generation X is comparatively having less technical prowess. At times older generation, takes millennials as a challenge for their jobs. Because of the fear of loosing the job, they try to keep distance from the new methods of work, and advocates the older work art of completing the task. Figure 2, represent the process of reverse mentoring.

Eliminating these barriers in the organizational culture leads to enhancement of various management practices such as, improved organizational structure, effective leadership style, cynosure on short-term innovation goals & long-term capabilities and keenness to abandon the investment and knowledge. Asian countries is in urgent need to unlearn and re-learn in order to expand their productivity.

Figure 2.

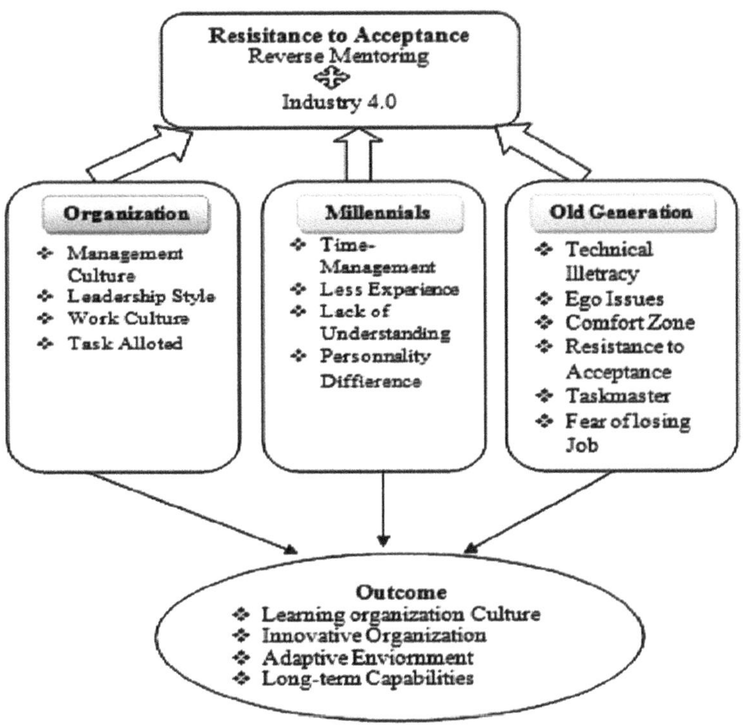

CONCLUSION

The economic and social success of the Asian countries depends on its Gross Domestic Product and Per capita Income. Countries where industries are technologically magnificent, the productivity is way higher. This in turn increasing the competitive index and happiness quotient of the people. Other Asian countries should move from Industry 2.0 to Industry 4.0, this would accelerate their growth and advancement.

Industry 4.0 proposes the espousal of new technology, which is providing a cutting edge to the ICT industry. The rise of enthusiasm for industry 4.0 has expanded in recent years because of the conviction that the present development is denoting a noteworthy change in history. Industry 4.0 is no longer a ' Future Pattern'. For some undertakings, it is currently at the core of their strategic and research plan.

Reverse Mentoring is not a concept, it is a practice to change the existing culture of the organization. As the workforce would witness inclusion of the wide age-group, including the three generations. It becomes indispensable for them to work together, in harmony and compassion. Organizations must put human beings in the loop to maximizing the potential value of its technologies. This is possible by rearranging, re-training and re-constructing the work and people

Reverse mentoring would provide opportunity to the executives and leaders assimilate knowledge and skills from a different generation. This would save the additional cost of training. One of the crucial outcome of this, it would help organisation to become a self-learning organisation by engaging workforce across different levels. This will also motivate the millennials to cherish back in the organization.

Many organizations such as; IBM, Vodafone, Telstra, Proctor & Gamble, Cisco, and Time Warner have already embodied this modern concept and are reaping exceptional benefits. Moving with this global trend, many Indian corporations, such as Bharti Airtel, are traversing the possibilities of using this growing phenomenon to mentor seniors. IBM employs sizeable millennial workforce and is enduring group reverse mentors executives to grow more socially networked and digitally skilled.

As discussed in this chapter, the changing organization set-up, and the introduction of robotic technologies, machine learning and artificial intelligence even in the areas of regular or specialized task, a rift would forge in the intelligence ratio of humans and robots. To bridge the gap between the intelligence, emotions and cognizance, training will have imperative and important latitude. In the future, with the advancement of technologies and increasing amoebic structures of organization, the lower end needs of the employees would be challenged withal.

REFERENCES

Berawi, M. A. (2018). Utilizing big data in industry 4.0: Managing competitive advantages and business ethics. *International Journal of Technology*, *3*(1), 430–433. doi:10.14716/ijtech.v9i3.1948

Biss, J. L., & DuFrene, D. D. (2006). An Examination of Reverse Mentoring in the Workplace. *Business Education Digest*, (15).

Carr, K. (2002). Building bridges and crossing borders: Using service learning to overcome cultural barriers to collaboration between science and education departments. *School Science and Mathematics*, *102*(6), 285–298. doi:10.1111/j.1949-8594.2002.tb17886.x

Chen, Y. C. (2013). Effect of reverse mentoring on traditional mentoring functions. *Leadership and Management in Engineering*, *13*(3), 199–208. doi:10.1061/(ASCE)LM.1943-5630.0000227

Cotugna, N., & Vickery, C. (1998). Reverse mentoring: A twist to teaching technology. *Journal of the American Dietetic Association*, *98*(10), 1166–1168. doi:10.1016/S0002-8223(98)00270-3 PMID:9787725

Eby, L. T., & Allen, T. D. (2002). Further investigation of proteges' negative mentoring experiences: Patterns and outcomes. *Group & Organization Management*, *27*, 456–479. doi:10.1177/1059601102238357

Eby, L. T., Butts, M. M., Lockwood, A., & Simon, S. A. (2004). Proteges' negative mentoring experiences: Construct development and nomological validation. *Personnel Psychology*, *57*(2), 411–447. doi:10.1111/j.1744-6570.2004.tb02496.x

Gerpott, F. H., Lehmann-Willenbrock, N., & Voelpel, S. C. (2017). A phase model of intergenerational learning in organizations. *Academy of Management Learning & Education*, *16*(2), 193–216. doi:10.5465/amle.2015.0185

Giddens, D., & Phillips, R. (2009, September). Reverse Mentoring: Finding a new way of working while discovering Web 2.0, ALIA National Library and Information Technicians Conference, Adelaide Convention Centre, Adelaide, SA, Australia.

Glas, A. H., & Kleemann, F. C. (2016). The impact of industry 4.0 on procurement and supply management: A conceptual and qualitative analysis. *International Journal of Business and Management Invention*, *5*(6), 55–66.

Greengard, S. (2002). Moving forward with reverse mentoring. *Workforce*, *81*(1), 15.

Kaše, R., Saksida, T., & Mihelič, K. K. (2019). Skill development in reverse mentoring: Motivational processes of mentors and learners. *Human Resource Management*, *58*(1), 57–69. doi:10.1002/hrm.21932

Kram, K. E. (1983). Phases of the mentor relationship. *Academy of Management Journal*, *26*(4), 608–625.

Microsoft. (2015, July 23). Reverse mentoring: How millennials are becoming the new mentors (Blog post). Retrieved from https://news.microsoft. com/Europe/features/reverse-mentoring-how-millennials-are-becoming-the-new-mentors

Ropes, D. (2013). Intergenerational learning in organizations. *European Journal of Training and Development*, *37*(8), 713–727. doi:10.1108/EJTD-11-2012-0081

Sodexo. (2017). 2017 global workplace trends. Retrieved from http://www. sodexo.com/files/live/sites/sdxcom-global/files/PDF/Media/ Sodexo-2017-workplace-trends-report.pdf

Thomas, R. E. (2019). Strengthening the Leadership Platform Through Effective Mentoring Programs. *Conflict Resolution & Negotiation Journal, 2019*(1).

Xu, L. D., Xu, E. L., & Li, L. (2018). Industry 4.0: State of the art and future trends. *International Journal of Production Research*, *56*(8), 2941–2962. doi:10.1080/00207543.2018.1444806

Chapter 11
Social Media and Increased Venture Creation Tendency With Innovative Ideas:
The Case of Female Students in Asia

Syed Far Abid Hossain
Xi'an Jiaotong University, China

Xu Shan
Xi'an Jiaotong University, China

Mohammad Musa
Shaanxi Normal University, China

Preethu Rahman
Shaanxi Normal University, China

ABSTRACT

The purpose of this chapter is to ascertain the contemporary role of social media in increased venture creation tendency along with innovative ideas. The key objective of this study is to discover the tendency of female students' innovativeness in venture creation in China. A random sampling method was used to conduct a survey in different universities in China to identify the scenario of innovativeness in venture creation. Findings from primary data collection indicated that the female students in China are highly involved with social media marketing with innovative ideas. As a result, apart from traditional marketing, society is involved with contemporary marketing where innovativeness with social media and smartphones are the key factors. Innovative ideas in venture creation may generate additional earning for people with low income in society. Future studies with mixed methodology and respondents who use different social media as a tool to innovate new venture may shed light on the undiscovered phenomenon of social media marketing in the context of the mobile phone.

DOI: 10.4018/978-1-7998-0357-7.ch011

INTRODUCTION

Although entrepreneurial tendency with the help of social media which can be expressed as media entrepreneurship, is poorly investigated so far (Khajeheian, 2013), it has potential market due to social networking, trust, and online reviews. A recent survey revealed that 88% of total online buyers either review, read or trust online reviews (Chen, Luo, & Wang, 2017) which enhance the probability to make a final buying decision. Systematic mobile device usage (Hossain et al., 2019) recently is the main motivation for innovative business ideas. The broad objective of this chapter is to ascertain the contemporary role of social media in increased venture creation tendency along with innovative ideas. The specific objective of this study is to discover the tendency of female students' innovativeness in venture creation with a special focus of China because of its large population and recent development in online shopping. The role of mobile phones in media entrepreneurship is investigated with social networking (Hossain, 2019). However, the role of social media in generating innovative business idea is an undiscovered phenomenon. This study is an attempt to shed light on it. This chapter focused on related data from significant and relevant literature and finally provided a basis for further research in the next sections. The following sections describe the methodology of the study, Review of literature, result, discussion, limitations, conclusion and further research direction.

BACKGROUND OF THE STUDY

Social media and venture creation have a strong relationship. Social media helps to create new ventures. The venture creation phenomenon heavily influenced by the social media and media industries as long as they, in their very nature, fall into the culture and creativity-related businesses. The crucial distinctiveness of the venture creation activities such as creation, innovation, and original ways of thinking are critical in making business success. "Media particularly social media also plays an important role in influencing the venture creation phenomenon, by making a talk that transmits qualities and pictures credited to venture creation, by giving a transporter advancing innovative practices, and by empowering a pioneering soul in the society. Through these methods, media and venture creation have a reciprocal impact" (Khajeheian, 2013). The qualities of the social media items are particularly adjusted to the elements of the pioneering procedure, for example, self-sufficiency, ingenuity, hazard taking, star liveliness, and focused forcefulness. These measurements speak to the enterprising introduction of the firm, which can be characterized as the processes, practices, and decision-making activities that lead firms to choose to enter another market or dispatch another product. Media and companies are urged to be particularly risk-taking and innovative. The entrepreneurial approach they have to develop is without a doubt extremely important (Hang & VanWeezle, 2007). The goal of social media in venture creation must be to build a bridge between the general discipline of venture creation and the specificities of the media industry and social media (Achtenhagen, 2008).

As expressed by Fruhling and Digman (2000), online life can upsurge the client base and piece of the overall industry, which thus can encourage the development methodologies of a business. Coherently, a two-path correspondence between the clients and the business can give data and thoughts, to upgrade the nature of the business' market contributions and to urge them to improve. The stage in this way can give chances to draw in potential clients and hold existing ones, constructing a more grounded connection between the gatherings included (Mangold & Faulds, 2009). Social media can help to provide

Table 1. Electronic database used for the literature review

Source	online access
Science Direct	http://www.sciencedirect.com/
Emerald Insight	http://www.emeraldinsight.com/
Scopus	http://www.scopus.com/
Springer	http://link.springer.com/
Web of Science	https://apps.webofknowledge.com
ACM Digital Library	http://dl.acm.org/

"promoting that is increasingly compelling, new correspondence and dissemination channels, shorter time to showcase, redid items, 24-hour online specialized help, and online intuitive network" (Mukolwe & Korir, 2016). The reason organizations are exploiting the long-range informal communication destinations is that there is another age of purchasers whose purchasing conduct is altogether different from the past gatherings of clients. This development of a completely new economy on the Internet is the result of purchasers looking for more comfort in shopping, better correspondence, greater commitment, and capacity to choose (Tigo, 2012; Perju, 2015; Tosifyan & Tosifyan, 2017). Business people consider internet-based life a helpful device since it energizes the ID of chances in the business condition (Park & Sung, 2017). There are changing perspectives about "pioneering opportunity"; where Schumpeter (1932) expressed that one must search for new data accessible in the market so as to make a chance, Kirzner (1997) then again contended that a business must utilize the current data to find a chance. Web-based life is such a stage, that has now empowered business visionaries to find and make openings by surveying both existing and new data by conveying and interfacing with friends on the system (Park & Sung, 2017) and variety seeking (Hossain et al., 2019)

RESEARCH METHODOLOGY

First, the study utilized a literature review in the arena of social networking, innovation, media entrepreneurship and device usage in shopping. In addition, the authors used a case study from a real entrepreneur to represent the real scenario in China. Finally, a total number of 42 female students were interviewed in China. They all are involved directly with media entrepreneurship with the help of social networking. Most of the students are from 18-25 years old, studying at the university level in business, social science, and engineering majors. The total number of 42 students were asked questions about the perception of social media, value co-creation, target customers and difficulty handling.

REVIEW OF LITERATURE ON SOCIAL MEDIA AND VENTURE CREATION

Throughout the decades, various patterns have been knowledgeable about the business condition and got shriveled in the blink of an eye by any stretch of the imagination. Once in a while, the entry of certain marvels holds the ability to change and impact the business condition, as it were, one such wonder is web-based social networking, which is frequently reciprocally utilized with the term Web 2.0 (Kadam &

Ayarekar, 2014). There are assortments of internet-based life stages extending from informal organizations, private interpersonal organizations, to web journals and miniaturized scale online journals (Shabbir et al., 2016). A portion of the well-known and generally-utilized, long-range informal communication destinations include Facebook, LinkedIn, Instagram, Twitter, Skype, WhatsApp, Viber, WordPress, YouTube, Flikr, Google+, Pinterest, Quora, Reddit and Snapchat (Bajaj, 2017; Mehra, 2017; Maina, 2018). In spite of its origin in 1997 (Shabbir et al., 2016), there was a blast in internet-based life in the year 2000, when a noteworthy increment was found in the quantity of long-range interpersonal communication destinations (Kadam & Ayarekar, 2014). Thus, the manner in which clients and businesspeople impart changed in light of the fact that this stage enabled organizations to have more noteworthy access to a more extensive scope of target groups of onlookers; grasp their shifting needs and needs; improve and enhance items and benefits; and support client commitment with the business (Smith & Taylor, 2004; Jagongo & Kinyua, 2013). Be that as it may, among all the web-based life destinations, the most generally utilized online life stage by organizations and advertisers is Facebook (Driver, 2018). As indicated by the reports of Statista (2018), the main web-based life stages for advertisers everywhere throughout the world incorporate Facebook (94%), Instagram (66%), Twitter (62%), LinkedIn (56%), YouTube (half), Pinterest (27%) and Snapchat (8%). So also, in Bangladesh, most of the online retailers go through Facebook (Gilchrist, 2018). Because of the different points of interest of working together via web-based networking media, it has reshaped the regular promoting techniques in the nation. Presently many, beginning from understudies to housewives can initiate a start-up via web-based networking media since it doesn't require physical space or tremendous labor (Farhin, 2018). As expressed by Fruhling and Digman (2000), online life can upsurge the client base and piece of the pie, which thusly can encourage the development procedures of a business. Sensibly, a two-route correspondence between the clients and the business can give data and thoughts, to upgrade the nature of the business' market contributions and furthermore to urge them to improve. The stage can, in this way, give chances to draw in potential clients and hold existing ones, constructing a more grounded connection between the gatherings included (Mangold & Faulds, 2009). Web-based life can help give "promoting that is increasingly compelling, new correspondence and circulation channels, shorter time to showcase, modified items, 24 hours online specialized help, and online intelligent network" (Mukolwe & Korir, 2016). The reason organizations are exploiting the long-range informal communication destinations is on the grounds that there is another age of purchasers whose purchasing conduct is altogether different from the past gatherings of clients. This rise of a completely new economy on the Internet is the result of customers looking for more comfort in shopping, better correspondence, greater commitment and capacity to choose (Tigo, 2012; Perju, 2015; Tosifyan & Tosifyan, 2017). Business visionaries consider web-based social networking a valuable apparatus since it empowers distinguishing proof of chances in the business condition (Park & Sung, 2017). There are shifting perspectives about "enterprising chance"; where Schumpeter (1932) expressed that one must search for new data accessible in the market so as to make a chance, Kirzner (1997) then again contended that a business must utilize the current data to find a chance. Web-based life is such a stage, that has now empowered businesspeople to find and make openings by evaluating both existing and new data by conveying and collaborating with companions on the system (Park & Sung, 2017).

Social media has without a doubt opened up windows of new opportunities for people (Mukolwe & Korir, 2016). Another type of online businesspeople has surfaced who are not just reassuring new companies on

Internet-based life, but at the same time are utilizing the stage to upgrade the current organizations and make systems of clients more than ever (Fischer & Reuber, 2011). There is no satisfactory data, on what number of individuals are enlisting their organizations via web-based networking media every year, or how are their organizations performing, however. In any case, numerous analysts investigating the field of enterprise, unveil that internet-based life has made better approaches for working together and organizing feasible for people (Genç & Oksüz, 2015; Cesaroni et al., 2017; Mukolwe & Korir, 2016; Upkere et al., 2014; Melissa et al., 2013; Oke, 2013). A portion of the common advantages have been separated from the discoveries of investigates led on the connection between online networking and business visionaries in the creating nations (Melissa et al., 2013; Upkere et al., 2014; Vivakaran & Maraimalai, 2016; Cesaroni et al., 2017), which are referenced as pursues:

- Social media has enabled people to perceive pioneering opportunity through selling on the web and understand that their time can be utilized in profitably to acquire benefit from an enormous potential market.
- There are a low venture and working costs engaged with an internet-based life start-up that in the long run propels youthful businesspeople.
- Since numerous ladies can maintain their organizations from home and at their own comfort, this has helped them to have a superior work-life balance, where they can successfully oversee both their expert and individual lives at the same time.
- It is leeway that in most of the organizations via web-based networking media does not require an abnormal state of innovative proficiency, which decreases the start-up intricacy.
- Some different advantages incorporate practical advancement; access to a more extensive target group of onlookers; item improvement and upgrade; the foundation of outer networks of a client who might share their encounters; and the extent of better teaching clients (Baghdadi, 2013; Brengman & Karimov, 2012; Cesaroni et al., 2017).

As opposed to the preferences examined, subjective research led by Genç and Oksüz (2015) uncovered a few challenges business visionaries face while working together via web-based networking media. Some of the serious issues recognized were: impersonation of work, proficient disappointment, trouble in distinguishing explicit target gathering of people, more extensive spread-out of piece of the pie, and unreasonable challenge. In spite of the fact that a couple of creators have investigated the impact of web based life on businesspeople with regards to creating nations (Cesaroni et al., 2017), there is next to no scholarly proof on how internet based life have quickened enterprise. Various writing has been separated in importance to businesspeople (Afroze et al., 2014; Ahammad & Moudul-Ul-Huq, 2013; Islam & Ahmed, 2016), yet none of those showed the effect online life has on them, or the degree to which it has added to the development of potential business visionaries.

TENDENCY OF FEMALE STUDENTS USING SOCIAL MEDIA IN ASIAN REGION

Needless to say, in many regions of Asia, females are deprived and have less opportunity for work due to innumerable reasons. Sometimes, the work opportunity is not sufficient enough and sometimes culture is a factor which restricts women to do a certain type of work. For example: from a traditional Muslim family from Pakistan or Bangladesh, women may not work as a sales executive or in the marketing department. This is not the case in all the Asian region, however, this is a common phenomenon in many Asian regions. That is why now a day, women are using social media to do something for themselves.

Figure 1. Venture creation opportunity with social media

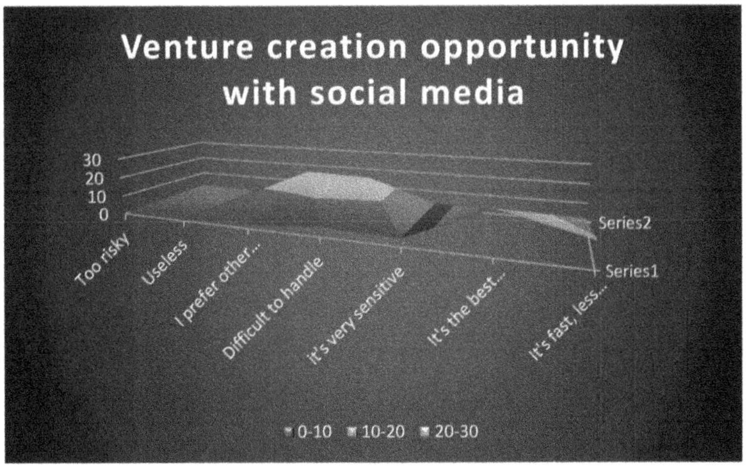

Till date, many of them are conducting a kind of little business by themselves with very short manpower and limited resources. However, the attraction or intention to conduct business is very high.

VENTURE CREATION OPPORTUNITY WITH SOCIAL MEDIA

Regardless of the type of social media, venture creation usually depends on the circle of friends of a motivated entrepreneur. For example: If someone wants to start selling goods via an online platform, the important issue is to determine the social media platform used by his or her circle of friends and well-known people. This is the first stage of motivation and individual can be pursued or initiated by a kind of self-confidence level that the circle of friends may help. In real life, this is true most of the time, however, if the seller can't maintain the reputation or fail to serve properly, the circle of friends can't help for a long time. So, there is tremendous opportunity to create venture due to the availability and affordability of social media and smartphones in particular. This process even works faster rather than other traditional techniques of selling. For example: From production to consumption, there are the complexity of intermediaries and promotional activities as well as cost, so, usually, it takes time to make the product known to the potential customers. But, in social media people trust friends and known-people and they can decide faster without hesitation. This is one of the reasons why social media is attracted by many people in the contemporary era.

CASE STUDY

Although social media is used by so many people, the innovativeness is always unique. Especially, in China the market is huge, the buyers are unlimited and the competition is tremendous. As a result, innovativeness is a crucial issue in China. A few months ago, one of the authors of this study was invited for lunch by a Chinese language teacher. The author had no idea about any business talk. The Chinese teacher who invited the author was actually a new entrepreneur with a social media marketing platform and she was trying to enhance her channel. She cordially ordered some dishes with due respect to her guest and started a normal conversation. All on a sudden, she revealed the main purpose of that meeting. Actually, she was involved with a new kind of social media platform where members can sell goods to their known people and earn profit without investing anything. The platform is simply a mobile Apps. Anyone can register with a phone number and enjoy shopping. The only difference between regular online shopping and this innovative new platform is earning profit by the person who introduces others to buy from this platform. Although it is hard to say that it is not a kind of real entrepreneurship where people conduct business activities by themselves, however, this is a platform generated by the third party, sellers enjoy the advantage of mobile technology and conduct social media marketing activities to boost-up the sales. This is not a job because there is no work time and no salary. So it is purely a kind of start-up from a different perspective which is an example of innovative entrepreneurship practice in China.

THE BLESSING OF TECHNOLOGY IN INNOVATION

The importance of social networking was acknowledged by Ardichvili et al. (2000), who claimed that social networks are one of the main factors influencing the core process of opportunity recognition. Fischer & Reuber (2011) also stated that, in order to answer complex business issues that arise, business industries are likely to interact with other people to discuss their options to come to a decision as well as cooperative partnerships. For these, they often lead to new business insights and therefore new goals to attain. Social networks can be identified with the more popular term, social media. The role of social media has been researched in various fields already (Fischer & Reuber, 2011; Greve & Salaff, 2003). Therefore, in this book tries to discover the influence of national culture on new business processes, there is also the aim to discover what kind of role there is for social media platforms in the business decisions. The research of Fischer and Reuber (2011) discovered that social media platforms such as Twitter can help businesses to create and follow opportunities. Fischer and Reuber (2011) aim to propose social media as a "corridor for opportunity creation and exploration", yet they state that there are several factors to be taken into account.

De Carolis, Litzky, and Eddleston (2009) state that research has indicated the importance of networks and social capital during the process of creating a new venture. More specific, new venture creation appears to be the result of the social network of the entrepreneur combined with their cognitive biases. De Carolis et al. (2009) found out that cognitive bias could even explain why social capital has a greater effect on the progress of creating a new venture for some. However, research on the intersection of the concepts of social networks, new venture creation, and effectuation is lacking. Fischer and Reuber (2011) claim that the use of effectual thinking processes by businesses can increase the efficacy of new ventures. Also, costs of business failure have proven to be reduced, due to earlier failure and lower levels of investment of effectual firms compared to causational ones. Besides, the level of expertise correlates with the usage of effectual logic by industries. Prior research has indicated that using of effectual logic is forced to make decisions within an uncertain business environment, relating to the fact that they shape the market instead of treating the market as a given fact. This relates inevitably to one of the most important concepts in business processes, namely the recognition of opportunities.

NEW CONCEPT DEVELOPMENT FOR VENTURE CREATION

The technological innovation of new products and concepts can be seen as a key driver of competitive advantage and commercial success. The number of worldwide social media users is increasing every day, which becomes a great opportunity for businesses to reach their online audience through social networks. One of the advantages of the social network is that it enables businesses to reach a worldwide customer population so that customers can survey, select, and purchase products and services from businesses around the world (Al Kailani & Kumar, 2011). In particular, peer communication through social media, a new form of consumer socialization, has profound impacts on consumer decision making and thus marketing strategies. Consumer socialization theory predicts that communication among consumers affects their cognitive, effective, and behavioral attitudes (Ward, 1974). Nowadays, the analysis of consumer behavior is central to marketing success, especially since most potential consumers are using the internet and different online socializing tools. The online audience is a booming market worldwide, however giving its globalized nature a level of segmentation is needed cross-culturally. The unique aspects of social media and its immense popularity have revolutionized marketing practices such as advertising and promotion (Hanna, Rohm, & Crittenden, 2011). Social media has also influenced consumer behavior from information acquisition to post-purchase behavior such as dissatisfaction statements or behaviors (Mangold & Faulds, 2009) and patterns of Internet usage (Ross et al., 2009; Laroche et al., 2012). Other functions of social media involve affecting and influencing perceptions, attitudes and end behavior (Williams & Cothrell, 2000) while bringing together different like-minded people (Hagel & Armstrong, 1997). In an online environment, Laroche (2012) pointed out that people like the idea of contributing, creating, and joining communities to fulfill needs of belongingness, being socially connected and recognized or simply enjoying interactions with other like-minded members.

In recent years, social networking sites and social media have increased in popularity, at a global level. For instance, Facebook has more than a billion active users (as of 2012) since its beginning in 2004. Indeed, online social networks have profoundly changed the propagation of information by making it incredibly easy to share and digest information on the internet (Akrimi & Khemakhem, 2012). Marketing through social media can help increase brand recognition, it is cost-effective, and it helps improve brand dependability and power. We can talk about Facebook marketing here. Nearly any kind of brand can be promoted through Facebook, turning potential customers into active fans who follow news of promotions and developments and share the news with their own circle. For successful marketing through Facebook, a business needs a clear goal and strategy, stay active in online with regular posts, encourage comments and reply quickly, nurture relationships with customers and frequent online promotions.

LOWER RISK AND LIMITED INVESTMENT USING SOCIAL MEDIA

Social networks can be used as a great tool and effective platform (Jain & Sinha, 2018) for people who typically share a common interest or activity. They provide a variety of ways for users to interact with each other. Every person, who wants to join a social networking site he must create his own profile. This profile describes his interests, needs, and wishes. Through that profile, we can know his friends who have similar interests. These networks offer a unique opportunity for highly targeted marketing. The use of social network can contribute to the success of the company. The Internet-based applications have the advantage that they are actively working with the customers and can get feedback directly from there.

Figure 2. New concept development

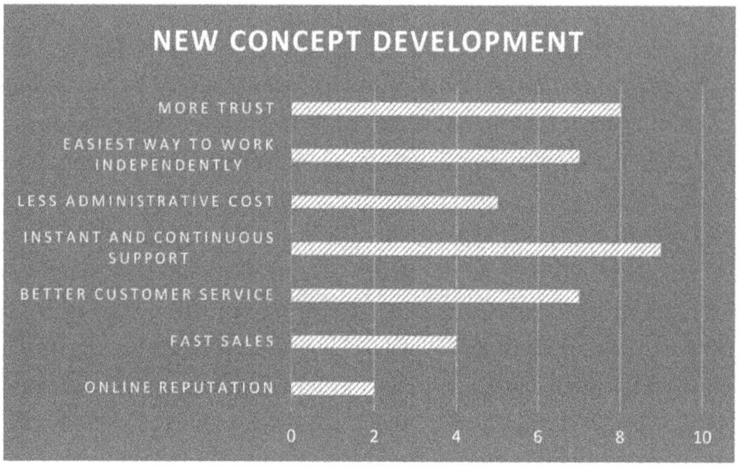

Social network marketing can be very advantageous for businesses as it has lower risk and limited investment options for investors. Establishing communities around products and services is a potential strategy to build brand loyalty, establishing exit barriers, and facilitating viral marketing through self-emergent customer testimonials. Such communities can also be a source of innovation by soliciting consumer input, and customer suggestions. Social networking can find new customers, and help conduct brand intelligence and market research with that the company can make its advertising in a social network. The communication in the social network gets shoppers to listen to one another, review ratings for products and services, and provide product knowledge and personal information. Social networks protect users from interaction with the outside world and keep information and interaction away from strangers. Social marketing can be an inexpensive way to promote a company rather than putting together a huge marketing team or a prohibitive budget. While using social media as a business promoter business must be especially careful in the market research and advertising laws not to cross legal borders. The advertisement in a social net brings danger to product brands. Social networks are user-generated contents where end users about their experiences with products, services, and customer service, etc. The contents sometimes can be critical to the product, and societies have very little control in which end users share in their social network. The supervision of the general perception of brands of the online community and directing problems discussed online is extremely important for brands of the product and service. We must not forget the most important thing through social marketing there is an opportunity to collect consumer data; however, these compilations of data cannot just be trapped in it. Social network users are more careful regarding sharing their private data. To have access to the consumer Information, a business must build products and applications which will require that a user divides at least some personal data.

OPPORTUNITY TO EXPAND BUSINESS

In recent years, a change in the relationship between companies and customers has revealed. The customer has received more and more control over and through regarding the company and its products through the goal to achieve sustainable business development (Oskam, Bossink & de Man, 2018). The core of

Figure 3. Handling difficulty

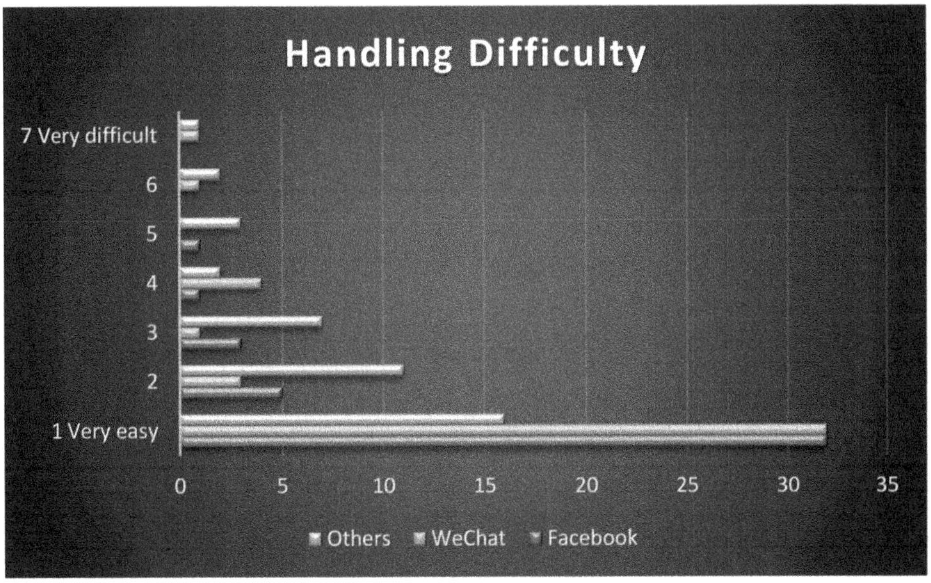

any business is the customers, and social networking represents an opportunity to build even closer and more profitable relationships with customers, which brings unlimited opportunity to expand the business. Therefore, the company must respond to this change. In fact, companies can gain benefits through using social networking in marketing: they can achieve a better understanding of the customer needs and can build better relationships with customers. For companies to achieve measurable commercial benefits, they must plan their activities in social networks for better control and measurement. The correct behavior can also change the way in how the companies consider their customers. This is the area for the use of customer relationship management to intersect with social networks, and customers, and social networks, which are looking for ways to deal with the companies in connection. Whenever the coordination between social networking and marketing is achieved, companies can more easily follow their clients, achieve their requirements, control, and measure their activities. Social-networking tools may provide a better introduction mechanism for accounts with higher conversation rates. Social networking sites are the source of almost inexhaustible views of clients and situations, and the challenge is to control this information in an appropriate manner and in a meaningful way for the company and that brings real benefits for them. Social networking is also a suitable framework for core activities in marketing on the Internet. Strategically, establishing communities around products and services has been a well-known method of building brand loyalty, establishing exit barriers, and facilitating viral marketing through self-emergent customer testimonials. One of the key success factors of social marketing is involving its customers and determining their needs on a personal level by encouraging consumers to participate enthusiastically and listening to their desires. Burt (1992) states that relations with colleagues, friends, and clients can create opportunities that the business can turn into a success by successfully using their financial and human capital (to finance and create the product). It is, therefore, the final arbiter of competitive success. Coleman (1988) agrees with this point of view. He defends that there are different functions and purposes assigned to the concept of social capital. He adds that an important form of social capital is

Figure 4. Target customers

the potential for information, which gives the access to the information that may make investors exploit more opportunities or create more opportunities and find ways to expand business gradually.

IMPLICATIONS, LIMITATIONS AND FUTURE RESEARCH

Feedback from the customers is a tool to come up with innovativeness in social media marketing and it also can help to make a good relationship (Lin, Luo, Cheng, & Li, 2019). Customers sometimes have latent demand which is complicated to understand sometimes for the marketer. As the social media marketing procedure is very direct and contact with the customer is easily reachable, sellers can get more feedback from buyers and come up with more innovative ideas to satisfy them and retain them for a longer period of time. Also, the feedback from the customers works as a tool to boost up the business or bring new customers. Word-of-mouth or e-word of mouth is a very crucial factor in this regard. It is true that all the customers are not satisfied with the same product always, however, the majority is an issue. If the majority of customers provide positive feedback, the others could take it positively and reach a final decision without enough hesitation. In China, people usually read comments from other users before buying a particular product online. As there is no chance to touch or feel the product in real, comments or feedback from other users motivate the potential buyers to go ahead and make purchases. As a result, feedback from customers is a very crucial factor in social media marketing in achieving innovativeness in business. This study was limited to collect data from only one Asian country. Future research may go in depth with more scientific research in broad scale.

CONCLUSION

In the contemporary era, people are fascinated with change and focus on sustainable business development (Naranjo-Valencia et al., 2018). New and innovative ideas attract people and excite them as well. Innovativeness is observed in various industries in the market especially in fashion and clothing, jewelry, travel, film, and food industry. Innovativeness is appreciated by consumers over the years and made them happy and satisfied. If the new venture is innovative, people will be automatically attracted to it. They will be curious to know about it. Apart from that, innovative venture creation is a tool to compete with the other existing competitors within the same industry in the market. In some restaurants, innovative interior decoration attracts customers. Some people go there just for the innovative interior or service such as eating inside an airplane or customers enjoy the variation when waitresses serve the food wearing a nurse uniform in some restaurants. It is easily arguable that all the innovative venture may not be accepted by the customers, however, innovativeness could be either a primary or secondary tool to face the competitors and survive in the market. In addition, a venture can grow fast due to innovativeness.

REFERENCES

Achtenhagen, L. (2008). Understanding entrepreneurship in traditional media. *Journal of Media Business Studies*, *5*(1), 123–142. doi:10.1080/16522354.2008.11073463

Afroze, T., Alam, K., Akther, E., & Jui, S. N. (2014). Women Entrepreneurs in Bangladesh-Challenges and Determining Factors. *Journal of Business and Technology (Dhaka)*, *9*(2), 27–41. doi:10.3329/jbt.v9i2.26194

Ahammad, I. & *Moudud-Ul- Huq, S.* (2013). Women Entrepreneurship Development in Bangladesh Challenges And Prospects. *International Journal of Innovative Research and Development*, (pp. 41–48). Retrieved from http://www.ijird.com/index.php/ijird/article/view/36096/29238

Akrimi, Y., & Khemakhem, R. (2012). What Drive Consumers to Spread the Word in Social Media? *Journal of Marketing Research & Case Studies*, 1–14. doi:10.5171/2012.969979

Al Kailani, M., & Kumar, R. (2011). Investigating Uncertainty Avoidance and Perceived Risk for Impacting Internet Buying: A Study in Three National Cultures. *International Journal of Business and Management*, *6*(5), 76–92. doi:10.5539/ijbm.v6n5p76

Baghdadi, Y. (2013). From E-commerce to social commerce: A framework to guide enabling cloud computing. *Journal of Theoretical and Applied Electronic Commerce Research*, *8*(3), 12–38. doi:10.4067/S0718-18762013000300003

Bajaj, R. (2017). Top 10 Most Popular Social Sites and Apps in 2017. Retrieved from https://www.linkedin.com/pulse/top-10-most-popular-social-networking-sites-apps-2017-rajiv-bajaj

<bok>Mukolwe, E., & Korir, J. (2016). Social Media and Entrepreneurship: Tools, Benefits, and Challenges. A Case Study of Women Online Entrepreneurs on Kilimani Mums Marketplace on Facebook. International Journal of Humanities and Social Science, 6(8), 248–256.</jrn>

Brengman, M., & Karimov, F. P. (2012). The effect of web communities on consumers' initial trust in B2C e-commerce websites. *Management Research Review*, *35*(9), 791–817. doi:10.1108/01409171211256569

Burt, R. S. (1992). The social structure of competition. In N. Nohria, & R. Eccles (Eds.), *Networks and Organizations: Structure, Form and Action* (pp. 57–91). Boston, MA: Harvard Business School Press.

Cesaroni, F. M., Demartini, P., & Paoloni, P. (2017). Women in business and social media: Implications for female entrepreneurship in emerging countries. *African Journal of Business Management, 11*(14), 316–326. doi:10.5897/AJBM2017.8281

Chen, K., Luo, P., & Wang, H. (2017). An influence framework on product word-of-mouth (WoM) measurement. *Information & Management, 54*(2), 228–240. doi:10.1016/j.im.2016.06.010

Coleman, J. S. (1988). Social Capital in the Creation of Human Capital. *American Journal of Sociology, 94,* Supplement: Organizations and Institutions: Sociological and Economic Approaches to the Analysis of Social Structure, S95-S120.

De Carolis, D. M., Litzky, B. E., & Eddleston, K. A. (2009). Why Networks Enhance the Progress of New Venture Creation: The Influence of Social Capital and Cognition. *Entrepreneurship Theory and Practice, 33*(2), 527–545. doi:10.1111/j.1540-6520.2009.00302.x

Driver, S. (2018, Oct. 15). Social Media for Business: A Marketer's Guide. Business News Daily. Retrieved from https://www.businessnewsdaily.com/7832-social-media-for-business.html

Farhin, N. (2018, Jan. 9). How small businesses use Facebook to promote products and services. Dhaka Tribune. Retrieved from https://www.dhakatribune.com/ business/2018/01/09/small-businesses-use-facebook-promote-products-services

Fischer, E., & Reuber, A. R. (2011). Social interaction via new social media: (How) can interactions on Twitter affect effectual thinking and behavior? *Journal of Business Venturing, 26*(1), 1–18. doi:10.1016/j.jbusvent.2010.09.002

Fruhling, A. L., & Digman, L. A. (2000). The Impact of Electronic Commerce on Business-Level Strategies. *Journal of Electronic Commerce Research, 1*(1), 13–22.

Genç, M., & Oksüz, B. (2015). A fact or an Illusion: Effective social media usage of female entrepreneurs. *Procedia: Social and Behavioral Sciences, 195*, 293–300. doi:10.1016/j.sbspro.2015.06.345

Gilchrist, K. (2018, July 17). Facebook and 3 millennials are changing the start-up scene in Bangladesh. CNBC. Retrieved from https://www.cnbc.com/2018/07/17/shopup-bangladesh-start-up-uses-facebook-to-help-micro-entrepreneurs.html

Greve, A., & Salaff, J. W. (2003). *Social Networks and Entrepreneurship, Entrepreneurship Theory and Practice*, (Fall): 1–22. doi:10.1111/1540-8520.00029

Hagel, J., & Armstrong, A. G. (1997). *Net gain: Expanding markets through virtual communities*. Boston, MA: Harvard Business School Press.

Hang, M., & Van Weezle, A. (2007). Media and entrepreneurship: A survey of the literature relating both concepts. *Journal of Media Business Studies, 4*(1), 51–70. doi:10.1080/16522354.2007.11073446

Hanna, R., Rohm, A., & Crittenden, V. L. (2011). We're all connected: The power of the social media ecosystem. *Business Horizons, 54*(3), 265–273. doi:10.1016/j.bushor.2011.01.007

Hossain, S. F. A. (2019). Social Networking and Its Role in Media Entrepreneurship: Evaluating the Use of Mobile Phones in the Context of Online Shopping–A Review. [JMME]. *Journal of Media Management and Entrepreneurship*, *1*(1), 73–86. doi:10.4018/JMME.2019010105

Hossain, S. F. A., Nurunnabi, M., Hussain, K., & Saha, S. K. (2019). Effects of variety-seeking intention by mobile phone usage on university students' academic performance. *Cogent Education*, *6*(1). doi:10.1080/2331186X.2019.1574692

Hossain, S. F. A., Ying, Y., & Saha, S. K. (2019). Systematic Mobile Device Usage Behavior and Successful Implementation of TPACK Based on University Students Need. In *Science and Information Conference* (pp. 729-746). Cham, Switzerland: Springer.

Islam, N., & Ahmed, R. (2016). Factors Influencing the Development of Women Entrepreneurship in Bangladesh. doi:. doi:10.2139srn.2851786

Jain, S., & Sinha, A. (2018). Social Network Analysis: Tools, Techniques, and Technologies. In Social Network Analytics for Contemporary Business Organizations (pp. 1-18). Hershey, PA: IGI Global.

Kadam, A., & Ayarekar, S. (2014). Impact of Social Media on Entrepreneurship and Entrepreneurial Performance: Special Reference to Small and Medium Scale Enterprises. *SIES Journal of Management*, *10*(1), 3–11.

Khajeheian, D. (2013). New Venture Creation in Social Media Platform; Towards a Framework for Media Entrepreneurship. Handbook of Social Media Management Value Chain and Business Models in Changing Media Markets, 125-142. doi:10.1007/978-3-642-28897-5_8

Khajeheian, D. (2013). New venture creation in social media platform; Towards a framework for media entrepreneurship. In *Handbook of social media management* (pp. 125–142). Berlin, Germany: Springer. doi:10.1007/978-3-642-28897-5_8

Kirzner, I. M. (1997). Entrepreneurial discovery and the competitive market process: An Austrian approach. *Journal of Economic Literature*, *35*(1), 60–85.

Laroche, M., Habibi, M. R., Richard, M. O., & Sankaranarayanan, R. (2012). The effects of social media based brand communities on brand community markers, value creation practices, brand trust and brand loyalty. *Computers in Human Behavior*, *28*(5), 1755–1767. doi:10.1016/j.chb.2012.04.016

Lin, J., Luo, Z., Cheng, X., & Li, L. (2019). Understanding the interplay of social commerce affordances and swift guanxi: An empirical study. *Information & Management*, *56*(2), 213–224. doi:10.1016/j.im.2018.05.009

Maina, A. (2018). 20 Popular Social Media Sites Right Now. Retrieved from https://smallbiztrends.com/2016/05/popular-social-media-sites.html

Mangold, W. G., & Faulds, D. J. (2009). Social media: The new hybrid element of the promotion mix. *Business Horizons*, *52*(4), 357–365. doi:10.1016/j.bushor.2009.03.002

Mehra, G. (2017). 105 Social Networks Worldwide. Retrieved from https://www.practicalecommerce.com/105-leading-social-networks-worldwide

Melissa, E., Hamidati, A., & Saraswati, M. S. (2013). Social Media Empowerment: How Social Media Helps to Boost Women Entrepreneurship in Indonesian Urban Areas. *The IAFOR Journal of Media,* Communication and Film, 1(1), pp. 77-90.

Naranjo-Valencia, J. C., Calderón-Hernández, G., Jiménez-Jiménez, D., & Sanz-Valle, R. (2018). Entrepreneurship and innovation: Evidence in Colombian SMEs. In Handbook of Research on Intrapreneurship and Organizational Sustainability in SMEs (pp. 294–316). Hershey, PA: IGI Global. doi:10.4018/978-1-5225-3543-0.ch014

Oke, D. F. (2013) The Effect of Social Network on women entrepreneurs in Nigeria: A case study of Ado-Ekiti Small scale Enterprise. *International Journal of Education and Research, 1*(11), pp. 1-14.

Oskam, I., Bossink, B., & de Man, A. P. (2018). The interaction between network ties and business modeling: Case studies of sustainability-oriented innovations. *Journal of Cleaner Production, 177,* 555–566. doi:10.1016/j.jclepro.2017.12.202

Park, J. Y., & Sung, C. S. (2017). Does Social Media Use Influence Entrepreneurial Opportunity? A Review of its Moderating Role. *Sustainability, 9*(1593), 1–16. doi:10.3390u9091593

Perju, A. (2015). Gender Differences in Modeling the Influence of Online Marketing Communication on Behavioral Intentions. *Procedia Economics and Finance, 27,* 567–573. doi:10.1016/S2212-5671(15)01034-5

Schumpeter, J. A. (1932). *The Theory of Economic Development.* London, UK: Transaction Publishers.

Shabbir, M. S., Ghazi, M. S., & Mehmood, A. R. (2016). Impact of Social Media Applications on Small Business Entrepreneurs. *Arabian Journal of Business and Management Review*6(203), pp. 1-3. doi:. doi:10.4172/2223-5833.1000203

Smith, P., & Taylor, J. (2004). Marketing Communications: An Integrated Approach. London, UK: Kogan Page.

Jagongo, A., & Kinyua, C. (2013). The Social Media and Entrepreneurship Growth (A New Business Communication Paradigm among SMEs in Nairobi). *International Journal of Humanities and Social Science, 3*(10), 213–227.

Tigo, M. (2012). Revisiting the Impact of Integrated Internet Marketing on Firms' Online Performance: European Evidences. *Procedia Technology, 5,* pp. 418-426.

Tosifyan, M., & Tosifyan, S. (2017). A Research on the effect of social media on tendency to entrepreneurship and business establishment (Case Study: Active Iranian Entrepreneurs in Social Media). *Italian Journal of Science & Engineering, 1*(1), pp. 43-48.

<unknown>Assaad, W., & Gómez, J. M. (2011). Social Network in marketing (social media marketing) opportunities and risks. International Journal of Managing Public Sector Information and Communication Technologies (IJMPICT), 2(1). Retrieved from http://www.seokursu.com.tr/social-network-in-marketing.pdf</eref>

<unknown>Statista. (2018). Leading social media platforms used by marketers worldwide as of January 2018. Retrieved from https://www.statista.com/statistics/259379/social-media-platforms used-by-marketers-worldwide/</eref>

<unknown>Upkere, C. L., Slabbert, A. D., & Upkere, W. I. (2014). Rising Trend in Social Media Usage by Women Entrepreneurs across the Globe to Unlock Their Potentials for Business Success Mediterranean Journal of Social Sciences, 5(10), 551–559.</jrn>

Vivakaran, M. V., & Maraimalai, N. (2016). Feminist pedagogy and social media: A study on their integration and effectiveness in training budding women entrepreneurs. *Gender and Education,* pp. 1-21. doi:10.1080/09540253.2016.1225008

Ward, S. (1974). Consumer Socialization. *The Journal of Consumer Research, 1*(2), 1–14. doi:10.1086/208584

Williams, L., & Cothrell, J. (2000). Four smart ways to run online communities. *Sloan Management Review, 41*, 81–91.

Chapter 12
Study of Technology-Based Innovations in Supply Chain Management Function of Indian Firms:
Strategic Imperatives

Som Sekhar Bhattacharyya
https://orcid.org/0000-0003-3259-0319
National Institute of Industrial Engineering, India

Bibhash Laik
National Institute of Industrial Engineering, India

Raunak Jaiswal
National Institute of Industrial Engineering, India

ABSTRACT

Supply Chain Management (SCM) has gained importance in recent years. Innovation and technological interventions in SCM would be required to remove inefficiencies. It has become imperative for firms to undertake new innovations in SCM to remain competitive. This chapter focuses on physical and digital innovation in Indian market context in the context of SCM . The authors explore the strategic imperative of technology-based SCM innovation by performing detailed literature review regarding new automated technological innovations in SCM to understand the new set of business gains to be incurred from SCM. The authors then carried out, through a semi-structured questionnaire, in-depth personal interviews of the 24 SCM experts in the study. Thematic content analysis was done. The main finding of the study was that physical innovation in SCM has occurred at a slower pace as compared to digital innovation. Digital innovation was perceived to be helping firms more than physical innovation in SCM. The major challenge has been the integration of the new system with the existing SCM system.

DOI: 10.4018/978-1-7998-0357-7.ch012

INTRODUCTION

Supply Chain Management (SCM) has become a critical business function (Jain et al.,2010). The concept of SCM encompassed the upstream and downstream flow of products, services, finances, information and other customers (Felea & Albastroiu, 2013). This has also been true for the reverse flow as well (Govindan, Soleimani, & Kannan, 2015). Recent developments like globalisation and its increasing competitive pressures brought a great change in the mindsets of top management team of organizations towards organizational SCM function (Hervani et al., 2005). This was done by mainly focussing upon increasing the efficiency and profitability of SCM functions through sourcing and increased just in time logistics (Li, 2014). To enhance the effectiveness of SCM functions, upgradation and introduction of new technologies would be the need of the hour (Ivanov & Sokolov, 2010). These would provide a competitive edge to bring about a transformation in the management of SCM (Reddy, 2005). Proper management of SCM function would ensure an increased focus on quality, quantity, and timely delivery (Fish, 2011). The main aim of SCM function has been to increase the SCM surplus, both towards value creation and addition (Kim, Jeong, & Jung, 2014). This has been enabled through integration of technology in SCM function (Jain et al., 2010). Combination of SCM functions and technologies have become essential to remain competitive (New, 2010). There has been a clear inclination towards greater application of SCM technologies (Russel & Taylor, 2008). With the advancement achieved in technologies like in automation, firms could adopt and implement SCM technologies to protect market share and improve market penetration by spotting new trends ahead of competitors (Christensen, 2013). In the present-day context, consumers required varieties of products in a shorter time frame (Vonderembse et al., 2006). For firms this would reduce the product life cycle and the time to market so as to provide higher levels of customer service (Lee, 2002). As a result, various SCM firms have entered competitive marketplace to provide SCM-related technological and business solutions (Li et al., 2006).

Application of technology has helped the simplification of SCM function (Chou, Tan, & Yen, 2004). This enabled business firms to operate more efficiently, provide transparency in process, control firm level inventory and help to reduce logistical operational costs (Gupta & Kohli, 2006). Additionally, through a more stable and efficient SCM, firms could greatly enhance customer satisfaction and customer retention (Kim, 2006). Application of information technology in SCM has been viewed as a means to enhance business competitiveness and performance (Trkman et al., 2010). In developed countries, many new SCM technologies have been used, whereas in India the rate of adoption of new technologies in SCM has been slow (Govindan et al., 2014). By the late 2010s, adoption of technology had gained momentum as the competitive pressure has been building up after prolonged globalisation (Chopra & Meindl, 2016). 'SCM innovation' could be simply defined as any new idea, device, method or such initiatives which changed the existing ways of doing work and made work more efficient and simpler (Devila, Epstein, & Shelton, 2012; Kumar, 2012; Chopra & Meindl, 2016). SCM innovations reduced the consumption of resources such as time, money and such other transactional efforts (Davenport, 1993; Chopra & Meindl, 2016). In present day business environment innovation of organizational process and product has been a big challenge and has been viewed as crucial for firm level success (Teece, 2010). According to Dittmann (2017) innovation in SCM was classified into two broad categories namely 'Physical innovation' which involved technologies that were in physical form like '3-D printing', 'Drones', 'Robots', and 'Wearable technologies.' Whereas, the second classification was 'Digital/analytical innovation'. It involved technologies that were in digital form which included 'Big Data', 'Cloud Computing', 'Internet of Things (IOT)' and such others. Insights would be drawn from the data stored on SCM for better decision making

(Dittmann, 2017; Verma, Bhattacharyya, & Kumar, 2018). These categories are interlinked. Physical innovation must be connected to the digital side through appropriate compatible software for best results (Barton & Court, 2012). This connectivity would generate data for machines to learn and grow its range of application exponentially rather than remaining constrained to a fixed environment (Akter et al., 2016). Literature has already been available which indicates how in different industries (Nayak & Bhattacharyya, 2019) and different market segments like the Bottom of the Pyramid (Verma & Bhattacharyya, 2016) technology has been a game changer. Bhattacharyya & Shrey (2019) had talked about how technology has been applied by entrepreneurs in the domain of SCM in India. However, a detailed study on how technology innovations in SCM has been altering the SCM landscape in emerging economies like in India has been lacking. In this study the authors attempt to address this gap.

LITERATURE REVIEW

The authors have undertaken Literature Review (LR). The authors conducted a conceptual LR in 'SCM Innovation' & 'SCM Strategy' (Rocco & Plakhotnik, 2009). The authors conducted the LR search on EBSCO, Proquest, Emerald Insight, Jistor and such other journals databases. The authors presented the LR in two parts namely 'innovation in SCM' and 'SCM & strategic benefits'. The authors undertook a conceptual LR regarding both 'SCM Innovation' as well as for 'SCM Strategy'. Firstly, a review of extant research on the wide aspects of recent innovations on SCM has been presented. The authors presented the different SCM technologies, its benefits, and its adoption challenges. This area of research under SCM innovations has been tabulated in Table 1.

In Table 2, the authors discussed an integrated perspective on SCM Strategy (SCMS), Corporate Strategy (CS), Competitive Strategy (CMS) and its relationships, contribution and impact on each other.

RESEARCH GAPS AND METHODOLOGY

In the literature survey conducted by authors, it was noticed that there has been a lack of availability of literature on drivers, challenges and strategic value of new SCM innovation from an Indian perspective. Most of the studies have been conducted in countries like USA, UK & other European countries (Alicke, 2016). An Indian perspective was found to be missing in this context given the fact that the Indian economy was evolving and complex (Bhattacharyya, 2011). The authors in this paper have attempted to bridge this gap. The objective of this research was to understand the strategic value of new technology-based SCM innovations in India. An exploratory study was conducted by using a semi structured in-depth open-ended questionnaire (Creswell & Poth, 2017). Interview of experts from different types of organisations where SCM played a prominent role was done (Rowley, 2012). Research was also conducted with a motive to understand the current level of adoption and prospects of new SCM innovations in India (Silverman, 2013; Rowley, 2012) done like in (Bhattacharyya & Jha, 2015) . Further, the study also assessed the strategic value which was created by the new innovations in SCM for the companies adopting it. The authors also researched to find out the drivers of new technology-based SCM innovation and to assess the importance of each driver. In the context of India thus the research questions were -

1. What are the new technology based innovations in SCM practices occurring?

Table 1. Literature review on SCM innovation

S.No	Author(s)	Insights
1	Heutger& Kückelhaus (2015)	Unmanned Aerial Vehicle (UAV) use in logistics was discussed. UAV could be used in urban first and last mile delivery, rural delivery, and intralogistics. Apart from the benefits, the challenges in terms of regulatory pressure in adopting and public acceptance of UAV were discussed.
2	Agarwal, Modgil, Patyal, & Maddulety (2012)	The role of RFID (Radio Frequency Identification) in minimizing reverse logistics was discussed. In present day context, how RFID could minimize reverse logistics costs was discussed. The impact and effectiveness of RFID in minimizing reverse logistics was explained with an algorithm.
3	Diwan (2016)	Internet of Things (IOT) in logistics for autonomous logistics & smart logistics entities was studied. The application of IOT in logistics industry and its impact on logistics by improving inventory management, transparency and business process optimization was explained. It was illustrated with a used case how IOT could be an enabler for autonomous logistics.
4	Ratnajeewa & Bandara (2015)	Green logistics distribution practices and its significance was discussed. The benefits a firm could get by going green like reduction of CO_2 emission, cost savings, SCM optimization as well as boosted business performance was described. Intangible benefits associated with green logistics like image and reputation enhancement was demonstrated. Profitability and environmental consideration were both simultaneously required to be practiced in a firm.
5	Angeleanu (2015)	Emerging technology trends especially on cloud logistics and 3D printing were focussed in the context of SCM. These technology trends would improve the performance, efficiency and lower costs. The logistics service provider could migrate its logistics process in the cloud by adopting Software as a Service (SaaS) solution. This would help in improved capital expenditure and faster deployment.
6	Brar, Rabbat, & Raithatha (2015)	Drones would be successful in last mile delivery because of lower cost, value of fast delivery and convenience. The major disadvantage of drone technology has been the poor battery life which affected the hour of operation and also drones has been heavier in dead weight. The major barrier for adoption were mainly regulation and societal factors like privacy and safety.
7	Gibson and Matthews (2013)	Technological changes and how the act of doing business within organizations and its role on SCMS have altered were discussed. Both public and private firms were evaluated. The major challenge was in integrating new technology with the existing SCMS systems.
8	Palanivelu & Dhawan (2015)	The growing concern for environment and how this factor had pushed companies to adopt sustainable process and technologies were discussed. The drivers of green logistics like climate change, GHG emission and rising energy costs were also mentioned.
9	Tiwari, Wee, & Daryanto (2018)	The role of big data analytics in SCM was discussed. Big data analytics could capture and analyse huge amount of data and could give meaningful insights which could help in the decision making in SCM. Big data analytics could play an important role across all supply chain activities which could help in process improvement and sustainable supply chain development.
10	Buyukozkan & Gocer (2018)	This article discussed the importance of digital supply chain for an organization. Digital supply chain could take advantages of new approaches with new technological innovation and analytical methods. This article also discussed the various challenges and issues in implementing digital supply chain like lack of proper planning, lack of collaboration among members and lack of flexibility of SCM.
11	Daya, Hassini, & Bahroun (2017)	This paper discussed the role and impact of Internet of Things (IOT) in SCM. Here the article discussed the role and impact of IOT in each of the major four process that is source, make, deliver and return. Some of the major impacts were improved visibility and reduced lead time and overall costs. The author also discussed about the challenges of IOT implementation on SCM. The major challenges would be lack of framework and security and privacy issues.
12	Tjahjono et al. (2017)	The article discussed about the impact of Industry 4.0 (fourth industrial revolution) on the supply chain. Warehouse, procurement, logistics and order fulfilment were the functions of supply chain that have been considered. It was noticed that order fulfilment and logistics were most affected by the introduction of Industry 4.0. The author also discussed about the benefits after introduction of Industry 4.0. Some of the benefits were increased efficiency and increased flexibility which could help to meet customer demands effectively.
13	Merlino & Sproge (2017)	The authors explored the major technological changes that have been impacting and transforming the supply chain processes. Some of the technologies have been artificial intelligence, robotics and big data. The authors also discussed their benefits along the supply chain. The author also found out the major obstruction in deployment of new technologies have the human ones like change management.

Table 2. Literature review on SCM strategy and corporate strategy

S.No	Author(s)	Insights
1	Prasad, Subbaiah, & Rao (2012)	Alignment of the CMS with SCMS through Quality Function Deployment (QFD) was discussed. If these functions were not aligned it might lead to actions not being consistent with customer needs. CMS of firms had to be defined and SCMS had to be designed in aligning with CMS to meet customer satisfaction. SCM index was computed using QFD and utility-based optimization for choosing the optimum set of SCM objectives.
2	Feizabadi, Singh, & Motlagh (2014)	The objective was to find out the circumstances under which SCM could get more attention to corporate level. It was found out that when a corporation sought related diversification strategy, SCM capabilities could be exploited as it would play a prominent role. These included low clock speed, high interdependency between Original Equipment Manufacturer (OEM) and supplier, resource pooling in terms of purchasing and logistics activities.
3	Hofmann (2010)	The relationships between CS and SCM was discussed. The nature of SCMS was discussed and a framework was developed to understand the linkages between CS and SCM. It focussed on linking between CS and business unit strategies with SCMS and capabilities, especially at the network level. Fit between CS and SCM positively impacted the performance of a firm.
4	Tompinks supply chain consortium (2014)	The importance of aligning SCM with Business Strategy (BS) was discussed with a survey study. Findings were that many managers believed that SCM has been an enabler of BS. Even the non-SCM leaders felt there has been a higher alignment between SCM and BS. The results indicated that well aligned companies had an interconnected BS and SCMS.
5	Kodali & Soni (2011)	Strategic fit between CMS and SCMS in Indian manufacturing industry was examined. The choice of CS and SCMS affected business and SCM performance. The major hurdles in implementing SCM practices were overcoming traditional practices followed by external support needed, insufficient knowledge and others
6	Li, Nathan, & Rao (2006)	The relationship between SCM practices, competitive advantage and organizational performance was explored. It was found that to outperform competitors and to maintain a robust competitive position, higher level of SCM practice was required. Enhanced organizational performance provided increased capital to a firm for implementing SCM practices.
7	Sillanpaa & Sillanpaa (2014)	The different SCMS which were present in Europe and Asia were discussed. SCMS framework that covered business environment (high or low volume), CS (cost leadership or differentiation), SCM demand (predictable or unpredictable demand) and SCMS (efficiency or responsive) was developed. It was found that in both the SCM contexts business volume was low, CS was differentiation, and SCM was unpredictable.
8	Pozo, Tachizawa, Akabane, & Soares (2015)	SCM could be used as a CMS for costs reduction was focussed. Implementation and impact of SCM in small sized firms and whether it affected performance flexibility and cost reduction was discussed. By properly implementing SCM and improving strategic alliances, the lead time of the products was reduced as a result the service level became more efficient.

2. What are the challenges in implementing the new technology based SCM innovations?
3. What are the drivers for this technology based SCM innovations?
4. How technology based SCM innovations were adding strategic business value to firms?

For this exploratory study, a total of 24 interviews were conducted with SCM experts. The data was collected between January 2018 to March 2019. All the experts had experience (especially regarding SCM and technological innovations) in the Indian context for a minimum of 36 months. All the managers had expertise in SCM functions. The In-depth interviews conducted by the authors explored the research questions both as leading and probing questions (Rowley, 2012). The data was collected as physical and telephonic interviews (average duration was 45 minutes). It was summary transcribed and thematic content analysed for findings (Hsieh & Shannon, 2005; Mayring, 2004). The first author undertook intracoder reliability (value – 95), within the prescribed quality limit (Lombard, Snyder-Duch, & Bracken, 2002).

FINDINGS

The findings of the research have been tabulated in Table 3. Representative responses of respondent for each question have been presented.

Based on the answers by the experts, certain common themes were identified from their responses. This was carried out by thematic content analysis. Table 4, illustrated the common challenges faced by firms in India in implementing SCM innovation.

Table 5, tabulated the important themes pertaining to drivers of SCM innovations in India which were identified based on interviews conducted.

Table 6, tabulated the strategic business value of SCM innovation in India which were identified through interviews conducted.

DISCUSSION

The authors found out that there were various challenges in implementing technology based SCM innovations in India. In technological innovations in the domain of SCM, technology becomes the central pivot (Bhattacharyya & Shrey, 2019) .However, the authors noted that in non-technological innovations in the SCM space market segmentation, way of targeting market and pricing are generally altered (Bhattacharyya & Shrey, 2019). From technology point of view the authors found out that the major challenge was in integrating new system with the existing ones. This had been explored by Delaney & Agostino (2015). There remained a common consensus amongst experts that the reliability of new technology

Table 3. Representative interview responses

New-innovation	Challenges	Drivers	Strategic Value
"Internet Of Things (IOT), Big data analytics, Cloud Enterprise Resource Planning (ERP), Transportation Management System (TMS), Warehouse Management System (WMS), and Blockchain would be some of the prominent new innovations in SCM."	*"High Cost of innovation would be one of the major challenges for choosing a particular technology in India." "Integration of data with the existing system would be another challenge." "The labour cost in India is cheap. The benefits of innovation should surpass the cheap labour cost to be selected." "Resistance of the employees in adopting the change that is change management required." "Lack of capabilities of the organisation in terms of resources for deployment of big data would be a hindrance."*	*"Reducing human error and redundant task would be greatest drivers." "Big data analytics could help in properly forecasting customer demands and help to carry optimal inventory." "Increasing competition is also one of the drivers of SCM innovations which had been forcing firm for innovation." "Becoming market leader in this cut-throat competition is the main reason which is driving these innovations." "Drivers to innovation in SCM are the continuous strive to increase SCM surplus."*	*"These would help to optimize inventory and reduce working capital required for a firm." "This would help the company to reduce their OPEX (operational expenditure) as the firm need not own the infrastructure." "There would be new opportunity for revenue growth." "As manpower would be free for value added work. It would also help in increasing the productivity." "Under-utilization and over utilization of capacity would not happen reducing the wastage, efficiency, human intervention."*

Table 4. Challenges in implementing new technology based SCM innovations in India

S. No	Challenges
1	Cost of innovation was relatively higher than available options.
2	Integration of new system with the existing ones was challenging.
3	Total cost of ownership has been higher compared to the cost of product.
4	Training people has been more time consuming.
5	Building infrastructure has been costly.
6	Immaturity of the organisation in terms of technological gap was inhibiting adoption of new technology.
7	Implementing new technology in relatively smaller sized firm was found to be difficult because of resource constraint.
8	Change management was challenging as the employees resisted to new adoption.
9	Cost of labour in India was less compared to cost of new innovation which inhibited the adoption of new technology.
10	Quality of data was poor, and this was challenging.
11	Regulatory norms and policies of government was generally not supportive in nature.
12	Affordability, user knowledge, acceptance and customer awareness of new innovation would be important for fast adoption.
13	Reliability of the new technology was perceived as uncertain.
14	Deviation from plan while implementation has been a challenge.
15	Lack of clarity regarding usage of 'Big Data' through SCM has been a hindrance.
16	Choosing the right solution from the varieties of solutions available based on the requirement of firm has been a daunting task.
17	SCM sector has mostly been unorganised in nature.
18	Quality of internet provided has been poor.

was perceived as uncertain which had also been argued by Ghosh & Bhowmick (2014). Implementing new technology in relatively smaller sized firms thus would continue to be challenging. This would be primarily because of resource constraints. Similar results were found by Farsi & Toghraee (2014) but this was in the context of Iran. As per Uyar (2014) the total cost of ownership of a product affected firm profitability. This factor should be considered primarily by a firm rather than considering product cost for choosing a particular innovation. The same was observed in this study. Amongst the various solution present in the market choosing the right kind of solution which fitted a firm's requirement would be of great importance. Similar results were found by Ganguly (2016). As per Hall & Khan (2003), user knowledge and acceptance led to faster diffusion of new innovation. The same was observed by the authors. From the point of view of employees of an organisation the major challenges were imparting training to employees as it had been more time consuming. Further, change management has been challenging in nature which had also been argued by Umble, Haft, & Umble (2003) as the critical success factor for implementation. Similar results were found out by researchers. This was because employees resisted to new adoption-based changes in organizational process. The macro economic factors which had been creating challenge in technology based SCM innovation implementation was lower labour cost in India. Labour cost had been cheap in India as compared to other countries like USA and UK. The benefits from new innovation had to be more than cheap labour cost. Regulatory norms and policies of government in India was generally not supportive in nature which also created a hindrance in implementation. This had not been explored before by the researchers.

Table 5. Drivers of technology based SCM innovations in India

S. No	Drivers
1	Enhanced market competitiveness.
2	Provided a firm the image of a market leader.
3	Reduced human error.
4	Optimized inventory level.
5	Redundant tasks were automated.
6	Improved serviceability of SCM functions occurred.
7	Reduced forecasting error in SCM planning occurred.
8	Customer specific demand which required high level of customization and flexibility were being fulfilled.
9	Competing on cost, quality and service level because of increased globalisation became easier.
10	Rising income level of consumers increased the demand from customers for better SCM and this demand resulted more quality conscious SCM.
11	Infrastructure was getting developed because of government policies.
12	Shrinking product life cycles was forcing firms to go for new innovations.
13	Large amount of data was available which could be used for drawing insights.
14	Machine-machine and human machine interaction had become ubiquitous.
15	For firms to become customer centric had become the need of the hour.
16	Reduce problem of counterfeiting by implementing blockchain technology was occurring.

The major driver of technology based SCM innovation in India from market perspective had been that it had been an enabler of market competitiveness. Similar results were found by Momaya & Ajitabh (2005). This helped the firms lead in its respective industry and eventually become market leader which had also been argued by Simpson, Siguaw, & Enz (2006). There remained a common consensus amongst experts that competing on cost, quality and service level was becoming easier which had been argued by Sabbaghi & Sabbaghi (2004). Rising income level of consumers increased the demands from customers for better SCM, forcing firms for technological applicability with regard to SCM function. Similar results were found by Sheth, Sisodia, & Sharma (2000). Infrastructure was getting developed in the Indian market due to government initiatives like dedicated freight corridors. This has not been explored before by researchers for India. Human factors also contributed to SCM innovation. Reduced human error and redundant tasks which were time consuming were automated because of new innovations in SCM. Similar results were found by Huber, Michael, & McCathie (2007) but it was in the context of RFID adoption in SCM. From the firm level perspective, inventory would be optimized and forecasting error would be reduced due to adoption of technology based new innovations. There would be greater savings for firms. This has been explored by Chase Jr (2014) with respect to predictive analytics. Customer specific demand which required high level of customization and flexibility were being fulfilled. Innovation would help the supply chain of firm to be responsive which had been argued by Gunasekaran, Lai, & Cheng (2008). As per the researcher consumers demand heterogeneity reduced product life cycles period and similar results were found by Adner & Levinthal (2001). Industry 4.0 have made human machine and machine-machine interaction ubiquitous as pointed out by Zhou, Liu, & Zhou (2015). Large amount of

Table 6. Strategic business value of SCM innovation in India

S. No	Strategic business value
1	Reduced human dependency to reduce the element of human error.
2	Total supply chain cost was reduced.
3	Optimized inventory and reduced operational expenditure were achieved.
4	The working capital required would be reduced.
5	New opportunities for revenue growth was being explored.
6	Productivity would be increased as non-value-added activities would be reduced and employees would be more engaged in value added work.
7	Product availability would be increased as time to market for the product would be reduced.
8	Capacity requirement could be optimized and would help in rationalisation of products.
9	Reduced wastage at raw material processing stage was achieved.
10	Higher return on investment was achieved.
11	Through use of cloud technologies relatively smaller firms were able to compete with its bigger counterpart as it is was based on pay per use basis.
12	Reduction in lead time of firm and its supplier was achieved.
13	Helped in better planning and coordination within supply chain functions thereby increasing efficiency.
14	Created new market with greater market share.
15	Generated marketing value for the firm as it would help a firm stand out in the market and differentiated it from its competitors.
16	Customer centricity helped gain market share and become market leader.
17	Business with slack capacity enabled to transact with other sister counterparts which were in a state of scarcity of resources.

data was available which could be used for drawing insights and helped in better decision making which had been argued by Marr (2015). This was also corroborated in this study.

The strategic value of technology based SCM innovation in India from market perspective would be that it created new market space. Firms which would implement it at the earliest would have greater market share achieved. This had been generally explored by Kim & Mauborgne (1999) for any strategic initiative. There remained a common consensus amongst the experts that technological innovation would generate marketing value for the firm which had been argued by Hoflinger, Nagel, & Sandner (2018). From the firm perspective in the initial stage new technology based SCM innovation would increase the cost of working capital and SCM costs but once it was fully developed then the Return On Investment (ROI) would be high. This had not been explored earlier by researchers for Indian context. New opportunities would be created which would help in revenue & economic growth. Similar results were found by Gerguri & Ramadani (2010). Capacity requirement would be optimized, and this would help in rationalization of products. This had also not been explored earlier. Lead time of firm and its supplier would be reduced with the help of new SCM innovation which would increase the transparency & coordination in firm supply chain. This has been explored by Patterson, Grimm, & Corsi (2003). The new set of innovations would help business with slack capacity to transact with other sister counterparts which were in state of scarcity of resources. This aspect had not been studied earlier. From the employee perspective the new innovation in SCM would reduce the human dependency, reduction of human error and reduction of non-value adding activities. Similar results were found by Blecker (2014). From the supply chain perspective, the new innovation would reduce the total SCM costs post its implementation. Similar results were found by Bhandari (2014). There would be improvement in product availability which would satisfy the ultimate goal of being customer centric. This had not been explored earlier by researchers in the context of Indian SCM function perspective. It is important for Indian SCM managers to facilitate the drivers for technology adoption in Indian firms and reduce the barriers for SCM adoption. This has been discussed in detail in the next section.

CONCLUSION AND IMPLICATIONS FOR MANAGEMENT

In this study, the authors explored the strategic imperatives, drivers & challenges of automated technologies applied in SCM function of firms in Indian industries. Semi structured open-ended in-depth interviews were conducted. It can be concluded from the findings that a lot of development was going regarding the digital innovation side. But the advancement of innovation in physical innovation has been relatively slow and it would take a decade or so down the line to gain proper foothold. Firm management would resort to innovation when the firm had achieved full efficiency point and there was no more scope of increasing efficiency with current level of SCM technology. Innovation especially digital innovation has been gaining importance because of competitive pressure and market forces. New technologies especially in the area of demand planning and forecasting, big data analytics, IOT, Software as a Service (SAAS), Enterprise Resource Planning (ERP), Transportation Management System (TMS) & Warehouse Management System (WMS) had been changing the way companies did business. These technologies would be enabling firms to stay ahead of competition if applied for SCM function. Digital transformation could be viewed as a journey and these solutions were evolving continuously. Firms were finding new ways to improve revenue and cut down costs. Digital innovation would be helping firms to reduce its Operational Expenditure (OPEX) and improve the bottom line that is profit. It would also reduce the human intervention and reduce human errors. Integrated supply chain had become the need of the hour as the extent of SCM of a firm was broadening and becoming complex. With throughout integration of SCM, the transparency would increase for a firm. Wastage of resources would reduce and resources like skilled worker would be free for undertaking value adding activities. In India, a major concern before implementing any innovation initiative at an organization has been the ROI. One could argue that most employees were aware of the outcome of the innovation but the process to go through often lacked clarity. Most employees in organizations experimented to get to the desired result. The major challenges in implementing digital innovation had been the high cost of innovation, high total cost of ownership, integration of new system with the existing system (which has been very difficult), poor quality of input data, lack of proper infrastructure and difficulty in implementing change management.

There has not been much advancement occurring in the adoption of physical innovation in SCM domain. This has been because of the cheap labour availability in India. Further, technology has been in its nascent stage and often not reliable. RFID has been gaining importance especially in the logistics and ecommerce domain. Physical innovation would take time to mature in the Indian market for widespread use. The market condition of India has been unique as it had been a developing market. Western researchers didn't mention much about the hindrances an Indian firm (or for that matter any emerging economy firm) would confronting in implementing such technologies in SCM domain. One could argue regarding the strong role of technology in implementing effective supply chain strategy. This study was carried out in India in the domain of SCM on automated technology. The scope of this research was limited to various experts from different types of Indian organizations where SCM played a prominent role. It was a qualitative research based on interview techniques. India being one of the fastest growing economy in world lacked various infrastructural and technological advancement. But with Foreign Direct Investment (FDI) coming and also due to various government policies there has been a boost in the technological advancement in the field of SCM. Also, India has been one of the biggest potential market for any company in the world and to gain more and more market share companies were investing heavily in the application of technology in SCM.

SCMS has been an enabler of BS. Increasing the efficiency of SCM would automatically aid to increase the efficiency of BS. ROI for a firm would increase when CMS and SCMS achieved right match. So high degree of alignment of SCM and BS was important. Total cost of ownership of the product has been very important in choosing the right product or solution. The cost of product might be less initially, but total cost of ownership should be considered for choosing the product. The technology landscape in India has not been reliable from supply side. It has been in its nascent stage. So, a pilot project should be compulsory before implementing the solution in full scale. Firms should invest in training of skilled workers to reap the full benefits of new innovation in SCM. This would also help in the change management function to adopt to the new changes. The authors provided an integrated point of view of technology SCM & Strategic benefits. As evident in table 1 & 2, the extant literature was piece meal & contextual. The authors integrated three fields of studies namely technology management (Schiederig, Tietze, & Herstatt, 2012), SCM functions (Tarn et al., 2002; Croxton et al., 2001), and strategic gains (Spekman, Kamauff Jr., & Myhr, 1998; Towill, 1997). This was the main theoretical contribution of the study. In future researchers could undertake comparative analysis between different country contexts & technology context in the domain of technology-based innovations in SCM functions.

REFERENCES

Adner, R., & Levinthal, D. (2001). Demand heterogeneity and technology evolution: Implications for product and process innovation. *Management Science*, *47*(5), 611–628. doi:10.1287/mnsc.47.5.611.10482

Agrawal, T. (2012). Role Of Rfid To Minimize Reverse Logistics: A Case Study Perspective. *IOSR Journal of Business and Management.*, *1*(4), 44–51. doi:10.9790/487X-0144451

Akter, S., Wamba, S. F., Gunasekaran, A., Dubey, R., & Childe, S. J. (2016). How to improve firm performance using big data analytics capability and business strategy alignment? *International Journal of Production Economics*, *182*, 113–131. doi:10.1016/j.ijpe.2016.08.018

Angeleanu, A. (2015). New technology trends and their transformative impact on logistics and supply chain processes. *International Journal of Economic Practices and Theories*, *5*(5), 413–419.

Barton, D., & Court, D. (2012). Making advanced analytics work for you. *Harvard Business Review*, *90*(10), 78–83. PMID:23074867

Ben-Daya, M., Hassini, E., & Bahroun, Z. (2017). Internet of things and supply chain management: A literature review. *International Journal of Production Research*, 1–24. doi:10.1080/00207543.2017.1402140

Bhandari, R. (2014). *Impact of technology on logistics and supply chain management. IOSR Journal of Business and Management.*

Bhattacharyya, S. S. (2011). Reflections on strategic insights for winning in the complex emerging market of India. *International Journal of Business Excellence*, *4*(1), 15–43. doi:10.1504/IJBEX.2011.037247

Bhattacharyya, S. S., & Jha, S. (2015). Mapping micro small and medium enterprises from the resource-based view and dynamic capability theory perspectives and innovation classification. *International Journal of Entrepreneurship and Small Business*, *25*(3), 331–350. doi:10.1504/IJESB.2015.069700

Bhattacharyya, S. S., & Shrey, A. (2019). Supply Chain Startups in India: A Cross Case Comparative Analysis. In Handbook of Research on Corporate Restructuring and Globalization (pp. 320-344). Hershey, PA: IGI Global. doi:10.4018/978-1-5225-8906-8.ch016

Blecker, T. (Ed.). (2014). Innovative methods in logistics and supply chain management: current issues and emerging practices, (Vol. 19), epubli.

Brar, S., Rabbat, R., Raithatha, V., Runcie, G., & Yu, A. (2015). *Drones for Deliveries*. Berkeley, CA: University of California Berkeley, Sutardja Center for Entrepreneurship & Technology.

Büyüközkan, G., & Göçer, F. (2018). Digital supply chain: Literature review and a proposed framework for future research. *Computers in Industry*, *97*, 157–177. doi:10.1016/j.compind.2018.02.010

Chase, C. W. Jr. (2014). Innovations in business forecasting: Predictive analytics. *The Journal of Business Forecasting*, *33*(2), 26.

Chopra, S., & Meindl, P. (2016), Supply chain management: Strategy, planning, and operation.

Chou, D. C., Tan, X., & Yen, D. C. (2004). Web technology and supply chain management. *Information Management & Computer Security*, *12*(4), 338–349. doi:10.1108/09685220410553550

Christensen, C. M. (2013). The innovator's dilemma: when new technologies cause great firms to fail. Brighton, MA: Harvard Business Review Press.

Creswell, J. W., & Poth, C. N. (2017). *Qualitative inquiry and research design: Choosing among five approaches*. New York, NY: Sage.

Croxton, K. L., Garcia-Dastugue, S. J., Lambert, D. M., & Rogers, D. S. (2001). The supply chain management processes. *International Journal of Logistics Management*, *12*(2), 13–36. doi:10.1108/09574090110806271

Daghfous, A., & Barkhi, R. (2009). The strategic management of information technology in UAE hotels: An exploratory study of TQM, SCM, and CRM implementations. *Technovation*, *29*(9), 588–595. doi:10.1016/j.technovation.2009.05.007

Davenport, T. H. (1993). Process innovation: reengineering work through information technology. Brighton, MA: Harvard Business Press.

Davila, T., Epstein, M., & Shelton, R. (2012). Making innovation work: How to manage it, measure it, and profit from it. Upper Saddle River, NJ: FT Press.

Delaney, R., & D'Agostino, R. (2015). The Challenges of Integrating New Technology into an Organization.

Dittmann, P. (2017). New Supply Chain Technology Best Practices. Available at https://haslam.utk.edu/sites/default/files/GSCI%20InnovationPaper%20FIN4-4-17.pdf

Diwan, M. (2016, March). Internet of Things In Logistics: Towards Autonomous Logistics & Smart Logistics Entities. Paper presented at International Maritime transport & Logistic Conference. Available at https://marlog.aast.edu/archive/2016/pdf/Papers/s06p02.pdf

Durga Prasad, K. G., Venkata Subbaiah, K., & Narayana Rao, K. (2012). Aligning the competitive strategy with supply chain strategy through QFD. *Journal of Advances in Management Research, 9*(2), 189–198. doi:10.1108/09727981211271931

Farsi, J. Y., & Toghraee, M. T. (2014). Identification the main challenges of small and medium sized enterprises in exploiting of innovative opportunities (Case study: Iran SMEs). *Journal of Global Entrepreneurship Research, 4*(1), 4. doi:10.1186/2251-7316-2-4

Feizabadi, J., Singh, M., & Motlagh, S. A. (2014). Contribution of supply chain to corporate strategy: A case study in agriculture machinery industry. *International Journal of Logistics Systems and Management, 18*(4), 473–499. doi:10.1504/IJLSM.2014.063981

Felea, M., & Albăstroiu, I. (2013). Defining the concept of supply chain management and its relevance to Romanian academics and practitioners. *Amfiteatru Economic Journal, 15*(33), 74–88.

Fish, L. A. (2011). Supply chain quality management. In Supply Chain Management-Pathways for Research and Practice. *IntechOpen.*

Ganguly, P. (2016, January). Selecting the right IoT cloud platform. In *2016 International Conference on Internet of Things and Applications (IOTA)* (pp. 316-320), IEEE. 10.1109/IOTA.2016.7562744

Gerguri, S., & Ramadani, V. (2010). The impact of innovation into the economic growth.

Ghosh, S., & Bhowmick, B. (2014, October). Technological uncertainty: Exploring factors in indian start-ups. In *IEEE Global Humanitarian Technology Conference (GHTC 2014)* (pp. 425-432), IEEE. 10.1109/GHTC.2014.6970317

Gibson, E. C., & Matthews, D. (2013, July). The impact of rapidly changing technology on the supply chain. In *2013 Proceedings of PICMET'13: Technology Management in the IT-Driven Services (PICMET)* (pp. 1944-1950), IEEE.

Govindan, K., Kaliyan, M., Kannan, D., & Haq, A. N. (2014). Barriers analysis for green supply chain management implementation in Indian industries using analytic hierarchy process. *International Journal of Production Economics, 147*, 555–568. doi:10.1016/j.ijpe.2013.08.018

Govindan, K., Soleimani, H., & Kannan, D. (2015). Reverse logistics and closed-loop supply chain: A comprehensive review to explore the future. *European Journal of Operational Research, 240*(3), 603–626. doi:10.1016/j.ejor.2014.07.012

Gunasekaran, A., Lai, K. H., & Cheng, T. E. (2008). Responsive supply chain: A competitive strategy in a networked economy. *Omega, 36*(4), 549–564. doi:10.1016/j.omega.2006.12.002

Gupta, M., & Kohli, A. (2006). Enterprise resource planning systems and its implications for operations function. *Technovation, 26*(5-6), 687–696. doi:10.1016/j.technovation.2004.10.005

Hall, B. H., & Khan, B. (2003). Adoption of new technology (No. w9730). *National bureau of economic research.*

Hervani, A. A., Helms, M. M., & Sarkis, J. (2005). Performance measurement for green supply chain management. *Benchmarking International Journal (Toronto, Ont.), 12*(4), 330–353.

Heutger, M., & Kückelhaus, M. (2015). Unmanned Aerial Vehicle in Logistics. Available at http://www.dhl.com/content/dam/downloads/g0/about_us/logistics_insights/DHL_TrendReport_UAV.pdf

Hoberg, K., & Alicke, K. (2016). How SC4. 0 will Enhance the Customer Experience. *Supply Chain Management Review*, (September/October), 28-37.

Höflinger, P. J., Nagel, C., & Sandner, P. (2018). Reputation for technological innovation: Does it actually cohere with innovative activity? *Journal of Innovation & Knowledge, 3*(1), 26–39. doi:10.1016/j.jik.2017.08.002

Hofmann, E. (2010). Linking corporate strategy and supply chain management. *International Journal of Physical Distribution & Logistics Management, 40*(4), 256–276. doi:10.1108/09600031011045299

Hsieh, H. F., & Shannon, S. E. (2005). Three approaches to qualitative content analysis. *Qualitative Health Research, 15*(9), 1277–1288. doi:10.1177/1049732305276687 PMID:16204405

Huber, N., Michael, K., & McCathie, L. (2007, September). Barriers to RFID adoption in the supply chain. In *2007 1st Annual RFID Eurasia* (pp. 1–6). IEEE. doi:10.1109/RFIDEURASIA.2007.4368128

Ivanov, D., & Sokolov, B. (2010). Conceptual Frameworks for Supply Chain Management. *Adaptive Supply Chain Management*, 19-33.

Jain, J., Dangayach, G. S., Agarwal, G., & Banerjee, S. (2010). Supply chain management: Literature review and some issues. *Journal of Studies on Manufacturing, 1*(1).

Kim, K., Jeong, B., & Jung, H. (2014). Supply chain surplus: Comparing conventional and sustainable supply chains. *Flexible Services and Manufacturing Journal, 26*(1-2), 5–23. doi:10.100710696-012-9163-2

Kim, W. C., & Mauborgne, R. (1999). Creating new market space. *Harvard Business Review, 77*(1), 83–93. PMID:10345394

Kumar, V. (2012). *101 design methods: A structured approach for driving innovation in your organization*. Hoboken, NJ: John Wiley & Sons.

Lee, H. L. (2002). Aligning supply chain strategies with product uncertainties. *California Management Review, 44*(3), 105–119. doi:10.2307/41166135

Li, S., Ragu-Nathan, B., Ragu-Nathan, T. S., & Rao, S. S. (2006). The impact of supply chain management practices on competitive advantage and organizational performance. *Omega, 34*(2), 107–124. doi:10.1016/j.omega.2004.08.002 PMID:17876965

Li, X. (2014). Operations management of logistics and supply chain: Issues and directions. *Discrete Dynamics in Nature and Society*.

Lombard, M., Snyder-Duch, J., & Bracken, C. C. (2002). Content analysis in mass communication: Assessment and reporting of intercoder reliability. *Human Communication Research*, *28*(4), 587–604. doi:10.1111/j.1468-2958.2002.tb00826.x

Marr, B. (2015). *Big Data: Using SMART big data, analytics and metrics to make better decisions and improve performance*. Hoboken, NJ: John Wiley & Sons.

Mayring, P. (2004). Qualitative content analysis. *A companion to qualitative research, 1*, 159-176.

Merlino, M., & Sproģe, I. (2017). The augmented supply chain. *Procedia Engineering*, *178*, 308–318. doi:10.1016/j.proeng.2017.01.053

Momaya, K., & Ajitabh, A. (2005). Technology management and competitiveness: Is there any relationship? *International Journal of Technology Transfer and Commercialisation*, *4*(4), 518–524. doi:10.1504/IJTTC.2005.006702

Nayak, B., & Bhattacharyya, S. S. (2019). Integrating Digital Wisdom and Human Capital. *Journal for Quality and Participation*, *41*(4), 20–23.

New, S. (2010). The transparent supply chain. *Harvard Business Review*, *88*, 1–5.

Palanivelu, P., & Dhawan, M. (2015). Sustainability and Green Logistics in Supply Chain. Available at https://www.academia.edu/28094615/Green_Logistics_Whitepaper

Patterson, K. A., Grimm, C. M., & Corsi, T. M. (2003). Adopting new technologies for supply chain management. *Transportation Research Part E, Logistics and Transportation Review*, *39*(2), 95–121. doi:10.1016/S1366-5545(02)00041-8

Pozo, H., Souza, C. P., & Tachizawa, T. (n.d.). Supply chain management as a competitive strategy for costs reduction: a case study in two small manufacturing companies.

Ratnajeewa, D., & Bandara, J. (2018, November). A Review of Research on Green Logistics Distribution Practices. Paper presented at the KDU International Research Conference, Sri Lanka. Available at http://ir.kdu.ac.lk/bitstream/handle/345/1552/msh-042.pdf?sequence=1&isAllowed=y

Reddy, J. S. (2005). Gaining competitive advantage through Supply chain management. *Indian Journal of Marketing*, *35*(6).

Rocco, T. S., & Plakhotnik, M. S. (2009). Literature reviews, conceptual frameworks, and theoretical frameworks: Terms, functions, and distinctions. *Human Resource Development Review*, *8*(1), 120–130. doi:10.1177/1534484309332617

Rowley, J. (2012). Conducting research interviews. *Management Research Review*, *35*(3/4), 260–271. doi:10.1108/01409171211210154

Russell, R. S., & Taylor-Iii, B. W. (2008). *Operations management along the supply chain*. Hoboken, NJ: John Wiley & Sons.

Sabbaghi, A., & Sabbaghi, N. (2004). Global supply-chain strategy and global competitiveness. *International Business and Economics Research Journal*, *3*, 63–76.

Schiederig, T., Tietze, F., & Herstatt, C. (2012). Green innovation in technology and innovation management–an exploratory literature review. *Research Management*, *42*(2), 180–192.

Sheth, J. N., Sisodia, R. S., & Sharma, A. (2000). The antecedents and consequences of customer-centric marketing. *Journal of the Academy of Marketing Science*, *28*(1), 55–66. doi:10.1177/0092070300281006

Sillanpää, I., & Sillanpää, S. (2014). Supply Chain Strategy: Empirical Case Study in Europe and Asia. *Management, 9*(2).

Silverman, D. (2013). Doing qualitative research: A practical handbook. New York, NY: Sage.

Simpson, P. M., Siguaw, J. A., & Enz, C. A. (2006). Innovation orientation outcomes: The good and the bad. *Journal of Business Research*, *59*(10-11), 1133–1141. doi:10.1016/j.jbusres.2006.08.001

Soni, G., & Kodali, R. (2011). The strategic fit between "competitive strategy" and "supply chain strategy" in Indian manufacturing industry: An empirical approach. *Measuring Business Excellence*, *15*(2), 70–89. doi:10.1108/13683041111131637

Spekman, R. E., Kamauff, J. W. Jr, & Myhr, N. (1998). An empirical investigation into supply chain management: A perspective on partnerships. *Supply Chain Management*, *3*(2), 53–67. doi:10.1108/13598549810215379

Tarn, J. M., Razi, M. A., Yen, D. C., & Xu, Z. (2002). Linking ERP and SCM systems. *International Journal of Manufacturing Technology and Management*, *4*(5), 420–439. doi:10.1504/IJMTM.2002.001459

Teece, D. J. (2010). Business models, business strategy and innovation. *Long Range Planning*, *43*(2-3), 172–194. doi:10.1016/j.lrp.2009.07.003

Tiwari, S., Wee, H. M., & Daryanto, Y. (2018). Big data analytics in supply chain management between 2010 and 2016: Insights to industries. *Computers & Industrial Engineering*, *115*, 319–330. doi:10.1016/j.cie.2017.11.017

Tjahjono, B., Esplugues, C., Ares, E., & Pelaez, G. (2017). What does industry 4.0 mean to supply chain? *Procedia Manufacturing*, *13*, 1175–1182. doi:10.1016/j.promfg.2017.09.191

Tompkins, B. (2014). Aligning Supply Chains with Business Strategy. Available at http://archive.tompkinsinc.com/wp-content/uploads/2014/04/Supply-Chain-Alignment-with-Business-Strategy.pdf

Towill, D. R. (1997). The seamless supply chain-the predator's strategic advantage. *International Journal of Technology Management*, *13*(1), 37–56. doi:10.1504/IJTM.1997.001649

Trkman, P., McCormack, K., De Oliveira, M. P. V., & Ladeira, M. B. (2010). The impact of business analytics on supply chain performance. *Decision Support Systems*, *49*(3), 318–327. doi:10.1016/j.dss.2010.03.007

Umble, E. J., Haft, R. R., & Umble, M. M. (2003). Enterprise resource planning: Implementation procedures and critical success factors. *European Journal of Operational Research*, *146*(2), 241–257. doi:10.1016/S0377-2217(02)00547-7

Uyar, M. (2014). A Research on Total Cost of Ownership and Firm Profitability. *The International Institute for Science, Technology and Education*, *5*(1), 9–16.

Verma, S., Bhattacharyya, S. S., & Kumar, S. (2018). An extension of the technology acceptance model in the big data analytics system implementation environment. *Information Processing & Management*, *54*(5), 791–806. doi:10.1016/j.ipm.2018.01.004

Verma, S., & Sekhar Bhattacharyya, S. (2016). Micro-foundation strategies of IOT, BDA, Cloud Computing: Do they really matter in bottom of pyramid? *Strategic Direction*, *32*(8), 36–38. doi:10.1108/SD-06-2015-0093

Vonderembse, M. A., Uppal, M., Huang, S. H., & Dismukes, J. P. (2006). Designing supply chains: Towards theory development. *International Journal of Production Economics*, *100*(2), 223–238. doi:10.1016/j.ijpe.2004.11.014

Wook Kim, S. (2006). Effects of supply chain management practices, integration and competition capability on performance. *Supply Chain Management*, *11*(3), 241–248. doi:10.1108/13598540610662149

Zhou, K., Liu, T., & Zhou, L. (2015, August). Industry 4.0: Towards future industrial opportunities and challenges. In *2015 12th International Conference on Fuzzy Systems and Knowledge Discovery (FSKD)* (pp. 2147-2152). IEEE. 10.1109/FSKD.2015.7382284

Chapter 13
The Role of Digital Economies in the Development and Growth in Asian Business Models

Kinza Yousfani
Isra University, Pakistan

Farhana Khowaja
Isra University, Pakistan

ABSTRACT

This chapter discusses the development of information and communication technology across Asian economies. Digital technology is the presentation of information in bits that emphasize digital technology which covers all business, financial, social, and cultural events that are supported by the web and other digital communication technologies. Technology has minimized the cost of storage, and flow of information. In the last 15 years, digital technology has affected human lives and the chapter examines how digital technology changes economic activity. There are three principal segments: e-business, e-business framework, and e-commerce. The digital economy is known as the web economy because of its dependence on the network. Modern technologies, cloud computing, mobile app, and social media influence the business landscape, reshaping the idea of work, boundaries of enterprises, and the obligations of business pioneers. Thus, the digital economy features the opportunity for organizations and people to execute existing tasks on the PC more frequently than before.

Introduction To Digital Economy (De)

The "Digital economy" is a term for those financial procedures, exchanges, communications, and exercises that depend on online technologies. The DE is one aggregate term for every single financial exchange that happens on the web. It is otherwise called the Web Economy or Internet Economy. With

DOI: 10.4018/978-1-7998-0357-7.ch013

the upcoming development or technology and the method of globalization, the digital and traditional economies are converging into one. While the DE is all the more comprehensively dependent on any of the different online devices applied in the present financial world. The term was first begotten in a book "The Digital Economy: Promise and Peril in the Age of Networked Intelligence" by author Don Tapscott in 1995. A widely recognized understanding of the DE is its activities close by the digital world. Thomas Mesenbourg (2001) has given three principal components to Digital Economy, specifically

1. E-Business framework (hardware, programming, telecoms, network, human capital, and so forth.),
2. E-Business (how business is led, any procedure that an association directs over PC – mediated systems),
3. E-Business (exchange of merchandise, for instance when a book is sold on the web).

The DE or web economy over the most recent 15 years is the enormous development of online platforms and their impact on our lives. Presently customers are influenced by things they see via web-based networking media, for example, Facebook, Twitter, Instagram, Google's Alphabet, Amazon, and other such famous sites YouTube, and so forth are from the Internet world. So this economy is an approach to misuse this opportunity. Presently it is incorporated into each part of the client's life – medicinal services, training, education, banking, entertainment, and so on. Online financial activities result from billions of online associations among individuals, organizations, business, data, gadget, information, and procedures. Furthermore, the establishment of the DE is as such hyper accessibility and connectivity which makes interconnectedness of individuals, organizations, and machines that rely upon Internet, mobile technology and the Internet of things. The internet of things, which is known as the infrastructure of the information society associates physical and tiny gadgets, building and numerous different things embedded with hardware, programming, sensors, and so forth to participate in the trading of information (Global System for Mobile Association, GSMA Report 2015). DE is well-known by the main place held by phenomena of growing products such as the more consumers and clients a company has, the more productive and dynamic it is and able to offer a superior service at the same price, which attracts new clients, and the process goes on (Arthur, 1996). This phenomenon is connected with network effects like the fineness of the service relies upon the scope of the network and number of users. Though network effects already happened in the traditional economies like hotels and transport system etc., they have been significantly increased by the digital economy. The rise in the number of organization and industries with network system which effects within the DE is clarified by the decrease of transaction costs.

Digital technologies made easier to authenticate the other party in a transaction and increase information of reputations; likewise, simpler correspondence and the reviewing of exchanges. Moreover, to facilitate the formation of trust among parties that don't have any acquaintance with one another (Dyer et al., 2003). This has gone to the presence of vast platforms, on which laypersons and semi-experts can discover clients under best and safe conditions and deliver them with services whose quality is occasionally higher than that presented by traditional professions. These transitional platforms work at a phenomenal level like Uber as personal transport and as close to the home vehicle. Increasing efficiency is also an outcome of Data Science which online companies use in order to repeatedly progress their performances such as cost, viability, excellence, and so forth., recognition to the gathering and processing of enormous information streams. Clients are enlisted by organizations in order to contribute to making the good or service known, set up client support. This nature of the DE strengthens network effects.

Background Of Digital Economy In Asia

The digital economy refers to an extensive scope of financial activities that utilize digitized data and information as a key factor of production. The web, cloud computing, big data, fintech, and other new online technology are utilized to gather, store, examine, and share data carefully and change social associations. The digitization of the economy makes advantages and efficiencies as digital technologies drive development and fuel openings job opportunities and monetary development. The DE similarly penetrate all parts of society, influence the manner people connect and achieving expansive sociological changes.

Asia's digital transformation is as of now massively affecting the locale's economies. Asia's online business exchanges represent 25% of the business to customer (B2C) market in the market, driven by the People's Republic of China (PRC), where organizations like Alibaba and Tencent have developed at a break-neck pace. The exchange volume of the PRC's retail online business market has expanded from (currency unit) CNY1.32 trillion out of 2013 to CNY5.33 trillion out of 2016, with an expected CNY7.57 trillion out of 2017 (ADB Report, 2018).

Financial technologies have also given rise to better methodologies for delivering financial services in Asia, particularly in assisting payment and lending; it promotes financial inclusion in numerous Asian nations. Fintech-based lending in Asia came to $102.8 billion in 2015 while the development of technologies furthermore enhanced the productivity of the payment system and reinforced Asia's position as the major payment market on the planet. Also, digitized, networked, and intelligent information and communications technologies (ICTs) enable modern economic activities to be more flexible, agile, and smart (Asian Development Bank Report, 2018).

Asia keeps on profiting by this internet revolution, understanding the DE remains a challenge due to its multifaceted nature. Internet modification is about vast information and platforms as well as how those online technologies can be used to utilized maximum opportunities, new business models and procedures, and smart products and services. Further, the computerized economy is empowering territorial organizations to move long way from the nearby and into worldwide, with respect to the long-term patterns toward market movement and reduced trade barriers. But, given the wide computerized skills gaps and changes in the levels of guidelines and substructure in Asia, not all nations can take the advantages offered by the digital economy. Normally, individuals in developing Asian countries don't approach to a basic online account, because of the absence of computerized gadgets, generally recognized forms of identifications or socioeconomic barriers.

To make more addition in the DE, there ought to be a significant comprehension of the distinctions in access and appropriation inside the populaces of several nations. New issues identified related to trust, security, transparency, and straightforwardness also needed to be addressed as Asia's online revolution increases. The DE can possibly change the social condition or environment and economic activities of Asia. It is now experiencing high development, quick advancement, and extensive application to other financial sectors. However, despite of gigantic opportunities exhibited by the DE, Asia has not yet totally grasped the capability of associating online technology for feasible development, due to poor ICT substructure, insufficient skills development, and financial hindrances that keep a lot of Asia's populace from participating in the DE (OECD, 2015).

Importance of the Digital Economy

The certain significance of availability, there is a rational need to ensure that the technology and infrastructure, mostly mobile framework, in a nation satisfies the needs of an internet society. This will be accomplished by eradicating hindrances to investment around access to range and the burden of the obligation. The Key partners, including governments and administrators, likewise need to collaborate on mindfulness campaign for online services, which ought to be easy to use and accessible through various channels and dialects that meet the necessities of nearby clients or users. The role of the government in setting up a computerized society does not stop at making an empowering environment. It should include a forceful push for the digitization of public facilities, which contact all people and organizations inside a nation and, accordingly, can fill in as a motivation for the take-up and utilization of online facilities by residents crosswise over various socioeconomics and income levels. Using the online society initiative or potentially economic goals of nations in the area including Pakistan, India, Indonesia, Thailand, and Bangladesh, we consider the fundamental factors that should be set up to build up an online society and the financial advantages thereof. Although a few organizations and people use technologies to just execute existing responsibilities on the PC, the DE is further developed than that. It isn't just utilizing a PC to perform assignments normally done physically or on simple gadgets. The internet or advanced economy is growing quickly around the world. It is the absolute most significant driver of advancement, competitiveness, and development, and it holds gigantic potential for business visionaries and small and medium-sized enterprise (SMEs). How organizations receive online technologies will be a key determinant of their future development.

Additionally, the term reveals the capacity to influence technologies to execute the task and take part in activities that weren't thinkable previously. Such opportunities for current entities to improve, to accomplish more, to do things creatively and to do a new thing that is covered in the related thought of online modification. New online trends, for example, mobile web service, cloud computing, smart grid, and online networking are essentially changing the business knowledge, redesigning the notion of work, the limits of activities and the obligations of business pioneers. These developments empower something beyond technological advancement. The outcome in business models, networking, and the exchange of information and access to worldwide markets. Two billion individuals are linked with the web and by 2016, and this number will surpass 3 billion practically 50% of the total populace (OECD, 2014). Organizations that neglect to grow digitally will become eluded from the worldwide market.

Advantages of Digital Economy

The DE has offered numerous new patterns and start-up thoughts. According to Toppr (2018) some important advantages of the DE are as under:

Encourages the Use of the Internet

Looking at this logically, the majority of your day by day work should today be possible on the web. The huge development of the internet and the web that started in the USA is presently an overall system. So there is a histrionic rise in the venture on everything interrelated to equipment, hardware, technological research, programming, facilities, and online correspondence, and so forth. Thus this has guaranteed that the economy with the web is digging in for the long term as are online organizations.

Escalation in E-Commerce

The organizations that modified and received the web and encompass online business in the most recent decade have prospered. The DE has pushed the web-based business division into overdrive. Direct selling, as well as purchasing, conveyance, promoting, making, offering, have all turned out to be simpler because of the digital economy.

The Online Goods and Services

Gone are the times of Movie DVD and Music CD's or records. Presently, these products are accessible to us online. There is no requirement for any substantial items any longer. Similar is true for services like banking, insurance, and so forth. There is no compelling reason to visit the bank on the off chance that you can do each exchange on the web or online transaction. So certain merchandise and services have been totally digitized in this online and advanced economy.

Transparency

Utmost exchanges and their payment in the DE occur on the web. Money exchanges are getting to be uncommon. This lessens the black money and exploitation in the market and makes the economy progressively crystal clear. Actually, during the demonetization, the legislature made a push for online trade to encourage the web economy.

Disadvantages of Digital Economy

Loss in the Employment

The more we rely upon the technology, the less we rely upon HR. The development of the computerized economy may rapid the loss of numerous professions. As the procedures get progressively computerized, the necessity for HR decreases. Take the case of internet banking itself.

Dearth of Professionals

The DE entails multifaceted procedures, advancements, and technologies. To formulate the phases and their support requires specialists and skilled experts. These are not voluntarily reachable, particularly in rustic and semi-provincial regions.

Substantial Investment

The DE involves a robust structure, advance and effective Internet, strong mobile systems and media transmission. The majority of this requires immense and overwhelming investment process. In a developing nation like Afghanistan, Bahrain, Kenya and so forth the expansion of the infrastructure and system is an exceptionally moderate, dull and expensive procedure.

The Difference Between Internet Economy Versus Sharing Economy Versus Token Economy

In its most initial days, the DE also called the web economy, which mentions to an economy that depends on digital computing technologies, which conduct business through market sectors dependent on the web, for example, Amazon.

In any case, financial experts and business pioneers state that the DE is furthermore developed and multifaceted than the web economy, which, below one meaning, essentially implies economic worth resulting from the web. The DE imitates the change from the third industrial revolution to the fourth modern transformation.

The third modern industrial revolution at times called the online revolution, which mentions to the progressions that occurred in the late twentieth century with the change from simple electronic and mechanical gadgets to online improvements. The fourth industrial revolution expands on the online modification as technologies currently to keep on crossing over the physical and online universes or cyber worlds (World Bank, 2018).

Key Features:

1. **Digitized Data/Work Processes:** The transaction can be completely made by online procedures given that the transaction interrelated data is as of now digitized, including service /product information and payment method, and so on.
2. **Logistics**: Internet Economy depends on effective logistics procedures to supply the product / service to finish the transaction.
3. **Light-Asset Based:** As long as the Internet Economy requires a disconnected or offline location to connect with clients, the Internet Economy is light-asset based and more cost-effective in valuing for rivalry or competition.

Sharing Economy likewise called Collaborative Consumption, which alludes to peer-to-peer (P2P) transactions reliant on the Internet. Such transactions are frequently encouraged by means of network-based online service. It varies from Internet Economy as in the sense that the stages only facilitate the online association of product/service consumer and providers. The platform does not sell the item or deliver the service nonstop.

Key Features:

1. **Peer-to-Peer Transactions**: Transactions in Sharing Economy are peer-2-peer, which can be B2C, C2C or B2B.
2. **Trust Authority**: The platform fills as a trusted authority to underwrite equally the customer and service/product provider.
3. **Unused Value to Share**: Sharing Economy to a great extent reliant on the ready to share unused value, which can be reprocessed or reused. For instance, cars are not utilized 92% of the time and the unused value can be a significant resource/chance for sharing economy car solutions, for example, Uber and Lyft.

Token Economy represents the system of incentives based on cryptographic that strengthen and shape necessary behaviors in the blockchain ecosystem. To shape the understanding in blockchain, it

Content	Internet Economy	Sharing Economy	Token Economy
Processed Information	Digitalized	Digitalized	Digitalized
Platform	Centralized	Centralized	Decentralized
Business Model	B2P2C	C2P2C	P2P
Trust Authority	Certainly	Certainly	Not any
Info Transparency	Not any	Not any	Certainly
Verification cost	Nearly	Nearly	Costless
Manual Operation	Nearly	Some	Streamlined

Source: Authors

involves mineworkers to deliver validation service for transactions (GSMA, 2015). Token Economics is the component to boost mineworkers to give better service on the system and network.

Examples:

- Blockchain Mechanisms, which make and confirm the transaction utilizing cryptography algorithms without centralized expert and authority;
- Token Economics, which boost the service (e.g., mining and validation) providing on the blockchain network.

In analogy, the blockchain system is the skeleton and the token economics is the nerve, muscle, and veins to carry life to the blockchain technology. In Rapid, Token Economics is essential to keep the system manageable in the long term.

Key Features:

- **Blockchain-Based**: Token works certainly with blockchain technology as a mediator of significant worth and trade. It does not involve a trusted expert to approve the P2P transactions on the system and network.
- **Multiple-Purpose**: Token is named as a utility token and security token. For utility tokens, it very well may be utilized for transaction, vote, and risk.
- **Multi-Factor Driven**: Depending on the application situation, the token financial matters can be modified to address exceptional attributes of the application scenario.

The following comparison precise the common and various features of Internet Economy, Sharing Economy and Token Economy.

Qualities of the Internet Economy

Over the previous decade important changes in how individuals and organizations link. The building of popular social networks, enterprise have launched their own business networks to link with supplier clients and internal frameworks. The outcome is increasing worldwide trade that is assessed to reach $65 trillion by 2020. To effectively adapt, one should initially comprehend the five key characteristics of the DE:

1. **Digitized and Tracked:** In a DE, analog objects produce online signals that can be measured, estimated, tracked, and examined for healthier decision making.
2. **Connected:** Connecting resources, suppliers, workers, and stakeholders by wireless correspondence permit people to make information-driven decisions, so improving security, viability, and visibility through the enterprise.
3. **Shared:** The DE functions on sharing. Shortly, companies will purchase only what is required and pay as they go. Buying what is required to minimize inventory costs, while buying usage as a service allows companies to pay just for the time used and value received.
4. **Personalized:** Additional feature of the DE is client personalization. Personalization means client get custom-made items and encounters from their preferred brands when and anywhere they need them.
5. **Direct:** The DE likewise permits bypassing the middleman, eradicating unnecessary mediators or networks and making a more direct relationship among purchaser and merchant. A simple ecosystem has less confrontation and lowers the barrier to entry for players in another part of the value chain. Foreign service monitoring is a good case of more direct operations. Leveraging foreign intelligence to track, monitor, manage, report, and resolve asset issues through the service lifecycle eradicates the requirement to have full-time local personnel.

The Role of Internet in Economic Development and Growth in Asia

Asia is apparently one of the most diverse states in the world, with a changing level of socioeconomic conditions and digital society growth and also signifies to an exceptional landscape of emerging, evolution and advanced online societies. (GSMA Digital Societies Report 2016), namely

1. Emerging internet social networks which include nations where digitization is fundamentally an instrument for financial improvement or socioeconomic development, especially in connection to improving social inclusion;
2. Transition online societies include nations where the emphasis is on personalization of services, for better commitment among people and institutions.
3. Advanced internet societies are those where it is likely to grow appropriately interconnected and interoperable procedures and services crosswise over sectors for profitability and proficiency gains. Asia incorporates various computerized and internet societies among the most advanced.

In the journey to grow the internet value chain, policymakers should concentrate on all components of an interoperable computerized ecosystem at the national dimension, comprising (I) encouraging computerized and online proficiency and reasonable gadgets; (ii) making it feasible for the new services to be given flawlessly; and (iii) supporting the development of the Internet of Things. Suitable regulatory settings support the development of a computerized society. Acceptance of a regulatory rule built on possible on an appropriate understanding of the marketplace that is to be regulated is the key.

The Significance of Small Businesses in Asia

1. **India**: Formally, more than 33.2 million small and medium-sized enterprises (SME) hire more than 101 million individuals and symbolize 40 percent of Indian export revenues.66 Informally, the figure is likely much greater.
2. **Indonesia:** 56.3 million micros, small, and medium-sized enterprises (SME), accounting for 99.9 percent of entire businesses, hire 107.7 million people (97.2 percent of the total staff) and symbolize nearly 57.1 percent of GDP.67
3. **Japan:** 4.2 million small and medium-sized enterprises (SME), embracing some 99.7 percent of all enterprises, hire 70 percent of the staff and symbolize 53 percent of GDP.
4. **Philippines**: Nearly 816,759 micros, small, and medium-sized enterprises (SME), accounting for 99.6 percent of all businesses, hire 3.87 million individuals and symbolize 35 percent of GDP.68
5. **Vietnam:** Nearly 303,729 micros, small, and medium-sized enterprises (SME), embracing more than 97 percent of all enterprises, hire 18.3 million individuals and symbolize 40 percent of GDP.69

Moreover, in Indonesia, less than 1 of every 10 private companies viewed themselves as having progressed online capacities, whereas 73% are offline or have basic online without abilities (Digital societies report, 2010). The potential advantages of internet technology for Indonesian SMEs contain:

1. Equal 80% advanced progress in income
2. One-and-a-half times the present-day probability of expanding employment
3. Seventeen times the present potential for advancement
4. More universal competitiveness

The study found that increasing SMEs' online commitment could build Indonesia's yearly financial development by 2%, the jump it desires to turn into a middle-income nation by 2025. For India's situation, the internet plan just received important support, as the government chose to demonetize huge denomination currency in an effort to encourage less cash. Early subjective proof focuses towards more noticeable acceptance of online approaches of payment by consumers and organizations even after re-monetization is nearly complete in the Indian economy (GSMA Intelligence, 2016).

Demonetization: Push to Additional Digitization?

The Indian government has questionably pulled off one of the most significant reform measures in the recent past by demonetizing high-value cash notes. This has triggered an unanticipated lack in currency flow, as around 86% of the currency had to be traded. Whereas the change had some adverse effect, there are some long-term positives too. A push to digitization in the payment stream was one such phenomenon. The verdict, in last 2016, to review banknotes from flow has fundamentally been a shot in the arm for fintech and another banking service in the country. The shortage of currency made it essential for all sectors of society to utilize the electronic cash. The outcome was a massive increase in online payment mechanisms even after the economy was entirely re-monetized. In fact, the economy has seen a 59% growth in transactions through digital networks in March compared to the first month after demonetization was proclaimed. It is important to take note that while a portion of this increase would be an early response, long-term structural factors show that India is equipped for an internet revolution and this occurrence will lead to everlasting convert in online. By 2021, the Indian Internet user base is forecast to reach 555.3 million which is 44 percent of the populace. Over the longer term, payment entries, cards, mobile wallets, online retail, payment banks, and e-marketplace industries are maybe to see net rises (World Bank, 2015).

Entrepreneurs in the Internet Economy

Various business/entrepreneurs grasped on the technology that energized the DE to make new organizations and business plan that couldn't have happened and occurred at the scope and scale they do today, in past generations.

These new organizations incorporate the ride-sharing stages such as Uber and Lyft; the home rental platform Airbnb; and content on request service, for example, Netflix and Spotify (Thomas et al, 2019).

Customer Engagement

- Internet is driving network in Asian nations, especially by means of the Mobile Internet.
- This ranges from mobiles connecting purchasers, to the increase of social media and opportunities exhibited by online banking, especially in nations where there is huge unbanked and rural populace.

People everywhere in the world over are grasping and driving transformation, at least where the technology is accessible and reasonable. Online technology can possibly open up a variety of opportunities for consumers, from availability and flexibility to access to social media and online banking, and this can encourage economic development and growth (Mahar, 2010).

Internet is driving network in Asian nations, from mobiles connecting customers in rural regions to improving livability and comfort in urban territories, especially as urbanization all through Asia results in blockage and natural difficulties. Asian customers are either previously grasping online or representing that they will when access improves and costs comfort, as confirmed by the strong acceptance of mobile.

Internet And Mobile Phone

Mobile tariffs are increasing quickly in the locale, driving across Internet acceptance and changing consumer behavior.

In Indonesia, for instance, the consumer has involved in the mobile Internet in a nation where the problematic region has delayed investment in fixed-line communication. There are over 1.3 mobiles per capita in Indonesia, and the tendency has been far more from 2G telephones near telephones that can get to the Internet. A study by Nielsen in 2011 indicate that Indonesians who do utilize the Internet were bound to get to it on their cell phones than in any of the other significant nations in Southeast Asia. There has been strong usage of social media in 2016, Indonesia was placed as fourth on the planet in term of having Facebook users around 78 million, behind India, USA, and Brazil.

Internet penetration in India has likewise developed quickly in the most recent couple of years, with around 432 million users in 2016, and a potential 750 million supplementary potential user. Additionally, 77 percent of urban and 92 percent of rural users consider mobile as the key gadget for accessing to the Internet (Mobile financial service, GSMA, 2014).

Figure 1.
Source: Economic Intelligent Unit

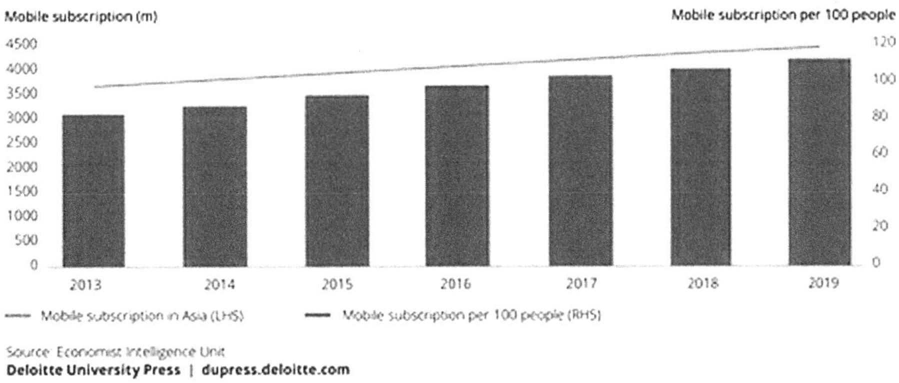

The Internet Economy Waves

The initial wave originated from the formation of the shared or common stock organization, where proprietors could spread the hazards and rewards of setting up new ventures.

Another wave emerged from the twin inventions of the telegraph and railroad, which made correspondences and coordination stage for enormous scale industry. The need to fund these activities propelled the capital markets in presence nowadays.

The third wave emphasis on a mutual stage—the web/internet. Mobility, Cloud computing, business knowledge, and social media support this change, which is occurring in equally developing and developed nations. This wave will change numerous parts of the worldwide from consumer behavior to new business models (Infinite information technology, 2019)

The digitization technology waves contain wave of digitization which is related with the overview and acceptance of what nowadays are measured "mature" technologies, for example, the management information system aimed at automating data processing and applied to monitor and reporting of business

Figure 2.
Source: Deloitte.com (2013)

	Europe	Arab states	CIS	Americas	Asia & the Pacific	Africa
Pre-paid handset-based	1.1	5.7	5.7	5.9	5.9	38.8
Post-paid handset-based	1.1	2.2	5.6	5.0	3.5	36.2

Source: ITU 2013, https://www.itu.int/en/ITU-D/Statistics/Documents/facts/ICTFactsFigures2013-e.pdf
Deloitte University Press | dupress.deloitte.com

performance, telecommunications technologies, for example, broadband and voice media communication which allow the remote access of information and data. PCs were introduced in the business environment during the 1960s and achieved 92.61% penetration among organizations in OECD nations just in 2014. Whereas Mobile phones were introduced in 1985 and accomplished around 99 percent universal dissemination by 2015. Personal PCs, produced in 1982, were embraced by 80.29 percent of OECD families in 2015. Additionally, fixed broadband was presented around in 1995 and has achieved 80.07 percent adoption within the same range. Whereas mobile broadband systems 3G or more achieved 84 percent of the worldwide populace in 2016 (OECD, 2014).

Also, digitization involves the dissemination of the Internet and its correspondence stages such as search engine, and marketplaces which allow the systems of the enterprise to consumer and enterprise between themselves for buying of supplies, and distribution of output. Moreover, diffusion start point can be somewhat randomly placed around 2010, which involves the acceptance of a numerous range of technologies which aimed at enhancing the information, processing and the quality of decision making, which further automating the routine tasks within business and governments. They are comprising of:

1. **Big Data Analytics:** The capacity for preparing very huge informational data sets to classify the pattern of relationships such as correlation, causality between information to be utilized in recognizing market trends, consumer behavior, and priorities. Most usual applications range from environmental or climate change research, as well as advertising and business structure within the private sector (UNCTAD, 2016).
2. **Internet of Things:** IOT includes stages that link different sensors and information device so as to create an entire vision of the behavior of an organization, framework, business activity, or some phenomena. Such as agriculture which comprises control fertilizers and regulates the most appropriate harvest. Smart cities which allow the control of traffic flows and cope energy use in public places. Telemedicine which monitors hospital patient's health. The adoption of the Internet of things is directly related to vertical applications and while these platforms are different from machine to machine (M2M) applications they are based on common components. Moreover, the Internet of things system is a platform that interconnects a variety of separate devices to provide a complete vision of certain phenomena. Therefore, M2M devices are components of an IOT networks.
3. **Robotics:** includes the application of online technology to the exhibition of monotonous manual responsibilities, for example, those required in vehicle assembly, and research in risky situations.
4. **3D Printing:** technology that permits the manufacture of items by means of progressive printing of adhesive materials such as polymers. While usage of 3D printing is widespread, its usage is genuinely common in product design such as medicinal prosthetics, architectural models, textile plan), as well as the development of extra parts in consumer electronics and industrial products.
5. **Machine Learning (ML) and Artificial Intelligence:** These two technologies are not alike; despite the fact, they share some basic conceptions. ML is an artificial intelligence application involving the development of programs that let a computer to learn procedures without being necessarily preprogrammed. The most extensively recognized application, the ML program alter itself after it starts processing data. The common ML application is product recommendation, internet platforms like Netflix, and Amazon, fraud revealing in credit card usage, and computation of consumer credit profile.

Figure 3.
Source: Statista

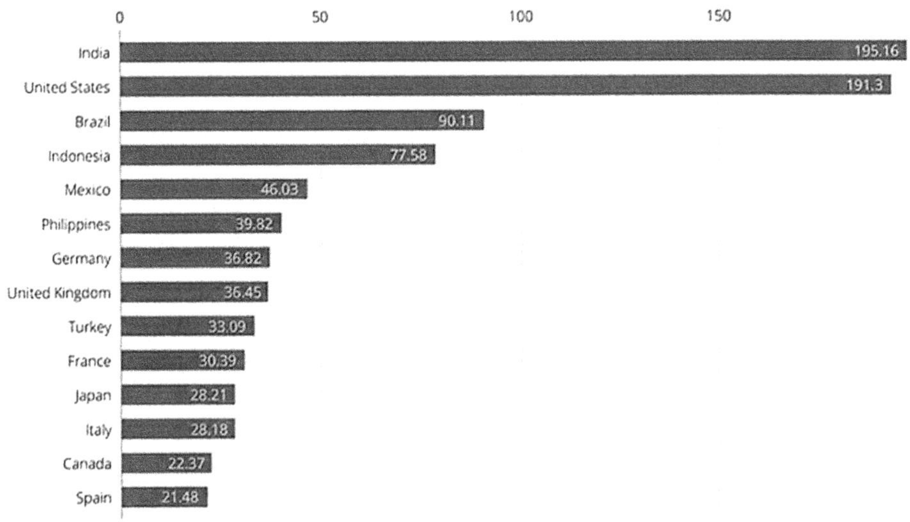

Source: Statista https://www.statista.com/statistics/268136/top-15-countries-based-on-number-of-facebook-users
Deloitte University Press | dupress.deloitte.com

Social Media

There is high involvement in social networks, with India primarily the highest number of Facebook users in the world with around 195 million in 2016. Whereas, Indonesia ranks 4th, with 78 million users. Facebook has been restricted in China, however, there are other social networks such as Renren social networking website and WeChat. When seeing about all types of social media and internet-based life, more than 1 billion people crosswise Asia were dynamic users of online networking in 2016, with 806 million in China alone an additional 130 million in India and 76 million in Indonesia. In Japan, LINE is the most widespread mobile messaging app, letting users send messages, share pictures, flicks, and tunes. LINE has 50 million active regular users, signifying 40 percent of the nation's populace. The under histogram shows the Facebook user in 2016 around the globe.

Online business is also growing in China, presently, the world's major web-based business. Virtual shopping is developing as a strong competitor to shopping in physical stores, with its accomplishment supported by sophisticated payment systems and well-organized and proficient logistics networks. A portion of the major online bazaars enables users to make accounts with a platform itself, so that overheads are not overdue while waiting for bank clearance, letting goods or products to be supplied quicker.

The Taboos Marketplace propelled in 2003 is the Alibaba Group's consumer-to-consumer web-based platform. The Alibaba Group likewise works the Tmall site launched in 2008, which is its business-to-consumer web-based platform. Whereas Taobao Bazaar facilities the transaction of products among private purchasers and suppliers, including small businesses, Tmall vendors are characteristically greater listed businesses. Tmall vendors are typically larger registered businesses. These two platforms had 443 million active purchasers as of the end of 2016,_which generally signifies 60 percent of all Internet users in China. Tmall.com and Taobao.com are presently the third- and fifth-most visited sites in China, correspondingly, and ninth and tenth worldwide (Hudson Kuteesa, 2018)

Furthermore, dock less bicycle sharing systems are a genuine case of China speed in real life. The principal supplier, Ofo, introduced in 2014 and within few years a large number of rental bicycles, unlocked by means of a cell phone application, were visible on Chinese lanes. The market consequently contracted drastically yet the speed and size of the rollout of an overall new urban mobility service were surprising. Also, Hangzhou is world's leading cashless cities and home to Alibaba Group Holding Ltd, a Chinese multinational conglomerate specializing in e-commerce, mobile payments, retail, internet, Artificial Intelligence, and technology established in 1999, the company delivers consumer-to-consumer, business-to-consumer and business-to-business sales service through web portals, as well as electronic payment services, shopping search engines, and cloud computing services (Marcus Fairs, 2019)

Examples: TechCrunch, a leading technology media property, Uber, the world's biggest transport or ride sharing organization. Facebook, the world's greatest well-known media proprietor, make no content. Alibaba, the most valuable retailer, has no stock and inventory. Additionally, Airbnb, the world's largest lodging supplier, claims no real estate.

Online Banking

Mobile banking encourages financial inclusion by providing access even in regions where the number of physical branches is limited.

Online banking incorporates cell phone and Internet banking. Cell phone or smartphone banking is making specific inroads in emerging Asia, with the penetration increasing in using smartphones increasing over fivefold in the three years from 2011 to 2014, going from 5% of the populace utilizing cell phone banking in 2011 leading to 26% in 2014. However, Mobile banking likewise improves monetary inclusion in the locale. For instance, in Papua Guinea, mobile network operators have been exempt under the banks and financial institution Act 2000 to lead cell phone money transfers. In certain nations, electronic top-ups enable people to send or receive airtime or broadcast, which can act as transfers like mobile money. Moreover, cell phone banking endorses financial inclusion by giving access even in areas where the quantity of physical branches is inadequate (Review of Economics, 2016).

For instance: India incorporate the usage of Unstructured Supplementary Service Data (USSD), which permits for cell phone banking transactions using the basic feature of phones without the need for mobile Internet network; Unified Payments Interface, which powers numerous bank accounts in a solo mobile app; and micro ATMs, where business correspondents (BCs) who may be local shop proprietor which act as "micro or small scale ATMs" to conduct prompt transactions. There is an important possibility for such technologies to have an influence on consumers and increase accessibility, especially as much of Asia's populaces remain unbanked.

Also, in China, there are now over 650 million smartphones users which are nearly double the whole population of the United States of America. The Chinese's nation is moving immense to adopt mobile payment technology. In 2017, Chinese consumers spent around 9 trillion dollars via mobile payments, compared with $112 billion in U.S. (Marcus Fairs, 2018).

Some of the key benefits of online platforms are as follow:

1. The broader customer reaches to online platforms significantly increase the prominence of partner companies because of their access to a huge customer pool.
2. Cost reduction Partnering with online platforms assists to decrease operation costs due to economies of scale and partition of functions that online platforms help to attain. For instance, restaurants can

supply food through GoJek (transportation network system company located in Jakarta, Indonesia) rather than employing their own delivery team. Web-based business platforms release partners from the hassle of logistics and payments by providing these services at practical costs.

3. Opportunities for improvement online stages, for example, Food panda. The Food panda is a German cell phone food delivery bazaar, the headquarter located in Berlin, Germany, which provide built-in analytics that helps partners to understand sales patterns and possibly improve their product offerings.

CASE STUDY ON KENYA

Summary

Kenya's major mobile operator Safaricom introduced its M-PESA mobile cash service in 2007, taking benefits of the increasing uptake of mobile services and low banking penetration in the nation to offer a solution that spread the financial services to large swathes of the populace without access to traditional financial institutions.

Throughout the years, M-PESA has evolved from a beginning person to person remittance service to a internet commerce suite, facilitating a wide range of online transaction among public institutions, private businesses and individuals. In 2015, Safaricom introduced M-PESA debit cards and point of sale terminals to allow users to pay for government services and taxes, in the line with the government's plan to use internet technologies to improve effectiveness and control bribery and corruption. The Kenyan government expects financial transactions between public institutions and citizens, including tax, bills, fines, and business registration payments, to be done electronically in the following couple of years.

Impact

Safaricom isn't just mobile money service provider in Kenya. Rivalry from other service providers is helping to drive development and make the service more reachable and accessible to consumers. In 2014 the total value of mobile money transactions in Kenya reached around KES 2.372 trillion ($24.7billion) across 86 million transactions. While online transfers and credit/debit card payment are on the rising country, the mobile money platform will play a key role in the internet commerce space given its possible to reach more consumers compared to other solutions.

Safaricom witness strong growth in person-to-business (64%) and business-to-business (83%) transaction in the half year to September 2014, compared to 20% for person-to-person transactions, featuring the potential of mobile-based internet commerce.

Reference: Central Bank of Kenya

CASE STUDY

Ma Te Sai, Lao PDR

Ma Te Sai is a social enterprise established in 2010 by Emi Weir and Clemence Pabion. They work with rural communities and people in Lao PDR to sell its handmade products through both its offline and online stores. Apart from exhibiting organization and product information on its independent website, Ma Te Sai also runs an official Facebook page for marketing purposes. The proprietor of the business shared that locals love Facebook and use it as an e-commerce platform. Ma Te Sai's Facebook page has more than 1,700 followers who often receive marketing feeds on new products and special campaigns.

Source: GSMA intelligence (2016)

CASE STUDY

IT Adoption Promotion Grant Program, Japan

The IT Adoption Promotion Grant Program run by the Ministry of Economy, Trade, and Industry, Japan (METI) serves as a good illustration of government support in helping MSMEs kick-start digitalization. A business owner (Client) can search for registered and qualified ICT vendors, consultants, or providers (support) which have been accepted by the authority on the program website. Apart from that, companies can also browse for appropriate ICT tools that fit their businesses. If the client managed to match with vendors successfully, the authority will disburse the grant straight to the client.

Source: ASEAN (2017)

CASE STUDY

Singapore: Skills Future

'Skills Future' is a national movement introduced by the Government of Singapore to deliver its citizens with the opportunities to grow themselves to their fullest potential through their lives, regardless of their starting points. Through this program, the skills, passion, and contributions of every individual will drive Singapore's next phase of growth towards a progressive economy and comprehensive society. As part of the national 'Skills Future' movement, the Government has rolled out the 'Skills Future Series' in partnership with the Institutes of Higher Learning (IHLs). The 'Skills Future Series' includes a list of modular, industry-relevant courses that pursue to equip working adults with specific skill sets to meet altering job demands in developing areas, allowing them to stay significant and competitive in the future. As a start, the initiative focused on eight parts: (i) data analytics, (ii) finance, (iii) tech-enabled services, (iv) online media, (v) cyber security, (vi) entrepreneurship, (vii) advanced manufacturing, and (viii) urban solutions. After some time, the IHLs will keep on developing further courses, and improve their focus to meet industry needs. Furthermore, the Skills Future Series, the Government of Singapore has also rolled out the 'Skills Future for Online Workplace'. This program helps every Singaporean adults and grown up, including those planning to return to the workforce, to understand emerging technologies and how they affect work, understand and use data, and accept a positive attitude for change, innovation, and resilience. For instance, applicants of this program will learn how to apply frequently used mobile apps in their daily life and perform basic cybersecurity activities, for example, setting up passwords to secure information and data.

CASE STUDY: ESTONIA

Summary

The government of Estonia has executed the most comprehensive online citizenship activity to date. Over the last 15 years, the e-Estonia initiative has effectively digitized key government services and residents' civic obligations. From e-elections, e-taxes, e-police to e-healthcare, e-identity and e-school, Estonians have essentially become online citizens. The 'X-Road' data exchange infrastructure launched by Estonia's Information System Authority links all the country's decentralized public and private databases and is one of the key elements behind the country's internet society initiative. The other key component is the chip-enabled 'eID' card – the nationwide standardized system of online verification and online authentication. With near universal mobile adoption, eID is increased by Mobile-ID, negating the need for carrying smart card readers. With this substructure in place, e-Estonia delivers a vast range of government services to all stakeholders, including residents and businesses.

Impact

The e-Estonia initiative has resulted in four noticeable impacts: 1) unprecedented levels of transparency and availability in government; 2) safe, suitable and flexible exchange of private, public and corporate data; 3) healthier, a well-educated populace with easy access to social services; 4) a wealthy atmosphere for business and entrepreneurship. Currently, more than 90% of the 1.3 million Estonians have E-ID cards. Apart from validation (m-ID), other examples of mobile e-government services progressively used by citizens in Estonia are paying parking charges (m-Parking) and purchasing tickets on public transport without cash (m-Ticket). Additionally, 94% of tax returns were filed through the e-Tax Board in 2011 however a quarter of the votes in the 2011 parliamentary elections were cast over the Internet. Estonia strategies to have 10 million total users (including residents and foreigners) of the e-Estonia platform by 2025 (Digital report, 2012).

CASE STUDY: SEOUL, SOUTH KOREA

Summary

Seoul is the capital of South Korea and the nation's biggest city with a populace of 10 million people. In June 2011, the city's authorities introduced 'Smart Seoul 2015', a smart city initiative to incorporate ICT in the city's development strategies. Supported by advanced ICT infrastructure, including free Wi-Fi in public places and mobile data networks, the city administration and other institutions developed a wide range of online services, comprising;

- **Smart Work Centers**: A framework that empowers government employees to work from 10 offices close to their homes.
- **Community Mapping:** Open governance model for increased citizen input in the administration of the city.
- **Smart Metering:** Productive improvement initiative to decrease the city's total energy use by 10%
- **U-Seoul Safety Service:** A project that integrates location-based services and CCTV tragedies involving kids, disabled, old aged, and those suffering from Alzheimer's disease.

Impact

Mobile technology is a key part of Smart Seoul. Through the m. Seoul apps, government offices provide the city's residents with 62 unique services reachable on more than 10 types of mobile devices. m.Seoul apps also support location-based services pinpointing near public offices, washrooms, hospitals, superstores and bus stations. Other services include live real-estate listings, daily job-search updates, and announcements of free cultural events. The average number of m. Seoul users improved from under 50,000 per month in 2007 to more than six million every month in 2014, as indicated Seoul Metropolitan Government, supporting Seoul's top 10 rankings among worldwide cities in a variety of indices, including e-Governance, labor efficiency, and city competitiveness.

Reference: Young (2016)

CONCLUSION

This chapter seeks to gain a fundamental understanding status and adoption of digital technology by business, or Enterprises, the common online tools used, and how deeply surrounded digitalization is in their business operations. Moreover, to gain healthier information and improved capacity of developing nations in understanding the DE and its cumulative effect in Asia. Additionally, develop and improve the discussion between government authorities and professionals on key issues correlated to the online economy. However, online networking and communication infrastructure offer a worldwide platform over which people and organization methodologies, plan, strategies, collaborate, communicate and search for information. It is commonly recognized that the development of DE has an extensive effect on the entire economy. Furthermore, DE has the prospective to completely change the social environment and economic activities of Asia. It is now encountering high development, quick innovation and advancement, and broad application to other economic divisions. Despite that gigantic opportunities exhibit by the online economy; Asia has not yet completely got the potential of connecting online technology for maintainable development, due to other poor ICT infrastructure, insufficient skills development, and socioeconomic hindrance that stop much of the Asia's populace from taking part in the online economy. The internet revolution isn't only big data and online platforms but also how those advanced technologies can be applied to increase opportunities for innovation, new business plans, procedures and smart product and services. More, the DE is enabling territorial organizations to move away from the local into the global, in keeping with the long term trends to market liberalization and reduced trade hindrances.

FUTURE RESEARCH PLAN

Substantially more research is required with concern to sophisticated online users and their influence on society and online markets. It has been proposed that people integrate the internet into their existing routine rather then creating a different way of life around the internet. Hence, integrating e-commerce into people's normal set of activities will rely on the level to which they pursue to achieve financial gain. For people who want to find the best deal or product that is perfect for them, online shopping will be attractive and more effectively coordinated into their everyday activities. This demonstrates that the already discriminating shoppers will be more likely to gravitate towards the online channel, thus by fueling some rich get richer phenomena. Selwyn et al. (2005) specify that the people making use of online business are more knowledgeable and repeated users, future stimulating this phenomenon. Research is compulsory that analyze the economic and social consequences of a small group of users accounting for a large portion of online business activity. Research may address the scope to which this small segment of the populace controls or effects online markets, pursue to better understand the demographics of these users, also analyze the economic benefits accumulating to these peoples.

REFERENCES

Albert. H. (2018). The Infinite information technology, *Digital Transformation*. Retrieved from http://www.infiniteinformationtechnology.com/digital-transformation-what-is-digital-economy

Arthur, W. B. (1996). Increasing Returns and the New World of Business. *Harvard Business Review*, *74*(4). PMID:10158472

Asian Development Bank. (2018, February). Understanding the Digital Economy: What Is It and How Can It Transform Asia? Event. Asian Development Bank. Retrieved from https://www.adb.org/news/events/understanding-digital-economy-what-it-and-how-can-it-transform-asia

Broadband Connectivity in South East Asia. (2018). World Telecommunication/ICT Indicators database. Retrieved from http://www.itu.int/en/ITU-D/Statistics/Pages/publications/wtid.aspx

Central Bank of Kenya. (2015). Bank supervision annual report of 2015. Retrieved from https://www.centralbank.go.ke/uploads/399346751_2015%20Annual%20Report.pdf

Deloitte. (2017). Role of digital in Asia's economic growth | Deloitte Insights, 2017. Retrieved from https://www2.deloitte.com/insights/us/en/economy/voice-of-asia/may-2017/digital-role-economic-growth.html

Deloitte. (2017). What is the digital economy? Deloitte Malta, 2017. Retrieved from https://www2.deloitte.com/mt/en/pages/technology/articles/mt-what-is-digital-economy.html

DevelopmentsS. M. E. in ASEAN. (2017). Retrieved from http://asean.org/asean-economic-community/sectoral-bodies-underthe-purview-of-aem/micro-small-and-medium-enterprises/overview/

Digital Efficiency Report, Cabinet Office. (2012, November). Retrieved from https://www.gov.uk/government/publications/digital-efficiency-report

Dubbah, M. M. (2010). *International and Comparative Competition Law* (p. 115). Cambridge, UK: Cambridge University Press. doi:10.1017/CBO9780511777745

Dyer, J. H., & Chu, W. (2003). The Role of Trustworthiness in Reducing Transaction Costs and Improving Performance: Empirical Evidence from the United States, Japan, and Korea. *Organization Science*, *14*(1), 57–68. doi:10.1287/orsc.14.1.57.12806

Garrison, G., Wakefield, R. L., & Kim, S. (2015). The effects of IT capabilities and delivery model on cloud computing success and firm performance for cloud supported processes and operations. [</ unknown>]. *International Journal of Information Management*, *35*(4), 377–393.

GSMA. (2014). State of the Industry: Mobile Financial Services for the Unbanked. Retrieved from https://www.gsma.com/mobilefordevelopment/wp-content/uploads/2015/03/SOTIR_2014.pdf

GSMA. (2016). Advancing Digital Societies in Asia. Available at http://www.gsma.com/newsroom/press-release/new-gsma-study-tracks-digitalsociety-progress-asia/, page 20 44.

GSMA. (2015). Competition Policy in the Digital Age. A practical handbook. Available at http://www.gsma.com/competition-policy-handbook

Kuteesa, H. (2018, Nov. 3). Hangzhou: Inside one of the world's leading cashless cities. The New Times, Rwanda.

Lin, J. (2011). Technological adaptation, cities, and new work. *The Review of Economics and Statistics*, *93*(2), 554–574. doi:10.1162/REST_a_00079

Fairs, M. (2019). China is fast becoming the world's creative superpower. Retrieved from https://www.dezeen.com/2019/04/17/china-design-power-opinion-marcus-fairs/

Medium. (n.d.). Internet Economy vs. Sharing Economy vs. Token Economy. Retrieved from https://medium.com/@dennis_z/internet-economy-vs-sharing-economy-vs-token-economy-13c69905946b

GSMA. (2014). Mobile Economy Report Sub-Saharan Africa. Available at http://www.gsma.com/mobileeconomy/archive/GSMA_ME_SubSaharanAfrica_2014.pdf, page 29.

OECD. (2014). The digital economy, new business models and key features. In *Addressing the Tax Challenges of the Digital Economy*. Paris, France: OECD Publishing. doi:10.1787/9789264218789-7-

OECD. (2015). *Measuring the Digital Economy, A New Perspective*. OECD Publishing.

Digital Societies Report. (2016). Indonesia and Thailand.

Digital Societies Report. (2016). Australia, Japan, and Singapore were selected as representative countries.

Share of Population in Selected Countries Who Are Active WhatsApp Users as of 3rd quarter. (2017). Retrieved from https:// www.statista.com/statistics/291540/mobile-internet-user-whatsapp/

Study on MSMEs Participation in the Digital Economy in ASEAN. (2018). pp. 1-28.

Thomas, A., Passaro, R., & Quinto, I. (2019). Developing Entrepreneurship in Digital Economy: The Ecosystem Strategy for Startups Growth. In Strategy and Behaviors in the Digital Economy. IntechOpen.

Toppr guide, Digital Economy, 2019.

<unknown>Ja-Young, Y. (2016, March 1). Korea shifting to a cashless society. Korea Times.

World Bank. (2010). ITU, *Info*Dev, IFC, *Telecommunications Regulatory Toolkit*, 10th Anniversary edition. Retrieved from https://openknowledge.worldbank.org/

World Trade Organization. (2013). *E-Commerce in Developing Countries: Opportunities and Challenges for Small and Medium-Sized Enterprises*. Geneva, Switzerland: World Trade Organization.

ADDITIONAL READING

Brynjolfsson, E., Hu, Y., & Smith, M. D. (2003). Consumer surplus in the digital economy: Estimating the value of increased product variety at online booksellers. *Management Science*, *49*(11), 1580–1596. doi:10.1287/mnsc.49.11.1580.20580

Brynjolfsson, E., & Kahin, B. (Eds.). (2002). *Understanding the digital economy: data, tools, and research*. MIT press.

Coyle, D. (1999). *The weightless world: strategies for managing the digital economy*. MIT press.

Ghosh, R. A. (Ed.). (2005). *CODE: Collaborative ownership and the digital economy* (pp. 275–276). Cambridge, MA: Mit Press.

Johansson, B., Karlsson, C., & Stough, R. (Eds.). (2006). *The emerging digital economy: entrepreneurship, clusters, and policy*. Springer Science & Business Media. doi:10.1007/3-540-34488-8

Malecki, E. J., & Moriset, B. (2007). *The digital economy: Business organization, production processes and regional developments*. Routledge. doi:10.4324/9780203933633

Pateli, A. G., & Giaglis, G. M. (2004). A research framework for analysing eBusiness models. *European Journal of Information Systems*, *13*(4), 302–314. doi:10.1057/palgrave.ejis.3000513

Raisinghani, M. S. (Ed.). (2003). *Business Intelligence in the Digital Economy: Opportunities, Limitations and Risks: Opportunities, Limitations and Risks*. Igi Global.

Tapscott, D. (1999). *Blueprint to the digital economy: Creating wealth in the era of e-business*. McGraw-Hill, Inc.

Tapscott, D. (2000). *Digital capital: Harnessing the power of business webs.* Harvard Business School Press.

Yoo, Y., Lyytinen, K. J., Boland, R. J., & Berente, N. (2010). The Next Wave of Digital Innovation: Opportunities and Challenges: A Report on the Research Workshop'Digital Challenges in Innovation Research'. *Available at SSRN 1622170.*

KEY TERMS AND DEFINITIONS

Big Data Analytic: Big data analytics enables data scientists to study large and complex varieties of data using predictive modeling, statistics and other analytics to discover unseen patterns, market trends, customer preferences, unknown correlations and other valuable information to help organizations improve their decision-making method, enabling more well-informed business decisions.

Digital Platform: Digital platform is the software and technology used to combine and streamline business operations and IT systems. A digital platform serves as a company's backbone for operations and customer engagement.

Digital Transformation: Digital transformation is the act of adopting and integrating technology into all aspects of business, creating a foundational shift that allows sustainable innovation and creative progress for an organization. It comprises of cultural changes, internal resource considerations, and product development that supports improved, technology-powered user experiences.

Digitization: Knowledge can be stored in digital form. Unlike the old economy where information was analog or physical, communication was only possible through the actual movement of people. In the new economy, information in digital form, eased by the digital devices that allows free movement of massive amounts of information in the shortest time possible among people in different parts of the world.

E-Business: contraction of 'electronic business', refers to the use of information and communication technology to support a business strategy.

E-Commerce: Electronic commerce is the purchasing and selling of goods and services on the internet, particularly the world wide web. e-commerce is one factor of e-business that has the potential to include monetary transactions.

Machine Learning (ML): ML is a branch of Artificial Intelligence that uses statistical techniques or algorithms to allow a computer to become better at what it does. The computer classifies patterns in behavior or speech, such as, it notes those differences to "learn" more about a specific user or set of users.

Mobile Internet Usage: Use of the internet away from home or work place on moveable computers or handheld devices through mobile phone networks.

Chapter 14
The Role Of Strategic Orientation In Export Performance Of China Automobile Industry

Muhammad Imran
https://orcid.org/0000-0001-7126-2831
The Islamia University of Bahawalpur, Pakistan

Jawad Abbas
Forman Christian College University, Turkey

ABSTRACT

During the last four years, the China automobile industry experienced a 49% drop in sales in the domestic and international markets. Company owners and the Government of China are exploring factors which could help them overcome the issues relating to sale, reputation, and brand image. Nonetheless, the investigation of company export performance factors in the automobile sector of China has largely been ignored. However, authors of this chapter conducted a literature review on factors of firm export performance. Therefore, the conceptual framework has found the factors of firm export performance such as total quality management (TQM), entrepreneurial orientation (EO), export market orientation (EMO), and brand orientation (BO) based on resource-based theory. This research believes that the proposed factors can increase the firm export performance of China automobile industry. The future studies should validate the proposed research framework empirically in the context of the Chinese automobile industry.

INTRODUCTION

Considering the social and industrial changes mainly caused by the technological revolution, a number of industries have reshaped their operational activities (Shahzad et al., 2019). The automobile industry is among the rapidly growing industries in the world and plays a significant role in economic development,

DOI: 10.4018/978-1-7998-0357-7.ch014

such as the creation of employment, export etc. (Thun, 2018). During the last two decades, China has dominated all countries with respect to economic growth (Habib, Abbas, & Noman, 2019). Because of its cheap labour and manufacturing cost, most companies have relocated their production units in China. According to Sachon (2016), until 1999, the Chinese automobile industry was producing less than two million vehicles. However, this industry has experienced a significant boom in the 20th century and, in 2015, this industry produced more than 24.5 million vehicles (Sachon, 2016). According to Bloomfield (2017), in 2010, China surpassed the United States and Japan in the automobile industry and become the biggest producer of automobiles in the world. The automobile industry has contributed more than seven percent to GDP of China (Chen, Lawell, & Wang, 2017).

Although China is the leading producer of the automobile's units, it exports only 3% of its total production (Huo, Gu, & Jiang, 2018). Moreover, during the last four years, this industry experienced a 49% drop in sale in the domestic and international market (Xie & Li, 2018). Considering this situation, it is imperative for Chinse automobile industrialists to investigate the factors which could help them to overcome the issues relating to sale, reputation and brand image. According to Knoke (2018) and Chaston (2017) quality and trade barriers are the two principal factors hindering the Chinese auto industry to expand further. According to De-Clercq and Zhou (2014), entrepreneurial orientation (EO), export market orientation (EMO) and brand orientation (BO) are the other possible factors which could significantly impact on the export performance of any country.

This study conceptualizes that quality management, EO, EMO and BO could enhance the export performance of the China automobile firm. To understand the link between these factors and their potential role in export performance resource-based View (RBV) theory is used. According to RBV, firms can create a competitive advantage by formulating strategies and capitalizing on resources which cannot be followed by any competitors (Barney, 1991). The success of an organization is primarily determined by internal resources (assets and capabilities) that owned by the firms as well as external resources (Abbas & Sağsan, 2019). These resources could be tangible as well as intangible, such as human resource, technological infrastructure etc. (Abbas et al., 2014).

Weerawardena and Coote (2001) stated that firm intangible resources that categorized in RBV element would lead to higher export performance. The RBV theory also anticipated a firm as an embodiment of distinctive packaged, concrete, and intangible resources (Li, 2018) that lead to higher performance. TQM, EO, EMO, and BO which are considered as internal capabilities or strategic resources of the firms would be the factors that trigger the firm performance. Therefore, TQM, EO, EMO, and BO are taking into account as a firm's resources which can lead to higher firm export performance. The remainder of this article discusses the literature pertaining to the study's variables, followed by a methodological overview and concluding remarks.

LITERATURE REVIEW

The advancement of technology, especially the internet, has turned the world into a global village (Abbas, Mahmood, & Hussain, 2015). Because of globalization, the geographical boundaries have been eliminated (Mahmood et al., 2014) and businesses are facing global competition. The globalization has also provided customers with more options to select the product, thus forcing firms to be more innovative and creative. In the present business atmosphere international business has become a key factor for the economic development of any country (ALI, 2017). According to Hennart, Majocchi, and Forlani (2017), there are several strategies which could help the organization to attain firm's internationalization, such as export, licensing and overseas presence (Seyoum, 2017). Export performance refers to the level of sale of one or more commodities in the form of goods or services to foreign countries (Hoekman & Shingal, 2017). It is also considered as the degree to which firm achieve its strategic and financial objectives with respect to exporting products to the international market (Cavusgil & Zou, 1994). Shoham (1998) defined export performance as the outcome of firm financial performance in the international markets such as export sales, export profitability and export growth.

The literature provides a number of factors which could impact the operational and financial performance of firms. In the view of Crick and Dana (2004), managers perception and commitment to enhancing organizational performance has a direct association with the operational performance of firms. In addition to this, the motivation level of workers also significantly impacts on the productivity of firms. Dana, Hamilton, and Pauwels (2007) said that organizational offshore production strategy, international orientation, domestic market condition and networks are also key elements which make a valuable impact on organizational import or export performance. In their study, Dana and Paulin (2008) highlighted the role of resources and organizational capabilities to capitalize on those resources as the leading factor for enhancing firm export performance. In another study, Etemad, Wilkinson and Dana (2010) said that government level support and incentives to promote export greatly impacts on organizational attitude and focus to promote export.

According to Imran et al., (2018), despite the importance of TQM in organizational performance, this factor has rarely been analyzed from export promotion aspect. Moreover, the elements of EO and EMO also have largely been ignored in this context (Boso et al., 2016; Chen, Sousa, & He, 2016). According to Chen et al. (2016) and İpek (2017), there are limited studies which have examined the relationship between EO, market orientation, BO and firm export performance. Therefore, it is critical to study these factors with respect to export promotion.

TQM AND FIRMS EXPORT PERFORMANCE

Total quality management (TQM) is a management system which greatly enhances organizational capabilities to manufacture high quality of goods (Mahmood et al., 2014) with least consumption of resources (Al-Qahtani, Alshehri, & Aziz, 2015). It is also well known for enhancing firms' efficiency and effectiveness. TQM is a comprehensive approach which focuses on managing quality from resources acquisition to the delivery of final product or service (Allur, Heras-Saizarbitoria, Boiral, & Testa, 2018; Abbas et al., 2014). It has a special focus on continuous improvement which enables firms to achieve efficiency in production activities, meeting customer needs and taking competitive advantage (Araújo, Tavares, Raupp de Vargas, & Rocha, 2015). A number of studies, such as O'Neill, Sohal, and Teng (2016), Psomas and Jaca (2016) and Cetindere, Duran, and Yetisen (2015) have stated that TQM significantly enhances firms operational and financial performance. However, a key concern in the literature is that most of the studies have focused on the domestic market (Imran, Aziz, & Hamid, 2017) and inadequate attention has been paid to the international market or expert factor (Imran et al., 2018).

The past literature has divided TQM's constructs into two groups, namely soft and hard aspects (Green, 2012; Hoang, Igel, & Laosirihongthong, 2010; Ingelsson, Eriksson, & Lilja, 2012). The Soft or management practices are related to the perceptions of management, employee aspects and principals (Dahlgaard-Park, 2012; Leavengood, Anderson, & Daim, 2014). Employee relations, customer focus, and leadership aspects of TQM are generally taken as soft aspects of TQM. The hard or technical element refers to the quality of equipment and skills, productions, and measurement (Chen, 2013; Vouzas & Psychogios, 2007). The literature has largely focused on the hard practices such as quality of data and reporting, process management, and product/service design. The analysis of literature also highlights some of quality models, such as the American Malcolm Baldridge National Quality Award (MBNQA), the European Foundation for Quality Management (EFQM), and Swedish Quality Award (SIQ) (Bou-Llusar, Escrig-Tena, Roca-Puig, & Beltrán-Martín, 2009; Shafiq, Lasrado, & Hafeez, 2017).

The EFQM model focus on all activities of firms at all levels through continuous improvement of processes (Oliveira, Corrêa, Balestrassi, Martins, & Turrioni, 2017). This model compromises of enablers and results sections. The enablers criteria contain five elements namely; leadership, strategy, people, resources and process. The outcome or results criteria of EFQM comprises customers results, people results, society results and business results. The EFQM focuses on the leadership aspect and states that to achieve and sustain success an organization must focus on strong leadership and clear strategic direction (EFQM, 2013). They need to develop and improve their people skills, resources and process to deliver value-adding products and services to their customers (Suárez, Calvo-Mora, Roldán, & Periáñez-Cristóbal, 2017). The EFQM approach focuses on meeting customers' needs and demands by taking into account their expectations (Sternad, Krenn, & Schmid, 2017). Most of the studies used the EFMQ enabler criteria for dimensions of TQM, which are leadership, strategy, people/employees, resources and process (Go & Govers, 2000; Nwachukwu et al., 2017; Shafiq, 2012; Shafiq et al., 2017).

The analysis of the literature indicates that most of the literature on TQM is focused on firm performance (Corredor & Goñi, 2011; Lages et al., 2009). For instance, Ng and Jee (2012) stated that TQM positively impacts on the operational performance of firms. Munizu (2013) investigated TQM from the financial perspective and concluded that TQM practices significantly boost the financial performance of firms. Mahmoodd, Qureshi, and Nisar (2014) and Yusr (2016) investigated the link between quality and firm innovation performance and found a significant positive association between them. Valmohammadi and Roshanzamir (2015) said TQM greatly enhances organizational learning capabilities. Yazdani, Attafar, Shahin, and Kheradmandnia (2016) analyzed TQM from the market perspective and said TQM greatly increases organizational strength to obtain and sustain competitive advantage. The above discussion clearly indicates that the role of TQM in promoting export performance has largely been ignored. For this reason, it is imperative to investigate this factor from the export performance perspective, particularly in the Chinese automobile industry. Considering the above discussion, the following proposition is proposed:

Proposition 1: Significant relationship exists between total quality management and export performance of Chinese automobile firms.

Entrepreneurial Orientation and Firm Export Performance

Entrepreneurial orientation (EO) is considered as the firm level processes, practices and decision-making styles (Lumpkin & Dess, 1996). It is considered as an important strategic tool to achieve competitive advantage and it enhances the firm's profitability (Zahra & Covin, 1995). In their study, Covin and Slevin (1989) identified three core components of EO, namely innovativeness, risk-taking and proactiveness. Innovativeness shows firm ability to accept new ideas, encourage experiments and favour change (Hurley & Hult, 1998). Innovativeness also represents the firm involved in the development of new ideas and the creative process to introduce new products and services in the market (Lumpkin & Dess, 1996).

Risk-taking shows the inclinations of an individual to take resource commitment (Miller & Friesen, 1978). Proactiveness, the third component of EO, refers to the firm's ability and willingness to make the development and take the first mover advantage against its competitors (Frank, Kessler, & Fink, 2010).

According to Covin and Miles (1999) and Miller (1983), most researchers have operationalized the EO via three dimensions namely; risk taking, proactiveness, and innovativeness. However, in regards to EO and firm export performance relationship, most of the researchers overlooked this relationship (Ajayi, 2016; Hernandez-Perlines, 2018). Therefore, there is a need to investigate the relationship between EO and firm export performance. Taking into account the above discussion, the following proposition is proposed:

Proposition 2: Significant relationship exists between entrepreneurial orientation and export performance of Chinese automobile firms.

Export Market Orientation and Firm Export Performance

Market orientation (MO) is essential for a firm because it captures ability to anticipate, address and take advantage of market changes to achieve superior performance (Styles, Gray, Kropp, Lindsay, & Shoham, 2006). Kohli, Jaworski and Kumar (1993) termed MO as a behavioural activity. From a behavioural perspective, MO refers to "the organization-wide generation of market intelligence pertaining to current and future customer need, dissemination of the intelligence across the department, and the organization-wide responsiveness to it" (Kohli & Jaworski, 1990; Kohli et al., 1993).

Narver and Slater (1990) categorized MO as an aspect of organizational culture. They classified MO into three groups, (1) customer orientation, (2) competitor orientation, and (3) inter-functional coordination. Customer orientation refers to the ability of the firm to classify customer future needs and provide value-added products and services. Competitor orientation refers to an act to understand the competitor's strengths, weakness and understand their abilities and strategies. Meanwhile, inter-functional coordination is an act of achieving the set of firm's objectives through different functional units' coordination (Narver & Slater, 1990; Slater & Narver, 2000).

If firms are having strong local MO it does not mean that they will have similar or higher degree export MO (Cadogan, Paul, Salminen, Puumalainen, & Sundqvist, 2001). This is because export MO of firms, specifically in international markets, are linked with export market behaviour, which is defined as an export-focused generation, dissemination, and responsiveness to export market intelligence in the international market (Cadogan et al. (2009). Likewise, Cadogan, Diamantopoulos, and Mortanges (1999) and Rose and Shoham (2002) presented that EMO behaviour is different from domestic market-oriented behaviour since international market stability is more complicated than the domestic market. Other than that, activities of EMO require more resources and investment than the domestic market. In regards to export performance, EMO would be a significant factor to be included in explaining China automobile export performance.

Considerable portions of literature have been carried out on examination of EMO and firm export performance. Among them, Singh and Mahmood (2013) and Samson and Mahmood (2015) found the contributory role of EMO in firm export performance. However, there is still a great need to investigate the relationship between EMO and firm export performance in a comprehensive manner. On the basis of the above discussion, the following proposition is proposed:

Proposition 3: Significant relationship exists between export market orientation and export performance of Chinese automobile firms.

Brand Orientation and Firm Export Performance

Brand orientation (BO) defined as a market-driven strategy or "an approach in which the processes of the organization revolve around the creation, development, and protection of brand identity in an ongoing interaction with target customers with the aim of achieving lasting competitive advantages in the form of brands" (Urde, 1999). The main objective of BO is to build a strong brand on the basis of values for the firm and acquiring competitive advantage (Anees-ur-Rehman, Saraniemi, Ulkuniemi, & Hurmelinna-Laukkanen, 2017).

Mutlu and Aksoy (2014) investigated the relationship between BO and firm export performance in Turkish companies' context and found a positive relationship between them. In their study, Laukkanen et al. (2013) concluded that BO makes a significant contribution to firm growth. Kaličanin, Veljković, and Bogetić (2015) study also found a significant relationship between BO and firm financial performance. Despite the importance of this concept in the local and international market, limited attention has been paid to further explore this concept, particularly in the Chinese automobile industry's export performance. Taking into account this discussion, the following proposition is proposed:

Proposition 4: Significant relationship exists between brand orientation and export performance of Chinese automobile firms.

PROPOSED RESEARCH FRAMEWORK

The present study's research framework is based on literature. This conceptualized study focuses on firm export performance as the dependent variables and introduces the strategic resources in the context of RBV, namely TQM, EO, EMO, and BO. According to RBV, strategic resources are valuable, rare, inimitable and non-substitutable (VRIN). These resources enable organizations to achieve a sustainable competitive advantage, which will lead to high export performance (Barney, 1991). TQM, EO, EMO and BO are intangible resources, which are valuable, rare, inimitable and non-substitutable in their nature.

RESEARCH METHODOLOGY

The present study is based on the conceptual model, which reviewed the conceptual and empirical studies to develop the research framework. The authors took researcher papers from the variety of databases such as Emerald, Elsevier, Taylor and Francis, Wiley, Springer, Sage and InderScience. The authors also considered newspapers and official's websites of various national and international institutions. These sources have been reviewed to have a comprehensive insight into any potential gaps in the previous studies.

CONCLUSION

In the current conceptualized study, the relationship between TQM, EO, EMO, BO and firm export performance has been proposed. Previous literature has shown that TQM, EO, EMO and BO have a positive relationship with SMEs export performance (Abiodun & Mahmood, 2015; Ajayi, 2016; Jin & Jung, 2016; Lages et al., 2009; Mutlu & Aksoy, 2014). However, the role of these variables in large industries,

Figure 1. Conceptual framework of the study

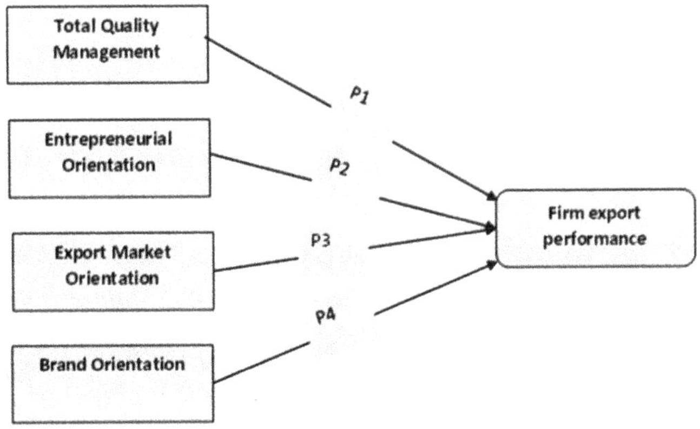

particularly the automobile industry in China has largely been ignored. The proposed framework has several critical contributions to firm export performance. If the proposed framework is validated, the finding of research will provide important information to owners/managers into the significant role of TQM, EO, EMO and BO in improving the export performance, specifically in the automobile industry in China. The finding of the proposed framework will also contribute to the country's GDP and employment creation that would be beneficial to the country as a whole.

REFERENCES

Abbas, J., Mahmood, H. K., & Hussain, F. (2015). Information security management for small and medium size enterprises. *Science International-Lahore*, *27*(3), 2393–2398.

Abbas, J., Muzaffar, A., Mahmood, H. K., Ramzan, M. A., & Rizvi, S. S. ul H. (2014). Impact of Technology on Performance of Employees (A Case Study on Allied Bank Ltd, Pakistan). *World Applied Sciences Journal*, *29*(2), 271–276.

Abbas, J., Muzaffar, A., Shoaib, M., & Mahmood, H. K. (2014). Do Business Schools Really Fulfill Industry Requirements? An Investigation of Industrial Performance of Business Graduates. *World Applied Sciences Journal*, *31*(7), 1378–1384.

Abbas, J., & Sağsan, M. (2019). Impact of knowledge management practices on green innovation and corporate sustainable development: A structural analysis. *Journal of Cleaner Production*, *229*, 611–620. doi:10.1016/j.jclepro.2019.05.024

Abdollahi, H., Razm, K., & Tan, H. (2014). TQM and market orientation's impact on SMEs & quot; performance. *Management Science Letters*, *4*(5), 887–892.

Abeykoon, M., & de Alwis, A. (2016). The Impact of Total Quality Management Practices on Export Performance of Apparel Exporters of Sri Lanka. *Kelaniya Journal of Human Resource Management*, *10*(1).

Abiodun, S. T., & Mahmood, R. (2015). Fostering Export Performance in SMEs: The Roles of Export Market Orientation and Learning Orientation in Turbulent Environment. *International Journal of Economic Perspectives*, *9*(2), 28.

Ajayi, B. (2016). The Impact of Entrepreneurial Orientation and Networking Capabilities on the Export Performance of Nigerian Agricultural SMEs. *Journal of Entrepreneurship and Innovation in Emerging Economies*.

Al-Qahtani, N. D., Alshehri, S. S., & Aziz, D. A. A. (2015). The impact of Total Quality Management on organizational performance. *European Journal of Business and Management*, *17*(36), 1–10.

Allur, E., Heras-Saizarbitoria, I., Boiral, O., & Testa, F. (2018). Quality and Environmental Management Linkage: A Review of the Literature. *Sustainability*, *10*, 1–15. PMID:30607262

Anees-ur-Rehman, M., Saraniemi, S., Ulkuniemi, P., & Hurmelinna-laukkanen, P. (2017). The strategic hybrid orientation and brand performance of B2B SMEs. *Journal of Small Business and Enterprise Development*, *24*(3), 585–606.

Antony, J., Fergusson, C., Warwood, S., & Hing Yee Tsang, J. (2004). Comparing total quality management success factors in UK manufacturing and service industries: Some key findings from a survey. *Journal of Advances in Management Research*, *1*(2), 32–45.

Araújo, C. A. S., Tavares, E., Raupp de Vargas, E., & Rocha, E. (2015). Developing learning capabilities through a quality management program. *Service Industries Journal*, *35*(9), 483–498.

Barney, J. (1991). Firm resources and sustained competitive advantage. *Journal of Management*, *17*(1), 99–120.

Berry, T. H. (1991). Managing the total quality transformation. New York, NY: McGraw-Hill.

Bloomfield, G. T. (2017). The world automotive industry in transition Restructuring the global automobile industry (pp. 19–60). Abingdon-on-Thames, UK: Routledge.

Boso, N., Oghazi, P., Cadogan, J. W., & Story, V. M. (2016). Entrepreneurial and market-oriented activities, financial capital, environment turbulence, and export performance in an emerging economy. *Journal of Small Business Strategy*, *26*(1), 1.

Bou-Llusar, J. C., Escrig-Tena, A. B., Roca-Puig, V., & Beltrán-Martín, I. (2009). An empirical assessment of the EFQM Excellence Model: Evaluation as a TQM framework relative to the MBNQA Model. *Journal of Operations Management*, *27*(1), 1–22.

Cadogan, J. W., Boso, N., Story, V. M., & Adeola, O. (2016). Export strategic orientation–performance relationship: Examination of its enabling and disenabling boundary conditions. *Journal of Business Research*.

Cadogan, J. W., Diamantopoulos, A., & Mortanges, D. (1999). A measure of export market orientation: Scale development and cross-cultural validation. *Journal of International Business Studies*, *30*(4), 689–707.

Cadogan, J. W., Kuivalainen, O., & Sundqvist, S. (2009). Export market-oriented behavior and export performance: Quadratic and moderating effects under differing degrees of market dynamism and internationalization. *Journal of International Marketing*, *17*(4), 71–89.

Cadogan, J. W., Paul, N. J., Salminen, R. T., Puumalainen, K., & Sundqvist, S. (2001). Key antecedents to "export" market-oriented behaviors: A cross-national empirical examination. *International Journal of Research in Marketing*, *18*(3), 261–282.

Cavusgil, S. T., & Zou, S. (1994). Marketing strategy-performance relationship: An investigation of the empirical link in export market ventures. *Journal of Marketing*, 1–21.

Cetindere, A., Duran, C., & Yetisen, M. S. (2015). The effects of total quality management on the business performance: An application in the province of Kütahya. *Procedia Economics and Finance*, *23*, 1376–1382.

Chaston, I. (2017). Technological Entrepreneurship: Technology-driven Vs Market-driven Innovation. Berlin, Germany: Springer.

<jrn>Chen, J., Sousa, C., & He, X. (2016). The determinants of export performance: A review of the literature 2006-2014. International Marketing Review, 33(5), 626–670.</jrn>

Chen, Y., Lawell, C.-Y. L., & Wang, Y. (2017). *The Chinese automobile industry and government policy*. Retrieved from Collis, D. J. (1994). Research note: How valuable are organizational capabilities? *Strategic Management Journal*, *15*(S1), 143–152.

Chen. (2013). Integrated analysis of the performance of TQM tools and techniques: a case study in the Taiwanese motor industry. *International Journal of Production Research, 51*(4), 1072-1083.

Corredor, P., & Goñi, S. (2011). TQM and performance: Is the relationship so obvious? *Journal of Business Research*, *64*(8), 830–838.

Covin, J. G., & Miles, M. P. (1999). Corporate entrepreneurship and the pursuit of competitive advantage. *Entrepreneurship Theory and Practice*, *23*(3), 47–47.

Covin, J. G., & Slevin, D. P. (1989). Strategic management of small firms in hostile and benign environments. *Strategic Management Journal*, *10*(1), 75–87.

Crick, D., & Dana, L.-P. (2004). 14. A comparative, exploratory investigation into the perceptions of internationalizing firms in Singapore and the UK. *Emerging Paradigms in International Entrepreneurship*, 319.

Dahlgaard-Park, S. M. (2012). Core values–the entrance to human satisfaction and commitment. *Total Quality Management & Business Excellence*, *23*(2), 125–140.

Dana, L. P. (2008). Handbook of research on European business and entrepreneurship: Towards a theory of internationalization. Cheltenham, UK: Edward Elgar Publishing.

Dana, L. P., Dana, T. E., & Spence, M. (2004). Public policy and international expansion of high-technology SMEs: A research agenda. *International Journal of Entrepreneurship and Innovation Management*, *4*(2-3), 116–123.

Dana, L. P., Grimwood, S., & William, G. (2009). Export incentives and international entrepreneurship in New Zealand firms. *Journal for International Business and Entrepreneurship Development*, *4*(1-2), 1–21.

Dana, L. P., Hamilton, R. T., & Pauwels, B. (2007). Evaluating offshore and domestic production in the apparel industry: The small firm's perspective. *Journal of International Entrepreneurship*, *5*(3-4), 47.

Dana, L. P., & Paulin, C. (2008). Internationalisation of the New Zealand nutraceutical industry. *Journal for International Business and Entrepreneurship Development*, *3*(3-4), 171–187.

De Clercq, D., & Zhou, L. (2014). Entrepreneurial strategic posture and performance in foreign markets: The critical role of international learning effort. *Journal of International Marketing*, *22*(2), 47–67.

Etemad, H., Wilkinson, I., & Dana, L. P. (2010). Internetization as the necessary condition for internationalization in the newly emerging economy. *Journal of International Entrepreneurship*, *8*(4), 319–342.

European Foundation for Quality Management. (2013). EFQM Model 2013. Retrieved from http://www.efqm.org/index.php/efqm-model-2013/

Frank, H., Kessler, A., & Fink, M. (2010). Entrepreneurial orientation and business performance-a replication study. *Schmalenbach Business Review*, *62*, 175–198.

Go, F. M., & Govers, R. (2000). Integrated quality management for tourist destinations: A European perspective on achieving competitiveness. *Tourism Management*, *21*(1), 79–88.

Green, T. J. (2012). TQM and organisational culture: How do they link? *Total Quality Management & Business Excellence*, *23*(2), 141–157.

Habib, M., Abbas, J., & Noman, R. (2019). Are human capital, intellectual property rights, and research and development expenditures really important for total factor productivity? An empirical analysis. *International Journal of Social Economics*. doi:10.1108/IJSE-09-2018-0472

Hennart, J.-F., Majocchi, A., & Forlani, E. (2017). The myth of the stay-at-home family firm: How family-managed SMEs can overcome their internationalization limitations. *Journal of International Business Studies*, 1–25.

Hernandez-Perlines, F. (2018). Moderating effect of absorptive capacity on the entrepreneurial orientation of international performance of family businesses. *Journal of Family Business Management*.

Hoang, D. T., Igel, B., & Laosirihongthong, T. (2010). Total quality management (TQM) strategy and organisational characteristics: Evidence from a recent WTO member. *Total Quality Management*, *21*(9), 931–951.

Hoekman, B., & Shingal, A. (2017). Aid for trade and international transactions in goods and services.

Huo, B., Gu, M., & Jiang, B. (2018). China-related POM research: Literature review and suggestions for future research. *International Journal of Production Economics*.

Hurley, R. F., & Hult, G. T. M. (1998). Innovation, market orientation, and organizational learning: An integration and empirical examination. *Journal of Marketing*, 42–54.

Imran, M., Aziz, A., & Hamid, S. (2017). Determinants of SME export performance. *International Journal of Data and Network Science*, *1*(2), 39–58.

Imran, M., Aziz, A., & Hamid, S. (2017). Total Quality Management, Export Market Orientation and Firm Export Performance: A Conceptual Framework. *International Journal of Academic Research in Business and Social Sciences*, *7*(9), 591–601. doi:10.6007/IJARBSS/v7-i9/3382

Imran, M., Aziz, A., Hamid, S., Shabbir, M., Salman, R., & Jian, Z. (2018). The mediating role of total quality management between entrepreneurial orientation and SMEs export performance. *Management Science Letters*, *8*(6), 519–532.

Imran, M., Hamid, S., & Aziz, A. (2018). The influence of TQM on export performance of SMEs: Empirical evidence from manufacturing sector in Pakistan using PLS-SEM. *Management Science Letters*, *8*(5), 483–496.

Ingelsson, P., Eriksson, M., & Lilja, J. (2012). Can selecting the right values help TQM implementation? A case study about organisational homogeneity at the Walt Disney Company. *Total Quality Management & Business Excellence*, *23*(1), 1–11.

İpek, İ. (2017). The Resource-Based View within the Export Context: An Integrative Review of Empirical Studies. *Journal of Global Marketing*, 1–23.

Jabeen, R., & Mahmood, R. (2015). The Effects of Total Quality Management and Market Orientation on Business Performance of Small and Medium Enterprises in Pakistan. *British Journal of Economics. Management & Trade*, *5*(4), 408–418. doi:10.9734/BJEMT/2015/14226

Jin, B., & Jung, S. (2016). Toward a deeper understanding of the roles of personal and business networks and market knowledge in SMEs' international performance. *Journal of Small Business and Enterprise Development*, *23*(3), 812–830.

Jing, Z., & Zhu, M. (2016). Market orientation, product innovation and export performance: Evidence from Chinese manufacturers. *Journal of Strategic Marketing*, *25*(5), 377–397.

Kaličanin, Đ., Veljković, S., & Bogetić, Z. (2015). Brand orientation and financial performance nexus. *Industrija*, *43*(1), 155–173.

Knoke, D. (2018). Changing organizations: Business networks in the new political economy. Abingdon-on-Thames, UK: Routledge.

Kohli, A. K., & Jaworski, B. J. (1990). Market orientation: The construct, research propositions, and managerial implications. *Journal of Marketing*, 1–18.

Kohli, A. K., Jaworski, B. J., & Kumar, A. (1993). MARKOR: A measure of market orientation. *JMR, Journal of Marketing Research*, 467–477.

Lages, L. F., Silva, G., & Styles, C. (2009). Relationship capabilities, quality, and innovation as determinants of export performance. *Journal of International Marketing*, *17*(4), 47–70.

Laukkanen, T., Nagy, G., Hirvonen, S., Reijonen, H., & Pasanen, M. (2013). The effect of strategic orientations on business performance in SMEs: A multigroup analysis comparing Hungary and Finland. *International Marketing Review*, *30*(6), 510–535.

Leavengood, S., Anderson, T. R., & Daim, T. U. (2014). Exploring linkage of quality management to innovation. *Total Quality Management & Business Excellence*, *25*(9-10), 1126–1140.

Li, W. (2018). Research on the Innovative Development Mode of Quality Education of College Students Based on the Perspective of Human Resource Management. *Educational Sciences: Theory and Practice*, *18*(5), 2447–2454.

Lumpkin, G. T., & Dess, G. G. (1996). Clarifying the entrepreneurial orientation construct and linking it to performance. *Academy of Management Review*, *21*(1), 135–172.

Mahmood, H. K., Hashmi, M. S., Shoaib, D. M., Danish, R., & Abbas, J. (2014). Impact of TQM Practices on Motivation of Teachers in Secondary Schools Empirical Evidence from Pakistan. *Journal of Basic and Applied Scientific Research*, *4*(6), 1–8.

Mahmood, K., Mahmood Ahmad Qureshi, I., & Nisar, A. (2014). An empirical study on measurement of performance through TQM in Pakistani aviation manufacturing industry. *International Journal of Quality & Reliability Management*, *31*(6), 665–680.

Mehralian, G., Nazari, J. A., Zarei, L., & Rasekh, H. R. (2016). The effects of corporate social responsibility on organizational performance in the Iranian pharmaceutical industry: The mediating role of TQM. *Journal of Cleaner Production*, *135*, 689–698.

Merrilees, B., Rundle-Thiele, S., & Lye, A. (2011). Marketing capabilities: Antecedents and implications for B2B SME performance. *Industrial Marketing Management*, *40*(3), 368–375.

Miller, D. (1983). The correlates of entrepreneurship in three types of firms. *Management Science*, *29*(7), 770–791.

Miller, D., & Friesen, P. H. (1978). Archetypes of strategy formulation. *Management Science*, *24*(9), 921–933.

Munizu, M. (2013). The Impact of total quality management practices towards competitive advantage and organizational performance: Case of fishery industry in South Sulawesi Province of Indonesia.

Mutlu, H. M., & Aksoy, H. (2014). Strategic Orientations, Firm Capabilities, and Export Performance: An Empirical Analysis in Turkey. *International Journal of Economic Practices and Theories*, *4*(2), 214–221.

Nabitz, U., Klazinga, N., & Walburg, J. (2000). The EFQM excellence model: European and Dutch experiences with the EFQM approach in health care. *International Journal for Quality in Health Care*, *12*(3), 191–202. PMID:10894190

Narver, J. C., & Slater, S. F. (1990). The effect of a market orientation on business profitability. *Journal of Marketing*, 20–35.

Ng, P. K., & Jee, K. S. (2012). Innovating TQM, CE and KM for productive manufacturing in a Malaysian firm. *Total Quality Management & Business Excellence*, *23*(9-10), 1089–1105.

Ngambi, M. T., & Nkemkiafu, A. G. (2015). The Impact of Total Quality Management on Firm's Organizational Performance. *American Journal of Management, 15*(4), 69.

Nwachukwu, C., Chladkova, H., & Zufan, P. (2017). Empirical assessment of microfinance banks in nigeria using efqm excellence model. *International Journal of Qualitative Research, 11*(2).

O'Neill, P., Sohal, A., & Teng, C. W. (2016). Quality management approaches and their impact on firms? financial performance – An Australian study. *International Journal of Production Economics, 171*, 381–393.

Oliveira, G., Corrêa, J., Balestrassi, P., Martins, R., & Turrioni, J. (2017). Investigation of TQM implementation: Empirical study in Brazilian ISO 9001-registered SMEs. *Total Quality Management & Business Excellence*, 1–19.

Pfau, L. D. (1989). Total quality management gives companies a way to enhance position in global marketplace. *Industrial Engineering (American Institute of Industrial Engineers), 21*(4), 17–25.

Psomas, E. L., & Jaca, C. (2016). The impact of total quality management on service company performance: Evidence from Spain. *International Journal of Quality & Reliability Management, 33*(3), 380–398.

Qhogwana, X. (2017). *The use of innovative strategies by automotive component manufacturers in Gauteng.*

Reijonen, H., Hirvonen, S., Nagy, G., Laukkanen, T., & Gabrielsson, M. (2015). The impact of entrepreneurial orientation on B2B branding and business growth in emerging markets. *Industrial Marketing Management, 51*, 35–46.

Rose, G. M., & Shoham, A. (2002). Export performance and market orientation: Establishing an empirical link. *Journal of Business Research, 55*(3), 217–225.

Sachon, M. R. J., Zhang, D., Zhang, Y., & Castillo, C. (2016). The Chinese Automotive Industry in 2016. Spain: Universidad de Navarra.

Sadikoglu, E., & Olcay, H. (2014). The effects of total quality management practices on performance and the reasons of and the barriers to TQM practices in Turkey. *Advances in Decision Sciences*, 2014.

Samson, A., & Mahmood, R. (2015). Fostering Export Performance in SMEs: The Roles of Export Market Orientation and Learning Orientation in Turbulent Environment. *International Journal of Economic Perspectives*, *9*(2), 28.

Seyoum, B. (2017). Export Controls and International Business: A Study with Special Emphasis on Dual-Use Export Controls and Their Impact on Firms in the US. *Journal of Economic Issues, 51*(1), 45–72.

Shafiq, M. (2012). Implementation of quality management systems and business excellence frameworks in Pakistani textile companies. *Journal of Quality and Technology Management, 7*(2), 11–23.

Shafiq, M., Lasrado, F., & Hafeez, K. (2017). The effect of TQM on organisational performance: Empirical evidence from the textile sector of a developing country using SEM. *Total Quality Management & Business Excellence*, 1–22. doi:10.1080/14783363.2017.1283211

Shahzad, M., Ying, Q., Ur Rehman, S., Zafar, A., Ding, X., & Abbas, J. (2019). Impact of knowledge Absorptive Capacity on Corporate Sustainability with Mediating Role of CSR: Analysis from the Asian Context. *Journal of Environmental Planning and Management*, 1–27. doi:10.1080/09640568.2019.1575799

Shoham, A. (1998). Export performance: A conceptualization and empirical assessment. *Journal of International Marketing*, 59–81.

Singh, H., & Mahmood, R. (2013). Determining the effect of export market orientation on export performance of small and medium enterprises in Malaysia: An exploratory study. *Advances in Management and Applied Economics*, *3*(6), 223.

Slater, S. F., & Narver, J. C. (2000). The positive effect of a market orientation on business profitability: A balanced replication. *Journal of Business Research*, *48*(1), 69–73.

Sternad, D., Krenn, M., & Schmid, S. (2017). Business excellence for SMEs: Motives, obstacles, and size-related adaptations. *Total Quality Management & Business Excellence*, 1–18.

Styles, C., Gray, S., Kropp, F., Lindsay, N. J., & Shoham, A. (2006). Entrepreneurial, market, and learning orientations and international entrepreneurial business venture performance in South African firms. *International Marketing Review*, *23*(5), 504–523.

Suárez, E., Calvo-Mora, A., Roldán, J. L., & Periáñez-Cristóbal, R. (2017). Quantitative research on the EFQM excellence model: A systematic literature review (1991–2015). *European Research on Management and Business Economics*, *23*(3), 147–156.

Tan, Q., & Sousa, C. M. (2015). Leveraging marketing capabilities into competitive advantage and export performance. *International Marketing Review*, *32*(1), 78–102.

Thun, E. (2018). Innovation at the middle of the pyramid: State policy, market segmentation, and the Chinese automotive sector. *Technovation*, *70*, 7–19.

Urde, M. (1999). Brand orientation: A mindset for building brands into strategic resources. *Journal of Marketing Management*, *15*(1-3), 117–133.

Valmohammadi, C., & Roshanzamir, S. (2015). The guidelines of improvement: Relations among organizational culture, TQM and performance. *International Journal of Production Economics*, *164*, 167–178.

Viet, B. N., Le Tan, B., Thanh, V. N., & Kim, N. V. (2017). Determinants of export performance: Case of seafood firms in Viet Nam. *Business and Economic Horizons*, *13*(5), 724–737.

Vouzas, F., & Psychogios, A. (2007). Assessing managers' awareness of TQM. *The TQM Magazine*, *19*(1), 62–75.

Weerawardena, J., & Coote, L. (2001). An empirical investigation into entrepreneurship and organizational innovation-based competitive strategy. *Journal of Research in Marketing and Entrepreneurship*, *3*(1), 51–70.

Wernerfelt, B. (1984). A resource-based view of the firm. *Strategic Management Journal*, *5*(2), 171–180.

Xie, Z., & Li, J. (2018). Exporting and innovating among emerging market firms: The moderating role of institutional development. *Journal of International Business Studies*, *49*(2), 222–245.

Yazdani, B., Attafar, A., Shahin, A., & Kheradmandnia, M. (2016). The impact of TQM practices on organizational learning case study: Automobile part manufacturing and suppliers of Iran. *International Journal of Quality & Reliability Management*, *33*(5), 574–596.

Yovovich, B. (1988). What is your brand really worth. *Adweek's Marketing Week*, *8*, 18–24.

Yusr, M. M. (2016). Innovation capability and its role in enhancing the relationship between TQM practices and innovation performance. *Journal of Open Innovation: Technology, Market, and Complexity*, *2*(1), 1.

Zahra, S. A., & Covin, J. G. (1995). Contextual influences on the corporate entrepreneurship-performance relationship: A longitudinal analysis. *Journal of Business Venturing*, *10*(1), 43–58.

Chapter 15
Transformational Human Resource Management: Crafting Organizational Efficiency

Ikramul Hasan
https://orcid.org/0000-0001-5510-1341
Independent University, Bangladesh

MD. Nazmul Islam
https://orcid.org/0000-0002-6778-508X
University of Malaya, Malaysia

Mohammad Ashraful Ferdous Chowdhury
https://orcid.org/0000-0001-8540-1353
Shahjalal University of Science and Technology, Bangladesh

ABSTRACT

Human Resource Management plays an essential role for attaining organizational goals. Nowadays, practitioners, researchers, and academicians around the world are emphasizing to transform and reshape the practice of human resource. However, very few research works have been done in the area of Transformational Human Resource Management (T-HRM). Hence, the aim of this chapter is to propose an integrated framework of T-HRM and organizational efficiency. In light of that, this study has proposed potential factors of the T-HRM. Secondly, this study presented positive effect of the factors of T-HRM on organizational efficiency management. Concept of knowledge management has introduced as a potential mediator, and ICT and organizational alignment has presented as a potential moderator of this study. Finally, knowledge of this study will provide better insights on T-HRM for ensuring organizational efficiency.

DOI: 10.4018/978-1-7998-0357-7.ch015

INTRODUCTION

Transformational human resource management (T-HRM) is currently a new concept in the area of human resource management. Now-a-days academicians, researchers and practitioners around the world in the area of organization behavior, management and human resource management are highlighting the importance of transformational human resource management. Alike transformational leadership (Shih & Orochena, 2016), T-HRM also considers employee as a partner in the organizational innovation and improvement. T-HRM is an outcome of cross-functional activities, which results impact on achieving organizational objective. T-HRM has potential impact on employee efficiency development. Being a modern HRM practices, T-HRM also confirms potency to develop employee through the effective knowledge management (Edvardsson, 2008; Oltra, 2005) training (Seeck & Diehl, 2017), leadership (Mehmood & Arif, 2011) talent management (Iles et al., 2010), information technology usage (Bourke & Crowley, 2015; Steijn & Tijdens, 2005) innovation and change (Santangelo & Pini, 2011; Lin & Sanders, 2017), social responsible human resource management (Barrena-Martínez et al., 2017; Milfelner et al., 2015) and organizational alignment (Shih & Chiang, 2005). However, due to the current changes in business industry and more importantly rate of failures or unsuccessful organizational changes or transformations in contemporary business era (Lewis, 2019; Habersang et al., 2019), reducing the rate of employee retention (Silva et al., 2019), job dissatisfaction (Moussa & Somjai, 2019; De Clercq et al., 2019), lack of trust (Mooijman et al., 2019), lack of commitment (Souza et al., 2019), disengagement (Wolff, 2019) and improper alignment (Davidson & Butcher, 2019) between individual employee skills, job attributes, organizational strategy as well as diversify workforce (Karim et al, 2019), global competition (Gershon, 2019) have provoked the notion of T-HRM. In order to attain organizational efficiency in highly competitive changing business environment and to ensure sustainable competitive advantage in organization, T-HRM has become one of the important factor for the organization. Transactional HR activities expressing employees are more likely clerical mundane work whereas Transformational HR activities are strategic.

In contemporary business world, involvement of HR division of an organization not only limited to the transactional functions but also align HR activities with organizational objective. Moreover, T-HRM is more involved with the activities that are strategic, forward thinking, proactive and comprehensive with more technology based to aid HR processes to make them effective, support organizational change and improvement (Manzoor et al., 2019). Hence, T-HRM is concerned with complete revamping and constant pruning of HR processes in order to make them best in class maximize efficiency as well as to ensure sustainable competitive advantage for the organization. However, very limited research work has been done in the area of T-HRM. With this background, formulation of this research is to identify factors of T-HRM and its impact of organizational effectiveness as well as to propose model for integrating factors of T-HRM and organizational efficiency management. Hence, this study will address the following questions:

- What are the important factors of transformational human resource management (T-HRM)?
- Are different factors of T-HRM influences organizational effectiveness?

In order to address the following questions, comprehensive literature review will be conducted to link potential relationships. This study also exhibits association to bring the key factors involved in T-HRM (knowledge management, training and development, organizational leadership, talent management, information technology, innovation and change, socially responsible human resource management and organizational alignment) in order to understand organizational efficiency management. Presenting the relationship among the constructs this study will postulate a number of effective relationships, which confirms the impact or influence on organizational efficiency management. Finally, this study will propose a conceptual framework. This study will contribute to HRM literature and will give new insights to the researchers and practitioners by adding the notion of T-HRM and its factors as an antecedent of employee development as well as organizational efficiency management.

This paper will be organized in four steps. First, this study will propose different factors of T-HRM. Secondly, this study will present the impact of the factors of T-HRM on organizational effectiveness. Then this study will present a conceptual model by connecting different factors of T-HRM as an independent, mediator and moderating variable (leadership, T&D, talent management, innovation and change, SR-HRM as independent variable, knowledge management as a mediator, ICT and organizational alignment as a moderator). Finally, this study will propose implications, future research and conclusion based on the discussion of the paper this study.

METHODOLOGY

The chapter aims to attain dual objectives of creating academic knowledge and develop assumption through the conceptual model in order to address and identify potential problems. In order to capture as many studies as possible and to limit biases caused by the study-identification process, the procedure was followed as proposed by Rosenbusch et al., (2011); Islam et al., (2018) and Hasan et al., (2018). A computerized keyword searches in the databases (e.g., Web of Science (ISI), Scopus, Web of Knowledge, ABI, Google Scholar) was conducted to explore the highly pertinent studies on HRM. In addition, most relevant journals in T-HRM elements, knowledge management, and organizational efficiency management, ICT (e.g., *The Leadership Quarterly, Management Information Systems Quarterly, Organizational Behavior and Human Decision Processes, Strategic Organization*) were searched manually. As a third step, the reference sections of the relevant articles are searched. Three keywords "antecedences of transformational human resource management to confer efficiency management" were used in the literature search.

LITERATURE REVIEW

Transformational Human Resource Management (T-HRM)

T-HRM is one of the contemporary areas in the field of human resource management. Due to the globalization, high level of competition, innovation and continuous change and up gradation in the technology or shift in market are the main forces for the transformation of human resource management (Beer, 1997). Moreover, being a key engine for the organization HR has drawn a considerable attention for ensuring organizational efficiency, effectiveness, success and sustainability, which drives HR to go for transformation. T-HRM addresses and fulfills the requirements and demands of internal and external stakeholders of the organization (Ulrich et al., 2009). Transformational Human Resource Management is more strategic, practical, flatter, faster, forward linking, less bureaucratic, more prompt and remove hierarchical barriers (Beer, 1997) than the traditional/conventional HR in order to achieve competitive advantages as well as ensures organizational efficiency and effectiveness. Moreover, T-HRM ensures alignment among the different departments in the organization, ensures commitment to continuous employee development, enhances competitiveness in organizational leadership and management level, ensures organizational innovativeness, and develops an environment for knowledge sharing and smooth communication in the organization (Beer, 1997).

Organizational Efficiency

Maintaining organizational efficiency is an important assignment for every organization. Efficiency measures how your organization transforms their input into output, moreover organizational efficiency can be realized by measuring the difference between input and output (Low, 2000). While discussing about the efficiency in their research Pinprayong and Siengthai (2012) presented two different types of efficiency such as business efficiency and organizational efficiency. In addition, authors highlighted the difference between two types of efficiency. Business efficiency represents the ratio of the input – output performance of the business, whereas organizational efficiency represents the enhancement of organizational internal process such as organizational culture, structure and community. Kumar and Gulati (2010) describe efficiency as allocating different resources of the organization for various uses. Similarly, Bartuševičienė and Šakalytė (2013) stated that organizational excellent performance is not known as efficiency whereas organizational excellence in operation is a more related to efficiency. Organizations are emphasizing to manage the efficiency level of the organization in order to optimize the organizational resource as well as to achieve organizational goals. Efficient organizations are more capable of competing with the other players or opponents and ensure organizational sustainability. Karlaftis (2004) stated that inefficient organization has no clear vision as well as inefficiency in an organization causes high employee turnover and costly organizational failure.

Conceptual Model for Integrating Factors of T-HRM and Organizational Efficiency

HR transformation is assumed as an integrated, aligned, innovative, and business focused approach that defines how HR is working within the organization so that it could fulfill the objectives of the customers, investors, and other stakeholders. While HR transformation, organization needs to consider factors related to both the internal and external environment of the organization. Ulrichet et al. (2009) have highlighted that organization need to transform HR in order to enhance organizational efficiency and it can be achieved not only by focusing on organizational profit maximizing but also identifying and fulfilling the needs of stakeholders of the organization. Transformational HRM will support organization to transform their activities in order to achieve organizational objective by identifying key stakeholders (shareholder, customer, supplier, competitors, employee, regulator and community) and their needs and expectations as well as by aligning HRM activities such as recruiting, training, motivating and rewarding etc., with the organizational objectives and stakeholder's expectation.

However, different factors may consider important for human resource transformation but in line with the previous literature related to the transformation, this study will consider knowledge management, organizational leadership, information and communication technology infrastructure, talent management, training, innovation and change management, SR-HRM and organizational alignment as important factors for managing human resource transformation. Notable, in capability aspects of the organization, Gold and his colleagues theorized knowledge management capabilities as multidimensional concepts and incorporate an infrastructural perspective as an important capability. They present it as knowledge infrastructure capabilities, which actually work as KM enablers too; Alavi and Leidner (2001); Lee and Choi (2003) also support these facts. In considering knowledge infrastructural capability, they comprise technology, organizational culture, and organizational structure as multiple composed dimensions (Gold et al., 2001). Handzic (2011) developed a socio-technical knowledge management model where she claimed that as a knowledge enabler various social and technical initiatives could facilitate the KM process and foster the development of organizational knowledge by emphasizing factors. Hence, this study propose mediating effect of knowledge management between the proposed factors of T-HRM and organizational efficiency. Consequently, leadership of the organization also refers to the process of influencing others to achieve some desired goals in which knowledge sharing occurs, as well as creating the incentives for doing so (Handzic, 2011; de Jong & Hartog, 2007; Kerr & Clegg, 2007). Also, modern day HR practices is concern about the importance of talent management (Ashton & Morton, 2005) for retaining and attracting talented and potential employee in organization (Deery, 2008; Christensen & Rog, 2008). Besides, contemporary training and development opportunity provided by the HRD in an organization improves the knowledge level and capacity of the employee of the organization. Similarly, T-HRM emphasizes on organizational innovation because HRM is also considered another antecedents of organizational innovation (Shipton et al., 2006). Hence, T-HRM integrates innovation and HR strategy in order to ensure organization competitiveness, effectiveness and sustainability (De Leede & Looise, 2005). Moreover, T-HRM underlines the importance of internal and external stakeholders of the organization, hence, socially responsible human resource management (SR-HRM) is another essential factor for T-HR for achieving organizational efficiency, goal and success (Nie et al., 2018; Milfelner et al., 2015). Because, organizational efficiency management determines the goals of the organization. Most of the literature also supports these facts and suggest IT as a strong influencer in today's organizational practice. In this study, the researchers follow the transformational side of the HR literature. Sharing among the employees is a social exchange process; social exchange theory supports this fact (Cabrera & Cabrera, 2005). Information technology is considered an important factor in transformation literature and confirm the flows in the areas of smooth's knowledge repositories, data mining, and decision support systems (Hahn & Subramani, 2000) by providing a platform for communication and sharing. Therefore, heading towards communication aspects as a facilitator, this study proposes the ICT infrastructure of the organization as a moderator between T-HRM and efficiency management. Furthermore, different studies also have presented the moderating effects of organizational alignment (Bezrukova et al., 2012; Byrd et al., 2006). Because organizational alignment connects organizational strategy, different departments and employee of the organization for achieving organizational objective and transforming organization (Henderson & Venkatraman, 1999). Hence, this study also considers organizational alignment as an important factor of T-HRM and propose a moderating effect of organizational alignment between knowledge management and organizational efficiency.

Figure 1 depicts the components of T-HRM developed in this study to analyze the influences on organizational efficiency management.

Figure 1. Components of T-HRM

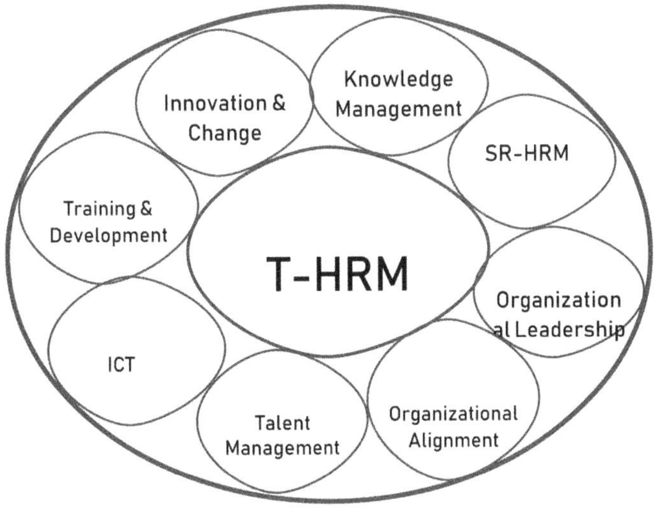

Knowledge Management

Knowledge is a valuable resource for business organizations. According to Grant (1996b) knowledge is considered as the firm's most important and precious asset to sustain its competitive advantage (Davenport & Prusak, 1998; Suppiah & Sandhu, 2011). Davenport and his colleagues found knowledge management as the improvement process of knowledge exercise for the organizations to achieve its organizational objectives (Davenport et al., 1998). Knowledge entails iteration and interaction between individuals in the workplace (Howell & Annansingh, 2013). A number of activities was recognized from the knowledge management literature to manage knowledge for the organizations. Heiseg (2009) explored 166 different terms in his study that are used to describe KM activities after analyzing 117 KM frameworks. Nevertheless, based on further classification, from the literature, knowledge management can be explored as the process of (1) knowledge acquisition (i.e., collecting and identifying useful information), (2) organizing knowledge (i.e., enabling employees to retrieve organizational knowledge), (3) knowledge leverage (i.e., exploiting and usefully applying knowledge), (4) knowledge sharing (i.e., disseminating knowledge through the whole organization), and (5) organizational memory (i.e., storing the knowledge in the repository; Kim & Lee, 2013; Nonaka & Takeuchi, 1995; Rowley, 2000). Nonetheless, Wang and Noe (2010) found the success of KM initiative in knowledge sharing because sharing the information to the right person on time to continue the action in ways that foster organizational performance. Consequently, knowledge sharing also is important in the area of innovation, organizational learning, the development of new skills and capabilities of the employee, increasing productivity, and maintaining a competitive advantage for organizations (Kim & Lee, 2012; Senge, 2006). Hence, knowledge management gains central attention in the area of HR management transformation to utilize employee efficiencies in generating new ideas and competencies to produce competitive advantage.

Organizational Leadership

Organizational leadership refers to the process of influencing others towards achieving some desired goals (de Jong & Hartog, 2007). Durcker (1996) has defined "a leader is someone who has followers." According to Packard (2009), in organizations, leadership connects organizational goal, aligns an employee with the vision, develop strategies, and motivates staff. Organizational and management literature also support this fact and studies report that leadership is a crucial element for organizational efficiency enhancement (Benson & Blackman, 2011; Carson, Tesluk, & Marrone, 2007). Considering effective leadership, Ozcelik et al. (2008) give more emphasis on cognitive tasks (e.g., planning, coordinating, organizing, and decision-making) since the practices aimed by the leaders to capitalize on the employees' emotional resources is a differentiating factor for organizational-level performance. Emphasizing leader-follower (employee) relationships, Epitropaki and Martin (2013) also place more importance on interpersonal influence, as it is thought to be a cornerstone factor for organizations. Contractor et al. (2012) also claimed in their study that the behavior of the leaders has a rigorous influence on collectives, including teams and units.

Information and Communication Technology Infrastructure

Technology comprises a crucial element for the organization in creating new knowledge. Using information and communication technology, an organization can integrate its previously stored information and knowledge (Gold et al., 2001). Consequently, apart from storing and retrieving data, it can improve the access of knowledge and eliminate temporal and spatial barriers between knowledge workers. Information and communication technology (ICT) can enhance knowledge sharing among the employees in different levels (Tohidinia & Mosakhani, 2010). The organizations invest in comprehensive IT infrastructure it can facilitates in rapid collection, storage, and exchange of knowledge among the employees (Lee & Choi, 2003). Representing ICT in an organizational perspective, Dalkir (2011) represented communication and collaboration technology. Though these technologies are invariably intertwined, communication technologies include the telephone, fax, video conferencing, teleconferencing, chatrooms, instant messaging, phone text messaging (SMS), Internet telephone (voice over IP or VOIP), e-mail, and discussion forums (Dalkir, 2011). Alternatively, groupware and collaboration technology represent a system that enhance the groups communication among colleagues (work groups) and involved them to a communication network (e.g., LAN) to organize their activities e.g. scheduling meetings and allocating resources, e-mail, password protection for documents, telephone utilities, electronic newsletters, and file distribution. Therefore, the ICT are usually integrated with some form of collaboration and work as a dissemination tool for knowledge sharing.

Talent Management

"The terms "talent management", "talent strategy", "succession management", and "human resource planning" are often used interchangeably" (Lewis & Heckman, 2006). Talent management deals with the talent "supply", "demand" and "flow" in the organization (Pascal, 2004). Transformation, technological up gradation and change in the current business world, organization are seriously struggling to develop perfect workforce for the organization (Panayiotou et al., 2019). In order to meet expectation of the modern era organizations are facing problem because of the growing number of skill or talent gap

(Hora, 2019). Now a day because of the slowdown of the employee turnover, it has become challenging for any business organization for attracting, and retaining talented employee. Traditional style of the HRD of the organization are not suitable to meet the challenge to overcome the issue related with the talent management. T-HRM as a modern approach of HRM with the help of the emotional intelligence, considers talent management as an essential element to achieve organizational efficiency (Laborde et al., 2019). T-HRM creates engine room for developing internal talent in an organization as well as ensures talent retention and external talent attracting strategy by developing proper organizational environment. Moreover, by offering competitive compensation and benefit for the both internal and external talented employee for managing organizational talent as well as ensuring organizational efficiency.

Training and Development

Training and Development (T&D) is considered as one of the important components of HR (Dhamodharan et al., 2010). Training and development involves with the gaining knowledge through improving skills of the employee for achieving organizational objective. Likewise, T-HRM also considers T&D as an important factor to achieve organizational goal and efficiency. Because, T&D not only improves employee knowledge level by providing training and coaching to their employee to achieve organizational objective but also enhances employee commitment (Brown & Sitzmann, 2011). Hence, in order to compete with the different competitors, also to ensure organizational transformation and innovation T&D will play very crucial role in the domain of T-HRM. Love and Singh (2011) in their research study have pointed out that T&D is one of eight aspects of HRD success. Because through T&D employee gets up-to-date and gain saleable knowledge which add value to their qualification and career progress and works as a strong motivating factor which results lower turnover in organization. Moreover, Kim and Ployhart (2014) stated that T&D not only motivates and develops employee competences but also improves organizational performance and competitiveness. Mainly, T-HRM will ensure accurate connection between organizational goal, employee skill gap and required training of the employee for achieving organizational efficiency. Alongside, training and development works as a predictor of employee and organizational performance. Training and development helps employee to enhance knowledge in their specialized area as well as enhance employee commitment and effort to achieve organizational goal and ensure organizational efficiency (Brown & Sitzmann, 2011).

Innovation and Change

Innovation and change is a modern-day topic, which focuses on importance and process of organizational change (Nilakant & Ramnarayan, 2006). In this modern and competitive business environment, it is difficult for any business organization to survive and out compete the rivalry organizations. Therefore, organization has to constantly looking forward to bring innovation and change in organization in order to ensure their sustainability. However, bring innovation and change in organization is a difficult process. Organization has to face numerous obstacles to bring innovation and change in organization (Sune & Gibb, 2015; Pascale et al., 1997). Most importantly, employee behavior during organizational change is another concerning issue for organizational dealing with change and transformation. Because, success and failure of the organization depends on the employee attitudes towards change, employee cynicism and employee readiness towards change (Choi, 2011; Bommer et al., 2005; Herscovitch & Mayer, 2002). Many business organizations are struggling in dealing with employee attitude and behavior during organizational change and innovation. Therefore, T-HRM will play important role for shaping employee attitude and behavior (Wood & De Menezed, 1998) through transformational HR practice and helps organization to innovate, change and transform. Moreover, T-HRM supports organization to become innovative, transformative and managing change successfully in order to ensure organizational effectiveness and efficiency (Linnenluecke & Griffiths, 2010; Lines, 2004).

Social Responsible Human Resource Management (SR-HRM)

SR-HRM is a combination of corporate social responsibility (CSR) and human resource management (HRM). The notion of SR-HRM involves with improving work-file quality of the employee in the viewpoint of the principles of CSR.SR-HRM focuses on organizational fairness, objectivity, employee empowerment and non-discrimination. Shen and Zhu (2011) classified SR-HRM in three sections, such as employee oriented human resource management (EO-HRM). EO-HRM focuses on employee empowerment, workplace democracy as well as employee personal development and employee family needs. Another section is known as legal compliance human resource management (LC-HRM), which works with the labor law and international standards such as ILO to ensure employee workplace safety and compliance, rules and regulation of employee wage etc. Similarly, general facilitation human resource management (GF-HRM) emphasizes on welfare of both employees and other stakeholders of the organization. SR-HR helps organization to achieve competitive advantage. Similarly, number of research studies have highlighted the importance of SR-HRM for improving employee behavior and organizational performance. Hence, T-HRM also considered SR-HRM as an important factor for achieving organizational goal and efficiency. Shen and Benson (2016) stated that SR-HRM ensures organizational short-term effectiveness and long-term sustainability. Therefore, T-HRM emphasizes SR-HRM for transforming HR of the organization in order to achieve organizational efficiency, effectiveness and sustainability (Shen & Zhu, 2011).

Organizational Alignment

Alignment is known as the fit and linkage between different departments or units of the organization. Organizational alignment is one of the contemporary areas in the field of strategic management (Tan and Tan, 2005). Moreover, alignment units' organizational culture, organizational strategy and structure for ensuring organizational efficiency. Organizational alignment works as a linkage between organizational strategy, culture and its structure. Porter (1996) describes organizational alignment as a common understanding of the objectives and goals of the organization between the managers of different levels and ladders of the organization. In order to achieve organizational competitiveness organization must align its instruments and resources. Nadler and Tushman (1988) in their studies also highlighted the importance of organizational alignment because appropriate alignment directs organization in right path for achieving organizational effectiveness and efficiency. Therefore, HR department of the organization also emphasizes aligning activities of the organization with the organizational goals and objectives for transforming human resources. Different scholars pointed that although it import for every organization to ensure alignment in order to achieve its organizational goal and efficiency but it is not an easy task to ensure organizational alignment (Hinings et al., 1996), hence transactional or traditional HR will not be suitable to achieve proper organizational alignment. Henceforth, T-HRM will transform HRD activities

Figure 2. Proposed model

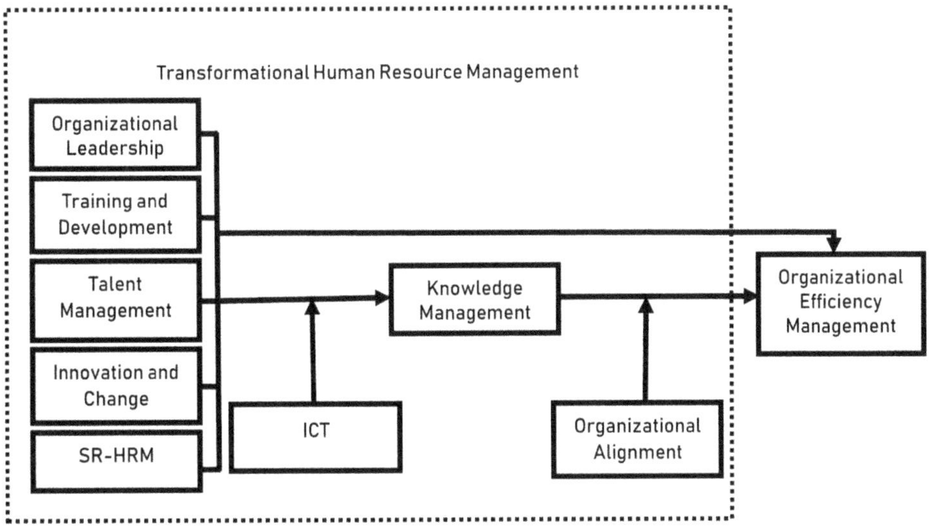

and link with organizational goal and plan strategy accordingly to ensure proper organizational alignment. Through proper alignment, T-HRM of the organization can ensure organizational effectiveness.

DISCUSSION

Limited research work has exhibited the factors of T-HRM and its effects on organizational efficiency management. This paper proposes a model to suggest the relationship between factors of T-HRM and organizational effectiveness. In this study, training and development, organizational leadership, talent management, change and innovation and social responsible human resource management serve as dependent variables, knowledge management serves as a mediating variable, both ICT and organizational alignment serve as moderating variables and organizational efficiency management serves as a dependent variable. The proposed model will help to link the factors of T-HRM, ICT, knowledge management and organizational alignment to ensure organizational efficiency management. The paper contributes the literature of T-HRM and organizational efficiency management theory and practices in numerous ways, which will provide new insights for both researchers and practitioners.

THEORETICAL AND PRACTICAL IMPLICATIONS

Proposed model of this study will contribute in the literature of T-HRM and organizational efficiency management in several ways. First, this study has identified a gap in the literature. Secondly, this study has satisfied the gap by suggesting several factors of T-HRM and integrating those factors of the T-HRM with organizational efficiency in order to see the effects of T-HRM on organizational efficiency management. Notably, to the best of the authors' knowledge such effects of T-HRM on organizational efficiency management not yet presented in the literature before.

In the practical point of view, the proposed conceptual model (Figure-2) could be useful for the organization for achieving organizational efficiency. This study sheds light on T-HRM, which is an essential and contemporary area in the field of HRM, which will provide proper guideline for organization as well as help leaders and managers of the organization to achieve competitive advantage, improve organizational performance, ensure efficiency and achieve organizational goal.

CONCLUSION

This paper proposed a number of influential factors of T-HRM, which have strong association with organizational efficiency management and open a space for further empirical investigation. Eventually, existing literature supports that proposed factors of T-HRM in this study, enhances organizational efficiency to achieve goals and objectives of the organization. Therefore, an empirical research on T-HRM and influence of its different factors as independent variable, mediating variable and moderating variable on organizational efficiency in an integrated model would offer new knowledge. However, proposed framework concluded that in order to achieve organizational efficiency, factors of T-HRM plays a vital role. Therefore, insights of this study will provide new scope for the researchers and practitioners around the world in the area of HRM and organizational study. Moreover, empirical rationalization of this proposed model will validate the strength of the model as well as will open a new area of further investigation.

REFERENCES

Alavi, M., & Leidner, D. (2001). Knowledge management and knowledge management systems: Conceptual foundations and research issues. *Management Information Systems Quarterly*, *25*(1), 107–136. doi:10.2307/3250961

Amburgey, T. L., Kelly, D., & Barnett, W. P. (1990, August). Resetting the clock: The dynamics of organizational change and failure. In Academy of Management Proceedings (Vol. 1990, No. 1, pp. 160-164). Briarcliff Manor, NY 10510: Academy of Management.

Ashton, C., & Morton, L. (2005). Managing talent for competitive advantage: Taking a systemic approach to talent management. *Strategic HR review*, *4*(5), 28-31.

Barrena-Martínez, J., López-Fernández, M., & Romero-Fernández, P. M. (2017). Towards a configuration of socially responsible human resource management policies and practices: Findings from an academic consensus. *International Journal of Human Resource Management*, 1–37. doi:10.1080/09585192.2017.1332669

Bartuševičienė, I., & Šakalytė, E. (2013). Organizational assessment: Effectiveness vs. efficiency. *Social Transformations in Contemporary Society*, *1*(1), 45–53.

Beer, M. (1997). The transformation of the human resource function: Resolving the tension between a traditional administrative and a new strategic role. *Human Resource Management: Published in Cooperation with the School of Business Administration, The University of Michigan and in alliance with the Society of Human Resources Management*, *36*(1), 49-56.

Benson, A. M., & Blackman, D. (2011). To distribute leadership or not? A lesson from the islands. *Tourism Management, 32*(5), 1141–1149. doi:10.1016/j.tourman.2010.10.002

Bezrukova, K., Thatcher, S., Jehn, K. A., & Spell, C. S. (2012). The effects of alignments: Examining group faultlines, organizational cultures, and performance. *The Journal of Applied Psychology, 97*(1), 77–92. doi:10.1037/a0023684 PMID:21744943

Bommer, W. H., Rich, G. A., & Rubin, R. S. (2005). Changing attitudes about change: Longitudinal effects of transformational leader behavior on employee cynicism about organizational change. *Journal of Organizational Behavior: The International Journal of Industrial, Occupational and Organizational Psychology and Behavior, 26*(7), 733–753. doi:10.1002/job.342

Bourke, J., & Crowley, F. (2015). The role of HRM and ICT complementarities in firm innovation: Evidence from transition economies. *International Journal of Innovation Management, 19*(05). doi:10.1142/S1363919615500541

Brown, K. G., & Sitzmann, T. (2011). Training and employee development for improved performance.

Byrd, T. A., Lewis, B. R., & Bryan, R. W. (2006). The leveraging influence of strategic alignment on IT investment: An empirical examination. *Information & Management, 43*(3), 308–321. doi:10.1016/j.im.2005.07.002

Cabrera, E. F., & Cabrera, A. (2005). Fostering knowledge sharing through people management practices. *International Journal of Human Resource Management, 16*(5), 720–735. doi:10.1080/09585190500083020

Carson, J. B., Tesluk, P. E., & Marrone, J. A. (2007). Shared leadership in Tams: An investigation of antecedent conditions and performance. *Academy of Management Journal, 50*(5), 1217–1234.

Choi, M. (2011). Employees' attitudes toward organizational change: A literature review. *Human Resource Management, 50*(4), 479–500. doi:10.1002/hrm.20434

Christensen Hughes, J., & Rog, E. (2008). Talent management: A strategy for improving employee recruitment, retention and engagement within hospitality organizations. *International Journal of Contemporary Hospitality Management, 20*(7), 743–757. doi:10.1108/09596110810899086

Contractor, N. S., DeChurch, L. A., Carson, J., Carter, D. R., & Keegan, B. (2012). The topology of collective leadership. *The Leadership Quarterly, 23*(6), 994–1011. doi:10.1016/j.leaqua.2012.10.010

Dalkir, K. (2011). Knowledge Management in Theory and Practice (2nded.). Cambridge, MA: The MIT Press.

Davenport, T. H., & Prusak, L. (1998). *Working Knowledge*. Boston, MA: Harvard Business School Press.

Davidson, S. L., & Butcher, J. (2019). Rural Superintendents' Experiences with Empowerment and Alignment to Vision in the Application of Principle-Centered Leadership. *Rural Educator, 40*(1).

De Clercq, D., Haq, I. U., Azeem, M. U., & Ahmad, H. N. (2019). The relationship between workplace incivility and helping behavior: Roles of job dissatisfaction and political skill. *The Journal of Psychology*, 1–21. PMID:30696391

De Jong, J. P., & Hartog, D. N. D. (2007). How leaders influence employees' innovative behaviour. *European Journal of Innovation Management, 10*(1), 41–64. doi:10.1108/14601060710720546

Deery, M. (2008). Talent management, work-life balance and retention strategies. *International Journal of Contemporary Hospitality Management, 20*(7), 792–806. doi:10.1108/09596110810897619

deLeede, J., & Looise, J. K. (2005). Innovation and HRM: Towards an integrated framework. *Creativity and Innovation Management, 14*(2), 108–117. doi:10.1111/j.1467-8691.2005.00331.x

Dhamodharan, V., Daniel, B. J. C., & Ambuli, T. V. (2010). An empirical study on assessing trainees' expectations and their perceptions. *International Business Research, 3*(2), 174. doi:10.5539/ibr.v3n2p174

Drucker, P. F. (1996). Your leadership is unique. *Leadership, 17*(4), 54.

Edvardsson, I. R. (2008). HRM and knowledge management. *Employee Relations, 30*(5), 553–561. doi:10.1108/01425450810888303

Epitropaki, O., & Martin, R. (2013). Transformational–transactional leadership and upward influence: The role of Relative Leader–Member Exchanges (RLMX) and Perceived Organizational Support (POS). *The Leadership Quarterly, 24*(2), 299–315. doi:10.1016/j.leaqua.2012.11.007

Gershon, R. A. (2019). Transnational Media and the Economics of Global Competition. *Global Communication. Multicultural Perspectives*, 37.

Gold, H. A., Malhotra, A., & Segars, H. A. (2001). Knowledge Management: An organizational Capabilities Perspective. *Journal of Management Information Systems, 18*(1), 185–214. doi:10.1080/07421222.2001.11045669

Grant, R. M. (1996b). Toward a knowledge-based theory of the firm. *Strategic Management Journal, 17*(S2), 109–111. doi:10.1002mj.4250171110

Habersang, S., Küberling-Jost, J., Reihlen, M., & Seckler, C. (2019). A Process Perspective on Organizational Failure: A Qualitative Meta-Analysis. *Journal of Management Studies, 56*(1), 19–56. doi:10.1111/joms.12341

Hahn, J., & Subramani, M. R. (2000). A framework of knowledge management systems: issues and challenges for theory and practice. *Proceedings of the International Conference on Information Systems, ICIS'2000*, 302-12.

Handzic, M. (2011). Integrated socio-technical knowledge management model: An empirical evaluation. *Journal of Knowledge Management, 15*(2), 198–211. doi:10.1108/13673271111119655

Hasan, I., Khan, A. N. M. S., Karim, M. A., Khan, S. R., Alam, S., & Sanjana, B. (2018). Health and safety compliance in the readymade garment sector of Bangladesh: Practices and Observations. *Independent Business Review, 11*(1-2), 25–32.

Heisig, P. (2009). Harmonisation of knowledge management–comparing 160 KM frameworks around the globe. *Journal of Knowledge Management, 13*(4), 4–31. doi:10.1108/13673270910971798

Heisig, P. (2009). Harmonisation of knowledge management – comparing 160 Km frameworks around the globe. *Journal of Knowledge Management, 13*(4), 4–31. doi:10.1108/13673270910971798

Henderson, J. C., & Venkatraman, H. (1999). Strategic alignment: Leveraging information technology for transforming organizations. *IBM systems journal, 38*(2.3), 472-484.

Herscovitch, L., & Meyer, J. P. (2002). Commitment to organizational change: Extension of a three-component model. *The Journal of Applied Psychology, 87*(3), 474–487. doi:10.1037/0021-9010.87.3.474 PMID:12090605

Hinings, C. R., Thibault, L., Slack, T., & Kikulis, L. M. (1996). Values and organizational structure. *Human Relations, 47*(7), 885–916. doi:10.1177/001872679604900702

Hora, M. T. (2019). *Beyond the skills gap: Preparing college students for life and work.* Cambridge, MA: Harvard Education Press.

Howell, K. E., & Annansingh, F. (2013). Knowledge generation and sharing in UK universities: A tale of two cultures? *International Journal of Information Management, 33*(1), 32–39. doi:10.1016/j.ijinfomgt.2012.05.003

Iles, P., Chuai, X., & Preece, D. (2010). Talent management and HRM in multinational companies in Beijing: Definitions, differences and drivers. *Journal of World Business, 45*(2), 179–189. doi:10.1016/j.jwb.2009.09.014

Islam, M. Z., Jasimuddin, S. M., & Hasan, I. (2018). Determinants that influence knowledge sharing: An integrated literature review. *International Journal of Knowledge Management Studies, 9*(4), 363–380. doi:10.1504/IJKMS.2018.096318

Karaa, D., Uysalb, M., Sirgyc, M. J., & Leed, G. (2013). The effects of leadership style on employee well-being in hospitality. *International Journal of Hospitality Management, 34*, 9–18. doi:10.1016/j.ijhm.2013.02.001

Karim, A. S., Zaki, A. R., & Mubeen, M. H. (2019). Managing Workforce Diversity in Multicultural Organizations. *Journal of European Studies, 35*(1), 79–91.

Karlaftis, M. G. (2004). A DEA approach for evaluating the efficiency and effectiveness of urban transit systems. *European Journal of Operational Research, 152*(2), 354–364. doi:10.1016/S0377-2217(03)00029-8

Kerr, M., & Clegg, C. (2007). Sharing knowledge: Contextualizing socio–technical thinking and practice. *The Learning Organization, 14*(5), 423–435. doi:10.1108/09696470710762646

Kim, T., & Lee, G. (2012). A modified and extended Triandis model for the enablers–process–outcomes relationship in hotel employees' knowledge sharing. *Service Industries Journal, 32*(13), 2059–2090. doi:10.1080/02642069.2011.574276

Kim, T. T., & Lee, G. (2013). Hospitality employee knowledge-sharing behaviors in the relationship between goal orientations and service innovative behavior. *International Journal of Hospitality Management, 34*, 324–337. doi:10.1016/j.ijhm.2013.04.009

Kim, Y., & Ployhart, R. E. (2014). The effects of staffing and training on firm productivity and profit growth before, during, and after the Great Recession. *The Journal of Applied Psychology, 99*(3), 361–389. doi:10.1037/a0035408 PMID:24377393

Kumar, S., & Gulati, R. (2010). Assessing the Effect of Ownership on the Efficiency of Indian Domestic Banks. *IUP Journal of Bank Management, 9*(3).

Laborde, Z. B., Burbano, K. B., Reinoso, V. G., Bangeppagari, M., Mulla, S. I., & Selvanayagam, M. (2019). Emotional Intelligence Models as Generators of Business Management Change in the Human Talent Area. *Journal of Artificial Intelligence, 12*(1), 1–10. doi:10.3923/jai.2019.1.10

Laborde, Z. B., Burbano, K. B., Reinoso, V. G., Bangeppagari, M., Mulla, S. I., & Selvanayagam, M. (2019). Emotional Intelligence Models as Generators of Business Management Change in the Human Talent Area. *Journal of Artificial Intelligence, 12*(1), 1–10. doi:10.3923/jai.2019.1.10

Lee, H., & Choi, B. (2003). Knowledge management enablers, process, and organizational performance: An integrative view and empirical examination. *Journal of Management Information Systems, 20*(1), 179–228. doi:10.1080/07421222.2003.11045756

Lewis, L. (2019). *Organizational change: Creating change through strategic communication*. Hoboken, NJ: Wiley-Blackwell. doi:10.1002/9781119431503

Lewis, R. E., & Heckman, R. J. (2006). Talent management: A critical review. *Human Resource Management Review, 16*(2), 139–154. doi:10.1016/j.hrmr.2006.03.001

Lin, C. H., & Sanders, K. (2017). HRM and innovation: A multi-level organisational learning perspective. *Human Resource Management Journal, 27*(2), 300–317. doi:10.1111/1748-8583.12127

Lines, R. (2004). Influence of participation in strategic change: Resistance, organizational commitment and change goal achievement. *Journal of Change Management, 4*(3), 193–215. doi:10.1080/1469701042000221696

Linnenluecke, M., & Griffiths, A. (2010). Beyond adaptation: Resilience for business in light of climate change and weather extremes. *Business & Society, 49*(3), 477–511. doi:10.1177/0007650310368814

Love, L. F., & Singh, P. (2011). Workplace branding: Leveraging human resources management practices for competitive advantage through "Best Employer" surveys. *Journal of Business and Psychology, 26*(2), 175–181. doi:10.100710869-011-9226-5

Low, J. (2000). The value creation index. *Journal of Intellectual Capital, 1*(3), 252–262. doi:10.1108/14691930010377919

Manzoor, F., Wei, L., Bányai, T., Nurunnabi, M., & Subhan, Q. A. (2019). An Examination of Sustainable HRM Practices on Job Performance: An Application of Training as a Moderator. *Sustainability, 11*(8), 2263. doi:10.3390u11082263

Mehmood, Z. U. I., & Arif, M. I. (2011). Leadership and HRM: Evaluating new leadership styles for effective human resource management. *International Journal of Business and Social Science, 2*(15).

Milfelner, B., Potočnik, A., & Žižek, S. Š. (2015). Social responsibility, human resource management and organizational performance. *Systems Research and Behavioral Science, 32*(2), 221–229. doi:10.1002res.2263

Mooijman, M., van Dijk, W. W., van Dijk, E., & Ellemers, N. (2019). Leader power, power stability, and interpersonal trust. *Organizational Behavior and Human Decision Processes*, *152*, 1–10. doi:10.1016/j.obhdp.2019.03.009

Moussa, M., & Somjai, K. (2019). *Job dissatisfaction and employee turnover: A qualitative case study in Thailand*. Sage Publications: Sage Business Cases Originals.

Nadler, D. A., & Tushman, M. L. (1988). Organizational frame bending: Principles for managing reorientation. *The Academy of Management Perspectives*, *3*(3), 194–204. doi:10.5465/ame.1989.4274738

Nie, D., Lämsä, A. M., & Pučėtaitė, R. (2018). Effects of responsible human resource management practices on female employees' turnover intentions. *Business Ethics (Oxford, England)*, *27*(1), 29–41. doi:10.1111/beer.12165

Nilakant, V., & Ramnarayan, S. (2006). *Change management: Altering mindsets in a global context*. Sage Publications India.

Nonaka, I., & Takeuchi, H. (1995). *The Knowledge Creation Company: How Japanese Companies Create the Dynamics of Innovation*. New York, NY: Oxford University Press.

Oltra, V. (2005). Knowledge management effectiveness factors: The role of HRM. *Journal of Knowledge Management*, *9*(4), 70–86. doi:10.1108/13673270510610341

Ozcelik, H., Langton, N., & Aldrich, H. (2008). Doing well and doing good: The relationship between leadership practices that facilitate a positive emotional climate and organizational performance. *Journal of Managerial Psychology*, *23*(2), 186–203. doi:10.1108/02683940810850817

Packard, T. (2009). Leadership and performance in human services organizations. Chapter 7 in The handbook of human services management, Sag5e, 143-164.

Panayiotou, A., Putnam, L. L., & Kassinis, G. (2019). Generating tensions: A multilevel, process analysis of organizational change. *Strategic Organization*, *17*(1), 8–37. doi:10.1177/1476127017734446

Pascale, R., Millemann, M., & Gioja, L. (1997). Changing the way we change. *Harvard Business Review*, *75*(6), 126. PMID:10174794

Pinprayong, B., & Siengtai, S. (2012). Restructuring for organizational efficiency in the banking sector in Thailand: A case study of Siam Commercial Bank. *Far East Journal of Psychology and Business*, *8*(2), 29–42.

Porter, M. E. (1996). What is strategy? *Harvard Business Review*, *74*(6), 61–78. PMID:10158474

Rehman, S., Zahid, M., Rahman, H. U., & Habib, M. N. (2019). A Partial Least Squares Approach to the Leadership Styles, Organizational Culture, and Employees' Productivity: A Case of Pakistan Banking Industry. *International Journal of Asian Business and Information Management*, *10*(1), 55–64. doi:10.4018/IJABIM.2019010104

Rosenbusch, N., Brinckmann, J., & Bausch, A. (2011). Is innovation always beneficial? A meta-analysis of the relationship between innovation and performance in SMEs. *Journal of Business Venturing*, *26*(4), 441–457. doi:10.1016/j.jbusvent.2009.12.002

Rowley, J. (2000). From learning organisation to knowledge entrepreneur. *Journal of Knowledge Management*, *4*(1), 7–15. doi:10.1108/13673270010315362

Santangelo, G. D., & Pini, P. (2011). New HRM practices and exploitative innovation: A shopfloor level analysis. *Industry and Innovation*, *18*(6), 611–630. doi:10.1080/13662716.2011.591977

Schweyer, A. (2010). *Talent management systems: Best practices in technology solutions for recruitment, retention and workforce planning*. Hoboken, NJ: John Wiley & Sons.

Seeck, H., & Diehl, M. R. (2017). A literature review on HRM and innovation–taking stock and future directions. *International Journal of Human Resource Management*, *28*(6), 913–944. doi:10.1080/09585192.2016.1143862

Senge, P. M. (2006). The Fifth Discipline: The Art and Practice of the Learning Organization. Currency, New York, NY.

Shen, J., & Benson, J. (2016). When CSR is a social norm: How socially responsible human resource management affects employee work behavior. *Journal of Management*, *42*(6), 1723–1746. doi:10.1177/0149206314522300

Shen, J., & Jiuhua Zhu, C. (2011). Effects of socially responsible human resource management on employee organizational commitment. *International Journal of Human Resource Management*, *22*(15), 3020–3035. doi:10.1080/09585192.2011.599951

Shih, C. P., & Orochena, O. D. C. P. (2016). Analyzing the Effect of Transformational Leadership on Innovation and Organizational Performance. *International Journal of Productivity Management and Assessment Technologies*, *4*(2), 11–27. doi:10.4018/IJPMAT.2016070102

Shih, H. A., & Chiang, Y. H. (2005). Strategy alignment between HRM, KM, and corporate development. *International Journal of Manpower*, *26*(6), 582–603. doi:10.1108/01437720510625476

Shipton, H., West, M. A., Dawson, J., Birdi, K., & Patterson, M. (2006). HRM as a predictor of innovation. *Human Resource Management Journal*, *16*(1), 3–27. doi:10.1111/j.1748-8583.2006.00002.x

Silva, M. R. A., de Amorim, J. C., & Dias, A. L. (2019). Determinants of Employee Retention: A Study of Reality in Brazil. In Strategy and Superior Performance of Micro and Small Businesses in Volatile Economies (pp. 44–56). Hershey, PA: IGI Global. doi:10.4018/978-1-5225-7888-8.ch004

Souza, J. G. S., Lages, V. A., Sampaio, A. A., Souza, T. C. S., & Martins, A. M. E. D. B. (2019). The absence of functional dentition is associated with the lack of commitment to oral functions among Brazilian adults. *Ciencia & Saude Coletiva*, *24*(1), 253–260. doi:10.1590/1413-81232018241.30432016 PMID:30698258

Steijn, B., & Tijdens, K. (2005). Workers and Their Willingness to Learn: Will ICT-Implementation Strategies and HRM Practices Contribute to Innovation? *Creativity and Innovation Management*, *14*(2), 151–159. doi:10.1111/j.1476-8691.2005.00335.x

Steijn, B., & Tijdens, K. (2005). Workers and Their Willingness to Learn: Will ICT-Implementation Strategies and HRM Practices Contribute to Innovation? *Creativity and Innovation Management*, *14*(2), 151–159. doi:10.1111/j.1476-8691.2005.00335.x

Sune, A., & Gibb, J. (2015). Dynamic capabilities as patterns of organizational change: An empirical study on transforming a firm's resource base. *Journal of Organizational Change Management, 28*(2), 213–231. doi:10.1108/JOCM-01-2015-0019

Suppiah, V., & Sandhu, M. S. (2011). Organisational culture's influence on tacit knowledge-sharing behaviour. *Journal of Knowledge Management, 15*(3), 462–477. doi:10.1108/13673271111137439

Tan, J., & Tan, D. (2005). Environment–strategy co-evolution and co-alignment: A staged model of Chinese SOEs under transition. *Strategic Management Journal, 26*(2), 141–157. doi:10.1002mj.437

Tohidinia, Z., & Mosakhani, M. (2010). Knowledge sharing behavior and its predictors. *Industrial Management & Data Systems, 110*(4), 611–631. doi:10.1108/02635571011039052

Ulrich, D., Allen, J., Brockbank, W., Younger, J., & Nyman, M. (2009). HR Transformation Building Human Resources from the Outside In. New York, NY: McGraw-Hill.

Wang, S., & Noe, R. A. (2010). Knowledge sharing: A review and directions for future research. *Human Resource Management Review, 20*(2), 115–131. doi:10.1016/j.hrmr.2009.10.001

Wolff, B. (2019). The Truth About Employee Disengagement. *Professional Safety, 64*(2), 24–24.

Wood, S., & De Menezes, L. (1998). High commitment management in the UK: Evidence from the workplace industrial relations survey, and employers' manpower and skills practices survey. *Human Relations, 51*(4), 485–515. doi:10.1177/001872679805100403

Chapter 16
Understanding Inter-Relational Dynamics of Different Factors Impacting the Cymbidium Orchids-Based Economy in Sikkim, India Using DEMATAL

Bibeth Sharma
Sikkim Manipal University, India

Saibal Kumar Saha
https://orcid.org/0000-0002-7842-698X
Sikkim Manipal University, India

Ajeya Jha
https://orcid.org/0000-0003-0491-5008
Sikkim Manipal University, India

ABSTRACT

Floriculture based economy is slowly increasing throughout the world. Orchids, which were once found only in the wild are now being widely cultivated and being sold in the market. In this chapter, an attempt has been made to understand the inter-relational dynamics of different factors impacting the cymbidium orchids-based economy in Sikkim, India using DEMATAL. A total of 14 factors were identified in this regard and based on expert opinion, these factors were rated and analysed using DEMATAL. The cause and effect relationship of the different factors was established in the process. It was found that policy, technology, e-commerce, floricultural parks, certification and infrastructure are the causes while cold chain, quality, pricing, promotion, market development, product development, entrepreneurs, and farming are the effects.

DOI: 10.4018/978-1-7998-0357-7.ch016

INTRODUCTION

Sikkim is distinct in being the second smallest state of India having a population marginally above 0.6 million. It is spread over an area of about 7096 Sq.Km. It has other unique distinctions - most important being, its proximity to both, the Tibetan Plateau in the north and the Bay of Bengal in the south. This manifests in a wide array of climatic conditions - from tropical moist forests in the south and cold desert in the north within a short distance of fewer than 100 kilometres. It harbours wholly a mountainous terrain and as a unit of eastern Himalayas, it is one of the top ten biodiversity hotspots all over the world. One important feature of its biodiversity are its splendid orchids. As many as 523 species of orchids (Including Cymbidium) are found in here. Sikkim possibly has the highest orchid's diversity ratio (Sudhizong Lucksom, 2007)

Sikkim traditionally has been an economically an underdeveloped region of the country. The twin challenge of economic under-development and fragile ecological status makes disruptive innovations in terms of economic choices vital for this tiny state. Most traditional economic sectors hardly make any sense in the context of this state. As it is exceptionally biodiverse as well as ecologically exceptionally fragile, it, therefore, does not permit any economic activity that can have even remotely severe ecological consequences. This results in very few options in terms of economic development. One of the core objectives of the State Industrial Policy is *maintenance of the Green State image while promoting industrial activities*. Cymbidiums are natural to this region and can be grown to add to the ecological richness rather than depleting it. This makes Cymbidium an attractive choice for economic development. Accordingly, in the year 2013 floriculture has been designated as an important economic sector for the state. Sikkim, consequently, has been declared a floriculture zone by the government of India. The growth in Floriculture Industry in Sikkim Himalayas is estimated to grow to 20% per annum by the year 2020. About 5,000 farmers are engaged in growing cymbidium in the state but considering the high growth of market space, more are expected to join. These opportunities, though welcome, yet may bring little cheer to the local economy as the accruing advantages at best are moderate. The approaches discussed so far are piece-meal. Jha A et al. (2018) have focussed upon the life-style segmentation of cymbidium consumers. S. Bibet et al. (2015) have studied Need assessment, Pricing and Availability perceptions of Sikkim Cymbidium. P. Kumar et al. (2018) have studied the feasibility of producing Cymbidium in Sikkim. Janakiram, T (2018) have focused upon flourientrepreneurship in the context of women empowerment and also in the context of Cymbidium.

What is needed, however, is to provide an entirely a revolutionary changeover to the traditional approach. A disruptive approach to the existing set-up is needed to meet the ecological and economic needs of the state. It, for example is found that most locally grown products are being sold locally, relatively at low prices. Very high values are possible if the product is sold internationally. This, however, is unthinkable now, primarily because of various reasons, one being that the state has poor connection with large markets.

In order to come up with a new economic vision in the context cymbidiums of Sikkim, it is imperative that a thorough analysis of macro environment of Sikkim Cymbidium is undertaken and which alone can help us to come up with an innovative approach that will help the state to harvest best of this opportunity. The aim of this study is to develop a conceptual system for developing better understanding of the macro environment and accordingly this study, therefore, is directed to understand the relationship between various macro-factors influencing the profitability of this sector in the state.

The motivation for formulating disruptive approach is to be found in literature. The significance if innovative systems approach to agriculture in developing countries has been studied by many researchers (Clark et al., 2016; Hall et al., 2004), Musau (2017). Iizuka et al. (2017) implemented the functions of innovation systems framework to comprehend and evaluate the process of system building for the successful transformation of the Ethiopian flower industry. Earlier, Bolo (2016) highlighted the relevance of the innovation systems (IS) approach to agriculture in developing countries and concluded that it is highly critical for shaping and defining the involvement of various policies, institutional frameworks and combinations of actors in innovative activity. He further underlines as how their interactions or rather lack of interactions, contribute to learning and innovation. Skvortsova et al. (2018) through their work have proposed ways to improve state policy of Russia in the context of innovation to bring about quality and safety in agricultural production.

This study contributes to the theory by implementing objective methods to understand macro-environment for nurturing an upcoming economic sector for an underdeveloped region. It is expected to provide tangible evidence to the policy makers and strategizing managers with an understanding to come up with innovative frameworks that can transform the economic potential into a realized vision. It promotes a systems approach for understanding and evaluating a macro business environment that facilitates defining problems and opportunities as also systems thinking. This study aspires to build a systems context for revolutionizing Cymbidium export potential for the given geographical region.

DEMATEL is a tool which effectively captures the phenomenon of disruptive innovation across various industries and potential businesses, in this study we would endeavour to observe the cause and effect perspective of disruptive innovation in nascent and highly promising business value chain of Cymbidium orchids.

LITERATURE REVIEW

Floriculture is production, processing and marketing of different types of flowers (Kadam, 2012). In the recent years there has been an increase in the consumption of floricultural products in developed and developing countries (Hanafi, 2012). In India the floriculture business is becoming a major source of employment (Sen & Raju, 2006) and has been growing at a compound annual growth rate (CAGR) of 25 percent over the past decade. Indian floricultural sector is small and unorganized but it is emerging as a vibrant sector in bio diversity hot spot of Northeast India (Singh, 2013). Due to globalization of the floriculture trade, orchids have become the most popular cut flower and potted floriculture crop (Mohanty et al., 2012).

In the context of Sikkim Cymbidium macro-factors have been identified through experts' interview as well as literature review. The identified factors are Government policy, Technological input, Quality Assurance, product Pricing, Promotional efforts, E-Commerce, market Development, Product Development, Floriculture Parks, Support Infrastructure, Certification (quality, organic character and phyto-sanitation), Entrepreneurship and farming.

Importance of Government policies in agriculture in general have been studied extensively. Even recent studies accept the hypothesis that Government policies are central to the growth of agricultural sector (Mozaffarian et al. (2018); Kay (2018); Resnick (2018); Klomp et al. (2018); England et al. (2018). In the context of Sikkim, policies related to Organic farming are significant. Recently Sikkim has been declared the first 100 percent-organic state by United Nations (Yadav et al., 2018; Bhardwaj,

2019). Further Sikkim Cymbidium are believed to be exclusive. But it does not enjoy any benefit of Geographical Indication (GI). Government Policies in this respect could also be critical. The economic validation for shielding geographical indications primarily originates from the fact that place of origin may be used as a quality indicator.

Technological input too historically has been identified as critical for innovative transformations. In the context of Floriculture pre and post-harvest technological inputs are important. Pre-harvest technological inputs have been the major focus often at the cost of post-harvest technological inputs. Cold-chain technologies Parrado-Moreno et al. (2018); Negi et al. (2018); Khan et al. (2018); Wang et al. (2018); Campos et al. (2018); Beasley et al. (2019); Wei et al. (2019) have been identified to be critical. Packaging technologies too have been studied for their importance Gawde (2018); Siddiqui et al. (2018); Tiwari et al. (2019). To what extent existing technological knowhow impact other major factors is always important to comprehend.

The characteristics that determine the quality of flowers range from colour, size, smell, freshness and longevity. Flowers have been known for being quality sensitive. Many researchers have focused on the quality aspects in floriculture (Souri et al.; Sajid et al. (2018); Mallahi et al. (2018); Prabhu et al. (2018); Grimes et al. (2018); do Nascimento Simões et al. (2018); Baker (2018); and Kongklom et al. (2018) and pre-harvest technology seems plays a vital role in ensuring the quality of flowers. To what extent technology and other factors affect quality and in turn how quality impacts actual commercialization of Sikkim Cymbidium needs exploration.

In last two decades' e-commerce has developed substantially as a means to reach out to the consumers directly and hence has emerged as an important macro-factor which needs to be included in any framework of factors that we study. This aspect is endorsed by two recent studies (Pudaruth et al. (2018), and D'souza et al. (2019). Pricing of flowers too is a critical factor because of price fluctuations due to geography, quality, occasion and so on. How pricing is affected and how it affects other factors need to be studied in a macro framework. Some recent studies on this aspect are by Dar (2015); Saripalle (2016); and Kumari et al. (2016).

Various certifications related to quality, geographic indication, organic origin, environmental integrity have become essential to ensure the flow of only the right kind of products through exports. Floriculture too has seen the growing significance of right certification and some of the recent studies on this include Gebreeyesus (2015); Roy (2015); Mwesige (2018); Vos et al. (2019). Which factors are essential for certification and how certification improves Cymbidium export also needs investigation.

Market development is essential factors determining the ultimate success of any endeavour. What role does it play? How is it affected by other macro-factors and how in turn it affects other factors need to be understood in the present context. Some recent work highlighting the importance of market development include those by Muraro, D. et al (2016); Rajeevan, P. K et al (2016).

No economic sector can develop until supported by new age entrepreneurs. Entrepreneurship however does not develop in isolation and needs to be nurtured. It is an outcome of may macro factors and in turn it influences other factors. This interplay is important to comprehend. Some recent works highlighting the significance of entrepreneurship on flower markets include those by Boruah et al. (2016); Khandave et al. (2017); Chhetri et al. (2018) and Kulkarni et al. (2019)

Literature review has helped us to identify the macro-factors influencing the overall success of transforming export of Sikkim Cymbidium. Interrelationship and interplay of these variables on each other needs to be determined. The objective of this paper, therefore is *to understand the inter-relational dynamics of different factors impacting the cymbidium orchids-based economy in Sikkim.*

To achieve this objective DEMATEL appears to be a useful tool. Decision-Making Trial and Evaluation Laboratory (DEMATEL) methodology is used for researching and solving complex and entwined problem groups as it can verify interdependence between variables. It also improves them by a chart which reflects interrelationships between variables. The DEMATEL process gives the Impact Relations Map which helps respondents to organize their actions in the world (Falatoonitoosi 2013). Graph theory is the foundation of DEMATEL. It helps to get an enhanced realization of casual relationships by dividing important and related issues to cause and effect (Chung-Wei & Gwo-Hshiung, 2009) and makes it possible to visualize the casual relationships of sub criteria and systems while forming casual diagrams. It may also demonstrate communication network and even control relationships between individuals (Jiann & Gwo-Hshiung, 2011; Wu & Lee, 2007; Chiu et al., 2006). This methodology also verifies interdependence among unpredictable features which may contain characteristic for a system or development trends and try to reflect the interrelationship between variables by improving the directed graph (Tamura et al., 2002; Hori & Shimizu, 1999). DEMATEL also helps to understand practical solutions, particular problems and cluster of complicated problems (Tzeng et al., 2007; Chiu et al., 2006; Huang et al., 2007; Liou et al., 2007).

DEMATEL technique has been applied in numerous situation like manufacturing planning and control, multi criteria decision making, Marketing strategy and customer performance (Chiu et al., 2006), administration control systems (Hori & Shimizu, 1999), safety and security measurement (Liou et al., 2007), modernization strategy set for Taiwan's SIP Mall (Huang et al., 2007), fuzzy approach and expert systems (Wu & Lee, 2007), industry material selection process (Shih-Chi et al., 2011), success factors of hospital service quality (Jiunn et al., 2010) etc. DEMATEL is more suitable in real-world applications than traditional methods for situations which assume independence among criteria (Chen et al., 2011; Hung et al., 2011; Liu et al., 2012).

METHODOLOGY

With help of DEMATEL an attempt has been made to understand the inter-relational dynamics of major factors that impact the cymbidium orchids-based economy in the state of Sikkim. The factors: Policy, Technology, Quality Assurance, Pricing, Promotional efforts, E-Commerce, market Development, Product Development, Floriculture Parks, Support Infrastructure, Certification (quality, organic character, and phyto-sanitation), Entrepreneurship and farming were identified to be influencing the economy of cymbidium orchids in Sikkim. Based on experts' opinion the level of dependence of these variables on each other were identified using paired comparison. Each respondent was asked to estimate the dependence of factors by a numeral score ranging between 0 to 5 with 0 implying no influence and 5 implying highest influence. The number of experts providing information stood at 10.

The steps of DEMATEL method calculation are described as follows:

Step 1: Generation of Average Matrix / Original Impact Matrix (A)
Step 2: Calculation of maximum values of all rows and columns of the impact matrix (A).
Step 3: Calculation of k
Step 4: Calculation of the Direct Impact Matrix (M)
Step 5: Creation of priority ranking of the Direct Impact Matrix (M).
Step 6: Calculation of Total Impact Matrix (T).
Step 7: Calculation of vector elements.

Table 1. Average matrix / original impact matrix (A)

A	Policy	Technology	Cold Chain	Quality	Pricing	Promo	E-com	Market Dev	Floric. parks	Product-Dev	Infrastr	Certific	Entrepr.	Farming	Σaij
Policy	0	2.1	0.8	2.8	2.1	0.2	0.8	1	0.4	1.1	0.2	4.4	0.5	4.5	20.9
Technology	4.3	0	3.6	0.8	0.3	0.5	0.2	0.5	2.3	0.5	3.4	2.2	2.2	2	22.8
Cold Chain	4.4	3.8	0	1.1	0.3	0.5	0.7	2.4	1.8	2.1	3.6	0.3	2.7	1.8	25.5
Quality	1.3	3.9	3.4	0	1	1.9	0.9	1.2	2.1	3.3	3.5	2.9	2.1	3	30.5
Pricing	1.4	2.6	2.3	3.6	0	1.3	2.3	2.8	1.1	3.6	1.3	4.3	1.2	0.7	28.5
Promotion	1.3	1.2	1.4	3.4	1.1	0	3.2	1.5	1.7	1.4	1	4.1	4.1	0.3	25.7
E-Commerce	3	1.4	0.3	1.3	1.3	2.1	0	1	0.2	0.6	2.3	1.3	1.9	0.2	16.9
Market Dev	5	1.2	1.1	3.2	3.9	4.1	3.8	0	0.9	3.8	3.6	4.3	1.3	0.9	37.1
Floriculture Parks	4.5	1.6	0.2	0.4	0.2	0.2	0.2	0.7	0	0.2	1.2	0.2	0.4	0.9	10.9
Product-Dev	3.2	4.6	0.5	1.4	0.9	1.1	1.3	2.4	3.2	0	3.1	1.2	2.8	3.4	29.1
Infrastructure	4.5	2.1	0.4	1.5	2.2	0.4	0.3	0.2	0.2	0.4	0	0.2	1	0.2	13.6
Certificate	4.6	3.8	2.1	2.2	0.4	0.2	0.7	1.8	1	2.4	3.1	0	1	1.1	24.4
Entrepreneurs	3.8	2.1	3.2	2.5	2.4	3.2	4.4	3.3	0.8	4.2	2.4	4	0	3.1	39.4
Farming	3.9	3.1	2.3	3.1	3.6	1.6	1.7	2.1	2.4	2.2	2.1	3.1	1	0	32.2
Σaij	**45.2**	33.5	21.6	27.3	19.7	17.3	20.5	20.9	18.1	25.8	30.8	32.5	22.2	22.1	

Step 8: Formation of Cause – Effect relationship diagram

RESULT AND DISCUSSION

DEMATAL calculations and analysis is as given below.

Step 1: Generation of Average Matrix / Original Impact Matrix (A)

The calculations of the Original Impact Matrix are conducted by pairwise comparisons of different dimensions to evaluate the perceived level of impact of each respondent regarding the dimensions.

From the Original Impact Matrix, it is seen that:

1. Policy has a high degree of impact on Certificate (4.6)
2. Technology has a high degree of impact on Product- Development (4.6)
3. Cold Chain has a high degree of impact on Technology (3.6)
4. Quality has a high degree of impact on Pricing (3.6)
5. Pricing has a high degree of impact on Market Development (3.9)
6. Promotion has a high degree of impact on Market Development (4.1)
7. E-Commerce has a high degree of impact on Entrepreneurs (4.4)
8. Market Development has a high degree of impact on Entrepreneurs (3.3)
9. Floriculture Parks has a high degree of impact on Product- Development (3.2)
10. Product- Development has a high degree of impact on Entrepreneurs (4.2)
11. Infrastructure has a high degree of impact on Cold Chain and Market Development (3.6)
12. Certificate has a high degree of impact on Policy (4.4)
13. Entrepreneurs has a high degree of impact on Promotion (4.1)
14. Farming has a high degree of impact on Policy (4.5)

Table 2. Direct impact matrix (M)

M=K*A	Policy	Technology	Cold Chain	Quality	Pricing	Promo	E-com	Market Dev	Floric. parks	Product-Dev	Infrastr	Certific	Entrepr.	Farming	Σaij column
Policy	0.000	0.046	0.018	0.062	0.046	0.004	0.018	0.022	0.009	0.024	0.004	0.097	0.011	0.099	0.4598
Technology	0.095	0.000	0.079	0.018	0.007	0.011	0.004	0.011	0.051	0.011	0.075	0.048	0.048	0.044	0.5016
Cold Chain	0.097	0.084	0.000	0.024	0.007	0.011	0.015	0.053	0.040	0.046	0.079	0.007	0.059	0.040	0.561
Quality	0.029	0.086	0.075	0.000	0.022	0.042	0.020	0.026	0.046	0.073	0.077	0.064	0.046	0.066	0.671
Pricing	0.031	0.057	0.051	0.079	0.000	0.029	0.051	0.062	0.024	0.079	0.029	0.095	0.026	0.015	0.627
Promotion	0.029	0.026	0.031	0.075	0.024	0.000	0.070	0.033	0.037	0.031	0.022	0.090	0.090	0.007	0.5654
E-Commerce	0.066	0.031	0.007	0.029	0.029	0.046	0.000	0.022	0.004	0.013	0.051	0.029	0.042	0.004	0.3718
Market Dev	0.110	0.026	0.024	0.070	0.086	0.090	0.084	0.000	0.020	0.084	0.079	0.095	0.029	0.020	0.8162
Floriculture Parks	0.099	0.035	0.004	0.009	0.004	0.004	0.004	0.015	0.000	0.004	0.026	0.004	0.009	0.020	0.2398
Product-Dev	0.070	0.101	0.011	0.031	0.020	0.024	0.029	0.053	0.070	0.000	0.068	0.026	0.062	0.075	0.6402
Infrastructure	0.099	0.046	0.009	0.033	0.048	0.009	0.007	0.004	0.004	0.009	0.000	0.004	0.022	0.004	0.2992
Certificate	0.101	0.084	0.046	0.048	0.009	0.004	0.015	0.040	0.022	0.053	0.068	0.000	0.022	0.024	0.5368
Entrepreneurs	0.084	0.046	0.070	0.055	0.053	0.070	0.097	0.073	0.018	0.092	0.053	0.088	0.000	0.068	0.8668
Farming	0.086	0.068	0.051	0.068	0.079	0.035	0.037	0.046	0.053	0.048	0.046	0.068	0.022	0.000	0.7084
Σaij	0.9944	0.737	0.4752	0.601	0.433	0.381	0.451	0.46	0.398	0.5676	0.678	0.715	0.488	0.486	

Step 2: Calculation of maximum values of all rows and columns of the impact matrix (***A***).
Max of Σaij (column and rows) = 45.2
Step 3: Calculation of k

$$k = Min\left(\frac{1}{\max\limits_{1\le i\le n}\sum_{f=1}^{n}|a_{ij}|}, \frac{1}{\max\limits_{1\le j\le n}\sum_{i=1}^{n}a_{ij}}\right), i, j = 1, 2, 3, ..., n$$

Putting the value of Max of Σaij (column and rows) = 45.2 in the above equation, we get

k = 0.022

Step 4: Calculation of the Direct Impact Matrix (M)
Step 5: Creation of priority ranking of the Direct Impact Matrix (M). It is done by taking the summation of row sum and column sum.

From the Direct Impact Matrix, the priority table is formed. It is seen that Policy has the highest impact priority which is followed by Entrepreneurs and Market Development. Promotion, E-Commerce and Floriculture Parks has the least impact priority.

Step 6: Calculation of Total Impact Matrix (T). It is done by using the following formula:
$T = M (I-M)^{-1}$
Where I is the Identity matrix.
$T = M (I-M)^{-1}$

Table 3. Priority table

Σaij column	Σaij row	Row sum + column sum	Priority	
0.4598	0.9944	1.4542	**1**	Policy
0.5016	0.737	1.2386	**6**	Technology
0.561	0.4752	1.0362	**10**	Cold Chain
0.671	0.6006	1.2716	**4**	Quality
0.627	0.4334	1.0604	**9**	Pricing
0.5654	0.3806	0.946	**12**	Promotion
0.3718	0.451	0.8228	**13**	E-Commerce
0.8162	0.4598	1.276	**3**	Market Dev
0.2398	0.3982	0.638	**14**	Floriculture Parks
0.6402	0.5676	1.2078	**7**	Product-Dev
0.2992	0.6776	0.9768	**11**	Infrastructure
0.5368	0.715	1.2518	**5**	Certificate
0.8668	0.4884	1.3552	**2**	Entrepreneurs
0.7084	0.4862	1.1946	**8**	Farming

Step 7: Calculation of vector elements.
C = column sum vector
R = row sum vector
R - C = row and column difference vector

Table 4. (I – M) matrix

I - M	Policy	Technology	Cold Chain	Quality	Pricing	Promo	E-com	Market Dev	Floric. parks	Product-Dev	Infrastr	Certific	Entrepr.	Farming
Policy	1.00	-0.05	-0.02	-0.06	-0.05	0.00	-0.02	-0.02	-0.01	-0.02	0.00	-0.10	-0.01	-0.10
Technology	-0.09	1.00	-0.08	-0.02	-0.01	-0.01	0.00	-0.01	-0.05	-0.01	-0.07	-0.05	-0.05	-0.04
Cold Chain	-0.10	-0.08	1.00	-0.02	-0.01	-0.01	-0.02	-0.05	-0.04	-0.05	-0.08	-0.01	-0.06	-0.04
Quality	-0.03	-0.09	-0.07	1.00	-0.02	-0.04	-0.02	-0.03	-0.05	-0.07	-0.08	-0.06	-0.05	-0.07
Pricing	-0.03	-0.06	-0.05	-0.08	1.00	-0.03	-0.05	-0.06	-0.02	-0.08	-0.03	-0.09	-0.03	-0.02
Promotion	-0.03	-0.03	-0.03	-0.07	-0.02	1.00	-0.07	-0.03	-0.04	-0.03	-0.02	-0.09	-0.09	-0.01
E-Commerce	-0.07	-0.03	-0.01	-0.03	-0.03	-0.05	1.00	-0.02	0.00	-0.01	-0.05	-0.03	-0.04	0.00
Market Dev	-0.11	-0.03	-0.02	-0.07	-0.09	-0.09	-0.08	1.00	-0.02	-0.08	-0.08	-0.09	-0.03	-0.02
Floriculture Parks	-0.10	-0.04	0.00	-0.01	0.00	0.00	0.00	-0.02	1.00	0.00	-0.03	0.00	-0.01	-0.02
Product-Dev	-0.07	-0.10	-0.01	-0.03	-0.02	-0.02	-0.03	-0.05	-0.07	1.00	-0.07	-0.03	-0.06	-0.07
Infrastructure	-0.10	-0.05	-0.01	-0.03	-0.05	-0.01	-0.01	0.00	0.00	-0.01	1.00	0.00	-0.02	0.00
Certificate	-0.10	-0.08	-0.05	-0.05	-0.01	0.00	-0.02	-0.04	-0.02	-0.05	-0.07	1.00	-0.02	-0.02
Entrepreneurs	-0.08	-0.05	-0.07	-0.06	-0.05	-0.07	-0.10	-0.07	-0.02	-0.09	-0.05	-0.09	1.00	-0.07
Farming	-0.09	-0.07	-0.05	-0.07	-0.08	-0.04	-0.04	-0.05	-0.05	-0.05	-0.05	-0.07	-0.02	1.00

Table 5. (I – M) inverse matrix

(I - M) inverse	Policy	Technology	Cold Chain	Quality	Pricing	Promo	E-com	Market Dev	Floric. parks	Product-Dev	Infrastr	Certific	Entrepr.	Farming
Policy	1.0783	0.108	0.061	0.108	0.08	0.032	0.048	0.058	0.043	0.0703	0.061	0.15	0.047	0.137
Technology	0.1708	1.06	0.1131	0.064	0.042	0.035	0.034	0.046	0.079	0.0529	0.122	0.101	0.081	0.088
Cold Chain	0.1837	0.145	1.0443	0.077	0.05	0.043	0.052	0.089	0.074	0.0922	0.135	0.073	0.098	0.091
Quality	0.139	0.164	0.1251	1.062	0.068	0.076	0.062	0.074	0.09	0.1263	0.148	0.133	0.097	0.121
Pricing	0.1333	0.136	0.1004	0.136	1.043	0.066	0.091	0.106	0.067	0.1342	0.103	0.162	0.077	0.072
Promotion	0.121	0.098	0.0783	0.127	0.063	1.037	0.109	0.076	0.072	0.0857	0.088	0.153	0.131	0.058
E-Commerce	0.1209	0.074	0.0376	0.067	0.056	0.067	1.027	0.049	0.027	0.0476	0.088	0.076	0.07	0.039
Market Dev	0.2249	0.125	0.0852	0.148	0.137	0.131	0.134	1.058	0.07	0.1509	0.16	0.186	0.091	0.089
Floriculture Parks	0.1308	0.061	0.0228	0.033	0.023	0.016	0.018	0.03	1.014	0.0235	0.048	0.035	0.024	0.044
Product-Dev	0.1715	0.17	0.063	0.091	0.067	0.06	0.07	0.094	0.108	1.0543	0.133	0.101	0.105	0.128
Infrastructure	0.1387	0.081	0.0352	0.063	0.068	0.025	0.026	0.027	0.024	0.0365	1.03	0.045	0.043	0.037
Certificate	0.1819	0.145	0.0868	0.096	0.047	0.033	0.047	0.074	0.056	0.0952	0.123	1.061	0.061	0.075
Entrepreneurs	0.2168	0.152	0.1343	0.139	0.114	0.119	0.152	0.133	0.074	0.166	0.147	0.186	1.069	0.14
Farming	0.1915	0.151	0.1051	0.133	0.123	0.072	0.081	0.095	0.096	0.1094	0.121	0.147	0.075	1.062

R + C = row and column sum vector

Higher value of ***(R + C)*** means that the mutual effects of the factors are greater (Seyed-Hosseini et al., 2005). From the table below it is seen that ***(R + C)*** is highest for Policy (3.285), which means that the mutual effect of Policy and other factors is the greatest. The difference vector (***R - C***) gives the net impact of the Total Impact Matrix. If (***R - C***) is > 0, it means that the factor has greater impact on other

Table 6. Total impact matrix (T)

T	Policy	Technology	Cold Chain	Quality	Pricing	Promo	E-com	Market Dev	Floric. parks	Product-Dev	Infrastr	Certific	Entrepr.	Farming
Policy	0.078	0.108	0.061	0.108	0.080	0.032	0.048	0.058	0.043	0.070	0.061	0.150	0.047	0.137
Technology	0.171	0.060	0.113	0.064	0.042	0.035	0.034	0.046	0.078	0.053	0.122	0.101	0.081	0.088
Cold Chain	0.184	0.145	0.044	0.077	0.050	0.043	0.052	0.089	0.074	0.092	0.135	0.073	0.098	0.091
Quality	0.139	0.164	0.125	0.062	0.067	0.076	0.062	0.074	0.090	0.126	0.148	0.133	0.097	0.121
Pricing	0.133	0.136	0.100	0.136	0.043	0.066	0.091	0.106	0.067	0.134	0.103	0.162	0.077	0.072
Promotion	0.121	0.098	0.078	0.127	0.063	0.037	0.109	0.076	0.072	0.086	0.088	0.153	0.131	0.058
E-Commerce	0.121	0.074	0.038	0.067	0.056	0.067	0.027	0.049	0.027	0.048	0.088	0.076	0.070	0.039
Market Dev	0.225	0.125	0.085	0.148	0.137	0.131	0.134	0.058	0.070	0.151	0.160	0.186	0.091	0.089
Floriculture Parks	0.131	0.061	0.023	0.033	0.023	0.016	0.018	0.030	0.014	0.023	0.048	0.035	0.024	0.044
Product-Dev	0.172	0.170	0.063	0.091	0.067	0.060	0.070	0.094	0.108	0.054	0.133	0.101	0.105	0.128
Infrastructure	0.139	0.081	0.035	0.063	0.068	0.025	0.026	0.027	0.024	0.037	0.030	0.046	0.043	0.036
Certificate	0.182	0.145	0.087	0.096	0.047	0.033	0.047	0.074	0.056	0.095	0.123	0.061	0.061	0.075
Entrepreneurs	0.217	0.152	0.134	0.139	0.113	0.119	0.152	0.133	0.074	0.166	0.147	0.186	0.069	0.140
Farming	0.192	0.151	0.105	0.133	0.123	0.073	0.081	0.095	0.096	0.109	0.121	0.147	0.075	0.062

Table 7. Cause effect relationship table

R	C	R+C	R-C	Factors	
2.204	1.082	3.285	1.122	Policy	Cause
1.670	1.089	2.759	0.581	Technology	Cause
1.092	1.249	2.341	-0.157	Cold Chain	Effect
1.343	1.486	2.829	-0.143	Quality	Effect
0.981	1.425	2.406	-0.444	Pricing	Effect
0.812	1.298	2.110	-0.486	Promotion	Effect
0.952	0.845	1.797	0.108	E-Commerce	Cause
1.010	1.790	2.799	-0.780	Market Dev	Effect
0.896	0.524	1.420	0.371	Floriculture Parks	Cause
1.245	1.415	2.660	-0.170	Product-Dev	Effect
1.506	0.680	2.186	0.826	Infrastructure	Cause
1.609	1.182	2.791	0.427	Certificate	Cause
1.068	1.942	3.010	-0.874	Entrepreneurs	Effect
1.182	1.564	2.746	-0.381	Farming	Effect

factors. Hence, it is called the cause. If ***(R - C)*** is < 0, the factor has a smaller impact on other factors. Hence, it is referred to as the effect.

Step 8: Formation of Cause – Effect relationship diagram

The analysis of DEMATAL can be summarized as:

1. The Policy is influenced by farming (acceptance of farmers), certification, and quality of cymbidium that can be grown in Sikkim and availability/acceptance of technology and in turn it influences certification, creation of infrastructures and floriculture parks and application of technology and creation of cold chains.
2. Technology is influenced by three major factors namely policy, availability of cold chain and infrastructure. In turn it influences product development, quality, cold chain and certification.
3. Cold chain depends primarily on policy, infrastructure, technology and entrepreneurship and in turn it influences technology, quality, and entrepreneurship, pricing and farming.
4. Quality depends mainly upon technology, cold chain, product development, infrastructure, certification, farming and entrepreneurship and has relatively greater influence upon pricing promotion, market development, farming and entrepreneurship.
5. Pricing is a factors that finds greatest influence of certification, product development, quality, market development and technology. It substantially influences market development, farming and entrepreneurship.
6. Promotion can be best served by certification, entrepreneurship, quality and e-commerce. It, in turn, can influence market development and entrepreneurship.
7. E-commerce can be considered to be primarily under the influence of policy and infrastructure but it does influence, entrepreneurship, promotion and market development.

Figure 1. Cause – effect relationship diagram

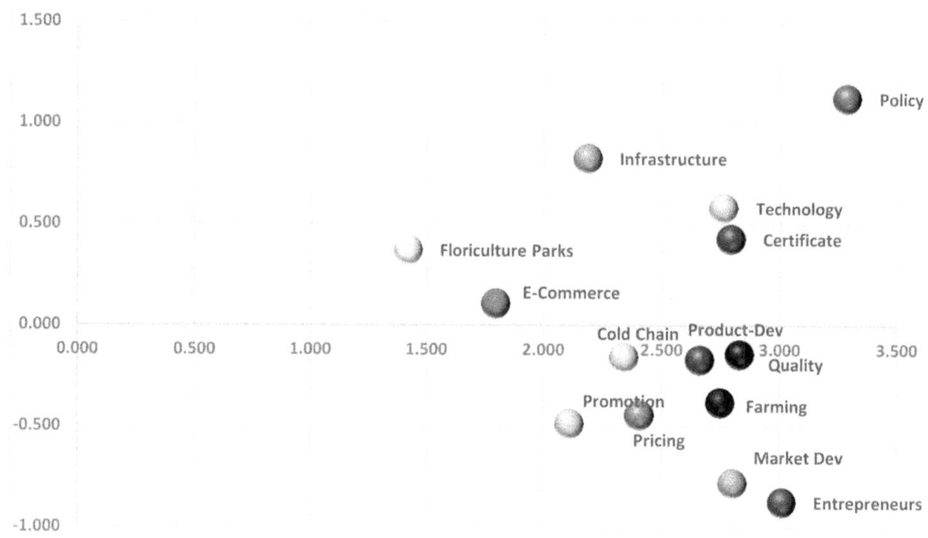

8. Market development that is one of the major desirable factors can be best served through policy, promotion, pricing, quality, e-commerce, product development, and certification. It finds influence over cold chain, pricing, entrepreneurship and product development.
9. Floriculture parks primarily depend upon policy alone but it is found to have substantial impact upon technology, product development and farming.
10. Product development depends primarily upon technology, policy, floriculture parks, infrastructure, entrepreneurship and farming. It finds major influence over pricing quality, market development, entrepreneurship and certification.
11. Infrastructure appears to be under the sole influence of policy although technology too influences it. It, however, in a major way impacts technology, quality, cold chain, market development, entrepreneurship and product development.
12. Certification yet again emerges as a factor that depends mainly upon policy, technology, and Infrastructure and product development. It, however, has considerable influence over policy, pricing, promotion, market development, entrepreneurship and farming.
13. Entrepreneurship, that emerges as the most prominent effect depends upon a large number of related factors including product development, e-commerce, certification, market development, farming, policy, cold chain, quality technology and infrastructure. It will influence promotion, product development, quality and cold chain.
14. Farming, the last factor, appears to be impacted most by policy, technology, cold chain, quality, market development, floriculture parks, product development, infrastructure and certification. It in turn impacts policy, product development, quality and entrepreneurship.

Implications for the government are:

1. The policy framework must be drawn from farmers' needs and aspirations, international certification dynamic requirements, availability/acceptance of technology and quality potential (and uniqueness) of Sikkim cymbidiums.
2. The policies should be specific on issues involving certification for the product, creation of required infrastructure and creation of floriculture parks
3. The policies should be framed with due assessment of technologies, transferring capabilities and application feasibility.
4. Cold-chains should be created to preserve the flowers for longer duration.
5. In state of Sikkim entrepreneurship is still to take strong roots. Hence, Government should create required environment and infrastructure for potential and existing entrepreneurs.
6. Opportunities for creating certification centres, operating cold chains, market development and product development should be created by the Government.

Implications for the farmers (growers) are:

1. Their understanding must remain consistent with the overall vision and policies of the State Government.
2. They should make use of technology to enhance quantity and quality of their products.
3. They should focus on quality and certification requirements and make use of government sponsored technology parks.
4. They must play active role in product development and in supporting allied entrepreneurs.

CONCLUSION

Sikkim Cymbidium provides the state of Sikkim in India with opportunities that are distinctly encouraging and feasible. The efforts so far are just traditional and perhaps the vision is narrow in this respect. In order to realize its true potential an innovative framework needs to be formulated based on a greater insight into the dynamics of the macro factors influencing the environment. This paper has been conceived to develop a better insight of the interrelational dynamics of various macro factors with a view to determine the cause and effect relationship between these factors as also the quantum of the relationships these share with each other. To achieve this target DEMATEL technique has been used and which is known for bringing objectivity to otherwise a largely subjective setting.

Indian farmers are advancing from growing just traditional crops to non-traditional crops and are making an attempt to improve their livelihood by increasing the portfolio of their crops production. Floriculture business has gained momentum in India along with the rest of the world and Indian farmers are making active efforts for meeting the demand of these plants' blossoms. After analysis of the different factors affecting the cymbidium orchids based economy in Sikkim it is found that flower growers need a lot of support in terms of policies, technology, infrastructure, marketing skills and certification (Geographical Indication, Organic Status, and Phytosanitary certification) for being able to grow, market and sell cymbidium orchids as per the world standards. A robust framework and infrastructural support from the Government in terms of certifications, technology, cold chains and entrepreneurial support can help in the development of the cymbidium orchids sales and marketing. The relationship between diverse factors (cause and effects) with their magnitude using DEMATEL has been studied and found that policy, technology, e-commerce, floricultural parks, certification and infrastructure are the causes while cold chain, quality, pricing, promotion, market development, product development, entrepreneurs, and farming are the effects. A collaborative support from the Government, NGOs, entrepreneurs, marketers and farmers will help to increase the cymbidium orchids-based economy in Sikkim.

Implications for practicing managers are substantially discernible. Practicing managers need to function closely with the policy-makers as well as the growers. Sikkim Cymbidium can acquire an international image due to its uniqueness and for which geographical indication needs to be established. This along with the UN certification of being the first 100% organic state will endow a distinct position for Sikkim Cymbidium and which will enhance its acceptance as well as make it less sensitive to the price-competition. Policies are also needed by the government to provide infrastructure for a smooth supply chain functioning. Cymbidium co-operatives need to be established to ensure that growers are in a position to bargain more effectively and are also confident of the sale of their products. Government must also formulate policies regarding capacity building of local potential entrepreneurs, specifically in the context of technology, marketing capabilities and e-commerce. Local entrepreneurs are also needed for having capability and recognition to certify the cymbidiums for their organic origin, geographical identity, quality and environmental integrity.

Product development and market development are two areas where professional managers can intervene substantially. Sikkim Cymbidium so far being sold as cut flowers. Is it possible to expand the product? Are there alternate uses of Sikkim Cymbidium? Sikkim cymbidium traditionally has been used as a delicacy as well as for its medicinal properties. These alternate uses can be further explored by the managers. Market development requires evolving a customer profile. Who are the people who buy cymbidium heavily? What life-style characteristics are shared by cymbidium consumers? Why do they buy it? Does Cymbidium have utilitarian purpose or do they just have an emotional value? When do people buy these? How can Cymbidium be integrated with various cultures in terms of colour symbolism? What values does cymbidium evoke also needs to be answered by managers. Quality has emerged as an important factor as it has tremendous impact on acceptance and pricing. Which quality parameters put Cymbidium on a high pedestal vis-à-vis other competing floral products? How can these characteristics be highlighted for its promotion. Cymbidium holds significant gift value. How is rated against other gift-products? What gift combos have greater acceptability by the buyers? Cymbidiums have a long vase life and it gives an opportunity to the marketers to sell it online? Is it being done? What are the challenges in this respect? Which e-commerce models are best for Cymbidiums?

As final concluding remarks it may be stated that export of Cymbidiums provides a worthwhile opportunity to the state of Sikkim in India. This exploratory work can help in coming up with innovative approaches to realize this potential. Transformation of Sikkim Cymbidium from a locally sold product to an export-oriented entity would provide a disruptive breakthrough.

REFERENCES

Baker, J. E. (2018). Preservation of cut flowers. In Plant growth regulating chemicals (pp. 177–191). Boca Raton, FL: CRC Press. doi:10.1201/9781351075749-10

Beasley, M., Leshuk, J. A., Yasar, M., & Cywilko, M. E. (2019). *U.S. Patent Application No. 16/077,090.*

Bhardwaj, M. (2019). Growth And Performance Of Organic Farming In India: What Could Be The Future Prospects? *Growth, 20*(01).

Bibeth, S., & Ajeya, J. (2015). Case Study: Need assessment, Pricing and Availability perceptions: A study of Cymbidium Orchids of Sikkim Himalayas. *Advances in Management, 8*(1), 26.

Bolo, M. (2016). Innovation Systems and Capability Building Among Smallholders: Lessons And Insights From Kenya's Flower Farmers. *Innovation Systems: Towards Effective Strategies in support of Smallholder Farmers*, 74.

Boruah, R., Borua, S., Deka, C. R., & Borah, D. (2016). Entrepreneurial behavior of tribal winter vegetable growers in Jorhat district of Assam. *Indian Research Journal of Extension Education, 15*(1), 65–69.

Campos, Y., & Villa, J. L. (2018, November). Technologies applied in the monitoring and control of the temperature in the Cold Chain. In *2018 IEEE 2nd Colombian Conference on Robotics and Automation (CCRA)* (pp. 1-6). IEEE. 10.1109/CCRA.2018.8588118

Chen, F. H., Hsu, T. S., & Tzeng, G. H. (2011). A balanced scorecard approach to establish a performance evaluation and relationship model for hot spring hotels based on a hybrid MCDM model combining DEMATEL and ANP. *International Journal of Hospitality Management, 30*(4), 908–932. doi:10.1016/j.ijhm.2011.02.001

Chhetri, K., & Ramswamy, R. R. (2018). Socio-economic Determinants of Agripreneurship: Daramdin Floriculture Cluster In Sikkim. *Small Enterprises Development, Management & Extension (Sedme). Journal, 45*(1).

Chowdhery, H. J. (1998). *Orchid of Arunachal Pradesh* (pp. 1–824). India.

Chung-Wei, L., & Gwo-Hshiung, T. (2009). Identification of a threshold value for the DEMATEL method using the maximum mean deentropy algorithm to find critical services provided by a semiconductor intellectual property mall. *Expert Systems with Applications, 8*(1), 9891–9989.

D'souza, D. J., & Joshi, H. G. (2019). E-Commerce Framework for Strategic Marketing of Udupi Jasmine. *AGRIS On-Line Papers in Economics and Informatics, 11*(1), 17–26. doi:10.7160/aol.2019.110102

Dar, M. A. (2015). *Impact of Floriculture Development Programme on Registered Flower Growers of Central Kashmir* (Doctoral dissertation, SKUAST Kashmir).

do Nascimento Simões, A., Diniz, N. B., da Silva Vieira, M. R., Ferreira-Silva, S. L., da Silva, M. B., Minatel, I. O., & Lima, G. P. P. (2018). Impact of GA3 and spermine on postharvest quality of anthurium cut flowers (Anthurium andraeanum) cv. Arizona. *Scientia Horticulturae, 241*, 178–186. doi:10.1016/j.scienta.2018.06.095

Du Puy, D., & Cribb, P. (1988). *The genus Cym-bidiwn*. Portland, OR: Timber Press.

England, M. I., Stringer, L. C., Dougill, A. J., & Afionis, S. (2018). How do sectoral policies support climate compatible development? An empirical analysis focusing on southern Africa. *Environmental Science & Policy, 79*, 9–15. doi:10.1016/j.envsci.2017.10.009

Falatoonitoosi, E., Leman, Z., Sorooshian, S., & Salimi, M. (2013). Decision-Making Trial and Evaluation Laboratory, Research Journal of Applied Sciences. *Engineering and Technology*, *5*(13), 3476–3480.

Forest Survey of India. (2009). State of Forest Report. Dehra Dun, India Forest Survey of India, Ministry of Environment and Forests, Government of India.

Gawde, P. (2018). Effect of packaging materials on vase life of tuberose (Polianthes tuberose) cv. Shringar. *IJCS*, *6*(5), 2011–2014.

Gebreeyesus, M. (2015). Firm adoption of international standards: Evidence from the Ethiopian floriculture sector. *Agricultural Economics*, *46*(S1), 139–155. doi:10.1111/agec.12203

Gebreeyesus, M., & Sonobe, T. (2009). Governance of global value chain and firms' capability in African floriculture, 1st Draft.

Grimes, S., Phillips, T., Hahn, V., Capezzone, F., & Graeff-Hönninger, S. (2018). Growth, Yield Performance and Quality Parameters of Three Early Flowering Chia (Salvia hispanica L.) Genotypes Cultivated in Southwestern Germany. *Agriculture*, *8*(10), 154. doi:10.3390/agriculture8100154

Hall, A., Carberry, P., Djikeng, A., Roy-Macauley, H., Pengelly, B., Njoya, A., ... Keating, B. (2016). The Journey to R4D: An institutional history of an Australian initiative on food security in Africa. In J. Francis, L. Mytelka, A. van Huis, & N. Röling (Eds.), *Innovation Systems: Towards Effective Strategies in Support of Smallholder Farmers. Technical Centre for Agricultural and Rural Cooperation (CTA) and Wageningen University and Research (WUR) Convergence of Sciences-Strengthening Innovation Systems (CoS-SIS) programme* (pp. 183–201). Wageningen, The Netherlands.

Huang, C. Y., Shyu, J. Z., & Tzeng, G. H. (2007). Reconfiguring the innovation policy portfolios for Taiwan's SIP Mall industry. *Technovation*, *27*(12), 744–765. doi:10.1016/j.technovation.2007.04.002

Hung, Y. H., Chou, S. C., & Tzeng, G. H. (2011). Knowledge management adoption and assessment for SMEs by a novel MCDM approach. *Decision Support Systems*, *51*(2), 270–291. doi:10.1016/j.dss.2010.11.021

Iizuka, M., & Gebreeyesus, M. (2017). Using Functions of Innovation Systems to Understand the Successful Emergence of Non-traditional Agricultural Export Industries in Developing Countries: Cases from Ethiopia and Chile. *European Journal of Development Research*, *29*(2), 384–403. doi:10.105741287-016-0004-0

India. Planning Commission. (2008). Sikkim Development Report. Academic Foundation.

Janakiram, T., Jain, R., Vyas, A. K., & Srilekha, R. (2018). Florienterpreneurship in India: New Avenue of Economic Empowerment of Women. *ASCI Journal of Management, 47*.

Jha, A., Sharma, B., & Kumar, J. (2018). Life-Style Distinction Between Customers and Non-Customers of Sikkim Cymbidium. In Management Strategies and Technology Fluidity in the Asian Business Sector (pp. 14–26). Hershey, PA: IGI Global. doi:10.4018/978-1-5225-4056-4.ch002

Jiann, L. Y., & Gwo-Hshiung, T. (2011). An integrated MCDM technique combined with DEMATEL for a novel cluster-weighted with ANP method. *Expert Systems with Applications*, *7*(2), 1417–1424.

Kay, C. (2018). Rural Latin America: exclusionary and uneven agricultural development. In Capital, Power, and Inequality in Latin America (pp. 21-52). Abingdon-on-Thames, UK: Routledge. doi:10.4324/9780429501869-2

Khan, M. S., Khan, M. M. I., & Haleem, A. (2018, August). Towards effective management of cold chain: A DEMATEL approach. [IOP Publishing]. *IOP Conference Series. Materials Science and Engineering*, *404*(1). doi:10.1088/1757-899X/404/1/012019

Khandave, S., Prajapati, M. R., & Patel, V. T. (2017). Entrepreneurial Attributes Of Nursery Growers. *Gujarat Journal of Extension Education*, 418.

Klomp, J., & Hoogezand, B. (2018). Natural disasters and agricultural protection: A panel data analysis. *World Development*, *104*, 404–417. doi:10.1016/j.worlddev.2017.11.013

Kongklom, N., Chuensangjun, C., Chisti, Y., & Sirisansaneeyakul, S. (2018). Improved keeping quality of Dendrobium "Bom" orchids using nutrients entrapped in a biodegradable hydrogel. *Scientia Horticulturae*, *234*, 184–192. doi:10.1016/j.scienta.2018.02.031

Kulkarni, N. P., & Jahagirdar, K. A. (2019). Entrepreneurial Behavior and Constraints Faced by the Rose Growers. *Asian Journal of Agricultural Extension. Economia e Sociologia (Evora, Portugal)*, 1–8.

Kumar, P., & Rai, S. C. (2018). Agricultural Diversities and Its Sustainability in Sikkim Himalaya: An Analysis. *Political Economy Journal of India*, *27*(1-2), 91.

Kumari, M., Mehta, P., & Raina, K. K. (2016). Farmers' Perceptions towards Marketing Problems and Challenges in Floriculture in Solan District of Himachal Pradesh, India. *International Journal of Economic Plants*, *3*(4), 143–149.

Liou, J. J. H., Tzeng, G. H., & Chang, H. C. (2007). Airline safety measurement using a hybrid model. *Journal of Air Transport Management*, *13*(4), 243–249. doi:10.1016/j.jairtraman.2007.04.008

Liu, C. H. G. H., Tzeng, G.-H., & Lee, M.-H. (2012). Improving tourism policy implementation— The use of hybrid MCDM models. *Tourism Management*, *33*(2), 239–488. doi:10.1016/j.tourman.2011.05.002

Lucksom, S. (2007). *The Orchids of Sikkim and North East Himalaya,* (pp. 1–965). Open Library: S. Z. Lucksom.

Mallahi, T., Ramezanian, A., Saharkhiz, M. J., Javanmardi, J., & Iraji, A. (2018). Antimicrobial activities of Asafoetida and Shirazi thyme essential oils improve the vase life of gerbera cut flowers. *Acta Ecologica Sinica*, *38*(3), 228–233. doi:10.1016/j.chnaes.2017.08.009

Mittermeier, R. A., Gils, P. R., Hoffman, M., Pilgrim, J., Brooks, T., Mittermeier, C. G., ... Da-Fonseca, G. A. B. (Eds.). (2004). *Hotspots Revisited. Earth's Biologically Richest and Most Endangered Terrestrial Ecoregions*. Arlington, VA: Conservation International.

Mozaffarian, D., Angell, S. Y., Lang, T., & Rivera, J. A. (2018). Role of government policy in nutrition—Barriers to and opportunities for healthier eating. *BMJ (Clinical Research Ed.)*, *361*, k2426. doi:10.1136/bmj.k2426 PMID:29898890

Muraro, D., Negrelle, R. R., Cuquel, F. L., & Anacleto, A. (2016). Market management: The impact on the development of an ornamental plants supply chain in Curitiba, Brazil. *Ciencia e Investigación Agraria*, *42*(3), 453–460.

Musau, S. M. (2017). *The Role of Strategic Management Practices on Competitiveness of Floriculture Industry in Kenya: A Case of Kiambu County* (Doctoral dissertation, United States International University-Africa).

Mwesige, P. (2018). *Value Chain governance and standards compliance in fresh fruit and vegetable (FFV) export sector in Uganda* (Doctoral dissertation, Makerere university business school Institutional repository).

Negi, S., & Anand, N. (2018). Wastage and Cold Chain Infrastructure Relationship in Indian Food Supply Chain: A Study From Farm to Retail. In Supply Chain Management Strategies and Risk Assessment in Retail Environments (pp. 247-266). Hershey, PA: IGI Global. doi:10.4018/978-1-5225-3056-5.ch014

Obara-Okeyo, P., Fujii, K., & Kako, S. (1997). Enzyme polymorphism in Cymbidium orchid cultivars and inheritance of leucine aminopeptidase. *HortScience*, *32*(7), 1267–1271. doi:10.21273/HORTSCI.32.7.1267

Parrado-Moreno, C. A., Ricardo-Hernández, R. E., Velásquez-Arredondo, H. I., Lopera-Castro, S. H., & Hasenstab, C. (2019). An Environmental Evaluation of the Cut-Flower Supply Chain (Dendranthema grandiflora) Through a Life Cycle Assessment. *Revista EIA*, *16*(31), 27–42. doi:10.24050/reia.v16i31.747

Prabhu, G., Thamaraiselvi, S. P., Aruna, P., & Sudhakar, R. (2018). Evaluation of chrysanthemum (Dendranthema grandiflora Tzelev.) genotypes for loose flower production under Coimbatore conditions. *IJCS*, *6*(4), 1618–1621.

Pudaruth, S., & Busviah, D. (2018). Developing and Testing a Pioneer Model for Online Shopping Behavior for Natural Flowers: Evidence from Mauritius. *Studies in Business and Economics*, *13*(1), 128–147. doi:10.2478be-2018-0011

Rajeevan, P. K., Geetha, C. K., & Rajendran, P. (2016,). Orchid-centric floriculture development in Kerala, India. In *International Symposium on Succulents and Other Ornamentals 1165* (pp. 15-26).

Resnick, D., Haggblade, S., Babu, S., Hendriks, S. L., & Mather, D. (2018). The Kaleidoscope Model of policy change: Applications to food security policy in Zambia. *World Development*, *109*, 101–120. doi:10.1016/j.worlddev.2018.04.004

Roy, T. N., Das, K. K., & Rai, D. (2015). Floriculture in West Bengal in Augmenting Income and Export. In *Diversification of Agriculture in Eastern India* (pp. 215–223). New Delhi, India: Springer. doi:10.1007/978-81-322-1997-2_19

Sajid, M., Rab, A., & Khan, I. A. (2018). The pre-harvest foliar application influenced the flower quality and vase life of chrysanthemum cultivars. *Horticult Int J*, *2*(4), 145–152.

Saripalle, M. (2016). Jasmine cultivation in Tamil Nadu: Market structure and pricing. *World Development Perspectives*, *1*, 12–14. doi:10.1016/j.wdp.2016.05.004

Seyed-Hosseini, S. M., Safaei, N., & Asgharpour, M. J. (2005). Reprioritization of failures in a system failure mode and effects analysis by decision making trial and evaluation laboratory chnique. *Reliability Engineering & System Safety*, *91*(8), 872–881. doi:10.1016/j.ress.2005.09.005

Shih-Chi, C., Sun, C. C., & Herchan, A. (2011). The DEMATEL approach applied to solar cell industry material selection process in Taiwan. *Session Interd. Manage. Sem.*, *15*(13), 253–267.

Siddiqui, M. W., Rahman, M. S., & Wani, A. A. (Eds.). (2018). Innovative Packaging of Fruits and Vegetables: Strategies for Safety and Quality Maintenance. Boca Raton, FL: CRC Press.

Singh, B. (2013). Given the rupee fall, Floriculture of Northeast India facing heat over rising cost of imported planting materials, ET Bureau, Aug. 23.

Skvortsova, T. A., Denisova, I. P., Romanenko, N. G., & Sukhovenko, A. V. (2018). Innovations and Support for Quality in Agriculture: A Case Study. *European Research Studies*, *21*, 423.

Souri, M. K., Goodarzizadeh, S., Ahmadi, M., & Hatamian, M. (2018). Characteristics of postharvest quality of chrysanthemum cut flowers under pretreatment with nitrogenous compounds. *Acta Scientiarum Polonorum. Hortorum Cultus*, *17*(3), 83–90. doi:10.24326/asphc.2018.3.8

Tamura, M., Nagata, H., & Akazawa, K. 2002. Extraction and systems analysis of factors that prevent safety and security by structural models. In *41st SICE Annual Conference*, Osaka, Japan. 10.1109/SICE.2002.1196584

Tiwari, A. K., & Nigam, V. K. (2019). Recent Bio-Processing Technologies for Value Added Horticultural Products. In *Applied Microbiology and Bioengineering* (pp. 57–67). Cambridge, MA: Academic Press. doi:10.1016/B978-0-12-815407-6.00004-6

Tzeng, G. H., Chiang, C. H., & Li, C. W. (2007). Evaluating intertwined effects in e-learning programs: A novel hybrid mcdm model based on factor analysis and DEMATEL. *Expert Systems with Applications*, *32*(4), 1028–1044. doi:10.1016/j.eswa.2006.02.004

Vos, J., van Ommen, P., & Mena-Vásconez, P. (2019). 12 To certify or not to certify. *Sustainability Certification Schemes in the Agricultural and Natural Resource Sectors: Outcomes for Society and the Environment*, 259.

Wang, K. Y., & Yip, T. L. (2018). Cold-Chain Systems in China and Value-Chain Analysis. In Finance and Risk Management for International Logistics and the Supply Chain (pp. 217–241). Amsterdam, The Netherlands: Elsevier. doi:10.1016/B978-0-12-813830-4.00009-5

Wei, J., & Lv, S. (2019, February). Research on the Distribution System of Agricultural Products Cold Chain Logistics Based on Internet of Things. In *IOP Conference Series: Earth and Environmental Science* (Vol. 237, No. 5, p. 052036). Bristol, UK: IOP Publishing 10.1088/1755-1315/237/5/052036

Wu, W. W., & Lee, Y. T. (2007). Developing global managers' competencies using the fuzzy DEMATEL method. *Expert Systems with Applications*, *32*(2), 499–507. doi:10.1016/j.eswa.2005.12.005

Yadav, A., Avasthe, R. K., & Dutta, S. K. (2018). Sikkim organic horticulture: Scope, challenges and prospects. *Progressive Horticulture, 50*(1, 2), 82-91.

Chapter 17
Employee Welfare Provisions:
An Administration Tool to Enhance Employee Efficiency

Chandra Sekhar Patro
https://orcid.org/0000-0002-8950-9289
Gayatri Vidya Parishad College of Engineering (Autonomous), India

ABSTRACT

Employee welfare is a prerequisite element for the success and growth of any form of organisation. The provision of welfare facilities improves the relations among the employees and the management of an organisation. These provisions boost the competence levels and value of the employees. The balance between employees'quality of life at the workplace and home is vital, as employees are the pillars of any organisation. The central aim of any organisation in adopting the welfare schemes is to secure the workforce by providing a proper work environment and minimising its hazardous effect on the employees' work life. The basic purpose of employee welfare is to enrich the lives of employees and to keep them happy and conducted. The provision of employees' welfare may be regarded as a wise investment as these would bring a profitable return in the form of greater efficiency. The chapter focuses on determining the various employee welfare provisions adopted by different private organisations and its influence on the employee's satisfaction and effectiveness.

INTRODUCTION

Welfare is a comprehensive term that refers to the physical mental, moral and emotional well-being of an individual (Aswathappa, 2010). Labour is every physical or intellectual activity applied in engineering production and the one who performs it as an employee. The concept of employee welfare is flexible and elastic. It extensively changes with times, regions, industry, country, social values, customs, beliefs, the degree of industrialization, social economic development of people and political ideologies prevailing at that moment. Coventry and Barker (1988) state that employee welfare includes provision of social club and sports facilities, supervising staff, canteens, running sick clubs, savings schemes, dealing with

DOI: 10.4018/978-1-7998-0357-7.ch017

superannuation, pension funds, leave grants, making loans on hardship cases, making long service grants, assistance to staff transferred to another area, providing fringe benefits, arranging legal aid and giving advice on personal problems.

Employee welfare has two negative aspects and positive aspects. On the negative side, employee welfare is concerned with counteracting the baneful effects of the large-scale industrial system of production especially capitalistic, and so far as India is concerned with the person or family, and social life of an employee. On its constructive part, welfare deals with the provision of opportunities to the employees and their families for a good life (Moorthy, 1968). Welfare measures should be provided by the organizations, whether public or private sector as it raises the morale of employees, reduce the risk and insecurity, eliminate turnover and absenteeism, and increase the production and productivity. Thus, improving the quality of working life by providing the employee welfare facilities would go a long way in achieving the goals of an organization (Patro, 2017a).

Employees have always been central to organizations, and their strategic importance is growing in today's knowledge-based industries. Employee welfare is an important fact of industrial relations, the extra dimension, giving satisfaction to the worker in a way that even a good wage cannot. Employee welfare is an important aspect of any industry. It acts as social security measures that contribute to improving the condition under which workers are employed. Employee welfare facilities in the organization affect the behavior of the employees as well as on the productivity of the organization. The management should provide good facilities to all employees in such a way that employees become satisfied and they work harder and more efficiently and effectively (Choudary, 2017). In the current scenario, the welfare schemes are provided to the employees to retain talented and skilled employees. Thus, employee welfare is very comprehensive and embraces activity provided by employers, state and central governments, trade unions and other agencies to help employees and their families to lead a contented work life. The welfare schemes change the way employees lead their quality life at the workplace as well as at home. The important question is whether the welfare schemes adopted by the companies have an impact on the employees' work life and their satisfaction (Patro, 2017b). Thus, employee welfare is a comprehensive term including various services, benefits, and facilities offered to employees by the employers. Through such generous fringe benefits, the employer makes life worth living for employees. The welfare amenities are extended by in addition to normal wages and other economic rewards available to the employees as per legal provisions. The welfare measures need not be monetary but in any kind/form. This includes items such as allowances, housing, transportation, medical insurance, and food. Employee welfare also includes monitoring of working conditions, the creation of industrial harmony through infrastructure for health, industrial relations and insurance against disease, accident, and unemployment for the workers and their families.

The chapter focuses on identifying the various welfare measures adopted by the companies for motivating the employees. Further, examines the influence of welfare provisions on employees' satisfaction and its impact on the effectiveness of the organization. Therefore, assortments of both statutory and non-statutory welfare provisions that can be provided by the companies according to the industrial acts are discussed for a better understanding.

BACKGROUND

Employee welfare provision is an imperative facet of organizational relations giving satisfaction to the worker in a means that even a good wage cannot. With the growth of industrialization and mechanization, welfare provisions have gained additional importance in the organization (Monappa, 1995). The report of the Committee on Labour welfare (GOI, 1969) includes the services such as adequate canteens, rest and recreation facilities, sanitary and medical facilities, arrangements for travel to and from work and for the accommodation of workers employed at a distance from their homes and such other services, amenities and facilities including social security measures as contribute to improve the conditions under which worker is employer. An employee cannot survive with the pace of modern life amid minimum available amenities. He needs added motivation to keep body and soul together at the workplace. The management has realized the importance of their role in providing these additional amenities to the employees (Patro, 2012). Many organizations have adopted welfare schemes as a strategy for improving the efficacy of employees since work-related troubles can lead to poor quality of work life and a decline in their performance (Manzini & Gwandure, 2011). Welfare facilities can be adapted to secure the workforce by providing appropriate human working conditions by minimizing the hazardous effect on the life of the employees and their family members. Employees' are the assets of any organization, so, the desires and necessities of the employees need to be satisfied to meet the goals of the organization (Patro, 2016a).

Any organization can operate effectively only when there is a high degree of coordination and cooperation between the employees and the management. The structure of a welfare state rests on its social security fabric. Government, employers, and trade unions have done a lot to promote the betterment of worker's conditions (Joseph et al., 2009). Patro (2017c) suggested that the existing welfare schemes may be improved further as they enrich the employee's standard of living and their satisfaction levels. New schemes and facilities should be added to existing ones to improve the efficiency and quality of work life of the employees by the management of all organizations. Therefore, welfare measures are considered as critical indicators that influence the performance of an employee. Satyanarayna & Reddi (2012) stated that the welfare measures which enrich the employees' standard of living and increase their satisfaction levels are to be necessarily provided by the organizations. Parul & Kumar (2013) concluded in terms of proving that the different welfare provisions provided to the employees working in an organization under the Factories Act, 1948 are having a positive relationship with employee satisfaction.

Patro (2015) in a comparative analysis of welfare measures in the public and private sector found that an employees' welfare facility is the key dimension to a smooth employer-employee relationship. The management provided all the health safety and welfares to the employees that helped to produce better performance in the work and working environment. The companies with strong employee relations initiatives will benefit because their workforce is highly motivated to expend their best efforts it involves providing fair and consistent treatment to all employees so that they will be committed to the organization (Gomez-Mejia et al., 2001). Jain (2016) study revealed that the employees were satisfied with existing welfare facilities like allowances, medical facilities, quality of work life, safety and security. Overtime allowances and canteen facilities were the two main areas where improvements are required. Culpepper (2009) found that implementing a pay for performance system has been shown to resolve organizational problems because it aligns the preferences of firms and employees. Also, creating pay for performance system serves as a sorting mechanism to identify and attract the most capable employees. This type of system has shown that individual pay incentives significantly improve productivity.

Nanda & Panda (2013) stated the Rourkela steel plant has adopted a better class of welfare activities that create an effective working environment and thus, better productivity. There are different kinds of welfare schemes like medical allowance, death relief funds, insurance, housing, transportation, and recreation club facilities, etc., are provided by the company to the employees to maintain better industrial relations. Neetha (2001) focused on the temporal changes in the labor process and the division of labor with a special emphasis on gender in the structure of the knitwear industry. Further, slowly the production relations are becoming more and more informal and the process has led not only to informalization of the workforce but also to the feminization and disorganization of the workforce. Logasakthi & Rajagopal (2013) found that the employees are not only satisfied with their jobs but also enjoyed the various facilities provided by the companies. The employees also extended their maximum support for the improvement of the company and also retain the employees for a long period (Patro & Raghunath, 2016). Sahoo & Sahoo (2018) identified that the variables interpersonal relationships, a safe and healthy work environment, and employee welfare were significant predictors of cordial employee relations.

Patel et al. (2017) found that there is no significant relationship between the satisfaction level of the employee having different age and designation level with welfare measures. There should be more emphasis on employees' welfare and it must be in a continuous manner, such measures will motivate employees and also increased their involvement in the job due to which Industrial growth will possible. Kumar and Yadav (2002) revealed that the overall satisfaction level of workers on welfare schemes was low in both the private and State sugar factories. The key behaviors which included belief in the organisation, desire to work to make things better, understanding of the business context and the 'bigger picture', being respectful of and helpful to colleagues, willingness to 'go the extra mile' and keeping up to date with developments in the field were found to be associated with employee engagement (Robinson et al., 2006). Ahmed (2018) stated that some welfare facilities are better in the public sector and some welfare facilities in the private sector. The canteen facility, latrine, and urinal facility, recreational facility, library facility, gymnasium facility, cultural activity, and training facility are better in the public sector compared to the private sector. On the other hand, restroom facility, drinking water facility, medical and first aid facility, washing facility, seating facility, transportation facility, housing facility, and educational facility are better in the private sector compared public sector organizations.

Motivating employees begins with recognizing that to do their best work, people must be in an environment that meets their basic emotional drives to acquire, bond, comprehend, and defend (Nohria et al., 2008). Sailesh (2012) stated that employee welfare measures increase the productivity of the organization and promote healthy industrial relations thereby maintaining industrial peace. Organizations provide welfare facilities to their employees to keep their motivation levels high. Business houses provide many such statutory and non-statutory things policies to maintain a satisfactory level of their employees. When they get better canteen facilities, good water to drink, clean restrooms, clean and hygienic washrooms, and bathrooms, regular medical checkups, health insurance, employee assistance program, grievance handling department, better facilities to sit or good workplace gives the employee a high level of satisfactory level. Kumar and Kumar (2018) study identified that the majority of the workers are satisfied by the various statutory welfare measures implemented by the organization. There is a relationship between Welfare measures implemented and workers satisfaction.

Patro (2019) stated that a satisfied employee is a key factor, who acts as the organization's ladder for success. Thus, good employee relations help in developing a satisfied, committed and productive workforce that leads towards the overall effectiveness of an organization.

NEED FOR EMPLOYEE WELFARE PROVISIONS

The Indian companies have greater need and importance of adopting employee welfare provision than any other country. Certain deficiencies in the Indian workforce are not found in other countries. The reason why employees' rights and welfare are very significant is that employees need to work in places where the environment is devoid of tension and rancor to be productive. The welfare provisions should also be provided based on the feeling of care and compassion to the employees considering that some of them will spend the whole of their life working for the organization (Hartman et al., 2004). The conducive, safe and healthy work environment will enable workers to give their best and perform effectively and efficiently (Jaskiewicz & Tulenko, 2012). While some organizations are taking the responsibility to provide welfare facilities and services, others are avoiding them. The perception of those companies not providing welfare facilities for their employees is that it is an unnecessary burden to them. The provision of welfare facilities is however mandatory under most company's law and must be statutorily provided.

Even today in some economies, the employees are not united into a class and the employee union associations are still in its infancy. If the employees' union does exist, there is a lack of enlightened leadership and there is no unity between these unions themselves. In the absence of well-organized employee unions, the workers can place their demands effectively neither before the employer nor can systematically think of their interests. Due to this lacking force, it is essential to focus on the employee welfare provisions by the employers and the Government. Due to the inadequate level of wages, the employees cannot comfortably settle down in one place or region, which may, in turn, give rise to absenteeism. The employees' migratory tendency can be curbed by providing them with adequate facilities and conditions, and this also shows a positive effect over the absenteeism in the organization. Healthy recreation should also be given priority as the efficiency of the workforce is diminished by energy being employed in all kinds of undesirable channels. Hence, it is essential that the recreation and enjoyment of a healthy kind be provided.

As the welfare of an employee at the workplace will not only enhance their work efficiency but also improve the willingness to perform their job involving themselves within the work environment. The welfare provisions help to maintain health and skills, avoids exploitation of the worker by the employer, increases the happiness of workers, and motivates the workers at the workplace. In India, most of the companies are not adopting the welfare measures properly at their workplace. The companies compel workers to work in uncomfortable/ adverse conditions. Their uncomfortable conditions lead to serious damages to the employee. Therefore, maintenance of employee welfare previsions is a must in the identity premises.

CATEGORIES OF EMPLOYEE WELFARE PROVISIONS

Modern and civilized societies contemplate workers as human beings entitled to the protection of the law and government (Steiner, Alston, & Goodman, 2008). The employees are deserved to be treated with dignity and respect irrespective of their position in the workplace. Employees are not slaves but are free born using their skills to contribute to the growth and development of an organization (Drescher, 2004). Therefore, they deserve ample welfare facilities and services which should be provided by the employers in addition to wages and salaries. The committee of experts on welfare facilities for employees categorized welfare measures (International Labour Organisation,1963) as statutory measures (intramural facilities) and non-statutory measures (extramural facilities), and mutual facilities. It is very difficult to classify the welfare activities into certain broad categories. Some of the facilities under statutory and non-statutory welfare measures are shown in Table 1.

Statutory Welfare Measures

Under this category, welfare facilities are provided according to the labor legislation passed by the Government. The nature and coverage of these facilities vary from country to country. These facilities must be provided by all the employers and cannot be ignored. Any contravention of the statutory provisions shall render the employer punishable under the Act concerned. According to the National Commission of Labour, the statutory measures are divided under two distinct heads:

- Facilities have to be provided irrespective of the size of the establishment.
- Facilities are to be provided subject to the employment of a specified number of persons.

The provisions provided inside the premises of the establishments are known as intramural facilities. These facilities include activities relating to minimization of industrial fatigue, provision of safety measures like fencing and covering of machines, good layout of the plant and machinery, sufficient lighting conditions,provision of first aid appliances, sanitary facilities, crèches, rest shelters, canteens, drinking water, prevention of fatigue, health services including occupational safety, administrative arrangements, uniforms and protective clothing, shift allowance, etc. Provisions of such facilities are also obligatory in all industrial establishments all over the world.

Non-Statutory Welfare Measures

Under this category, it is not mandatory for the companies to provide the welfare facilities to the employees. The provision of these facilities is voluntary. Earlier, due attention was not given to the provision of extramural facilities to the workers but now it is realized that these facilities are very important for the general welfare and upliftment of the employees. The welfare provisions offered to the workers outside the establishment are known as extramural facilities. They include better housing accommodations, educational facilities, maternity benefit, social insurance measures, sports, and cultural activities, library or reading room, leave travel facilities, workers co-operative credit societies, welfare programs for women, youth and children and transport facilities, etc. No doubt, the activities under this category ultimately lead to an increase in the efficiency of workers.

Mutual Welfare Facilities

These facilities are usually outside the scope of the statutory facilities. These activities are voluntarily undertaken by the workers themselves for their interests. As such the employer has no say in it.

Table 1. Categories of welfare provisions

Statutory Facilities	Non-Statutory Facilities
Drinking Water Canteens Toilets Crèche Facilities for storing and drying clothing Facilities for sitting First-aid appliances Shelters, restrooms, and lunch rooms Washing facilities Occupational Safety Uniform and protective clothing Shift Allowance Welfare officers	Economic Services Recreational Services Facilitative Services Housing Facilities Education Facilities Leave Travel Concession Social Insurance (gratuity, pension, PF) Benevolent Fund Maternity Benefits Health and Medical Facilities Vocational Training Workers Cooperatives

FACETS OF WELFARE PROVISIONS

Employees play an essential role in the development of any organization. Therefore, the management of the companies has to adopt welfare measures to get benefited in accomplishing the objectives. The benefits of employee welfare measures are more direct to the employees than the organization. However, it is important for the company to generate various employee welfare schemes to retain the employees (Patro, 2014). The different welfare facets identified are as shown in Figure 1.

1. **Education Schemes:** Education schemes play a very important role in motivating and enabling the progress of employees for their mental and physical development. The educational needs in any organization are two-fold i.e., adult/employee education schemes and schools/colleges for their

Figure 1. Welfare facets

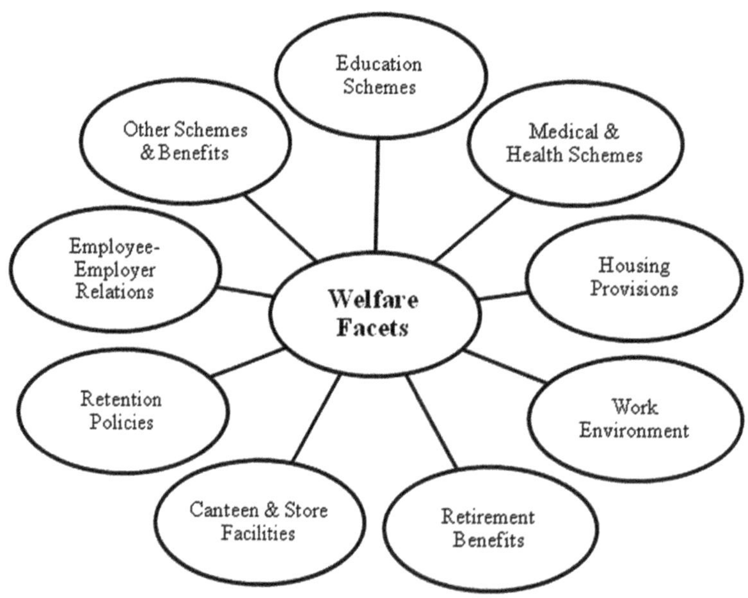

children. The educational benefits that can be provided by the private companies include, sponsoring for employees' higher education or permission for part-time education, reimbursement of children tuition fees, schools or colleges in the organization premises, and training & development programs (Patro, 2017d).

2. **Medical and Health Schemes:** The employees' health should be given utmost importance, as it also results in the development of an organization. Health care for employees helps to reduce the incidence of sickness, absenteeism and increases their efficiency and productivity (Patro & Raghunath, 2018). The organizations can provide medical facilities to the employees and their family members also. The management can offer facilities like medical fees reimbursement, maternity leaves or benefits, provision of emergency services, and frequent medical checkups, to the staff members. These facilities create a healthy work-life for the employees.
3. **Housing Provisions:** Housing facility has a direct bearing on the health and efficiency of the employees. The provision of quarter facilities at low cost with basic amenities such as adequate space, ventilation, electricity facilities, water facility, transport facility, maintenance facilities (roads, lighting, drainage systems, lavatories), and other sanitary arrangements. The organizations may also provide a housing allowance to the employees (Patro, 2017c). These facilities enhance the quality of life of the employees.
4. **Work Environment:** A positive work environment makes employees feel good about coming to work, and this provides the motivation to sustain them in the organization. The management should provide information and resources to assist their employees to make healthy lifestyle choices and to achieve and maintain good health. It promotes work-life balance by creating a healthy physical, social and psychological work environment as a core business goal (Rao et al., 2015). A healthy work environment makes the employees' more satisfied and improves their work efficiency. The companies can provide good environments such as a safe and secure environment, autonomy at work, restroom and locker facilities, workload and Job stress.
5. **Retirement Benefits:** The retirement benefit is one the important welfare scheme, which is to be provided to the employees in any organization. These benefits increase the satisfaction level of the employees towards the organizations and motivate them to achieve the organizational objectives. The companies can attempt to provide benefits such as a pension, gratuity and provident fund benefits, retirement or death compensation, and other post-retirement benefits (Patro, 2017c). These facilities will make the staff members motivated towards the management and in increased efficacy.
6. **Canteen and Store Facilities:** Cafeteria or canteens shall be provided by the employer so, as to provide hygienic and nutritious food to the employees. It is a method of creating more satisfactory working and living conditions for the employees working in an organization. This welfare measure would greatly reduce turnover and absenteeism among workers and improves workers efficiency to a greater extent and also create a permanently settled workforce by making service attractive to the employee (Patro, 2017c). This facet includes the facilities like the subsidized cost of food items and stores, precise and clean drinking water, maintenance of cleanliness and quality of food items that can be made available to the staff members.
7. **Retention Policies:** Employee retention has become the most critical issue for all types of organizations as a result of the shortage of skilled labor, economic growth, and employee turnover. Retention of employees is an important element in providing welfare schemes. Employee retention involves taking measures to encourage employees to remain in the organization for a maximum period. The

retention policy helps the management of the organization in reducing the cost of turnover, loss of company knowledge, interruption of work, regaining the efficiency and increases the productivity of the organization (Patro, 2016b). This policy includes better compensation packages, trust on employees' capability, provide growth opportunities, promotion and incentive Policies.

8. **Employee-Employer Relation:** There is a common saying that 'Employees leave their bosses, and not their jobs', which is true because the relationship with the management and the peers becomes often the reason for an employee to leave the organization. The management is sometimes not able to provide an employee with a supportive work culture and environment in terms of personal or professional relationships (Patro, 2019). The facet includes the factors like transparent and open communication, cooperation by management or superior, recognition and appreciation of hard work, grievance settlement forums.
9. **Other Schemes and Benefits:** The other welfare schemes or benefits must be provided to the employees, which in turn will help them to increase their satisfaction and quality of work life. These facilities motivate the employees towards management and increase their work efficiency. The other benefits may include recreation facilities, post office, and bank facilities, cooperative or credit societies, and group insurance schemes. These facilities also enhance the employees' satisfaction and lead quality work life.

FRAMEWORK AND HYPOTHESES

The welfare provisions change the way employees lead their work life in the companies and as well as at home. The question is would the welfare provisions have an impact on the employees' work life. Therefore, an inquiry into the employees' response to the various welfare provisions adopted in the companies has become essential. The various statutory and non-statutory welfare provisions adopted by the private companies according to the Industrial Acts are discussed in the framework as shown in Figure 2.

For the purpose the following hypotheses are framed:

Figure 2. Welfare provisions framework

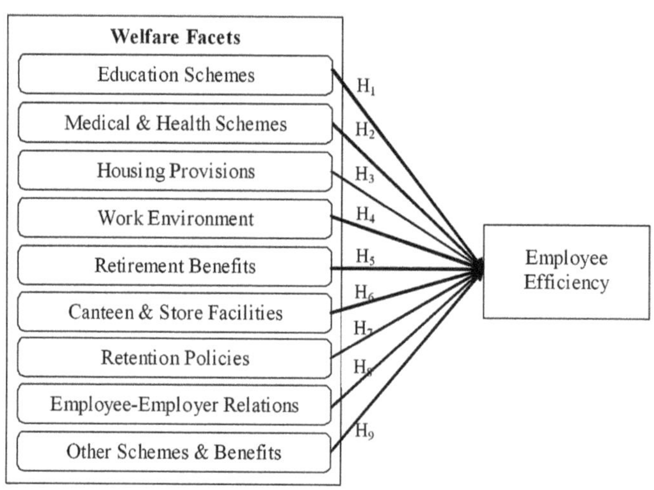

H_1: Education schemes do not show a significant effect on employees' efficiency.
H_2: Medical & Health schemes do not show a significant effect on employees' efficiency.
H_3: Housing provisions do not show a significant effect on employees' efficiency.
H_4: Work environment does not show a significant effect on employees' efficiency.
H_5: Retirement benefits do not show a significant effect on employees' efficiency.
H_6: Canteen & store facilities do not show a significant effect on employees' efficiency.
H_7: Retention policies do not show a significant effect on employees' efficiency.
H_8: Employee-employer relations do not show a significant effect on employees' efficiency.
H_9: Other schemes & benefits do not show a significant effect on employees' efficiency.

METHODOLOGY

The chapter enquires into the welfare provisions adopted by different private sector companies and the satisfaction experienced by the employees working in these organizations. The identified welfare facets having an impact on the employees' satisfaction at the workplace and their work life are examined. The required data were collected from both primary and secondary sources. The secondary data was collected from books, journals, magazines, unpublished thesis and institution websites to know the various welfare provisions adopted in the private sector companies. The primary data resources are collected using a survey method from the employees working in different private companies of Visakhapatnam city, India. The survey was conducted for a period of one month during March 2016. To get the required data, a structured questionnaire was designed and distributed to the employees through emails and some respondents are approached personally. The purpose of the survey was explained to the respondents and was assured of the confidentiality of their responses. A total of 290 valid responses were considered for statistical data analysis.

To measure the satisfaction level of the employees, a 5-point Likert rating scale technique was used ranging from Highly Satisfied (5) to Highly Dissatisfied (1). For evaluating the responses inferential statistics are used to make a conclusion based on the results and significance. Descriptive analytical tools such as mean, standard deviation, standard error, and t-values were used to summarize the respondents' opinions. The nine facets were tested for their relationship with employee's satisfaction using correlation and the hypotheses are tested using ANOVA analysis.

STATISTICAL DATA ANALYSIS

To examine the employees' perception towards the welfare provisions adopted by the companies, nine imperative welfare facets like housing schemes, education schemes, medical schemes, work environment, retirement schemes, canteen facilities, retention schemes, employee-employer relation, and other schemes are identified. The descriptive statistic results regarding the employees' satisfaction towards welfare provisions are shown in Table 2. The respondents are 59 percent (172) male and 41 percent (118) female. The mean scores, standard deviation, t-values, factor loadings, average variance extracted, composite reliability and Cronbach's' alpha are computed as shown in the table. The Cronbach's' alpha value reveals that all the variables are having internal consistency as they are above 0.90 falling in the category of superb. The composite reliability values also show the validity of the variables is above 0.80.

The statistics show that the overall mean values are in the range of 2.20 – 3.41 which indicates that the satisfaction level of the employees towards welfare provisions is average. The standard deviation values range from 0.729 – 1.113. In the case of education schemes, training & development programs (3.23) got the highest rating followed by the provisions sponsoring for employees' higher education (2.21), reimbursement of children tuition fees (1.70), and schools or colleges in organization premises (1.68). Regarding medical & health schemes the provisions frequent medical checkups (3.08) secured the highest rating from the respondents followed by the provision of emergency services (2.96), maternity leaves or benefits (2.51), and reimbursement of medical expenses (1.73). Concerning housing provisions, transport facility or allowances (3.40) got higher rating followed by the provisions housing allowances (2.86), hygiene or maintenance services (2.76), and provision of staff quarters (2.03) by the respondents. In the case of the work environment, restroom and locker facilities (3.54) scored the highest rating from the respondents followed by the safe and secure environment (3.34), autonomy at work (3.10), and workload and job stress (2.86).

The respondent's opinion on retirement benefits reveals that the provisions gratuity & provident fund benefits (3.15) secured the highest rating from the respondents followed by retirement or death compensation (2.01), post-retirement benefits (1.91), and pension benefits (1.71). In the case of canteen & store facilities the provisions precise and clean drinking water (3.85) got the highest rating followed by quality of food items (3.50), store facilities (2.83), and subsidized cost of food items (2.79) by the respondents. About retention policies, promotion and incentive policies (2.99) secured the highest rating by the respondents followed by the provisions trust on employees' capability (2.83), better compensation packages (2.76), and providing growth opportunities (2.28). Regarding employee-employer relations, cooperation by management or superior (2.59) got a higher rating from the respondents followed by the provisions recognition and appreciation of hard work (2.53), transparent and open communication (2.29), and grievance settling forums (2.20). In the case of other schemes & benefits, post office and bank faculties (2.76) secured the highest rating from the respondents followed by the provisions group insurance schemes (2.71), recreation facilities (2.51), and cooperative/credit societies (2.36).

The respondents' opinion on the employees' efficiency reveals that welfare provision helps in overall growth of employees (3.96) got the highest rating followed by the variables welfare provisions to increase the competencies of employees(3.86), welfare provisions are effectively adopted by the company (3.15), and satisfied with the welfare provisions of the company (2.65).

AVE-Average Variance Extracted, CR-Composite Reliability, CA-Cronbachs' Alpha

From Table 2, it can be observed that all the t-values are significant at ($p < 0.01$) significance level (2-tailed). The analysis show that the respondents are highly satisfied with the canteen & store facilities (3.24) followed by the provisions work environment (3.21), housing provisions (2.76), retention policies (2.72), other schemes & benefits (2.59), medical & health schemes (2.57), employee-employer relations (2.40). The low satisfied welfare provisions are education schemes (2.21) and retirement schemes (2.20). The respondents rating towards the employees' efficiency 3.41 shows a positive rating towards the welfare provisions adopted by the private companies. The factor loadings ranged from 0.633 – 0.894

Table 2. Summary of responses on employees' efficiency (n=290)

Welfare Facets		Mean	Std. Deviation	t	Factor Loadings	AVE	CR	CA
Education Schemes		***2.21***	***0.738***	***25.64***		***0.644***	***0.878***	***0.92***
	Sponsoring for employees' higher education	2.21	1.027	19.26	.786			
	Reimbursement of children tuition fees	1.70	0.604	25.18	.825			
	Schools or Colleges in organization premises	1.68	0.497	30.12	.888			
	Training & Development Programmes	3.23	1.031	27.98	.699			
Medical & Health Schemes		***2.57***	***0.958***	***23.26***		***0.549***	***0.829***	***0.95***
	Reimbursement of medical expenses	1.73	0.616	25.06	.796			
	Maternity leaves or benefits	2.51	1.114	20.18	.766			
	Provision of Emergency services	2.96	1.163	22.78	.686			
	Frequent medical checkups	3.08	1.100	25.01	.711			
Housing Provisions		***2.76***	***0.893***	***27.79***		***0.526***	***0.816***	***0.97***
	Provision of Staff Quarters	2.03	0.856	21.15	.747			
	Housing Allowances	2.86	0.990	25.85	.703			
	Hygiene or Maintenance services	2.76	0.889	27.78	.714			
	Transport facility or allowances	3.40	0.836	36.38	.737			
Work Environment		***3.21***	***0.859***	***34.65***		***0.576***	***0.844***	***0.96***
	Safe and secure environment	3.34	0.913	32.69	.726			
	Autonomy at work	3.10	0.880	31.50	.786			
	Restroom and locker facilities	3.54	0.680	47.87	.819			
	Workload and Job stress	2.86	0.964	26.55	.698			
Retirement Benefits		***2.20***	***0.729***	***27.08***		***0.531***	***0.817***	***0.94***
	Pension Benefits	1.71	0.508	30.15	.851			
	Gratuity & Provident Fund Benefits	3.15	0.873	32.28	.656			
	Retirement or Death compensation	2.01	0.819	21.98	.736			
	Post retirement benefits	1.91	0.715	23.92	.654			
Canteen & Store Facilities		***3.24***	***0.889***	***34.95***		***0.582***	***0.846***	***0.93***
	The subsidized cost of food items	2.79	0.910	27.40	.633			
	Precise and Clean Drinking water	3.85	0.618	55.69	.762			
	Quality of food items	3.50	0.928	33.74	.829			
	Store facilities	2.83	1.100	22.98	.812			
Retention Policies		***2.72***	***1.053***	***23.20***		***0.571***	***0.841***	***0.96***
	Better compensation packages	2.76	1.245	19.84	.769			
	Trust on employees' capability	2.83	0.965	26.19	.677			
	Providing growth opportunities	2.28	0.954	21.32	.864			
	Promotion and Incentive Policies	2.99	1.049	25.46	.699			
Employee-Employer Relations		***2.40***	***1.013***	***21.32***		***0.741***	***0.920***	***0.98***
	Transparent and Open communication	2.29	0.903	22.66	.876			

continued on following page

Table 2. Continued

Welfare Facets		Mean	Std. Deviation	t	Factor Loadings	AVE	CR	CA
	Cooperation by Management or Superior	2.59	1.110	20.84	.877			
	Recognition and Appreciation of hard work	2.53	1.147	19.69	.839			
	Grievance settling forums	2.20	0.892	22.07	.851			
Other Schemes & Benefits		***2.59***	***1.130***	***20.59***		***0.709***	***0.907***	***0.98***
	Recreation facilities	2.51	1.091	20.60	.894			
	Post Office and Bank faculties	2.76	1.265	19.53	.777			
	Cooperative/Credit Societies	2.36	1.128	18.74	.838			
	Group Insurance Schemes	2.71	1.034	23.47	.856			
Employee Efficiency		***3.41***	***1.113***	***23.13***		***0.517***	***0.810***	***0.91***
	Welfare provisions are effectively adopted by the company	3.15	1.008	27.96	.664			
	Satisfied with the welfare provisions of the company	2.65	1.170	20.25	.676			
	Welfare provisions increase the competencies of employees	3.86	1.386	16.54	.812			
	Welfare provision helps in the overall growth of employees	3.96	0.889	27.78	.714			
Note: All the t-values are significant at the 0.01 level (2-tailed)								

reveal that all the provisions have a positive influence on the employees' satisfaction and efficiency. The average variance extracted (AVE) values shows that employee-employer relations (0.741) got the highest variance whereas, employee efficiency recorded (0.517) lowest variance.

Table 3. Summary of one-sample t-test

Welfare Facets	t	df	Sig. (2-tailed)	Mean Difference	95% Confidence Interval of the Difference	
					Lower	Upper
Education Schemes	26.706	289	.000	2.203	2.04	2.37
Medical & Health Schemes	23.984	289	.000	2.569	2.36	2.78
Housing Provisions	28.997	289	.000	2.763	2.57	2.95
Work Environment	35.472	289	.000	3.234	3.05	3.42
Retirement Benefits	28.727	289	.000	2.197	2.04	2.35
Canteen & Store Facilities	35.203	289	.000	3.241	3.06	3.42
Retention Policies	24.298	289	.000	2.713	2.49	2.93
Employee-Employer Relation	21.839	289	.000	2.400	2.18	2.62
Other Schemes & Benefits	21.131	289	.000	2.588	2.34	2.83
Employee Efficiency	33.851	289	.000	3.038	2.86	3.22

Note: All the t-values are significant at the 0.01 level (2-tailed)

Table 4. Inter-item correlation matrix

Welfare Facets	1	2	3	4	5	6	7	8	9	10
Education Schemes	1.000									
Medical & Health Schemes	.982	1.000								
Housing Provisions	.967	.974	1.000							
Work Environment	.952	.959	.968	1.000						
Retirement Benefits	.972	.958	.950	.922	1.000					
Canteen & Store Facilities	.956	.966	.968	.965	.926	1.000				
Retention Policies	.962	.974	.980	.962	.939	.979	1.000			
Employee-Employer Relation	.914	.928	.943	.910	.905	.944	.969	1.000		
Other Schemes& Benefits	.930	.951	.952	.919	.914	.956	.981	.975	1.000	
Employee Efficiency	.961	.966	.943	.955	.924	.954	.944	.867	.895	1.000
All Correlation values are significant at the 0.01 level (2-tailed)										

The summary of the t-test indicates that all the welfare facets have a significant relationship with the employees' satisfaction and efficiency. The t-values with the significance levels (2-tailed), mean differences, and a 99 percent confidence interval of difference at the upper and lower level are shown in Table 3. The analysis show that the highly influential welfare facet is canteen & store facilities (35.203) followed by the welfare facets work environment (35.472), employee efficiency (33.851), housing provisions (28.997), retirement benefits (28.727), education schemes (26.706), retention policies (24.298), medical & health schemes (23.984), employee-employer relation (21.839), and other schemes & benefits (21.131).

The inter-item correlation between the welfare facets was also examined and found that all the paired correlation measures are significant at $p< 0.01$ level (2-tailed), as shown in Table 4. Therefore, the validity of the research has been achieved to a satisfactory level.

Table 5. Model summary

Model	R	R Square	Adjusted R Square	Std. Error of the Estimate
1	.984[a]	.969	.965	.18663
a. Predictors: (Constant), education schemes, medical & health schemes, housing provisions, work environment, retirement benefits, canteen & store facilities, retention policies, employee-employer relation, other schemes & benefits				

Table 6. ANOVAa

Model	Sum of Squares	df	Mean Square	F	Sig.
Regression	76.234	9	8.470	243.193	.000b
Residual	2.438	280	.035		
Total	78.672	289			
a. Dependent Variable: Employee Efficiency					
b. Predictors: (Constant), education schemes, medical & health schemes, housing provisions, work environment, retirement benefits, canteen & store facilities, retention policies, employee-employer relation, other schemes & benefits					

Table 7. Coefficients[a]

Model	Unstandardized Coefficients		Standardized Coefficients	t	Sig.
	B	Std. Error	Beta		
(Constant)	.563	.175		3.222	.002
Education Schemes	.290	.193	.214	1.499	.038
Medical & Health Schemes	.569	.151	.546	3.772	.000
Housing Provisions	.271	.154	.231	1.760	.028
Work Environment	.157	.129	.129	1.222	.026
Retirement Benefits	.037	.138	.025	0.268	.038
Canteen & Store Facilities	.457	.142	.377	3.233	.002
Retention Policies	.691	.216	.691	3.204	.002
Employee-Employer Relation	.382	.107	.376	3.555	.001
Other Schemes & Benefits	.338	.131	.371	2.584	.012
a. Dependent Variable: Employee Efficiency					

Note: All the t-values are significant at 0.01 and 0.05 level

RESULTS AND DISCUSSION

The statistical significance of all welfare facets was examined to determine the validity of the hypothesized paths. The results of the model summary, ANOVA, and regression coefficient values for the relationship between employee welfare facets and employee efficiency are discussed.

The model summary Table 5 reveals that the linear regression coefficient (R=0.984) indicates that there is a maximum correlation between the dependent and independent variables. In terms of variability R-Square (0.969) shows that the independent variables education schemes, medical & health schemes, housing provisions, work environment, retirement benefits, canteen & store facilities, retention policies, employee-employer relation, and other schemes & benefits can predict 97 percent of the variance in the factor employee efficiency.

The results of the ANOVA test shown in Table 6 indicates that the dependent variable employee efficiency (F=243.193, p=0.000 < 0.01) show a significant relationship with the independent variables education schemes, medical & health schemes, housing provisions, work environment, retirement benefits, canteen & store facilities, retention policies, employee-employer relation, and other schemes & benefits.

The coefficient results shown in Table 7 reveal that the education schemes have a significant positive relationship with the employee's efficiency (β=0.214; p=0.038 < 0.05) providing strong support for the rejection of the hypothesis H_1. Medical & health schemes show a significantly positive influence on employee efficiency (β=0.546; p=000 < 0.01), not supporting the null hypothesis H_2. The housing provisions have a positive significant relationship with the employees' efficiency (β= 0.231; p=0.028 < 0.05) rejecting the null hypothesis H_3. Work environment shows a strong positive significant impact on employees' efficiency (β= 0.129; p=0.026 < 0.05) thus, H_4 is also not supported. Moreover, the retirement benefits are having a significant positive impact on employee's satisfaction and efficiency (β= 0.025; p=0.038 < 0.05), hence H_5 is not supported. The canteen & store facilities also have a strong positive significant relationship with the employee's efficiency (β= 0.377; p=0.002 < 0.05), rejecting the null hypothesis H_6. The retention policies adopted by the management also significantly influencing the employees' efficiency ($\beta = 0.691$; $p=0.002 < 0.05$), hence H_7 is not supported. Employee and employer relations were found significantly influencing an employee's efficiency ($\beta = 0.376$; $p=0.001 < 0.05$), so the hypothesis H_8 is not supported. Finally, other schemes & benefits ($\beta = 0.371$; $p=0.012 < 0.05$) also show a significant positive relationship with employees' satisfaction and efficiency, hence not supporting the hypothesis H_9. Thus, the regression analysis results provide a strong support for the rejection of the null hypothesis relating to the relationship between independent variables education schemes, medical & health schemes, housing provisions, work environment, retirement benefits, canteen & store facilities, retention policies, employee-employer relation, other schemes & benefits with the dependent variable employee efficiency.

The welfare provisions such as housing schemes, education benefits, retirement benefits, medical benefits, retention policies, and the working environment have a strong significant influence on the employees' efficiency. Therefore, the welfare provisions are considered as a critical factor in influencing the work life of the employees. Proper implementation of both statutory and non-statutory welfare measures increases the employees' work efficiency and therefore, helpful in the overall development of the organization.

ADOPTION OF WELFARE PROVISIONS

The theories of labor welfare have evolved over the years. Before, the government had to compel the industrial organizations to provide basic amenities to their employees (Erasmus, 2008). Such compulsion was necessary because the employers used to exploit employees and treated them unfairly. The basic objective of employee welfare is to enable workers to live a richer and more satisfactory life. Adopting the welfare facilities depend on the management of an organization. It is based on the approach of the organization on how best it has organized the schemes to suit and benefit the workers (Cotton et al., 2005). While statutory welfare facilities are compulsory, non-statutory welfare schemes need to be shaped and driven by a very sound company policy that allows employees to have substantial input into what and how the facilities should be provided without conflicting or undermining the organization stance and focus.

The employer and employee need to work together to make sure that the facilities are provided based on mutual respects and understanding. Welfare provisions are desirable for employees but they cannot be substituted for wages. Workers have the inherent rights to living wages as a result of the labor they put into economic and productive aspects of an organization. However, it is pertinent to mention that living wages are not enough and cannot create a healthy work environment (Verdon, 2002). Wages are earned and usually used for social economics activities outside the workplace but welfare is provided for by the organization for a healthy and sustainable environment in the workplace (Portney, 2005). A combination of adequate wages with ample welfare provisions will intensely yield and achieve good results for the organization.

Statutory welfare facilities are products of the laws regulating the provision and implementation of welfare facilities in the workplace (Ridley & Channing, 2008). They are obligatory and imperative. Therefore, there are consequences for failure to provide welfare facilities as stipulated in the statutes. The health and safety of the workers are paramount and the necessary safety gadgets must be provided for the workers in addition to other important facilities and services. If not provided and there is a mishap or accident in the workplace, the victim may institute legal action against the erring organization in courts and punitive damages awarded against the organization if found wanting. Nowadays, an organization is not only considered successful as a result of the profits it has declared but also based on the welfare benefits given to employees that were used to make the profits (Schaltegger & Lüdeke-Freund, 2013). The organizational policy now contains numerous welfare facilities and how they are to be provided for the benefits of the employees. The company's annual reports and financial statements now reflect the amount of money being spent on various welfare schemes to the employees. Welfare schemes are now statutorily enforceable and the company has a corporate social responsibility to provide for them.

The successful adoption of employee welfare provisions depends on the approach which has been taken into account in providing such activities to the employees. Employee welfare is dependent on certain factors that influence the successful implementation of welfare programs.

1. Employee welfare measures cannot be a substitute for salaries and have a right to get adequate compensation.
2. A combination of social welfare, emotional welfare, and economic welfare together would achieve good quality results and make him lead a healthy work life.
3. Programs for housing, education, and training, provision of balanced diet and family planning measures are some of the important facets of welfare which increases the efficacy of the employees.
4. The development of human personality should be the goal of organizations. Therefore, it is necessary to implement intramural and extramural employee welfare services both inside and outside the company.
5. Co-operation and agreement of employees in the formulation and adoption of welfare services are essential for their growth.
6. Employees develop a sense of pride when they are made to feel that welfare programs are designed and adapted for them.
7. The responsibility of welfare provisions should be shared by different groups such as employers, employees and also trade unions, to implement the welfare programs more effectively and healthily.
8. Assessment or evaluation of existing welfare provisions is necessary on a periodical basis so that one can judge and analyze the success of employee welfare programs.
9. Identify the employees' problems and find out the help required to solve the issues, and provide necessary resources in planning welfare programs. Timely action in the proper direction is essential in any kind of social work.
10. Finally, the fact that employee welfare must aim at helping the employee to motive them and increase efficiency in the long run.

Every effort should be made to give employees some voice in the choice of welfare activities so long as it does not amount to dictation from employees. This helps the employees to become more responsible and lead a healthy quality work life.

CONCLUSION AND MANAGERIAL IMPLICATIONS

The concept of employee welfare has been used by many organizations as a strategy of improving the efficiency of employees, especially in private companies since work-related problems can lead to poor quality of life for employees and a decline in performance. Welfare provisions are essential for the higher efficiency of the employees and their satisfaction. Without getting effective support of welfare provisions from the management, the employees may not be able to balance between work life and family life, as a result, it may lead to low productivity of the organization. The employees are satisfied with the welfare provisions adopted in the private sector organizations but are not effectively implemented. The management should take necessary steps to solve the problems in effectively adopting the welfare measures so that the employees can do their job more interestingly.

The objectives of the chapter have been achieved after going through the relevant analysis. It is encouraged to implement similar welfare provisions as it is useful in motivating the employees in increasing their work efficiency. The facets of employee welfare provisions especially education, retirement benefits, and other facilities are indicative of influencing the employees to work efficiently in private sector organizations. The companies need to take necessary efforts to find out the employees requirements and adopt those welfare provisions for the successful development of the organization. New schemes and facilities should be added to existing ones to improve the efficiency and quality of work life of the employees by the management of all organizations.

The importance of welfare schemes and services are increasingly being agitated for by employees and organized labor. Welfare measures are recognized by international labor organizations and they all enjoin each member state to the organization to ensure its proper implementation. Employees deserve more than only salaries or wages, hence the argument for the provision and improvement of existing welfare facilities in the workplace. The welfare provisions benefit employees as well as serves as an impetus for efficiency and effectiveness in the chain of productive activities in the workplace. The companies will, however, benefit more because efficiency and effective production will lead to huge output which would invariably impact on the profit and margins of the organization and make it a perpetually sustainable venture.

REFERENCES

Ahmed, N. (2018). Workers' Welfare: A Comparative Study Between Public and Private Industries in Bareilly, Uttar Pradesh. *Management and Labour Studies*, *43*(3), 192–204. doi:10.1177/0258042X18768307

Aswathappa, K. (2010). *Human resources management*. New Delhi, India: Tata McGraw Hill Publishing Company Limited.

Choudary, S. (2017). Employee welfare: A scheme of a wise investment. *International Journal of Advanced Education and Research, 2*(1), 01-06.

Cotton, A. P., Sohail, M., & Scott, R. E. (2005). Towards improved labour standards for construction of minor works in low income countries. *Engineering, Construction, and Architectural Management, 12*(6), 617–632. doi:10.1108/09699980510634164

Coventry, W. F., & Barker, J. K. (1988). *Management* (International Edition). Portsmouth, NH: Heinemann Professional Publishing.

Culpepper, W. L. (2009). *Incentive pay eligibility varies by job and location*. Retrieved from http://www.shrm.org/hrdisciplines/compensation/Articles/Pages/IncentiveEligibility.aspx

Drescher, S. (2004). *The mighty experiment: Free labor versus slavery in British emancipation*. Oxford, UK: Oxford University Press.

Erasmus, B., & Schenk, H. (2008). *South African human resource management: Theory & practice*. Cape Town, South Africa: Juta and Company Ltd.

Gomez-Mejia, L. R., Balkin, D. B., & Cardy, R. L. (2001). *Managing Human Resources* (3rd ed.). Upper Saddle River, NJ: Prentice Hall.

Government of India. (1969). *Report of the Committee on Labour Welfare*. India: Ministry of Labour& Employment.

Hartman, L. P., Shaw, B., & Stevenson, R. (2000). Human Resources Opportunities to Balance Ethics and Neoclassical Economics in Global Labor Standards. *Business & Professional Ethics Journal, 19*(3/4), 73–116. doi:10.5840/bpej2000193/42

International Labour Organisation. (1963). *Asian Regional Conference*. Labour laws and legislation, 5-6.

Jain, S. (2016). Study of employee welfare & benefit practices at Indian oil corporation limited (Lubes Plant, Vashi, Navi Mumbai). *NCRD's Business Review: e-Journal, 2*(2), 1-10.

Jaskiewicz, W., & Tulenko, K. (2012). Increasing community health worker productivity and effectiveness: A review of the influence of the work environment. *Human Resources for Health, 10*(1), 38. doi:10.1186/1478-4491-10-38 PMID:23017131

Joseph, B., Joseph, I., & Varghese, R. (2009). Labour Welfare in India. *Journal of Workplace Behavioral Health, 24*(1&2), 221–242. doi:10.1080/15555240902849131

Kumar, G. S. A., & Kumar, K. A. (2018). A Study on Labour Welfare Measures in Singareni Collieries Company Limited. *International Journal of Engineering Technology Science and Research, 5*(3), 1376–1382.

Kumar, S., & Yadav, S. S. (2002). Satisfaction level from labour welfare schemes in sugar factories of Gorakhpur division. *Indian Journal of Economics, 33*(329), 171–188.

Logasakthi, K., & Rajagopal, K. (2013). A study on employee health, safety and welfare measures of chemical industry in the view of Sleam region, Tamil Nadu, India. *International Journal of Research in Business Management, 1*(1), 1–10.

Manzini, H., & Gwandure, C. (2011). *The Provision of Employee Assistance Programmes in South Africa Football Clubs*. Johannesburg, South Africa: University of the Witwatersrand.

Monappa, A. (1995). *Industrial Relations (Ninth print).Quality Control. Tool for Worker Development*. New Delhi, India: Tata Mc-Graw Hill Publishing Company Ltd.

Moorthy, M. V. (1968). *Principles of Labour Welfare* (1st ed.). Visakhapatnam, India: Gupta Brothers Books.

Nanda, N., & Panda, J. K. (2013). Challenges and effectiveness of industrial relation environment in Indian Industries study on Rourkela Steel Plant. *International Journal of Financial Services and Management Research, 2*(6), 163–174.

Neetha, N. (2001). *Gender and Technology: Impact of flexible Organisation and Production on Female Labour in the Tiruppur Knitwear Industry*. Noida, India: V. V. Giri National Labour Institute, 82.

Nohria, N., Groysberg, B., & Lee, L. E. (2008). *Motivating Employees: A Powerful New Model. Harvard Business Review*. Boston, MA: Harvard Business Publishing.

Parul, P. B., & Kumar, A. M. (2013). Provision of Welfare under Factories Act & Its Impact on Employee Satisfaction. *Journal of Business Management & Social Sciences Research*, *2*(2), 57–69.

Patel, A., Gohil, A., & Shah, H. (2017). Study on Labour Welfare Measures and Social Security on Selected Engineering Unit of Ahmadabad. *IBMRD's. Journal of Management Research*, *6*(1), 19–26. doi:10.17697/ibmrd/2017/v6i1/111654

Patro, C. S. (2012). Employee Welfare Activities in Private Sector and their Impact on Quality of Work Life. *International Journal of Productivity Management and Assessment Technologies*, *1*(2), 18–29. doi:10.4018/ijpmat.2012040102

Patro, C. S. (2014). A Study on the Impact of Employee Retention Policies on Organisation Productivity in Private Sector. *International Journal of Asian Business and Information Management*, *5*(3), 48–63. doi:10.4018/ijabim.2014070104

Patro, C. S. (2015). Employee Welfare Measures in Public and Private Sectors: A Comparative Analysis. *International Journal of Service Science, Management, Engineering, and Technology*, *6*(1), 22–37. doi:10.4018/ijssmet.2015010102

Patro, C. S. (2016a). A Study on Adoption of Employee Welfare Schemes in Industrial and Service Organisations: In Contrast with Public and Private Sectors. *International Journal of Service Science, Management, Engineering, and Technology*, *7*(2), 16–33. doi:10.4018/IJSSMET.2016040102

Patro, C. S. (2016b). Influence of Retention Policies on Employee Efficiency and Organisation Productivity. In U. Aung, & P. Ordoñez de Pablos (Eds.), Managerial Strategies and Practice in the Asian Business Sector (pp. 124-149). Hershey, PA: Business Science Reference (IGI Global). doi:10.4018/978-1-4666-9758-4.ch008

Patro, C. S. (2017a). A Study on Adoption of Employee Welfare Schemes in Industrial and Service Organisations: In Contrast with Public and Private Sectors. In M. Khosrow-Pour (Ed.), Public Health and Welfare: Concepts, Methodologies, Tools, and Applications (pp. 809-824). Hershey PA: Business Science Reference (IGI Global). doi:10.4018/978-1-5225-1674-3.ch038

Patro, C. S. (2017b). Employee Welfare Measures in Public and Private Sectors: A Comparative Analysis. In M. Khosrow-Pour (Ed.), Public Health and Welfare: Concepts, Methodologies, Tools, and Applications (pp. 1026-1042). Hershey PA: Business Science Reference (IGI Global). doi:10.4018/978-1-5225-1674-3.ch047

Patro, C. S. (2017c). Espousal of Welfare Schemes: A Means for Employees' Satisfied Work Life with Reference to Pharmaceutical Companies. *International Journal of Asian Business and Information Management*, *8*(3), 36–51. doi:10.4018/IJABIM.2017070103

Patro, C. S. (2017d). Welfare Regime: A Critical Discourse. In B. Christiansen, & C. Harish Chandan (Eds.) Handbook of Research on Human Factors in Contemporary Workforce Development (pp. 110-131). Hershey PA: Business Science Reference (IGI Global). doi:10.4018/978-1-5225-2568-4.ch005

Patro, C. S. (2019). Welfare Programs as a Strategy of Promoting Employees' Economic Growth and Work Productivity. In B. Christiansen, I. Sysoeva, A. Udovikina, & A. Ketova (Eds.), Emerging Economic Models for Global Sustainability and Social Development (pp. 291-311). Hershey PA: Business Science Reference (IGI Global). doi:10.4018/978-1-5225-5787-6.ch016

Patro, C. S., & Raghunath, K. M. K. (2016). A Take on Employee Welfare Facilities and Employees' Efficiency. *International Journal of Asian Business and Information Management*, *7*(3), 54–70. doi:10.4018/IJABIM.2016070104

Patro, C. S., & Raghunath, K. M. K. (2018). Employee Welfare Measures: The Impact on Employees' Efficacy and Organisations Productivity. In P. Ordóñez de Pablos (Ed.), Management Strategies and Technology Fluidity in the Asian Business Sector (pp. 215-234). Hershey PA: Business Science Reference (IGI Global). doi:10.4018/978-1-5225-4056-4.ch013

Portney, P. R. (2005). *Corporate Social Responsibility*. Washington, DC: REF Books Publishers.

Rao, P. V., Patro, C. S., & Raghunath, K. M. K. (2015). Employee Welfare is the Key - An Insight. *International Journal of Business and Administration Research Review*, *3*(11), 40–47.

Ridley, J., & Channing, J. (2008). *Safety at work*. Oxford, UK: Butterworth-Heinemann. doi:10.4324/9780080557137

Robinson, A., Sparrow, P., Clegg, C., & Birdi, K. (2006). Forecasting Future Competency Requirements: A Three-Phase Methodology. *Personnel Review*, *36*(1), 65–90. doi:10.1108/00483480710716722

Sahoo, R., & Sahoo, C. K. (2018). Drivers of Cordial Employee Relations: The Study of a State-owned Public Sector Undertaking. *Management and Labour Studies*, *43*(1&2), 123–139. doi:10.1177/0258042X18754428

Sailesh, S. (2012). Role of Organisation in welfare measures for employees. *International Journal of Research in IT and Management*, *2*(9), 3640.

Satayanarayan, M. R., & Redhi, R. J. (2012). Labour welfare measure in cement industries in India. *International Journal of Physical and Social Sciences*, *2*(7), 257–254.

Schaltegger, S., & Lüdeke-Freund, F. (2013). Business cases for sustainability. Encyclopedia of corporate social responsibility, 245-252.

Steiner, H. J., Alston, P., & Goodman, R. (2008). *International human rights in context: law, politics, morals: text and materials*. USA: Oxford University Press.

Verdon, N. (2002). *Rural women workers in nineteenth-century England: gender, work and wages*. Rochester, NY: Boydell Press.

KEY TERMS AND DEFINITIONS

Allowance: An allowance is an quantity of money given or fixed usually at regular intervals for a specific work or purpose.

Employee Relations: The term employee relations refers to the organization's efforts to manage relationships between employers and employees.

Employee Satisfaction: Employee satisfaction describes the employees happiness and fulfillment of their desires and needs at work.

Leadership: Leadership is the ability of an individual or a group of individuals to influence and guide followers or other members of an organization.

Motivation: Motivation is an important aspect that encourages employees to give their best performance and help in reaching organizational goals.

Welfare: Welfare is referred to the various services, benefits and facilities offered to the employees by the employers.

Work Environment: Work environment describes the surrounding conditions in which an employee works and operates.

Workforce Productivity: Workforce productivity is an assessment of the efficiency of an employee or a group of employees.

Chapter 18
Sustainopreneurship:
A New World Order

Jawad Iqbal
The Islamia University of Bahawalpur, Pakistan

Shakeela Kousar
The Islamia University of Bahawalpur, Pakistan

ABSTRACT

The current millennium is characterized with several global problems including unemployment, poverty, hunger, starvation, social exclusion, and environmental degradation. These social, economic, and environmental trends require the changes in current world order to provide such global mechanism where basic needs of each person will be satisfied and each person will have equal rights to resources and opportunity to utilize his full human potential. Entrepreneurship is an important driver that can transform a society towards a more sustainable future. Sustainopreneurship has emerged as a mutual product of sustainability and entrepreneurship. Nations have included the entrepreneurship in their developmental agenda for achieving the sustainable development goals. The purpose of this chapter is to briefly discuss the concept of sustainable entrepreneurship, its definition, evolution of the concept, and transition from sustainable development to Sustainopreneurship as a mechanism for disruptive innovations.

BACKGROUND

It became very challenging for states when their highly educated people are unable to get any job. Especially, it's harder for developing or under-developed countries where the government has scarce resources to support the unemployed labour force(Ahmed et al., 2010). Moreover, sustainability has become a world level issue and the declaration of sustainable development goals (SDG) by United Nations shows the seriousness of world community to make the earth the best place to live for the present and future generations through its sustainable means (Huda, 2016).

Since the independence of Pakistan, 1947, each successive government has focused the attention towards the development of large scale industries to the neglect of policies required to boost up entre-

DOI: 10.4018/978-1-7998-0357-7.ch018

preneurship and small and medium enterprises in the country (Cordova, 2013). Political instability, terrorism and law and order situation has shattered the economic activities in the countries (Ahmed et al., 2010). The unemployment rate is drastically increasing due to which country is facing several economic, social and environmental issues.

To resolve the social, economic and environmental issues of the world community and particularly of Pakistan there is a need for such mechanism that can help in making the world a better place to live not only for the present but also for the future generations. Sustainopreneurship (Sustainable Entrepreneurship) is a universal solution to address the universal life claims of the world community. This chapter has written to put the spotlight on the important concept of "Sustainopreneurship" and it has discussed in detail that how the concept was developed and can be adopted as a "New World Order" to resolve the problems of the world community.

PEOPLE (HUMAN RESOURCE): THE MISSING ELEMENT IN DEVELOPMENT PLANNING OF NATIONS

In the economic theory, human capital is considered as the important player in determining the growth of national income among all other factors used for the production of goods and services in any economy. Although human resources are the focus of all economic activities for a longer period of time, this factor has never received any attention. In order to strive for a better standard of living and expansion of economic activities countries are required to realize the importance of the human factor in their economies which are being run by human beings. The economic development of a nation is the consequence of its human resource efforts; therefore the quality of these resources must match with desired economic objectives.

While discussing the causes of economic development economists have highlighted various factors i.e. savings and formation of capital, organization of market, legal and political institutions, entrepreneurship, the level of scientific and technological knowledge, distribution of income, physical resources, population growth rate and their skill and ability. Keeping in view the importance of human resource in the economic development of nations, one group of experts from United Nations Organization have suggested that underdeveloped countries are required to design such educational programs that will help in reducing the manpower bottleneck. There is a universal agreement on the fact that the lack of well-trained human resource will vanish all the developmental efforts of the country. A successful and rapid economic development can only be possible with the real power of a quality human resource (Yang, 1967)

According to (Ul Haq, 1995), only 30 years ago, there was a strong belief among people that development means an increase in national income. Economists while discussing the sources of development were only focused towards the financial capital where physical capital was at the focal point and all other factors of production were ignored. No considerations were given to the human capital nor did any qualitative or quantitative measurements performed for this capital. It has never got the required importance anywhere. There are many nations who were unable to develop in spite of abundant financial resources because human resources, human development institutions and skills were absent among them without which economic gains cannot be translated into real development. Countries with the typeof natural resources differ in their development due to differences in their human capabilities. Human resources were just considered as residual elements in the development planning of almost all societies (Ul Haq, 1995).

The rediscovery of the concept of human development is not an unknown term rather this invention is the result of the efforts of some early political and economic leaders. Aristotle (322-384 BC) has focused on the idea of "Human Good", he has stated, "wealth is evidently not the good we are seeking, for it is merely useful and for the sake of something else". Then 'Adam Smith' (1723-90) while discussing poverty said that it is something that went beyond continuing calories, an effort to bring about the poor in the mainstream of the community. Similar supporting comments are evidenced from the opinion of modern economic experts including "Robert Malthus", "Karl Marx" and "John Stuart Mill". According to these modern thoughts the purpose of development is to enhance the options for people(Ul Haq, 1995) and these choices include the education, health, political freedom, cultural identity, personal protection, community involvement, environmental protection and many other areas of human well-being (Mahbub Ul Haq Research Center, 2018).

Human Development Report & Birth of Human Development Index

The classical economic literature was focused towards creating wealth as only means while the real objective of developing the well-being of humans was missed. The fascination about industrialism and technological advancements has not changed the thoughts of economic experts towards true wellness of humanity. The Second World War resulted in many economic growth models and national income accounts have rapidly emerged but still the people the real agents of change and beneficiaries were even missed from development planning of nations (Ul Haq, 1995).

In the 1980s, it was realized by many countries, in spite of the expansion of economic activity, they are still facing a reduction in human lives. Some countries were achieving good levels of economic development even with average income levels. While some developed countries were suffering from severe human distress in the form of crimes, HIV/AIDS, pollution that result in the poor social fabric. Furthermore, high-income rates were not translated into improvement of human lives. The quality of the economic growth of these nations was questionable. As a result, many Human Costs of Structural Adjustment Programs were initiated and supervised by IMF & World Bank in some developing countries. The purpose of these programs was to ensure the availability of alternative policies for the weakest areas of society, increased pollution and emerging democratic forces has created the urge for people-centred development models (Ul Haq, 1995).

In such circumstances, a Human Development Report has presented in 1989, in which reported the idea of a candid human development to serve the interests of the global community. Moreover, in t the 1990s has offered a dramatic change by presenting the new ideology of treating the human resource as both the means and ends of development. Most of the recent discussions about human development are the consequence of "Human Development Report-1990" presented under the supervision of United Nations Development Program prepared by the team of well-known economists like Dr Mahboob Ul Haq (Pakistani Economist) and Amrtya Sen (Indian Economist) and other social scientists. Since 1990 the "Human Development Report" is presented each year by "United Nations Development Program (UNDP)" (Ul Haq, 1995).

According to this report, the purpose of development is to increase choices for people and income is not the only choice but it's the sum of life expectancy, good health, education, quality environment and freedom. The report emphasized the role of humans in changing the social, economic and ecological issues of the world community by following a new people-centred world order (Ul Haq, 1995).

This report monitors the progress of humanity through each country's ranking in the human development index. Human Development Index was first introduced in 1990 as an alternative to traditional methods of measuring the economic development i.e. Per capita income & economic growth. Although income has the potential to expand people choices it's an imperfect measure to check the human development success of a given country. The Human Development Index is considered as a powerful measure to check the welfare, social and economic development of a country on the basis of following three dimensions: "1) Health 2) Education 3) Living standards" (Zavaleta & Tomkinson, 2015). Below Figure 1 is explaining the HDI dimensions and its indicators:

HDI is an important indicator to see the ranking of various countries in the above shown dimensions of human development around the globe. In sum, to analyze the socio-economic progress of nations, they are required to check the distribution of wealth and its consumption in society and the way it is visible from the lives of its masses. If a country's score in Human Development Index is high than its per capita gross National product or income it shows that their socio-economic priorities are going in favourable direction and country is establishing an effective base for enhancing its people choices. But is the country's score in HDI is lower than it shows that income is not equally distributed among its people and country is moving towards the wrong direction for development?(Ul Haq, 1995). Since 1990, each year "Human Development Report" is published by "UNDP (United Nations Development Program)" in which HDI ranking of all member nations are shared to show their updated status in human development. Now the countries are more concerned to improve their ranking in HDI to get and enjoy the title of a truly developed nation.

According to the Human Development Report, human development is based on the following four pillars i.e "1) Equity 2) Sustainability 3) productivity 4) empowerment". These components distinguish human development model from older models of economic growth. The first pillar describes as the equity is about enlarging the people choices through providing equitable access to opportunities. While "Sustainability" is about the right of humans that people must have equal access to resources in present as well as for future generations. In the context of "Productivity" investment in people is needed and providing them with a suitable environment to utilize their maximum potential. The element of "Empowerment" does not refer to the concept of welfare rather it is about economic development through people where they will have equal access to all opportunities for a better standard of living.

Human Development Index and Sustainability

The "United Nations Development Program" is publishing the Human Development Reports since 1990 which also presents the Human Development Index (HDI) of each country to check their progress in economic development. The HDI is the more effective measure of development as compared to the old one-dimensional measure like "Gross Domestic Product" (Sagar & Najam, 1998) and a well-known yardstick of well-being (Klugman et al., 2011) . Since, 1990, a series of Human Development Reports have been published by UNDP each year on various themes of development. The purpose of publishing

Figure 1. HDI Dimensions and Indicators
Source: (Report-UNDP, 2010)

Figure 2. Sustainability matrix
Source: (UNDP, 2015)

Human Development report each year is to highlight the aspect of Human development in intellectual as well as policy discussions. So it can be said that HDR has spread the development debates outside the boundaries of old economic thoughts. 'The Human Development Index' is very initial and a significant step for adopting the broad idea of "Sustainability" to be used as a measure of development (Sagar & Najam, 1998).

The Human Development index is annually included as a part of Human Development Report; it shows a big opportunity for United Nations Development Program to foster its mission of "Putting the people at the centre of development" by increasing the public awareness(UNDP, 1995). From the beginning of 1990, importance was given to the environmental aspects and sustainable development (Neumayer, 2001).

So, it can be said that human development and sustainability are compulsory elements for achieving universalism claims of human life. This idea is similar as the definition of sustainable development provided by "World Commission on Environment and Development -1987" as "development that meets the needs of the present without compromising the ability of future generations to meet their own needs."(UNDP, 2015).

Sustainability requires a steadiness between the present and future needs, the balance between activities of private and public domains, between personal goals and societal welfare concerns. According to the concept of sustainability, countries are needed to bring change in their budgetary preferences by cutting useless spending on military and public enterprises. They are required to spend on human development, effectively utilize available resources and invest on environmentally safe technologies (Ul Haq, 1995).

Sustainable Development: As World Agenda

By the end of the millennium as a result of increasing environmental degradation and disparities in the development of the planet, the world community has implemented the sustainable development as a new and efficient development model (Carley & Christie, 2017). Although this concept of sustainable development has been adopted as a new growth model in the 20th century in reality, the origin of the term is very old. All types of issues and concepts related to the term "sustainability" i.e. the need of food, living, energy and all environmental aspects have their roots thousands year back throughout the human history (Waas et al., 2011)

After the World War II, economic growth has adopted a new optimistic approach about the expectations of rising living standards worldwide (Du Pisani, 2006) and many countries have shown remarkable progress in development (Bass & Dalal-Clayton, 2012; Waas et al., 2011). During the period of fastly moving economic, scientific and technological growth along with rapidly increasing population growth rate, it was the time when the world has started exceeding the environmental limits of this earth, destroying the well-being of current as well as of future generations (Waas et al., 2011). This awareness among people about the damaging effects of rapid economic growth has changed the views of world community about the basic assumption of economic growth and successful development (Du Pisani, 2006) which has asked for a move from unfair industrialism "Business as usual" approach to sustainable development (Carley & Christie, 2017).

During the decade of 1950 to 1960, economic growth and increase in economic output was the only concern of development. In the start of 1970s, due to the acute poverty conditions in developing countries and lack of equitable access to the benefits of economic growth has to boost up the activities for the distribution of equal welfare choices for people (Waas et al., 2011). In the 1980s, environmental protection was claimed as the third major objective of development. Hence, "sustainable development" is the consequence of the combination of different ideas related to environmental safety, economic growth and development that has evolved over centuries (Du Pisani, 2006).

In less than half of the century, sustainable development has emerged as a globally acknowledged and formally endorsed development model. The concept has become the important world agenda by all the member states of the United Nations Organization as a result of the following events (Waas et al., 2011):

1. "United Nations Conference on the Human Environment (UNCHE; 1972)
2. World Conservation Strategy (WCS; 1980)
3. Our Common Future (1987)
4. United Nations Conference on Environment and Development (UNCED; 1992)
5. United Nations Millennium Summit (2000)
6. Earth Charter (2000)
7. United Nations World Summit on Sustainable Development (WSSD; 2002)
8. Rio+20 United Nations Conference on Sustainable development (UNCSD; planned in 2012)
9. United Nations Sustainable Development Goals 2015(SDGs)"

The "United Nations Conference on Human Environment (UNCHE)" which is also named as "(Stockholm Conference)" was an international conference organized by UNO. It was held on Stockholm, Sweden from June 5-16, 1972. This was the UNO's very first and significant on problems related to the environment that has highlighted the "environment' for the first time as international political agenda(Quental et al., 2011; Waas et al., 2011). The "World Conservation Strategy (Living Resource Conservation for Sustainable Development)" was presented in 1980. It led towards the development of the concept of "Sustainable Development". It has provided the base for "World Commission on Environment and Development (Brundtland Commission 1987)". It results in the creation of various institutions for further definition and promotion of the principles of sustainability (R.D., n.d). The well-known report "Our Common Future" (WCED, 1987) was printed in 1987, by the "World Commission on Environment and Development (WCED)", formed in 1983 by the UN General Assembly and presided by Gro Harlem

Table 1. United Nations millennium development goals 2015

Goal 1	Eradicate extreme poverty and hunger
Goal 2	Achieve universal primary education
Goal 3	Promote gender equality and empower women;
Goal 4	Reduce child mortality
Goal 5	Improve maternal health;
Goal 6	Combat HIV/AIDS, malaria and other diseases
Goal 7	Ensure environmental sustainability
Goal 8	Develop a global partnership for development

Source: (UNO, 2010)

Brundtland, former PM of Norway. Later, the report also became famous by his name. The commission was established to formulate the "global agenda for change (Waas et al., 2011).

This United Nations report named "Our Common Future" (WCED, 1987) has provided the base for "United Nations Conference on Environment and Development (UNCED)".The Conference was held in Rio de Janeiro, in 1992. This conference has presented the official worldwide formal recommendation of sustainability as a new development model by adopting the "AGENDA-21 (a global action plan for sustainable development)" and Rio Declaration with 27 sustainability principles (Waas et al., 2011). This conference has proved a successful effort to promote sustainability agenda (Quental et al., 2011) as sustainable development became the universal slogan for countries, businesses, NGOs, labour unions, organizations working for development, academic spheres, people and other having any stake all over the world.

By the end of the Millennium United Nations organization has organized the Millennium Summit in the year 2000 to discuss a broader agenda based on both development and environmental aspects (Quental et al., 2011). This declaration was also a confirmation regarding the approval of the world community for UNO 'Agenda-21' and 'Rio Declaration' which leads towards the development of 'Millennium Development Goals'. These MDGs are the time-based target list that each of the member state of UNO is required to achieve till 2015 to eradicate acute poverty and fulfilling the needs of world's poor community. This declaration is the commitment of member nations towards sustainability for the world's most deserving people and it was a step towards building the decent partnership for better lives of people of this planet and for providing such environment that will contribute towards peace and security (UNO, 2010). Following is the list of 8 Millennium Development goals signed by all member states to achieve till 2015:

The UNOs "World Summit on Sustainable Development (WSSD)" was organized in Johannesburg; in 2002. The main objective of this conference was to establish the necessary mechanism for the real implementation of AGENDA-21(Quental et al., 2011). In 2012, UNO organized a second conference to follow up "Rio+20 United Nations Conference on Sustainable development (UNCSD)", again in Rio de Janeiro. The conference had two major themes "1) a green economy in the context of sustainable development and poverty eradication; and (2) the institutional framework for sustainable development" (Leggett & Carter, 2012).

United Nations Sustainable Development Goals: World Agenda 2030

The UNOs "Millennium Development Goals" have proved as an effective way of mobilizing important social priorities worldwide. By including these priorities into a set of eight Goals, MDGs help in promoting "global awareness, political accountability, improved metrics, social feedback, and public pressure". As acknowledged by "Bill Gates" "The MDGs have become a type of global report card for the fight against poverty for the 15 years from 2000 to 2015"(Sachs, 2012). So the purpose of MDGs was to end poverty as the basic dimensions of human development by 2015 (UNDP, 2015).

Developing countries have shown considerable growth in the achievement of Millennium Development Goals although this progress is different from country to country. Some countries have achieved all MDGs while some have achieved very few of these goals. This achievement gap of Millennium Development Goals is very high and serious among underdeveloped countries of the world. This is the reason that all countries are eager to implement some universal set of goals that will help in the achievement of MDGs. As a result of this, an expert global sustainability panel was appointed to successfully organize the "Rio+20 summit" in June 2012, in which recommendations were made to adopt a set of Sustainable Development Goals (SDGs).

Since 2013, a formal process by UNO was underway to devise post-2015 agenda and to extend and enhance the work under "Millennium Development Goals". As a result of this, all the member states of the United Nations have formulated Sustainable Development Goals in late 2014. The major aim of "Sustainable Development Goals" is to eradicate the deprivations and disparities in human progress. These goals are a commitment to make improvement in the living conditions and providing equal opportunities to every person in this world by 2030. The achievement of these goals will change the state of human development in the entire world (UNDP, 2015). The Sustainable Development Goals are an important milestone that can help the world in moving towards a sustainable trajectory (Sachs, 2012).

All the member states of UNO have agreed upon 17 goals and 169 targets that constitute the "Sustainable Development Goals". These goals focus on five basic areas 1) People 2) Planet 3) Partnership 4) Peace 5) Prosperity. Since January 2015, SDGs are under discussion at UN General Assembly to ensure their adoption by all the heads of states and governments at the UN "Sustainable Development Summit" in September 2015. These Sustainable Development Goals address five areas for humanity and the planet. The UNO has also developed a process to monitor the progress of SDGs at the global, regional and national level (UNDP, 2015). Below figure shows the focus areas of UNO SDGs:

The Case of Pakistan

The economic development of a country is reliant on factors of production as well as effective management and utilization of human resources (Chani & Shahid, 2012). So, Human Development is the growth engine of any economy. Progress in human development always remains an important issue for developing countries like Pakistan. The country is facing multiple challenges on the domestic front including lack of energy resources, inflation, lack of employment opportunities, pathetic social structure, extremism and security challenge".Almost all of these problems are due to the lack of proper planning for the present and future. The rapidly changing dynamics of the global world i.e. rapidly changing technology and innovations are changing the developmental models of the world which requires modifications in prevailing socio-economic and political trends changes in social, political and economic behaviours and structure of societies to bring about dramatic change in existing capabilities (Council, 2014).

Figure 3. United Nations sustainable development goals
Source:(UNDP, 2015)

At the time of independence in 1947, Pakistan inherited a very poor economy without the availability of proper industrial infrastructure. But just after one decade in 1960, Pakistan became the fastly growing economy of Asia and was in a position to become so developed like Japan. Due to the "1965 and 1971 wars" and political conditions again hampered the economic developments. Since 1947, as shown in the below table, it was unable to perform as per expectations (Council, 2014).

Pakistan is the 7th most populous country in the world. Its population is expected to increase to over 227 million by 2025 and the major proportion of which comprises young people (63% below the age of 30). Such types of forecasts are resulting in large number of problems for Pakistan. One of the major challenges is to create jobs for a growing number of youth who will soon enter into the labour force. About 1.5 million jobs are required to create each year to keep the unemployment rate at the existing rate (Council, 2014).

To what extent these demographic projections will be addressed is totally based on the investment directed towards the development of people including education, training and their health. Moreover, a huge amount of these investments are required for food and nutrition needs of the increasing population. The climate change and drainage of water resources will be the major issues Pakistan is expected to face in the coming future. According to the recent "National Nutrition Survey", it has noted that almost sixty percent of masses is not having access to food and 50% of women and children are malnourished. The rapidly increasing population will badly affect food security and also to the country's development and

Figure 4. Comparative per capita income of Pakistan
Source: (Council, 2014)

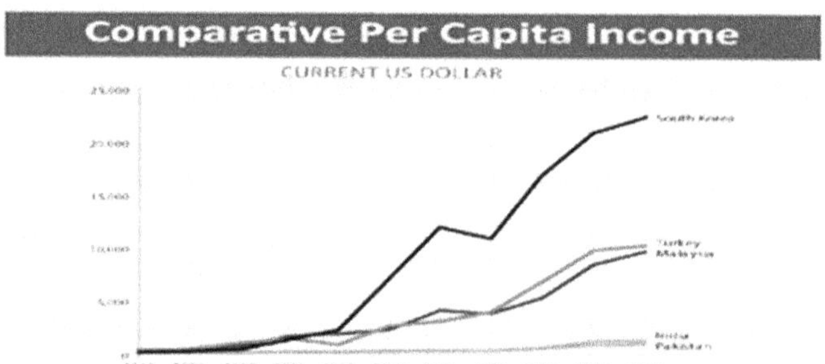

Figure 5. Population growth in Pakistan
Source: (Council, 2014)

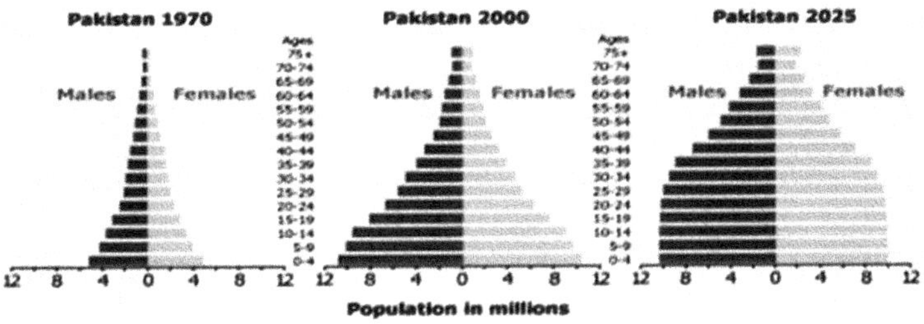

stability (Council, 2014).The development indicators of Pakistan of the last 68 years from 1947 to 2013 shows a very slow increase as given in the below table.

Being a developing country Pakistan's "Human development Index" is showing very poor condition. According to United Nations HDI of Pakistan is 0.5 which ranks it at 147 positions among 188 countries (UNDP, 2016).This poor condition effect the state of Human Resource Development in Pakistan which is considered the major element for organizational development which translates it into better business and better societies, make changes in the pattern of workplace according to globalization and help in the emergence of new knowledge and skills of individuals and society (Tabassum, 2015).

According to the United Nations report, Pakistan's position among its South Asian neighbours including India, Bangladesh, Bhutan and Nepal is not so good (UNDP, 2016). Pakistan has shown no improvements in its standing as compared to last years as its "Human Development Index" rank was at 148th position in the year 2014. According to the Human Development Report 2016, only 63% people in the country were seems satisfied with their living standards in 2014-15. Such position in Human development place Pakistan in the bracket of "Medium Human Development" where other neighbouring countries i.e. India, Bangladesh, Bhutan, Kenya, Myanmar and Nepal also stand (Akram, 2017).

United Nations Organization has presented Millennium Development Goals in the year 2000 which were signed by all 193 member states and 28 international organizations to achieve these eight goals

Table 2. Pakistan: 67 years of development

	1947	2013	Times increase
Population (Millions)	34.4	184.4	5.4
GDP(1999-2000 Rs. Bn)	267	9,863	37
Per capita GDP (1999-00 Rs.)	7,742	53,501	6.9
Industrial Value Added	7.8	20.9	2.7
Electric Generation (kw per capita)	5.0	517	103
Doctors per million people	30	800	27
Literate Population (%)	13.6	58.0	4.3
Primary School Enrollment (000)	544	18,748	35
Export (US $ Million)	444	24,754	56
Import (US$ Million)	319.0	39,822	125

Source: (Council, 2014)

Figure 6. HDI ranking of Pakistan
Source:(Akram, 2017)

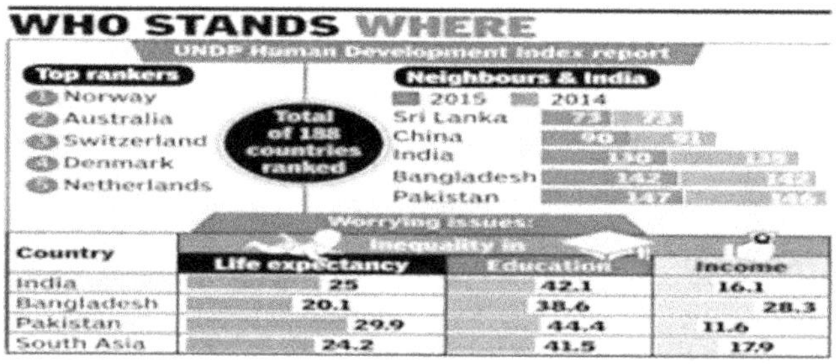

till the year 2015.Where most of the member states of UNO remained ineffective in achieving these goals, Pakistan also lagged behind in achieving these MDGs, which are now replaced with Sustainable Development Goals by United Nations Organization (Akram, 2017). The baseline of Pakistan against each SDG is explained below:

By signing the Sustainable Development Goals, Pakistan has aligned its Development Plan "Vision-2025" with the targets outlined in the United Nations Organization's 2030-Agenda. The purpose of these SDGs is to achieve inclusive growth and sustainable development (Dawn, 2016). In November 22, 2013 a National Consultative Conference was held to form the Vision-2025. Extensive consultation was conducted between all stakeholders including politicians, government officials, parliamentarians, national and private sector entrepreneurs, development partners and financial sector organizations, academia, think tanks, independent experts NGOs and civil society to take input from all sectors. Following are the elements of Pakistan vision 2025:

Pakistan Vision 2025 has adopted people-centric approach based on the agenda "Development has to be of people, for people, and by people". It is having 5+7 Model, with 5 enablers and seven pillars which provide the integrated formula for development and prosperity. The basic purpose of this vision is to put Pakistan on the fast track of development to transform it among top ten economies of the world till 2047 (its first century), till 2025 among top 25 economies and to place it at upper middle countries list. According to the economic indicators Pakistan is considered a middle-income country but its social indicators place it among the least developed countries.So, this Vision-2025 is the first stage towards the country's development journey by offering a strong and balanced platform for sustainable growth and development to help achieve the larger vision of transforming Pakistan as the next Asian tiger and high-income economy till 2047 (Council, 2014). Below given are the milestones to be achieved according to Pakistan Vision-2025.

Pakistan's Vision-2025 requires the focus on a number of priority areas. According to the demographic projections to achieve this vision, it requires the following actions:

- Fulfilling the basic needs of the increasing population by providing them access to basic needs i.e. health, education, energy, water, and sanitation.
- Empowering people by providing them the basic social, legal and physical infrastructure to ensure that they have needed facilities to live life with dignity, their basic human rights and lives are protected, full employment has achieved, women are empowered and poverty and hunger are eliminated.

Table 3. Sustainable development goals and Pakistan's baseline

SDGs		Pakistan's baseline
1	End poverty in all its forms	21.04% Population below poverty line $1.25 60.19% Population below poverty line $2
2	End hunger, achieve food security and improved nutrition, and promote sustainable agriculture	Ranked 76th among 107 countries on Global Food Security Index 58.1% food insecure households 169 kcal/person/day Intensity of Food Deprivation
3	Ensure healthy lives and promote wellbeing for all at all ages	88 <5y mortality rate (per1000livebirths) 170 Maternal mortality ratio (per100K live births)
4	Ensure inclusive and equitable Quality education and promote life-long learning	58% overall Literacy Rate 25.02 million Children (5-16) are not in school
5	Achieve gender equality and empower all women and girls	144th rank on women economic participation 132nd rank on women education attainment
6	Ensure availability and sustainable management of water and sanitation for All	35% doesn't have access to safe drinking water 52% doesn't has access to improved sanitation
7	Ensure access to affordable, reliable, sustainable, and modern energy for all	91.4% Population with access to electricity <1% Share of renewable energy in total mix
8	Promote sustained, inclusive and sustainable economic growth, full and productive employment and decent work for all	53.1% Labor force participation rate 6.2% Unemployment rate
9	Build resilient infrastructure, promote inclusive and sustainable industrialization and foster innovation	20.3% Industry's contribution in GDP 10.9% Population use Internet
10	Reduce inequality within and among countries	30 Income Gini coefficient 1.55 Palma Index(ratio of consumption of top 10% to the bottom 40%)
11	Make cities and human settlements inclusive, safe, resilient and sustainable	47% Urban population lives in 9 cities only 5.0 Cities population growth rate
12	Ensure sustainable consumption and production patterns	68.3% Electricity generation from fossil fuels 25% Energy losses in all sectors
13	Take urgent action to combat climate change and its impacts	310 (m.t of CO2 eq.) Total GHG emissions 6% Budget allocation for climate financing
14	Conserve and sustainable use the oceans, seas and marine resources for sustainable development	222 rank on global Ocean Health Index 350 (million gallon/day) raw sewage and untreated industrial waste flows into the Arabian Sea from Karachi
15	use of terrestrial ecosystems, sustainably manage forests, combat desertification, and halt and reverse land degradation and halt biodiversity loss	2.1% National forest cover 68 (million hectares) Land affected by desertification & degradation
16	Promote peaceful and inclusive societies for sustainable development, provide access to justice for all and build effective, accountable and inclusive institutions at all levels	9000 Avg. annual deaths from terrorist activities 3 million Cases pending before judiciary
17	Strengthen the means of implementation and revitalize the global partnership for sustainable development	

Source: (LEAD, 2016)

- There are a responsible and accountable government and public sector
- A vibrant entrepreneurial private sector that is able to generate the required number of jobs.

Figure 7. Elements of Pakistan vision-2025
Source: (Council, 2014)

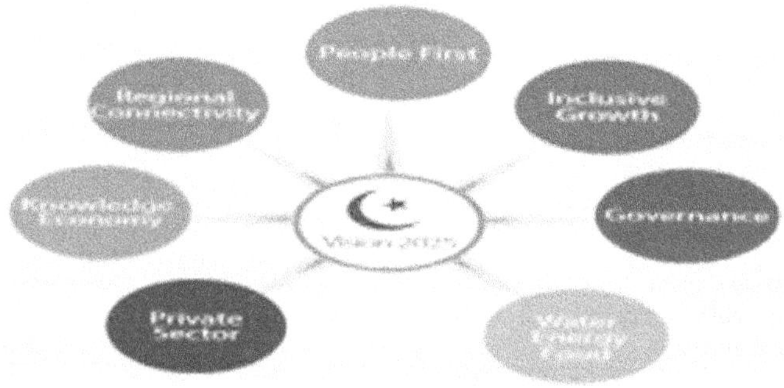

- Enabling the economy to effectively grabbing opportunities outside the borders of the country (especially regional opportunities) and getting their benefits.
- Enhancing and maintaining the competitiveness among the entire world that will transform the economy into knowledge-based.
- Protecting natural resources and effectively dealing with climate change issues.
- Focusing on the fastest GDP growth, in order to enhance the employment rate by emphasizing the "inclusive growth" that will discourage the social and regional inequalities that exist in our economy.

In order to achieve the desired development and growth goals, the Government of Pakistan is required to understand the changing global context. It has needed to take a lesson from developed and emerging economies. Based on the best practices from the entire world we need to develop our indigenous inclusive growth model a "Pakistani Model"(Council, 2014). The realization requires a simultaneous focus on various areas. The nationwide consultation of all sectors has identified 7 pillars each of these are integrated with Sustainable development Goals and pillar 1-4 with MDGs which is explained below:

Pakistan's vision 2025 is based on successful world pursuit of Millennium Development Goals and is a launching pad for the effective achievement of Sustainable Development Goals before its deadline of 2030 (Council, 2014). Currently, the UNOs Sustainable Development Goals have been adopted by all

Figure 8. Milestones of Pakistan vision-2025
Source: (Council, 2014)

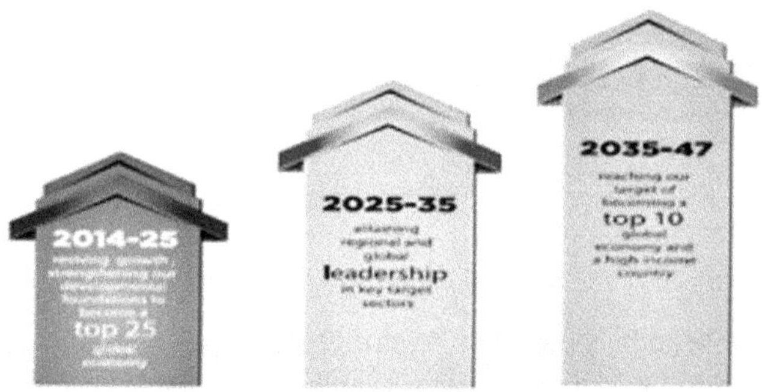

Table 4. 7 pillars of Pakistan's growth model, SDGs & MDGs

Sr.No.	Pillar	Link with MDGs	Link with SDGs
1	**People First:** Developing social and human capital and empowering women	This pillar encompasses poverty eradication (MDG1), access to health and education services (MDGs 2, 4, 5, and 6), and gender empowerment) (MDG 3).	SDGs 1 (poverty), 3, (health) 4 (education), and 5 (gender)
2	**Growth:** Sustained, indigenous, and inclusive Growth	While this was not an explicit goal, it was a key driver of MDGs, as well as being implicit in MDG 1 (equity, decent work), and MDG 6 (environmental sustainability).	The target is virtually identical to SDG 8, and also to SDGs 10, 12, 13, 14, 15
3	**Governance:** Democratic governance: institutional reform and modernization of the public sector	Again, while not an explicit goal, there is abundant evidence that shows the positive impact of good governance on the human development goals.	Again, the language is similar to that of SDG 16
4	**Security:** Energy, water, and food security	These are incorporated in MDG 1 (hunger), and MDG 6 (water)	Linked to SDG 2 (zero hunger), 6 (water security), 7 (energy security), and 11 (urban)
5	**Entrepreneurship:** Private Sector and entrepreneurship-led growth	-	This is linked to SDG 9 (foster innovation)
6	**Knowledge Economy**: Developing a competitive knowledge economy through value addition	-	SDG9(innovation), and 4 (education)
7	**Connectivity:** Modernizing transport infrastructure and regional connectivity	-	SDG 9 (infrastructure), and 17 (global partnership)

Source: (Council, 2014)

nations in 2015 that addresses the need of Human Resource Development for sustainable development of nations especially in Goal 4 Quality Education; Goal 8 Decent Work and Economic Growth; Goal 9 Industry, innovation and infrastructure; Goal 12 Responsible Consumption (Tabassum, 2015).

Entrepreneurship Led Sustainable Development

Entrepreneurial activity is becoming the major issue of concern for almost all governments around the world as well as in academia (Díaz-Casero et al., 2017). There is an inferred assumption that countries and industries with a large number of small businesses and startups are those where most innovative and high growth firms emerge. While countries with better institutional context and business-friendly policies have less number of low-quality firms but having high-quality entrepreneurs (Henrekson & Sanandaji, 2014). Entrepreneurship is a system through which firms with valuable innovations and well-organized firms as compared to their rivals in their respective markets enhance their share in the economy. As these firms grow they absorb the previously self-employed by offering them better options which result in a more prosperous economy and lower rates of self-employment (Henrekson & Sanandaji, 2014).

In the latest World Bank's ranking regarding ease of doing business, one country has shown a dramatic change in its position from 143rd in the list to the 67th. The country is Rwanda, who had faced serious destruction of its institutions and decrease in population due to genocide in the 1990s.Rwanda has moved up faster than its neighbouring countries Haiti, Liberia West Bank and Gaza on the world bank's list and it has moved past Italy, the Czech Republic, Turkey, and Poland. In another sub-index of study, Rwanda is placed on 11th position worldwide regarding ease of doing business. Rwanda's business revolution can also be realized at Costco, which is world's most demanding retail market for prestigious trade customers, where the coffee grown by even the small farmer of Rwanda is placed on the shelves. The economy of Rwanda is itself strong evidence of this dramatic growth as its per capita GDP has almost quadrupled since 1995. All such kind of dramatic change can only be experienced by a country through entrepreneurship. This importance of entrepreneurship was realized by Rwanda's president, Paul Kagame in the following words, he said that "Entrepreneurship is the surest way of development. "He is not alone who has realized the importance of entrepreneurship rather various economic studies from all over the world have consistently linked a variety of rapid growth outcomes to entrepreneurship i.e. creation of jobs, GDP growth rate and long term increase in productivity.

There are some other noticeable examples of outstanding entrepreneurial success stories on the Costco shelves just like Rwanda. The fresh fish of Chile is also placed at the shelves of Costco, which is ranked as second after Norway as a supplier of "Salmon". Evidence has noticed from Israel whose innovative entrepreneurs are supplying unique technological products to the entire world since the 1970s i.e. memory USBs. There is also an example of Iceland, whose manufactured drugs are sold by Costco Pharmacy. Iceland's drugs manufacturers have become global leaders just in 10 years. All these countries including Rwanda, Chile, Israel, and Iceland became the best place for the growth of entrepreneurship, and all credit goes to their governments. Although the products on the shelves of Costco are launched by innovative entrepreneurs of above-mentioned countries all these businesses are directly or indirectly supported by their governments who help in building such environment that helps in the nourishment and sustainability of entrepreneurship. So entrepreneurial ecosystem became the holy objective by all the governments around the world but unfortunately, some governments use a misguided approach to build an entrepreneurial ecosystem because they start pursuing some unattainable practices of the ecosystem by following the economies that are having a totally different context. But the governments are required to exploit all the available expertise and they should commit to the continuous experimentation. But government alone cannot do everything at its own; there is a need of combine efforts by private and non-profit sectors, corporate executives, and family business owners, universities, professional organizations, foundations, labour organizations, financers and also the entrepreneurs themselves have financed or even initiated the entrepreneurship education, conferences, research and policy guidance. An entrepreneurial ecosystem consists of a set of various elements i.e. leadership, culture, capital markets, and open-minded customers that are combined in complex ways. Separately working, all these elements are insufficient to sustain entrepreneurship while combine they can result in dramatic changes (Isenberg, 2010).

All over the world, most of the governments and agencies who are working with the help of international organization i.e. United Nations (UN), World Trade Organization (WTO), International Monetary Fund (IMF) and World Bank (WB) have failed by their all means to get rid of underemployment (İyigün, 2015).In this context, Sustainopreneurship is very crucial for the growth and development of developing nations. According to one recent research, in developing countries, SMEs (Small & Medium Enterprises) makeup to approximately 92% of businesses and providing employment to almost 80% of the workforce in those countries. These figures indicate that there is an immediate need for developing countries to commit themselves for preparing the next generation of entrepreneurs, innovators and leaders for achieving the sustainable development and growth (Cormier, 2003).

Among the academic spheres, the field of entrepreneurship, is getting greater importance as an important conduit for bringing about a transformation in society with sustainable solutions as well as a remedy for many environmental and social issues (Hall et al., 2010; Uslu et al., 2015). The idea of entrepreneurship is being utilized in the fields of economy and business administration for a long time, as according to J. B. Say, French Economist has called it the fourth factor of production. Hence it has been included in the basic production elements i.e. Labor, Capital & environment (Bilge & Bal, 2012; Uslu et al., 2015). Entrepreneurship cannot be considered as a luxury element rather it is a fundamental driver of sustainable economic performance (Cordova, 2013). In essence, Entrepreneurship is an important factor for improving living standards and the well-being of society. The creation of new entrepreneurial ventures is an important strategy for sustainable national economic growth and development as well as for achieving competitive economies at an international level (Salem, 2014).

The role of entrepreneurs as a vehicle for economic and social transformation of nations is not new in the economic literature (Youssef et al., 2018).In emerging economies around the world, interest in entrepreneurship is greater than ever to focus on the development of the youth population and the desire to move up the value chains (Cordova, 2013). Due to this increasing importance of entrepreneurship, United Nations General Assembly in its 2030 Agenda in September, 2015,referenced that entrepreneurship is not only crucial for achieving the Sustainable Development Goal (SDG)8 on decent work and economic growth rather it can also help in progress towards twin goal of prosperity and peace (Mahmoud et al., 2017; Naudé, 2013).

Development can be defined as a reduction in poverty, unemployment and inequality. While sustainability is about satisfying the needs of a nation for natural resources without compromising the generations (Luke, 2013). So, sustainable development can be defined as development that fulfil the needs of current generations without compromising the ability of future generations to fulfil their own needs (İyigün, 2015). Entrepreneurship appears as a precondition for sustainable development and a strategic tool to ensure an equality-based and non-discriminatory society which empowers its citizens to build a self-domain of economic activity through innovative business models (Huda, 2016).

Sustainopreneurship has emerged as a mutual product of sustainability and entrepreneurship. It has the ability to solve social and environmental sustainability issues, moreover, it converts these problems into opportunities with sustainable innovations (Abrahamsson, 2007). Sustainopreneurship requires such efforts and practices of sustainable entrepreneurship development that helps in promoting the sustainable development goals i.e. poverty eradication, child development, empowerment and curbing the life-threatening diseases. Such type of business models promotes creative problem solving the economic, social, and environmental sustainability (Huda, 2016). Sustainable entrepreneurs can act as a change agent from current economic conditions to a more sustainable economy. They fill the gaps that remained uncovered by businesses and government agencies in providing critical social and environmental goods and services by starting from a singer or sub-area of sustainability and slowly they cover all areas of sustainability(İyigün, 2015).

The business organizations are the major contributors for developing a sustainable world that comes through innovation and creative solutions (Hart, 2005). Therefore, creating the pro-sustainable organizations is a very challenging task which can only be achieved by practising the concept of sustainable development.

DEVELOPMENT OF THE CONCEPT: FROM SUSTAINABLE DEVELOPMENT TO SUSTAINOPRENEURSHIP

The concept of sustainable development has introduced due to the increasing awareness in the entire world about environmental problems, socio-economic issues, poverty, inequality and concerns about a healthy future for humanity. The concept of sustainable development focus on three areas environment, social and economic concerns (Hopwood et al., 2005).The concept has passed through various stages of its development which are as follows:

"Limit To Growth" (1974-86)

The concept of "sustainability" was first highlighted by (Meadows et al., 1972), in their model they have tried to discuss the economic growth and its side effects on the scarce resources of this earth. Their model was developed to address five important global trends 1. Increase in industrialization 2.rapidly increasing population 3. Prevalent malnutrition 4. Reduction in nonrenewable resources 5.Destruction in the environment. They have concluded their findings in their world model that if the current growth trends in above five areas remained unchanged then it will have resulted in the limit to the growth of this world within next one hundred years. Moreover, they have said that it is not impossible rather possible to bring about change in these growth trends to maintain a state of affairs that will be more sustainable in future. All this will result in such a global equilibrium where the basic needs of each person will be satisfied and each person will have equal rights to resources and opportunity to utilize his full human potential.

These findings were rejected by policymakers on the grounds that rapid technological innovations and advancements in markets will facilitate the economic growth of the world. Moreover, this destruction will be automatically controlled through the demand and supply mechanism, as the resources become limited there will be an increase in their prices that will lessen their demand (Bell & Stellingwerf, 2012). Although criticized but they have brought into light the ways that can help in reducing the destruction of scarce resources of this earth (Bell & Stellingwerf, 2012). So the book "Limit to Growth" by (Meadows et al., 1972) was the first attempt to highlight the concept to sustainability.

The Brundtland Commission (1987–96)

The term was first presented in 1972 at the United Nations Conference on Human Environment and later it gained importance through a report of the United Nations by the World Commission on Environment and Development (S. W. S. WCED, 1987) presided by Norwegian Prime Minister Gro Harlem Brundtland, the report was also referred to as "The Brundtland Report". According to this report, the term sustainable development is defined as "Sustainable development is the development that meets the needs of the present generation without compromising the ability of future generations' to meet their own needs." The idea that all natural systems have limitations and well beings of human requires living within those limits. Sustainable development infers that the resources that are renewable should be used wherever required and those that are not renewable must be reduced or recycled to enhance their sustainability for future generations (Hall et al., 2010).

According to (Balakrishnan et al., 2003) sustainability is basically at odds with the existing model of capitalism and its focus is on uncontrollable economic growth. They said that a major reduction of development if not the decentralization is required to achieve the goal of sustainability (Hall et al., 2010).

As a response to this criticism, many stands close to the possibility of large scale economic and societal transformation with the help of innovation. Many are looking towards established firms in providing leadership through innovation and delivering more sustainable products and services. Some thinkers have the opinion that existing business can't help in achieving sustainability rather change will be driven by entrepreneurs. An unstated assumption is that green, clean and low carbon entrepreneurs will help to avoid problems faced by industrial economies (Hall et al., 2010). According to the (Choi & Gray, 2008) successful entrepreneurs not only build a profitable enterprise but at the same time they effectively achieve multiple objectives of Triple Bottom line i.e. social, economic and environmental goals.

The Triple Bottom Line (TBL Or 3BL)

The term triple bottom line (TBL or 3BL) is associated with John Elkington who has first time introduced this term in 1995 (Bell & Stellingwerf, 2012; Majid & Koe, 2012). The purpose of introducing the concept was to identify such language that can help in understanding the use of sustainable values in business practices (Majid & Koe, 2012). Later in 1997, he has thoroughly elaborated the concept in his book titled "Cannibals with Forks: The Triple Bottom Line of 21st Century Business" in which he has presented 3 main value creating aspects of sustainable conduct "(1) Economic prosperity; (2) Environmental quality and; (3) Social justice". This concept was further developed into 3P elements including people, planet and profit (Majid & Koe, 2012).

ELEMENTS OF SUSTAINABILITY

There are three elements of sustainability 1) People 2) Planet 3) Prosperity, in order to be successful businesses are required to incorporate these 3P's in their practices (Bell & Stellingwerf, 2012). Each one of three P's is explained as under:

1. **People:** Valuing human resources is very crucial for sustainable businesses, selection, development and training of the right people helps in the achievement of business initiatives. The behaviour of business is explained through its workforce, so effective management of human resources is very critical in both social and ethical issues. A Triple Bottom Line based organization maintains a good working environment for its labour and the community in which it performs its operations. Such an organization does not endanger or exploit any of its stakeholders either they are children, suppliers, contractors etc. In short well-being of all stakeholders especially the employees are at the core of TBL based entities (Bell & Stellingwerf, 2012)
2. **Planet:** Sustainability is vital to the long life of the planet. An organization that follows the idea of Triple Bottom Line makes efforts to save the environment and they try to run their business in a way that is harmless to the environment as much as possible. Triple Bottom line based initiatives try to produce such businesses and products that do not damage the environment; they take care of the natural environment. Businesses that include the environmental protection into their mission they face problems in their profit maximization. Organizations that fully incorporate environmental protection they integrate their profit strategy with environmental protection initiatives. Government laws, regulations and individual self-regulation has a direct impact on profit related outcomes (Bell & Stellingwerf, 2012).

3. **Profit:** The measurement of the Triple bottom line based economic value is different from traditional accounting profits. In traditional organizations, profit is measured after deducting the cost of all direct or indirect expenses. But in Triple Bottom Line based initiatives this measurement is based on the real economic value that is enjoyed by the concerned society. Therefore, the Triple Bottom Line based measurement approach must consider the amount of profit that traditional entities understood as a social benefit (Bell & Stellingwerf, 2012).

1997 - Present

Since 1997, greater importance has given to the environment and social dimensions in academic research. This has highlighted the importance of sustainability and sustainable development and their role in defining sustainable entrepreneurship (Bell & Stellingwerf, 2012).

THE EMERGENCE OF VARIOUS FORMS OF ENTREPRENEURSHIP

Traditional or Commercial Entrepreneurship

Although a large amount of research is currently available about the field of entrepreneurship it was widely recognized as a meaningful discipline in the 1980s. Today, almost all business schools and business communities are familiar with this concept and they realize the importance of motivating the entrepreneurial behaviours within the commercial context (Schaper, 2002). According to (Schaper, 2002) "Entrepreneurs - in the traditional sense, play a decisive role as engines of change in market-based economies". Traditional or commercial entrepreneurs create social value by generating economic gains; they are responsible for introducing breakthrough innovations by adopting new ideas or bringing about change in existing market offerings. They identify gaps in the marketplace satisfy the unfulfilled needs and resolve the unsolved problems (Bell & Stellingwerf, 2012). The basic aim of entrepreneurial activities is to generate profit which helps entrepreneurs to build their personal wealth (Certo & Miller, 2008). So commercial entrepreneurship works for the single bottom line of 'economic gains'.

Social Entrepreneurship

It is not possible to change the world and to enhance the quality of life without being socially responsible. Social entrepreneurship is considered as an important element for sustainable development, it is therefore very crucial for defining and conceptualizing the sustainable entrepreneurship(Bell & Stellingwerf, 2012). There is increasing global attention to the area of social entrepreneurship (Sekliuckiene & Kisielius, 2015). It is very new and complex phenomena, having different definitions from different authors. The components of these definitions covers a wide variety of concepts including social justice, social value, viable socio-economic structures, forging a new equilibrium, employing innovation, entrepreneurial skills, market gaps, solving social problems, to social entrepreneur as a change agent (Zahra et al., 2009).

According to (Korosec & Berman, 2006) social entrepreneur is " individuals or private organizations that take the initiative to identify and address important social problems in their communities". As defined by (Greblikaite, 2012) "Social entrepreneurship encompasses the activities and processes undertaken to discover, define, and exploit opportunities in order to enhance social wealth by creating new ventures or managing existing organizations in an innovative manner". The focus of traditional entrepreneurship is economic results of activity, while social entrepreneurship comprises double results economic and social success and creation of equilibrium in society (Greblikaite, 2012). (Boschee & McClurg, 2003) explained two main differences between traditional and social entrepreneur: 1) Traditional entrepreneurs may donate their money to nonprofit organizations but their efforts are indirectly attached to social problems while social entrepreneurs differ in their income-earning strategies that are directly connected with their social mission 2) traditional entrepreneurs measure their efforts clearly in financial terms but social entrepreneurs measure them as a blend of social and economic results. Social Entrepreneurship works on double bottom line 1) Social Gains 2) Economic gains. This double bottom line is derived from Triple Bottom Line, but social entrepreneurship is much more than balancing economic and social gains (Bell & Stellingwerf, 2012).

Ecopreneurship

The decade of 1990s a new group of environmental friendly change agents have emerged which are called ecopreneurs. All those individuals and organizations that came up with environmentally friendly ideas are called ecological entrepreneurs (Pastakia, 1998). Ecopreneurs play an important role for the development of environmentally friendly systems by producing eco-friendly products, services and processes to achieve the sustainable development objective "development that meets the needs of the present generation without compromising the ability of future generations to meet their own needs" (Bell & Stellingwerf, 2012). No doubt, transition to sustainable development requires ecopreneurship on large scale (Pastakia, 1998).

Sustainable Entrepreneurship (Sustainopreneurship)

Entrepreneurship is an important driver that can transform society towards a more sustainable future (Belz & Binder, 2017; Hall et al., 2010). It is considered as a precondition for sustainable development and a strategic tool for a balanced and non-discriminatory society. It empowers the citizens to create a self-domain of economic activity through innovative business models (Huda, 2016). While, the term "Sustainable Development was first introduced at UN conference on Human Environment in 1972 and later it has gained popularity by the way of a report of UN World Commission on Environment and Development in 1987(also referred as The Brundtland Report).According to this report definition of sustainable development is "Sustainable development is a development that meets the needs of the present generation without compromising the ability of future generations to meet their own needs". This definition reveals the concept of sustainability by emphasizing that sustainable development is all about using the renewable resources wherever required and non-renewable resources should be recycled to extend their viability for future generations (Hall et al., 2010).

Despite this flourishing research domain and relevancy of established entrepreneurship literature, a very few numbers of research papers were published on sustainable development in the top entrepreneurship journals, for example, those listed on Thomson Reuters's Journal Citation Reports. Although sustainability-based research papers are scarce in the leading entrepreneurship journals but still a lot of material related to this area has published in non-entrepreneurship journals. Before the year 2002, only a few papers were there in the area of sustainable development and entrepreneurship but this trend is changing with the passage of time and the number of papers published is increasing. Most of the recent papers appeared in relatively new journals i.e. Greener Management International (which published two special issues on the topic), Business Strategy and the Environment (which published six articles) and the Journal of Cleaner Production (which published three articles). Another journal "Energy Policy" practitioners focused journal with an impact factor of 1.76 have published three papers but it did not include any details about the area of entrepreneurship. Most of these researchers were more focused the subject area of sustainable development rather than entrepreneurship. Although it's an emerging research area, the basic concepts of sustainable development and entrepreneurship still remained disconnected. Most of the literature is prescriptive than descriptive. Many researchers have studied the motivation of entrepreneurs to start sustainable ventures (Hall et al., 2010). According to (Dixon & Clifford, 2007) there is a strong relationship between entrepreneurialism and environmentalism. Some researchers have stated that ecopreneurs are a totally separate idea (Hall et al., 2010).

There is increasing awareness for the transformation of society in changing the way it uses natural resources to produce energy in order to cope up with severe environmental issues like degradation of ecosystem and change in global climate. In this context, entrepreneurship is considered an important mechanism to bring about a dramatic transformation to sustainable products and services. A large number of books that give serious warnings about environmental disaster mostly end up with optimistic closing note by concluding that civilization preservation depends upon the heroic social and environmental entrepreneurs (Dajian & Plan, 2003; Hover-Dixon, 2006). In spite of the importance of entrepreneurship for promoting sustainable development, uncertainty still exists about the nature of entrepreneurship role in the area of sustainability and how it evolved. Since a long time entrepreneurship has been considered as an important vehicle for societal transformation especially during the transition of an economy from one technological era to another (Schumpeter, 1942; J. A. Schumpeter, 1934).

Sustainable Entrepreneurship is based on the concept of the triple bottom line that is the concern about economic, social and ecological goals. Sustainable entrepreneurship process consists of six phases: 1) recognition of a social or ecological problem; 2) Identifying a social or ecological opportunity; 3) developing a double bottom line solution; 4) developing a triple bottom line solution; 5) financing and formulation of a sustainable enterprise; 6) creating or entering into a sustainable market (Belz & Binder, 2017). Sustainopreneurship is used as a solution to economic, social and environmental sustainability issues, it converts the problems arises into these areas as opportunities through their sustainable innovations (Huda, 2016).

Below Figure 9 shows the polycentric view by combining the concepts of economic, social and environmentally driven entrepreneurship. Based on the relevant entrepreneurial opportunities and intentions sustainable entrepreneurship is presented as an intersection of these three concepts (Schlange, 2009).

The three objectives of Sustainopreneurship are considered as mutually dependent in a way that a sustainable human society on our planet will completely on economic development which has a greater tolerance for the natural environment.But the question arises that whether the stakeholder framework of sustainable entrepreneurs can be considered as a combination of the other three variants. In this case, it can be seen located inside the intersection as shown in above figure 9.

There are various definitions of Sustainopreneurship, one group of scholars who are publishing in the sustainability journals they focus on the concept of sustainable development and triple bottom line and consider entrepreneurial activity at its second (Huda, 2016; Parrish & Foxon, 2009; Schlange, 2009). While another group who are writing in mainstream Entrepreneurial journals, they align the concept of the triple bottom line with entrepreneurship process (Cohen & Winn, 2007; Hall et al., 2010; Patzelt & Shepherd, 2011). So according to the later, Sustainopreneurship can be defined as the identification,

Figure 9. Sustainability-driven entrepreneurship as a concept of intersection
Source: (Schlange, 2009)

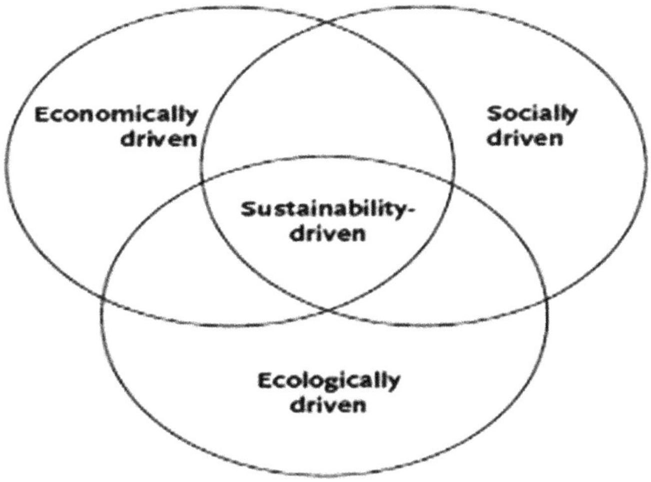

evaluation and exploitation of opportunities by individuals to produce such goods and services that bring economic, social, and ecological gains (Belz & Binder, 2017).

All these above definitions offer a valuable understanding of the development of sustainable entrepreneurship. The definitions highlight the important aspects and characteristics of sustainable entrepreneurship that have developed over time. Four defining attributes of the concept have emerged after studying these definitions " which are 1) Balancing environmental and social concerns 2) Economic gains, 3) Market failures and disequilibria, and 4) Transforming Sectors towards sustainability" (Bell & Stellingwerf, 2012).

According to (Shepherd & Patzelt, 2011), "Sustainable Entrepreneurs bring into existence future products, processes, and services for gain, where the gain is broadly construed to include economic and non-economic gains to individuals, the economy, and society".These -non-economic benefits are the differential point between traditional and sustainable entrepreneurship. Such type of gains is for the benefits of people and society while the economic gain is the base of traditional entrepreneurship where benefits of society are ignored (Bell & Stellingwerf, 2012). Sustainopreneurship is a complex phenomenon and development paths for sustainable ventures not only based on a single factor rather there are a large number of factors that play their role. Therefore explaining the development of sustainable development oriented ventures requires to focus on those interrelationships rather than just taking into account the individual factors (Muñoz & Dimov, 2015).

Sustainable entrepreneurship has considerable overlapping characteristics of other forms of entrepreneurship i.e. Social and ecological. Sustainopreneurship attempts to find the balancing state of social, economic and environmental concerns (Bell & Stellingwerf, 2012). Ecological entrepreneurship does not consider the non-economic gains for the larger interest of society and community (Shepherd & Patzelt, 2011); therefore ecopreneurship cannot be called Sustainopreneurship because former is concerned about the preservation of the natural environment. Hence, the sustainable entrepreneur is solving a particular environmental problem and transforming a sector into an environmentally sustainable condition. Sustainable entrepreneurship also comprises the characteristics of Social entrepreneurship which includes enhancing social impact, solving social problems and increasing social wealth (Bell & Stellingwerf, 2012).

Figure 10. Sustainable entrepreneurship as the mix of people, planet & profit
Source:(Bell & Stellingwerf, 2012)

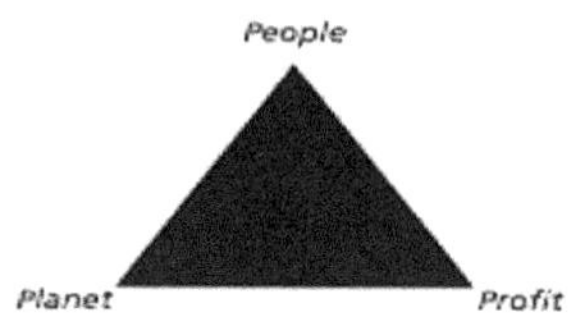

Besides environmental and social gains, sustainable entrepreneurs also work for economic gains or profit. As (Shepherd & Patzelt, 2011) clearly defined the concept of gain "Gain is broadly construed to include economic and non-economic gains to individuals, the economy, and society". Profit is also a very crucial element for sustaining a business. Hence sustainable entrepreneurship works for economic, social and environmental gains which differentiate it from social and ecological entrepreneurship which covers only a single aspect of sustainability that is social or ecological (Bell & Stellingwerf, 2012).

It can be concluded that any entrepreneurial activity will be said to be sustainable or will satisfy sustainable development goals if it incorporates a clear balance between 3 P's (Profit, People, and Planet) in their business initiatives. Sustainable entrepreneurship, a form of entrepreneurship which is derived from sustainable development, clearly consider and balance each P of sustainability, therefore sustainability can be called as the base of sustainable entrepreneurship (Bell & Stellingwerf, 2012).

Based on the definitions and literature about Triple Bottom Line, it can be said that sustainable entrepreneurship creates a balance between economic (Profit) as well as non-economic gains (social & environment) (Bell & Stellingwerf, 2012). Although, the concept of the triple bottom line was widely used by researchers to explain sustainable entrepreneurship, the concept is also having some limitations. As proposed by (Cohen & Winn, 2007) the link of the triple bottom line to sustainable entrepreneurship requires some further investigations. Based on the concept of the triple bottom line a revised model of Sustainopreneurship was presented by (Majid & Koe, 2012) including one additional dimension of "culture". The model is given as under:

A large number of researchers have emphasized the need to include a new dimension into the sustainable entrepreneurship framework. According to (Majid & Koe, 2012), culture must be included as the fourth pillar of sustainable development to create harmony among cultural, social, environmental

Figure 11. A revised model of sustainopreneurship
Source:(Majid & Koe, 2012)

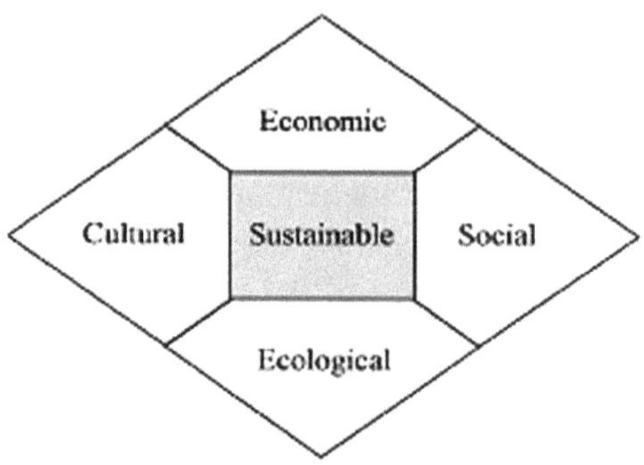

and economic dimensions. They said that equal weights must be given to each dimension in order to be a truly sustainable entrepreneur.

CONCLUSION

The need of entrepreneurship for sustainable development is having crucial for Asian countries because this Region is the home of 60% of world population having rich natural resources but their employment opportunities are very low which highlights the need for entrepreneurial education for the people of this Region (Rasool et al.,). The Government of Pakistan is facing several challenges to achieve Sustainable development Goals signed with UN in the year 2015, although the government has aligned its development plan Vision 2025 with the targets outlined in the UNs 2030 Agenda (Council, 2014). Given the weak socio-economic indicators, for Pakistan being a rapidly urbanized country in South Asia, achieving the developmental targets is an arduous and more challenging task due to migration from rural areas, poor standard of living, health, and water and sanitation facilities. Therefore, hard efforts are required to achieve a decent level of development(Akram, 2017).

Keeping in view the significance of human resource in the economy, Pakistan can't make sustainable development without addressing its human resource issues (Malik, 2003). At this current state of affairs self-employment or entrepreneurship is the only solution for economic development of Pakistan (Qaiser et al., 2014). Realizing this importance of entrepreneurship for sustainable development of the country, entrepreneurship is an area attracting a lot of attention from academician and researchers (Ahmed et al., 2010). Based on the given facts, while the government has included the entrepreneurship in its developmental agenda for achieving the sustainable development goals, it is the need of the hour to highlight the role of sustainable entrepreneurship as the only solution to address social, economic, environmental and cultural concerns simultaneously. So, it can be said that sustainable entrepreneurship can be adopted as the new world order for achieving the universal life claims of sustainable development by all nations, specifically by Pakistan.

Implications of the Study

The study has some important implications for policymakers, government, businesses and entrepreneurs. The study will highlight the role of Sustainopreneurship for the achievement of Sustainable Development Goals by all member nations of UNO, particularly by Pakistan. This will urge the policy makers and government to establish and incorporate such policies that will help in creating the awareness of Sustainable Entrepreneurship and for providing the conducive environment for the development and growth of sustainable business models. The awareness of this important phenomenon of Sustainopreneurship can help the Government of Pakistan in achieving its vision 2025 which is aligned with SDGs. The understanding of the concept will promote the sustainable business models which will help in enhancing the socio-economic and environmental condition of Pakistan which will result in improving the HDI ranking of the country. The sustainable business models will help in saving the resources for the present as well as for future generations. The understanding of the concept will help the entrepreneurs and businesses in converting the socio-economic and environmental issues into entrepreneurial opportunities. Located in the same context, the mechanism of Sustainopreneurship will help underdeveloped and developing

Asian countries (Region is the home of 60% of the world population having rich natural resources) to follow the same models to achieve their SDGs

Future Recommendations

Future research can be conducted on the role of higher education institutions for providing such entrepreneurial ecosystem that will help in the development of Sustainopreneurship in Pakistan as well as in Asian countries for the achievement of their Sustainable Development Goals.

Note: This Chapter is taken from an unpublished PhD thesis of Shakeela Kousar, PhD Scholar, Department of Management Sciences, The Islamia University of Bahawalpur, Pakistan working under the supervision of Dr. Jawad Iqbal.

REFERENCES

Abrahamsson, A. (2007). Sustainopreneurship-business with a cause: conceptualizing entrepreneurship for sustainability.

Ahmed, I., Nawaz, M. M., Ahmad, Z., Shaukat, M. Z., Usman, A., Rehman, W.-u., & Ahmed, N. (2010). Determinants of students' entrepreneurial career intentions: Evidence from business graduates. *European Journal of Soil Science*, *15*(2), 14–22.

Akram, S. (2017). *Pakistan's Development Dilemma*. Retrieved from Pakistan: In U. Balakrishnan, T. Duvall, & P. Primeaux, (2003) Rewriting the bases of capitalism: reflexive modernity and ecological sustainability as the foundations of a new normative framework. *Journal of Business Ethics*, *47*(4), 299–314.

Bass, S., & Dalal-Clayton, B. (2012). *Sustainable development strategies: a resource book*. Abingdon, UK: Routledge. doi:10.4324/9781849772761

Bell, J., & Stellingwerf, J. (2012). Sustainable entrepreneurship: The motivations and challenges of sustainable entrepreneurs in the renewable energy industry.

Belz, F. M., & Binder, J. K. (2017). Sustainable entrepreneurship: A convergent process model. *Business Strategy and the Environment*, *26*(1), 1–17. doi:10.1002/bse.1887

Bilge, H., & Bal, V. (2012). Entrepreneurship Aptitude: An Empirical Study On Undergraduate And Vocational High Scholl Students In Celal Bayar University. *Journal of Süleyman Demirel University Institute of Social Sciences*, *16*, 131–148.

Boschee, J., & McClurg, J. (2003). Toward a better understanding of social entrepreneurship: Some important distinctions.

Carley, M., & Christie, I. (2017). *Managing sustainable development*. Abingdon, UK: Routledge. doi:10.4324/9781315091525

Certo, S. T., & Miller, T. (2008). Social entrepreneurship: Key issues and concepts. *Business Horizons*, *51*(4), 267–271. doi:10.1016/j.bushor.2008.02.009

Chani, M. I., & Shahid, M. (2012). Human capital formation and economic development in Pakistan: an empirical analysis. *Актуальні проблеми економіки*(6), 486-495.

Choi, D. Y., & Gray, E. R. (2008). The venture development processes of "sustainable" entrepreneurs. *Management Research News*, *31*(8), 558–569. doi:10.1108/01409170810892127

Cohen, B., & Winn, M. I. (2007). Market imperfections, opportunity and sustainable entrepreneurship. *Journal of Business Venturing*, *22*(1), 29–49. doi:10.1016/j.jbusvent.2004.12.001

Cordova, D. (2013). Creating the environment for entrepreneurial success. *Center for International Private Enterprise.* Retrieved from http://www.cipe.org/creating-environment-entrepreneurial-success

Cormier, S. M. (2003). *Business incubation in inner-city emerging markets as an economic development tool.*

Council, N. E. (2014). *Pakistan Vision 2025.* Pakistan.

Dajian, Z., & Plan, B. (2003). Rescuing a planet under stress and a civilization in trouble. *Chinese Journal of Population, Resources and Environment*(6), 4.

Dawn. (2016). Pakistan's development challenges. *Dawn.*

Díaz-Casero, J. C., Fernández-Portillo, A., Sánchez-Escobedo, M.-C., & Hernández-Mogollón, R. (2017). The Influence of University Context on Entrepreneurial Intentions Entrepreneurial Universities (pp. 65–81). Berlin, Germany: Springer.

Dixon, S. E., & Clifford, A. (2007). Ecopreneurship–a new approach to managing the triple bottom line. *Journal of Organizational Change Management*, *20*(3), 326–345. doi:10.1108/09534810710740164

Du Pisani, J. A. (2006). Sustainable development–historical roots of the concept. *Environmental Sciences*, *3*(2), 83–96. doi:10.1080/15693430600688831

Greblikaite, J. (2012). Development of Social Entrepreneurship: Challenge for Lithuanian Researchers. *European Integration Studies*(6), 210-215.

Hall, J. K., Daneke, G. A., & Lenox, M. J. (2010). Sustainable development and entrepreneurship: Past contributions and future directions. *Journal of Business Venturing*, *25*(5), 439–448. doi:10.1016/j.jbusvent.2010.01.002

Hart, S. L. (2005). Capitalism at the crossroads: The unlimited business opportunities in solving the world's most difficult problems. London, UK: Pearson Education.

Henrekson, M., & Sanandaji, T. (2014). Small business activity does not measure entrepreneurship. *Proceedings of the National Academy of Sciences of the United States of America*, *111*(5), 1760–1765. doi:10.1073/pnas.1307204111 PMID:24449873

Hopwood, B., Mellor, M., & O'Brien, G. (2005). Sustainable development: Mapping different approaches. *Sustainable Development*, *13*(1), 38–52. doi:10.1002d.244

Hover-Dixon, T. (2006). The Upside of Down: Catastrophe, Creativity and the Renewal of Civilisation. London, UK: Souvenir Press.

Huda, K. N. (2016). Towards Sustainopreneurship Development at the Tertiary Level Education: A Case study on Southern University Bangladesh. *Journal on Innovation and Sustainability*, *7*(2), 3–16.

Isenberg, D. J. (2010). How to start an entrepreneurial revolution. *Harvard Business Review*, *88*(6), 40–50.

İyigün, N. Ö. (2015). What could entrepreneurship do for sustainable development? A corporate social responsibility-based approach. *Procedia: Social and Behavioral Sciences*, *195*, 1226–1231. doi:10.1016/j.sbspro.2015.06.253

Klugman, J., Rodríguez, F., & Choi, H.-J. (2011). The HDI 2010: New controversies, old critiques. *The Journal of Economic Inequality*, *9*(2), 249–288. doi:10.100710888-011-9178-z

Korosec, R. L., & Berman, E. M. (2006). Municipal support for social entrepreneurship. *Public Administration Review*, *66*(3), 448–462. doi:10.1111/j.1540-6210.2006.00601.x

LEAD. (2016). *Sustainable Development Goals (SDGs) (2015-2030)* Knowledge Hub on SDGs. Briefing Notes. Leadership for Environment and Development (LEAD), Pakistan. Islamabad, Pakistan. Retrieved from http://www.lead.org.pk

<conf>Leggett, J. A., & Carter, N. T. (2012). Rio+ 20: The United Nations Conference on Sustainable Development, June 2012.</conf>

Luke, T. W. (2013). Corporate social responsibility: An uneasy merger of sustainability and development. *Sustainable Development*, *21*(2), 83–91. doi:10.1002d.1558

Mahbub Ul Haq Research Center. (2018). History. Retrieved from https://mhrc.lums.edu.pk/history-3

Mahmoud, Y., Makoond, A., & Naik, A. (2017). *Entrepreneurship for Sustaining Peace*. International Peace Institute Issue.

Majid, I. A., & Koe, W.-L. (2012). Sustainable entrepreneurship (SE): A revised model based on triple bottom line (TBL). *International Journal of Academic Research in Business and Social Sciences*, *2*(6), 293.

Malik, I. (2003). *An Entrepreneurial Vacuum*. Pakistan.

Meadows, D. H., Meadows, D. H., Randers, J., & Behrens, W. W. III. (1972). *The limits to growth: a report to the club of Rome (1972)*. Google Scholar.

Muñoz, P., & Dimov, D. (2015). The call of the whole in understanding the development of sustainable ventures. *Journal of Business Venturing*, *30*(4), 632–654. doi:10.1016/j.jbusvent.2014.07.012

Naudé, W. (2013). *Entrepreneurship and economic development: Theory, evidence and policy*. Browser Download This Paper.

Neumayer, E. (2001). The human development index and sustainability—A constructive proposal. *Ecological Economics*, *39*(1), 101–114. doi:10.1016/S0921-8009(01)00201-4

Parrish, B. D., & Foxon, T. J. (2009). Sustainability entrepreneurship and equitable transitions to a low-carbon economy. *Greener Management International* (55).

Pastakia, A. (1998). Grassroots ecopreneurs: Change agents for a sustainable society. *Journal of Organizational Change Management*, *11*(2), 157–173. doi:10.1108/09534819810212142

Patzelt, H., & Shepherd, D. A. (2011). Recognizing opportunities for sustainable development. *Entrepreneurship Theory and Practice*, *35*(4), 631–652. doi:10.1111/j.1540-6520.2010.00386.x

Qaiser Gillani, D., Asghar, N., & Farooq, F. (2014). Socio-Economic Determinants of Self-Employment: Evidence from Southern Punjab (Pakistan). *Pakistan Journal of Social Sciences*, *34*(2).

Quental, N., Lourenço, J. M., & Da Silva, F. N. (2011). Sustainable development policy: Goals, targets and political cycles. *Sustainable Development*, *19*(1), 15–29. doi:10.1002d.416

R. D., S. (n.d). *Introduction to sustainable development: World conservation strategy of the international union for the conservation of nature and natural resources (iucn)*. York University, Toronto, Canada.

Rasool, F., Gulzar, A., & Naseer, S. (n.d.), Drivers of Entrepreneurship: Linking with Economic Growth and Employment Generation.

Report-UNDP, H. D. (2010). The real wealth of nations: Pathways to human development: United Nations Development Program New York.

Sachs, J. D. (2012). From millennium development goals to sustainable development goals. *Lancet*, *379*(9832), 2206–2211. doi:10.1016/S0140-6736(12)60685-0 PMID:22682467

Sagar, A. D., & Najam, A. (1998). The human development index: A critical review1. *Ecological Economics*, *25*(3), 249–264. doi:10.1016/S0921-8009(97)00168-7

Salem, M. I. (2014). The role of business incubators in the economic development of Saudi Arabia. *The International Business & Economics Research Journal (Online)*, *13*(4), 853.

Schaper, M. (2002). The essence of ecopreneurship.

Schlange, L. E. (2009). Stakeholder Identification in Sustainability Entrepreneurship. *Greener Management International*(55).

Schumpeter, J. (1942). *Capitalism, socialism and democracy*. New York, NY: Harper.

Schumpeter, J. A. (1934). *The schumpttr: Theory economic development*. Cambridge, MA: Harvard University Press.

Sekliuckiene, J., & Kisielius, E. (2015). Development of Social Entrepreneurship Initiatives: A Theoretical Framework. *Procedia: Social and Behavioral Sciences*, *213*, 1015–1019. doi:10.1016/j.sbspro.2015.11.519

Shepherd, D. A., & Patzelt, H. (2011). The new field of sustainable entrepreneurship: Studying entrepreneurial action linking "what is to be sustained" with "what is to be developed". *Entrepreneurship Theory and Practice*, *35*(1), 137–163. doi:10.1111/j.1540-6520.2010.00426.x

Tabassum, S. (2015). State of Human Resource Development (HRD) in Pakistan. *International Journal of Science and Research*, *6*(4).

Ul Haq, M. (1995). *Reflections on human development*. Oxford, UK: Oxford University Press.

UNDP. (1995). *Human Development Report-United Nations Development Program*. Retrieved from UNDP. (2015). *Human Development Report 2015*. Retrieved from UNDP. (2016). *Human Development Report 2016-Human Development for everyone*. Retrieved from UNO. (2010). *The Millennium Development Goals Report*. In Y. D. Uslu, Y. Hancıoğlu, & E. Demir (2015), Applicability to green entrepreneurship in Turkey: A situation analysis. *Procedia: Social and Behavioral Sciences*, *195*, 1238–1245.

Waas, T., Hugé, J., Verbruggen, A., & Wright, T. (2011). Sustainable development: A bird's eye view. *Sustainability*, *3*(10), 1637–1661. doi:10.3390u3101637

WCED. (1987). *Our common future*. Oxford, UK: Oxford University Press.

WCED. (1987). *S. W. S.* World Commission on Environment and Development.

Yang, W.-H. (1967). Human resource as the key factor of economic development.

Youssef, A. B., Boubaker, S., & Omri, A. (2018). Entrepreneurship and sustainability: The need for innovative and institutional solutions. *Technological Forecasting and Social Change*, *129*, 232–241. doi:10.1016/j.techfore.2017.11.003

Zahra, S. A., Gedajlovic, E., Neubaum, D. O., & Shulman, J. M. (2009). A typology of social entrepreneurs: Motives, search processes and ethical challenges. *Journal of Business Venturing*, *24*(5), 519–532. doi:10.1016/j.jbusvent.2008.04.007

Zavaleta, D., & Tomkinson, J. (2015). *Training Material for Producing National Human Development Reports*. UNDP Human Development Report Office.

Chapter 19
Crime Hotspot Prediction Using Big Data in China

Chunfa Xu
Tianjin University, China

Xiaoyang Hu
Tianjin University, China

Anqi Yang
Tianjin University, China

Yimin Zhang
Tianjin University, China

Cailing Zhang
Tianjin University, China

Yufei Xia
Tianjin University, China

Yanan Cao
Tianjin University, China

ABSTRACT

This chapter proves that utilizing big data and machine learning to predict crime is feasible in China. Researchers introduce five new machine learning algorithms into the field of crime prediction and compare them with four methods widely used in previous research. Using a weekly dataset in 213 street-level cells of Shanghai from April 2017 to March 2018, the researchers find new methods work better in predicting whether a specific cell will be a crime hotspot in next week. Five among nine methods can predict crime with more than 90 percent accuracy. These findings provide a scientific reference for urban safety protection. The research adds some significant evidence to a theoretical literature emphasizing that big data can predict crime.

DOI: 10.4018/978-1-7998-0357-7.ch019

INTRODUCTION

Crime now has been recognized as one of the most important social problems in the world, it is always related to such words like violence, unhappiness and insecurity. What is more, it will affect both the life quality of people and the economic development level of society, including public security, children development and adult socio-economic status and so on. According to some researches, crime tends to slower economic growth at both the national level (Mehlum et al., 2005) and the local level, such as cities and metropolitan areas (Cullen & Levitt, 2009).

Unfortunately, with the rapid development of the world, growing number of countries are facing increasingly serious crime problems. Since the early 1990s, the crime rate around the world has risen by an average of 5% a year. For example, according to BBC report, crime in England and Wales jumped by 13% in 2017, with a total of more than 5 million offences compared with 4.6 million last year, making it the fastest rise in crime in Britain for a decade. Beyond all that, the forms of crime are becoming more and more diverse, such as cyber-crime, due to the development of technologies. All of these, make it very hard to solve crime problems. Effective means of crime prevention are indispensable.

With the rapid development of network technology and the increasing speed of data transmission, people's daily life has been gradually separated into two levels: reality and network (Mcafee et al., 2012). At the network level, we have entered the "cloud" living environment. All kinds of basic data and behavioral data in real life will be uploaded and recorded to this "cloud" instantly. The huge database provides us with the possibility of various data analysis. With the advent of smartphones and wearable computing devices, every change in people's behavior, location, and even physical data becomes data that can be recorded and analyzed, an era of mass production, sharing, and application of data is beginning. With the continuous development of economy, the imbalance of national quality and cultural level induces the emergence of crimes. At the present stage, the total number of crimes in China presents a trend of increasing year by year, and the crime rate keeps rising (Kumar et al., 2018). Therefore, using big data to prevent crime is a necessary means for public security organizations to investigate crimes in the future.

As early as 2011, the crime prediction system has been put into operation in a large group of major cities in the United States and the United Kingdom, which achieved remarkable results (Xuemei, 2015). Time magazine even listed the crime prediction system based on big data as one of the top 50 inventions in 2012. According to the research report written by the Rand corporation in 2013, the prediction analysis of American crime intelligence is divided into four categories, namely, the method of predicting crimes, the method of predicting criminals, the method of predicting the identity of criminals and the method of predicting the victims of crimes (Andrey Bogomolov et al., 2014). Specific approaches are mainly based on low-complexity and small-scale historical crime data, and use Crime Mapping, Hot-spot Policing, Comp Stat and other analytical tools to conduct forecasts (Dumbill, 2013). The probability prediction and alarm prompt is obtained by computer program operation and deduction, through the analysis of historical crime data, alarm data, economic conditions and other small range of external data, combined with the crime map to determine the hot spots of crime (Inbaek et al., 2017).

Crime prediction is of great practical significance to the whole society. Scientific crime prediction methods and technologies can help the public security organs to effectively utilize the known data on criminal activities and their trends to predict the possible future criminal behaviors. And it is based on the predicted results to formulate the action deployment, in order to maximize the effectiveness of limited resources (Daichao et al., 2014).

Firstly, crime prediction can prevent crime in advance and realize accurate crime fighting. Crime prediction can provide preventive crime intervention, which makes it possible to kill crime in its cradle. On April 18, 2013, for example, Huairou district in Beijing launched the "Crime Data Analysis and Trend Prediction System," which collected data of more than 16,000 criminal cases in Huairou in the past 9 years and imported them into the system database through standardized classification. According to various prediction models established by mathematical experts, the probability and types of possible crimes in a certain period along with the region in the future were predicted automatically, providing guidance for fighting crimes. After the use of the prediction system, the criminal cases in the districts of Longshan, Quanhe and Huairou, which are prone to commit crime, dropped by 10.7%, 9.3% and 8.8% respectively in 2013. It can be considered that the Crime Data Analysis and Trend Prediction System has made great contributions to the prevention and prediction of crime.

Secondly, crime prediction can optimize the allocation of resources and improve work efficiency. If the crime prediction system can predict the probability of occurrence based on the terrain, time, weather and peripheral events on the basis of data, it will be conducive to the rational allocation of police force by the public security organs and optimize the allocation of resources. In 2013, for instance, Hushuguan police station, Huqiu branch of Suzhou public security bureau became the first pilot unit of crime prediction system. After three months of trial operation, 97 types of illegal crimes that can be controlled has dropped almost 20% month-on-month in the police station area, among which the road theft category dropped 45% month-on-month. In view of the key time and key areas identified by big data analysis, the police station increased the patrol intensity, changed the emphasis from quantity to quality, optimized the allocation of human resources within the station, and improved the patrol work efficiency.

Thirdly, crime prediction can deter criminals and maintain social stability. At present, monitoring is one of the most effective means to deter criminals. Compared with monitoring, the crime prediction system controls the crime before it happens. Due to the security responsibility, protection in place and other factors, criminals are even more afraid to act, which will better maintain social security.

Though crime prediction is valuable, it is undeniable that predicting crime event is really a challenging work (Yu et al., 2011). Fortunately, there are some relative work aiming to solve this problem. According to previous work, a key component of solving the crime problem is analyzing where and when the crime occurred. Firstly, crime is inherently geographical where it takes place in a certain place (Chainey & Ratcliffe, 2005). Some existing work in sociology and criminology has demonstrated that, regardless of the specific unit of analysis defined, there is a high concentration of crime at the micro geographic level, which can be explained by the interaction of the victim and the offender, as well as the existence of criminal opportunities at the micro geographic level (Branting & Branting, 2013; Weisburd & Green, 1994; Branting & Branting, 1984). In small geographic areas (such as streets), such clusters of crime are often referred to hotspots, and are not necessarily consistent with trends occurring at larger geographic levels (such as communities). In addition, many studies have taken into account the local variations of criminal behaviors on spatial and temporal scales, adding time into the consideration, and found the differences between geographical units within a specific time range (Pereira et al., 2015; Uittenbogaard & Ceccato, 2012; Jeong et al., 2010). Toole et al. (2011) analyzed that the spatiotemporal pattern of criminal records indicated that there was a complex relationship of multiple scales in the spatiotemporal structure of criminal records. Also, in the context of China, Hu et al. used space-time Bayesian model research development and the space-time characteristics of the crime (Hu et al., 2018). By studying the space and time feature of crime events, researchers are getting better at predicting crime.

In this study, the authors try to predict crime event using big data in a Chinese context. Specifically, what the researchers want to predict is whether a street-level cell of Shanghai would become a crime hotspot in the next week or not. The researches collect data that describe the features changing along with the crime time (e.g. weather/date of important festival/date of big activities), as well as the data showing the crime's space features (e.g. demographics/transportation and mobility/points of interest) to predict crime hotspots in specific street-level cells of China Shanghai. By comparing the effectiveness of 9 methods in crime prediction, the authors find the accuracy of five is higher than 90%, and Gradient Boosting is the best prediction algorithm with 91.2% accuracy rate. These findings prove that using big data and machine learning to predict crime is feasible in China.

The structure of this chapter is as follows: the BACKGROUND section describes background information and the relevant preliminary work in the field of crime prediction; MAIN FOCUS OF THE CHAPTER section describes the research questions of this chapter; DATA & DESCRIPTIVE STATISTICS section talks about the data that the authors used in the experiment; METHODOLOGY section provides the detailed information of the methods adopted; EXPERIMENTAL RESULTS section reports the detailed information about the experimental results; The authors talk about the limitations and future research directions in FUTURE RESEARCH DIRECTIONS section, and finally summarizes the study in CONCLUSION section.

BACKGROUND

About a decade ago, machine learning played a role in areas such as handwritten digit recognition, image processing, and item evaluation. However, in recent years, with the rise of big data, a large number of machine learning applications are highly coupled with big data, and machine learning is applied in a wider space.

Google has established an influenza warning model based on the most frequently searched entries by 50 million Americans and data from the US Centers for Disease Control and Prevention during the period of seasonal influenza transmission between 2003 and 2008. The model first predicted the outbreak of H1N1 flu in 2009 before the public health agency. Baidu Big Data Department collected the data of all national teams and clubs in the world from 2010 to 2013 and verified the 2010 World Cup knockout data, in order to build the prediction model of the event. This forecast model successfully predicted the results of all the knockouts in the World Cup in Brazil in 2014. In March 2016, Alpha Go, with Go World Champion and professional nine-segment player Li Shishi, played a man-machine battle with a total score of 4 to 1, which symbolizes that machine learning does not rely on exhaustive or iterative speculation, but on the real intelligent development. In May 2018, Google Voice Assistant made a haircut appointment by phone, which is no different from real people, opening a new path for machine learning and big data in natural language processing. In 2019, with the gradual promotion of 5G commercial applications, the eras of digitalization, intelligence and Internet of everything including crime prevention are coming.

Crime prevention has always been a hot topic for policy makers and academia. All countries in the world are committed to find more scientific means of criminal intervention. Recently, big data has been widely applied in crime predicting, for the use in reflecting of city dynamics. In a city with a high crime rate of 5 million people, this means saving up to 300 lives a year. Attacks, robberies, burglary and car theft can be reduced by 30% to 40%. The most important thing is to give residents the great benefits of freedom of movement and inner peace. Technology does not solve crime problems quickly, but intelligence agencies can use data to deploy scarce resources and people more effectively. For example, real-time crime maps use statistical analysis to highlight crime patterns, while predictive policing goes a step further, predicting crime and stopping it before the event occurs. The survey shows that some Asian big cities are outstanding in their awareness of smart technologies and the use of them. Given the shift in more and more people-centric smart cities, it is necessary to assess residents' perceptions of existing technologies in their environment. Asian cities perform best in terms of awareness, use and satisfaction, while European cities are lagging behind.

Human mobility, traffic flow and social media and other nontraditional data provide us with new insights to crime inference work. In general, scholars mainly predict crime from three perspectives, personal information-centric perspective, social relationship-centric perspective and place-centric perspective.

In personal information-centric perspective, research pays more attention to the people who are prone to crime. Fries et al. (2014) found they can use general file information such as offenders age and crime history to determine their crime rate. What's more, they can also infer their very complex criminal plans. On the other hand, Wang et al. (2013) aims to automatically identify crimes committed by the same person in the historical crime database. The proposed system, called a serial finder, seeks to identify and classify the M.O. As for the people who are inability to control oneself and take responsibility for their actions such as mentally ill offenders, Pflueger et al. (2015) made an analysis of 259 psychiatric patients combined with their criminal records showed that their average risk was 107 months. Based on random forest method, the risk factors of recidivism are deduced and classified. They used the most significant predictors of recidivism-age, type of index crime, diversity of criminal history, and substance abuse-to build a statistical method. And the prediction for reoffending was 58 to 95% accurate, as the same time that the overall crime prediction rate is 65 to 97%. Also for special groups, Grieger & Hosser (2012) contacted in three prisons for teenagers and young people. They found ADHD (Attention Deficit Hyperactivity Disorder) cannot predict criminal recidivism and only can partially explain ADHD is related to delinquency.

In social relationship-centric perspective, research focuses on the criminal profiling at community. At the community level, Wang et al. (2016) use extensive experiments to predict the crime rate at the neighborhood level. Compared with traditional studies, they added point of interest (POI) characteristics to assist in demographic characteristics, and taxi traffic supplemented geographical neighbors as hyperlinks, and observed significant improvements in the performance of crime rate inferences. In another paper, Traunmueller et al. (2014) used computational methods to prove various social theories on a large scale. They used mobile phone data belonging to London and saw people's movements as crime-related features. Aghababaei & Makrehchi (2016) proposed that social media which offered users more and more chances to voluntarily share their thoughts and concerns in large amounts of data may hidden variable important events when combined. They found despite only using the information from Twitter, the correlation between this information and crime trends was well documented when the types of crimes are highly correlated with shared content. In a further research, considering the development of social networks, in order to obtain hot dynamic location information, Huang et al (2015) combined fine-grained data related to social connections with geographical location information. They explained mining geographic and social characteristics from location-based social networks correlates with the frequency of different types of crime and developed and evaluated a range of features, ultimately found some information that was indicative of high-crime areas.

In the location-centered perspective, the research question is the location of the predicted crime (Wang et al., 2016). For example, based on that crime does not occur uniformly in different parts of the city, but in certain areas, Tayebi et al. (2014) built CRIMETRACER using a personalized random walk to predict the probabilistic activity space. Bogomolov et al. (2014) used human behavior data from a combination of mobile network activity and demographics, as well as open data related to criminal incidents, to predict crime hotspots in specific London communities. They trained various classifiers and found that the random forest (RF) algorithm produced the best performance, with an accuracy rate of nearly 70%. The difference is that Toole et al. (2011) used criminal records and the results of cross-correlation measurement, eigenvalue spectrum analysis and random matrix theory to identify multi-scale spatial and temporal patterns. Finally, they found significant spatial and temporal correlations between crime data. For the sake of accuracy, Wang et al. (2012) shows a preliminary survey of twitter-based crime prediction. Automated semantic analysis is used to understand natural language Twitter posts that report criminal incidents. They used feature-based models to predict hit-and-run crimes. Joel Caplan et al. (2015) used risk terrain modeling (RTM) to predict gun crime. Risk maps produced by RTM use a series of background information related to the shooting opportunity structure to estimate the risk of future shootings because they are distributed in different geographic areas.

However, these studies are based on the situation of foreign countries. Whether the big data are applicable to urban crime rate intervention in China has not been fully explained. In this paper, the researchers try to fill in the blank of this part of the research.

MAIN FOCUS OF THE CHAPTER

The research question of this chapter can be divided into three specific questions.

Firstly, since crime prediction plays a more and more important role in our daily life, researchers begin to pay more attention on this question. There are numerous studies exploring how to predict crime using different methods and data in different countries, especially in developed countries, but only a few of them were in a developing country context especially in Chinese context, which is more practical in Asia. Based on available information, the researchers propose the first research question.

RQ1: Whether predicting crime is feasible in China?

Secondly, as the metropolitan area in China, Shanghai is a typical modern city, it is of great practical significance to study the crime prediction in Shanghai. Using a weekly dataset in 213 street-level cells of Shanghai from April 2017 to March 2018, the researchers try to prove that using big data to predict crime is viable to in China. On the basis of available data and material, the researchers propose the second research question to solve the specific problem.

RQ2: Whether a street-level cell of Shanghai would become a crime hotspot in the next week or not?

At present, using big data to predict crimes is the most popular and effective way to predict crime, but there are many methods to use big data to predict crimes, and which method is the best in the context of this paper remains to be explored. Moreover, the researchers try to introduce five new methods to the field of crime prediction and compare the prediction results of these five machine learning algorithms with previous crime prediction models used widely in classic paper. Therefore, the researchers propose the third research question to testify which methods work better in predicting crime.

RQ3: Which methods work better to predict crime in the context of China?

DATA AND DESCRIPTIVE STATISTICS

The researchers compile a unique dataset of Shanghai for the period from April 2017 to March 2018 by manually collecting crime data and feature data. This research chooses Shanghai as the research object mainly considering the availability of data. Shanghai is the first provincial government to open data in China. Since officially opening data in May 2012, Shanghai has been ahead of other provinces in terms of data openness.

Crime Data

The crime dataset used in this paper is crawled from Shanghai Public Security Bureau, which provided daily data from April 2017 to March 2018, totaling 516042 pieces. Each criminal data record includes the criminal ID, the date of the crime, and the address of the crime, the administrative law enforcement units directly responsible for this crime.

Future Data

Learning from recent studies, this research selects feature variables from two dimensions of time and place (Bachner, 2013; Rumi et al., 2018). According to the principle of collecting as many features as possible under the condition of data availability, the researchers finally get 31 feature variables.

Time variables controls the characteristics that change with the time of crime. Weather is an important time variable for predicting crime. The correlation between weather and crime has been widely proved (Ranson, 2014; Carleton & Hsiang, 2016). Therefore, the researchers crawl weather data from Aggregate data platform API. Another time variable used in this paper is important festivals or activities, including important activities (such as the Olympic Games, Asian Games, National Games and vocal concert), traditional festivals, weekend, winter and summer vacations.

Place variables shows the characteristics of criminal areas. Place variables cover demographics features, transportation features, entertainment features, regional economic features and social Sentiment feature.

- **Demographics Feature:** The demographics of an area include numerous statistics (e.g., population density, age distribution, population flow) and are generally considered as one of the primary factors that affect its crime levels. The data is supplied by Shanghai Bureau of Statistics.
- **Transportation Feature:** Factors related to transportation and mobility characterize the nature and functionality of an area and, consequently, can affect its crime levels. These factors include various aspects, such as the density and type of the road network, the number and type of transportation nodes and the flow of passengers in a specific area. Thus, this study considers the number of subways and the number of buses. The data is crawled by the Autonavi Map API.

- **Entertainment Feature:** Another factor that is considered to be of high importance for determining the crime rates in an area is the number and type of various establishments, facilities. The researchers consider the number of hotel, restaurant, entertainment and shopping center. The source of the data is the Autonavi Map.
- **Regional Economic Feature:** Regional economic features reflect the level of economic development of one region. Generally, the crime rate in areas with good economic development is lower than that in areas with poor economic development. The researchers collect GDP, fixed asset investment, the proportion of tertiary industry and green space area. The data is supplied by Shanghai Bureau of Statistics.
- **Social Sentiment Feature:** The researchers believe that the higher the negative degree of social emotions, the greater the probability of crime (Botchkovar & Broidy, 2013). The researchers use python to crawl microblogs data from Sina weibo platform to examine people's social behavior and emotions. The researchers calculate the sentiment index of microblog by formula (1).

$$SentimentIndex_{it} = \frac{Negative_{it} - Positive_{it}}{total\ word_{it}} \tag{1}$$

Where $SentimentIndex_{it}$ represents negative emotions on microblog in district i week t; $Positive_{it}$ refers to the number of positive words in all microblogs in district i week t; $Negative_{it}$ refers to the number of negative words in all microblogs in district i week t; $total\ word_{it}$ is the number of words in all microblogs in district i week t.

In summary, the researchers summarize all the characteristic symbols, their meanings in Table1.

Descriptive Statistics

Before empirical analysis of the crime data, the researchers first examine the characteristics of Shanghai's crime data from two dimensions of time and space.

The researchers calculate the monthly and weekly distribution of crime in the time dimension. Figure 1 shows the monthly number of crimes. As can be seen from the Figure 1, the number of crimes committed in February was significantly smaller than that in other months. Except in February, the number of crimes in other months was more than 40,000. This may be because the Spring Festival, the most important festival in a year for Chinese people, is in February. Figure 2 records the average number of crimes committed on each day of one week. Compared with weekdays, the number of crimes on weekends is significantly lower. The reason for this may be that workers usually have weekends off. Workers usually choose to rest at home during holidays, which will reduce burglary and other crimes.

The researchers also count the distribution of crime in the street-level cell. Table 2 shows the top 10 street-level cells with the largest number of crimes during the sample period. The researchers find the street-level cell with the highest crime number is in Jing 'an district of Shanghai. Four of the top ten are in Jiading district.

Table 1. Characteristic data, meaning and administrative level

Dimension	Specific Dimension	Feature	Meaning	The administrative level
Time	Weather	Maxtemperature	Weekly maximum temperature	District
		Mintemperature	Weekly minimum temperature	District
		Windpower	Mean wind speed	District
		Maxtemfluctuation	Weekly maximum temperature fluctuation	District
		Mintemfluctuation	Weekly minimum temperature fluctuation	District
		Weatherflu	Weekly fluctuations in weather patterns	District
		Festival	The number of holidays	Street
		Activity	The number of important activities	City
Place	Demographics	Age0_17	The proportion of people aged 0 to 17	District
		Age18_34	The proportion of people aged 18 to 34	District
		Age35_59	The proportion of people aged35 to 59	District
		Age_60	Proportion of the population aged 60 and over	District
		Popdensity	The population density	District
		Floaingpopulation	Floating population	District
		Inpopulation	The population moved in from outside the city	District
		Outpopulation	To move out of the city	District
	Transportation	Subway	The number of subway stations	Street
		Bus	The number of bus stations	Street
	Entertainment	Restaurant	The number of resturants	Street
		Hotel	The number of hotels	Street
		Entertainment	The number of entertainment venues	Street
		Shopping	The number of shopping mall	Street
	Regional economic	GDP	GDP per capita	District
		Fixedinvestment	Fixed asset investment	District
		Thirdindustry	The proportion of the tertiary industry	District
		Greenarea	Proportion of green space	District
	Social Sentiment	SentimentIndex	The mood of the Microblog content	District
		Microblogcount	The amount of Microblog posts	District
		Microblogtranspond	Weibo forwarding amount	District
		Microbloglike	Thumb up on Microblog	District
		Microblogcomment	Comments of Microblog	District

METHODOLOGY

The researchers regard the prediction of crime hotspots as a binary classification problem. Specifically, what the researchers want to predict is which streel-level place of Shanghai would become a crime hot-

Figure 1. Monthly crime statistics of Shanghai (Shanghai Public Security Bureau, 2018)

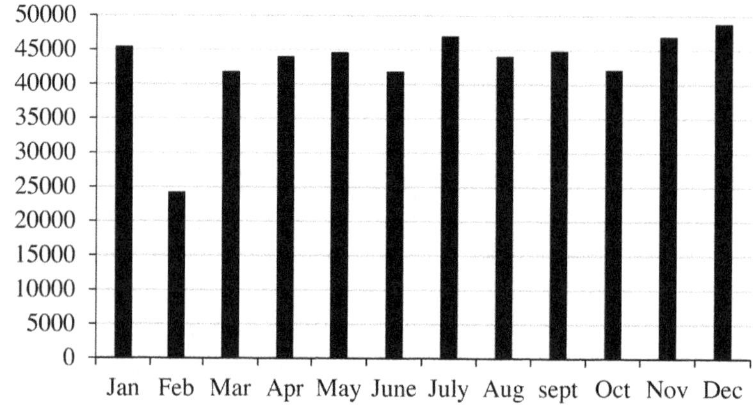

Figure 2. Weekly crime statistics of Shanghai (Shanghai Public Security Bureau, 2018)

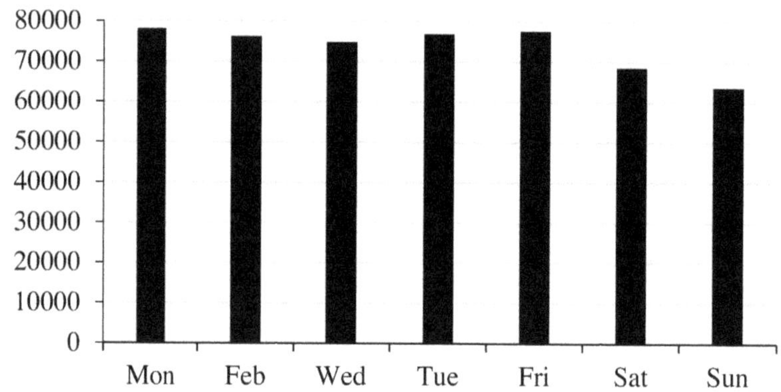

Table 2. Streets in the top 10 for the number of crimes of Shanghai

Street name	District name	Number of crime
Beizhan Street	Jing 'An	29866
Meilong Town	Minxing	18312
Anting Town	Jiading	12770
Pujiang Town	Minxing	10793
Malu Town	Jiading	10535
Chuansha New Town	Pudong	9664
Zhangjiang Town	Pudong	7530
Nanqiao Town	Fengxian	7311
Jiading Town	Jiading	7159
Fangsong Street	Songjiang	6326

Source: (Shanghai Public Security Bureau, 2018)

spot in the next week, and which not.

Data Preprocessing and Feature Extraction

Regionalization and Classification

Shanghai is divided into 213 street-level cells referring to the administrative plan of Shanghai in 2018. According to the information of the crime address, the researchers classify the crime data into different cells. The researchers calculate the longitude and latitude of the crime address and longitude and latitude boundaries of 213 street-level cells. If the latitude and longitude coordinates of one crime fall within the boundary of a street-level cells, the researchers believe that the crime occurred in the cell. Each crime event belongs to only one cell. Because of predicting crime hotspots with weekly frequency, researchers count the number of crimes in each cell by week.

Based on previous research on urban crime, the researchers define high crime areas as crime hotspots, where the number of crimes is higher than the median (Bogomolov et al., 2014). According to the median division rule, the researchers divide the data into two categories that high crime area is accounted for 49.11% and low crime area is 50.89%.

Feature Selection

All data in this paper are randomly divided into training data (80% data) and test data (20% data). The role of training data is feature selecting and fitting model. The test data is used to evaluate the prediction effect of the model.

The researchers calculate Pearson correlation coefficient among variables. The researchers find that there is a strong correlation between crime variable and demographic features, entertainment features. Next, based on the principle of average impurity attenuation of Random Forest algorithm, the researchers evaluate the importance of each features, and eliminate the traits of the bottom 10%. The top ten important features are shown in Table 3.

Model Building

In this subsection, the researchers compare the prediction results of nine models. In addition to previous crime prediction models used widely in classic paper, including Decision Tree (Nasridinov et al., 2013), Logistic Regression (Ludu, 2013), SVC (Bogomolovy et al., 2014), Random Forest (Sun et al., 2014), simultaneously the researchers applied some innovative methods widely used in machine learning algorithms to predict crime hotspot, containing Gradient Boosting, Ada Boost, Extra Trees, K Neighbors and Bagging on. The reason why the researchers try these new algorithms is that they have achieved good results in machine learning competitions.

Combined with these methods, the researchers trained various classifiers on the training data according to the above nine strategies to predict and estimate whether a specific cell will become a crime hotspot in the next week. According to the results, the researchers find Gradient Boosting modeled on algorithm yields the best performance in comparison of all other 8 classifiers.

The method Gradient Boosting is a machine learning technique for regression and classification problems that generates predictive models in the form of sets of weak predictive models (usually decision

Table 3. The top ten characters of importance

Feature	Importance	Rank
Entertainment	0.143	1
Subway	0.136	2
Restaurant	0.0899	3
Bus	0.0882	4
Hotel	0.0865	5
Popdensity	0.0367	6
Age35_59	0.0318	7
Outpopulation	0.0283	8
Microblogcount	0.0252	9
Microblogtranspond	0.0249	10

trees). As the same with other boosting methods, it builds models in a staged fashion and generalizes them by optimizing any differentiable loss function. The idea of Gradient Boosting originates from the observation of Leo Breiman, who believes that it can be interpreted as an optimization algorithm on an appropriate cost function (Breiman, 1997). Then Jerome H. Friedman (1999) developed an explicit regression gradient boosting algorithm, and Llew Mason et al. (1999) proposed a more general perspective of functional gradient boosting method. In the last two papers, the boosting algorithm is regarded as the iterative function gradient descent algorithm. In other words, the algorithm optimizes the cost function in function space by iteratively selecting a function (weak hypothesis) pointing in the direction of negative gradient. This view of boosting functional gradients has led to the development of boosting algorithms in many areas of machine learning and statistics in addition to regression and classification.

EXPERIMENTAL RESULTS

In this section, the researchers report and evaluate the experimental results of predictive effect gained by nine alternative models above. Binary classification is a wildly studied topic in statistical learning theory, machine learning and artificial intelligence (Proctor & Cho, 2006). Binary classification, which the researchers introduce to capture the cross-time dependence of crime prediction, which not only makes the feature space more compact, but also significantly improves the model performance indicators.

Binary Classification Problem and Evaluation

Table 4 describes a binary classification problem. Owing to our positive and negative samples, then there are four combinations of the predicted results and the real tags: TN, FN, FP, TP, as shown in table 4 below. The total number of samples is: TP+TN+FP+FN (Freeman & Moisen, 2008).

TN: predicted as 0 (Negative), and it was actually 0 (True-predicted correctly)

FN: predicted as 0 (Negative), but it was actually 1 (False-predicted wrong)

FP: predicted as 1 (Positive), but it was actually 0 (False-predicted wrong)

Table 4. Binary classification

		Real	
		0	**1**
Predict	0 (Negative)	TN (True Negative)	FN (False Negative)
	1 (Positive)	FP (False Positive)	TP (True Positive)

TP: predicted as 1 (Positive), and it was actually as 1 (True-predicted correctly)

The performance metrics and indicators that the researchers used to evaluate the predicted effect of 9 models above are as followed: Accuracy, Precision, Recall, F1 Score and AUC (Junker et al., 1999; Godbole et al., 2004).

Accuracy represents how many proportions of the samples are predicted correctly (Diebold & Mariano, 2002). The researchers calculate accuracy by formula (2).

$$Accuracy = \frac{TP+TN}{TP+TN+FP+FN} \tag{2}$$

Precision of a positive sample indicates how many of the samples predicted correctly are positive (Davis & Goadrich, 2006). Precision is calculated by formula (3).

$$Precision = \frac{TP}{TP+FP} \tag{3}$$

Recall for a positive sample describes how many samples that are actually labeled positive are predicted correctly (Junker et al., 1999). The researchers calculate recall by formula (4).

$$Recall = \frac{TP}{TP+FN} \tag{4}$$

F1 score is used to integrate precision and recall as an evaluation index, adding a weight coefficient which can assign different weights to recall and precision as required (Godbole et al., 2004). F1 score is calculated by formula (5).

$$F1score = \frac{2 \times recall \times precision}{recall + precision} \tag{5}$$

AUC is the area under the Receiver Operating Characteristic Curve (Robin, 2011).

Table 5. Metrics comparison

Model	Accuracy	Precision	Recall	F1score	AUC
Gradient Boosting	0.912	0.912	0.914	0.913	0.912
Random Forest	0.909	0.909	0.911	0.91	0.909
Ada Boost	0.904	0.901	0.909	0.905	0.904
Extra Trees	0.902	0.913	0.89	0.902	0.902
Bagging	0.901	0.91	0.892	0.901	0.901
Decision Tree	0.883	0.882	0.886	0.884	0.883
Logistic Regression	0.791	0.802	0.776	0.789	0.791
K Neighbors	0.739	0.751	0.72	0.735	0.739
SVC	0.514	0.509	0.999	0.674	0.511

Comparisons of Model Prediction Effectiveness

Table 5 includes all performance metrics gained by our models. The results of each evaluation indicator are obtained by taking the average value of 30 random running. The aim is to ensure the stability of the results.

As shown in the Table 5, all models except SVM achieve at least 70% prediction accuracy. Compared with traditional methods, the five methods introduced by the research work better in predicting whether a specific cell will be a crime hotspot in next week. The accuracy of the five new methods is more than 90% except K Neighbors. The Gradient Boosting Model achieves the best performance. Encouragingly, all the metrics showed by Gradient Boosting Model reach above 90%, which indicates that the model can relatively accurately forecast the crime hotspot with an ideal fitting result relative to those of previous crime prediction only reach 70% in other paper (Bogomolovy et al., 2014). These results prove that using big data to predict crime is viable to in China.

DISCUSSIONS OF PREDICTION RESULTS

The discussion in the previous section shows that Gradient Boosting is the best method to predict crime hotspot. In this section, the researchers further analyses the predictive effect of Gradient Boosting in practical application.

Unlike the previous division of training set and test set, the researchers use the data of the first 51 weeks as training set to predict 52th weeks of crime hotspot in this section. This approach is closer to our actual use because the researchers can use all the historical data as a training set in the actual prediction of crime hotspots. The accuracy based on the best predictive method Gradient Boosting performs is about 93.4%. Figure 3 and 4 present a comparison of the predicted results with the actual results.

The left graph in Figure 3 represents the real crime spots in Shanghai, while the right one describes the predictive crime spots. As is illustrated in the pictures, the black dots reflect crime hotspots and the black circles represent non-crime hotspots. As far as the entire city is concerned, the distribution of crime spots is roughly the same whose pattern shows a trend of radiation in both two subgraphs. In contrast, the crime hotspots are mainly scattered in the suburbs, rather than concentrated in the center of Shanghai. Due to the high density of the downtown area, the researchers observe the similarity between the two graphs from a more detailed perspective.

Specially, in order to obtain a clear and distinct view, the researchers zoom in on the central area of the image in Shanghai. Similar to Figure 3, the left graph in Figure 4 represents the real crime spots while the right one describes the predictive crime spots. The black dots refer to crime hotspots and the black circles represent non-crime hotspots. Figure 4 intuitively shows that crime predictive hotspots are almost identical to actual hotspots, only with delicate differences, which reflects that our Gradient Boosting algorithm provides us with excellent crime predictive effect.

FUTURE RESEARCH DIRECTIONS

The high prediction accuracy implies that using big data and machine learning is of great significance for crime prevention. The research in this field deserves further discussion. But the research does have some limitations: First, the time and place data that we used in the research is not very accurate, for example, the transportation data that we used is crawled by the Autonavi Map API, which is public API and doesn't provide a standard category of transportation information. So we crawl the transportation data using the keywords that we defined by ourselves, and it may lead to the incomplete data by this way. Second, the length of forecast period which is one week for now is not fine enough, this may cause that forecast is not timely. Third, our research is only about Shanghai, so it is geographically limited. And the model that we got good result in the research may not suitable for other cities in China. In future, the researchers plan to extend our research in two ways. For one thing, shorten the forecast period to predict more timely. Having access to crime and feature data aggregated on a daily or hourly basis would enable us to test our approach with finer times granularity. For another, explore the applicability of this method in more cities. Using data from one city as training dataset and other cities as test dataset is a good way to verify the model.

Figure 3. Comparison of crime hotspots in entire Shanghai

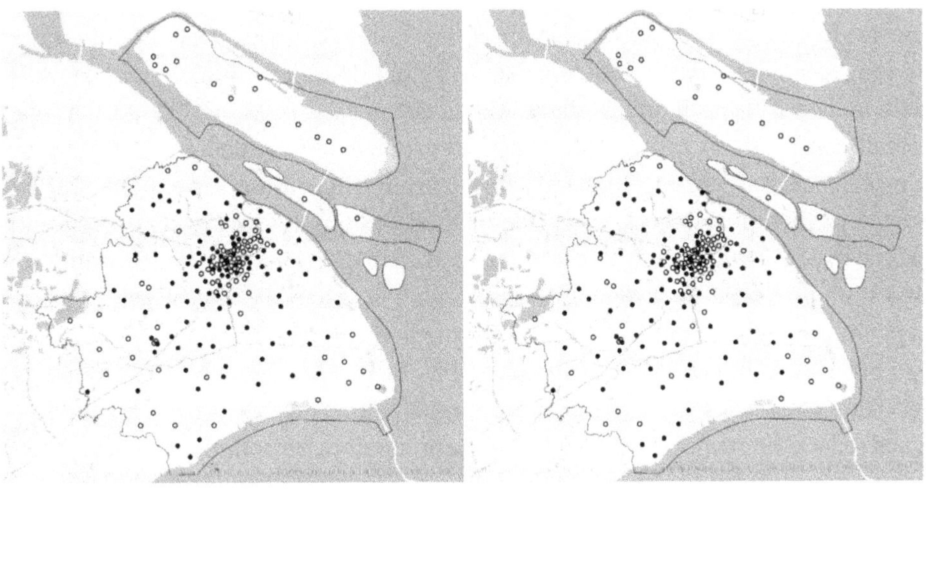

Figure 4. Comparison of crime hotspots in the center of Shanghai

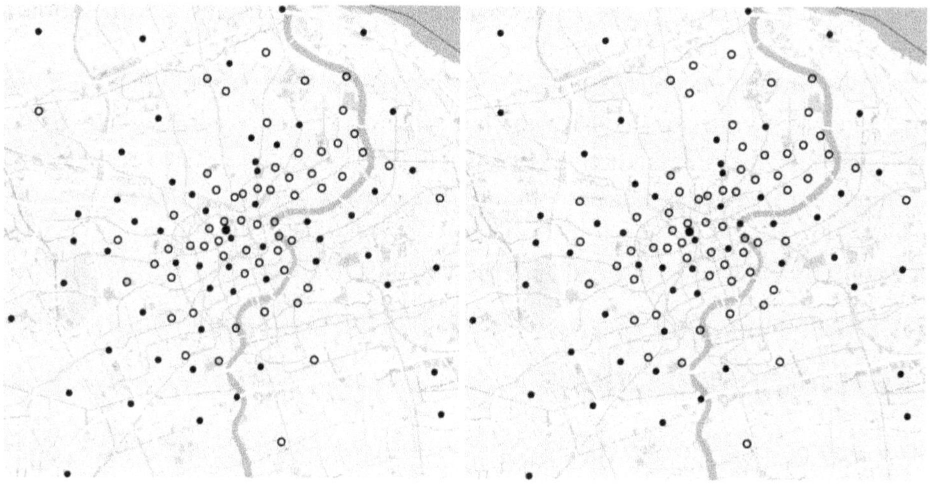

CONCLUSION

In this chapter the researchers collect both data that describe features changing along with the crime time and data showing the crime's space features to predict crime hotspots in specific street-level cells of China Shanghai. The researchers have compared the effectiveness of 9 machine learning algorithms in crime prediction, including Decision Tree, Logistic Regression, Support Vector Machine, Random Forest, Gradient Boosting, Ada Boost, Extra Trees, K Neighbors and Bagging on. The researchers find the accuracy of five algorithms is higher than 90%, and Gradient Boosting is the best prediction algorithm with 91.2% accuracy rate. These findings provide a scientific reference for urban safety protection. The research adds some significant evidence to a theoretical literature emphasizing that big data could predict crime. Compared with prior study, our main contributions from both theoretical and practical aspects are as follows.

This study has two main theoretical contributions. For one thing, this study broadens the application scenarios of big data theory. Despite the recent popularity of big data, little has been known regarding the effect of using big data to predict crime, especially in China. Most previous research about crime prediction are in a foreign context (Mehlum et al., 2005; Cullen & Levitt, 2009; Bogomolov et al., 2014), in this study the researchers try to predict crime using big data in a Chinese context. Although there are limitations in data collection, the result shows an accuracy exceeding 90% with a time interval of week. The result confirms that it's feasible to predict crime using big data in the Chinese context. For another, the researchers introduce five new machine learning algorithms to the field of crime prediction and compare the prediction results of these five methods with previous crime prediction models used widely in classic paper. Results show that new methods work better in predicting crime. The Gradient Boosting algorithm produced the best performance, with an accuracy rate of nearly 93%. This study, therefore, expands previous literature on traditional methods predicting crime.

Two major practical implications from this study. Firstly, this study provides a more scientific method for the supervision department to predict crime, and helps the supervision department to take timely precautions and reduce the crime rate; Secondly, crime prediction can optimize the allocation of resources and improve work efficiency. This study predict which streel-level place of Shanghai would become a crime hotspot in the next week, with an accuracy rate of nearly 93%, which is conducive to the rational allocation of police force by the public security organs and optimize the allocation of resources.

ACKNOWLEDGMENT

Thank seminar participants at Tianjin University for helpful discussions. The researchers especially thank Xi Zhang, Qi Wang, Xueyi Geng, Qiaoran Meng, Yuguang Yang, and Meng Chen for helpful comments on our research.

REFERENCES

Bachner, J. (2013). *Predictive policing: preventing crime with data and analytics*. Washington, DC: IBM Center for the Business of Government.

Bogomolov, A., Lepri, B., Staiano, J., Oliver, N., Pianesi, F., & Pentland, A. (2014). Once upon a crime: towards crime prediction from demographics and mobile data. In *Proceedings of the 16th international conference on multimodal interaction*. Istanbul, Turkey: ACM Press. 10.1145/2663204.2663254

Botchkovar, E., & Broidy, L. (2013). Accumulated strain, negative emotions, and crime: A test of general strain theory in Russia. *Crime and Delinquency*, *59*(6), 837–860. doi:10.1177/0011128710382346

Brantingham, P. J., & Brantingham, P. L. (1984). *Patterns in crime*. New York, NY: Macmillan.

Brantingham, P. L., & Brantingham, P. J. (1999). A theoretical model of crime hot spot generation. *Studies on Crime & Crime Prevention*, *8*(1), 7–26.

Breiman, L. (1997). *Arcing the edge (Technical Report 486)*. Berkeley, CA: Statistics Department, University of California. Retrieved from http://www.stat.Berkeley.EDU/users/breiman/

Caplan, J. M., Kennedy, L. W., Barnum, J. D., & Piza, E. L. (2015). Risk terrain modeling for spatial risk assessment. *Cityscape (Washington, D.C.)*, *17*(1), 7–16.

Carleton, T. A., & Hsiang, S. M. (2016). Social and economic impacts of climate. *Science*, *353*(6304), 9837. doi:10.1126cience.aad9837 PMID:27609899

Chainey, S., & Ratcliffe, J. (2005). *GIS and Crime Mapping*. Chichester, UK: John Wiley & Sons. doi:10.1002/9781118685181

Cullen, J. B., & Levitt, S. D. (1999). Crime, urban flight, and the consequences for cities. *The Review of Economics and Statistics*, *81*(2), 159–169. doi:10.1162/003465399558030

Daichao, L. I., & Sheng, W. U. (2014). Theme-Oriented Visual Analysis of Crime with Big Data. *Journal of Geo-Information Science*, *220*(1), 96–100.

Davis, J., & Goadrich, M. (2006). The relationship between Precision-Recall and ROC curves. In *Proceedings of the International Conference on Machine Learning*. New York, NY: ACM Press. 10.1145/1143844.1143874

Diebold, F. X., & Mariano, R. S. (1995). Comparing predictive accuracy. *Journal of Business & Economic Statistics*, *13*(3), 253–263.

Dumbill, E. (2013). Big data and thought crime: an interview with Jim Adler. *Big Data, 1*(1), 10-13.

Feifei, S., Zhuo, C., & Xiaolei, X. (2014). Application of an improved random forest based classifier in crime prediction domain. *Journal of Intelligence*, *33*(10), 148–152.

Freeman, E. A., & Moisen, G. G. (2008). A comparison of the performance of threshold criteria for binary classification in terms of predicted prevalence and kappa. *Ecological Modelling*, *217*(1-2), 48–58. doi:10.1016/j.ecolmodel.2008.05.015

Friedman, J. H. (2001). Greedy function approximation: A gradient boosting machine. *Annals of Statistics*, *29*(5), 1189–1232. doi:10.1214/aos/1013203451

Fries, D., Rossegger, A., Endrass, J., & Singh, J. P. (2013). The prediction of criminal recidivism using routinely available file information. *International Journal of Psychological Research*, *6*(2), 8–14. doi:10.21500/20112084.671 PMID:25374652

Godbole, S., & Sarawagi, S. (2004). Discriminative methods for multi-labeled classification. *Lecture Notes in Computer Science*, *3056*, 22–30. doi:10.1007/978-3-540-24775-3_5

Grieger, L., & Hosser, D. (2012). Attention deficit hyperactivity disorder does not predict criminal recidivism in young adult offenders: Results from a prospective study. *International Journal of Law and Psychiatry*, *35*(1), 27–34. doi:10.1016/j.ijlp.2011.11.005 PMID:22142896

Hu, T., Zhu, X., Duan, L., Guo, W., & Arcaute, E. (2018). Urban crime prediction based on spatio-temporal bayesian model. *PLoS One*, *13*(10). doi:10.1371/journal.pone.0206215 PMID:30379897

Huang, Y. Y., Li, C. T., & Jeng, S. K. (2015). Mining location-based social networks for criminal activity prediction. In *Proceedings of Wireless & Optical Communication Conference*. Taipei, China: IEEE. doi:10.1109/WOCC.2015.7346202

Inbaek, S., Kang, B. K., Doo, I. C., Mee, Y. E. S., & Park, H. (2017). Implementation of crime prevention system using public big data. *Advanced Science Letters*, *23*(10), 9574–9578. doi:10.1166/asl.2017.9750

Jeong, K. S., Moon, T. H., & Jeong, J. H. (2010). Hotspot analysis of urban crime using space-time scan statistics. *Journal of the Korean Association of Geographic Information Studies*, *13*(3), 14–28.

Junker, M., Hoch, R., & Dengel, A. (1999). On the Evaluation of Document Analysis Components by Recall, Precision, and Accuracy. In *Proceedings of International Conference on Document Analysis & Recognition*. Bangalore, India: IEEE. 10.1109/ICDAR.1999.791887

Kumar, R., & Nagpal, B. (2018). Analysis and prediction of crime patterns using big data. *International Journal of Information Technology*, (4), 1-7.

Ludu. (2013). Burglary Crime Analysis Using Logistic Regression. In S. Yamamoto (Ed.), *Human Interface and the Management of Information. Information and Interaction for Learning, Culture, Collaboration and Business*. Berlin, Germany: Springer.

Mason, L., Baxter, J., Bartlett, P., & Frean, M. (1999). Boosting algorithms as gradient descent. In *Proceedings of International Conference on Neural Information Processing Systems*. Cambridge, MA: MIT Press.

Mcafee, A., & Brynjolfsson, E. (2012). Big data: The management revolution. *Harvard Business Review*, *90*(10), 60–66, 68, 128. PMID:23074865

Mehlum, H., Moene, K., & Torvik, R. . (2005). Crime induced poverty traps. *Journal of Development Economics, 77*(2), 0-340.

Nasridinov, A., Ihm, S. Y., & Park, Y. H. (2013). A Decision Tree-Based Classification Model for Crime Prediction. In J. Park, L. Barolli, F. Xhafa, & H. Y. Jeong (Eds.), *Information Technology Convergence*. Dordrecht, The Netherlands: Springer.

Pereira, M. G., Caramelo, L., Orozco, C. V., Costa, R., & Tonini, M. (2015). Space-time clustering analysis performance of an aggregated dataset: The case of wildfires in Portugal. *Environmental Modelling & Software*, *72*, 239–249. doi:10.1016/j.envsoft.2015.05.016

Pflueger, M. O., Franke, I., Graf, M., & Hachtel, H. (2015). Predicting general criminal recidivism in mentally disordered offenders using a random forest approach. *BMC Psychiatry*, *15*(1), 62. doi:10.118612888-015-0447-4 PMID:25885691

Proctor, R. W., & Cho, Y. S. (2006). Polarity correspondence: A general principle for performance of speeded binary classification tasks. *Psychological Bulletin*, *132*(3), 416–442. doi:10.1037/0033-2909.132.3.416 PMID:16719568

Ranson, M. (2014). Crime, weather, and climate change. *Journal of Environmental Economics and Management*, *67*(3), 274–302. doi:10.1016/j.jeem.2013.11.008

Robin, X., Turck, N., Hainard, A., Tiberti, N., Lisacek, F., Sanchez, J.-C., & Müller, M. (2011). Proc: An open-source package for R and S+ to analyze and compare ROC curves. *BMC Bioinformatics*, *12*(1), 77. doi:10.1186/1471-2105-12-77 PMID:21414208

Rumi, S. K., Deng, K., & Salim, F. D. (2018). Crime event prediction with dynamic features. *EPJ Data Science*, *7*(1), 43. doi:10.1140/epjds13688-018-0171-7

Tayebi, M. A., Ester, M., Glasser, U., & Brantingham, P. L. (2014). CRIMETRACER: Activity space based crime location prediction. In *Proceedings of IEEE/ACM International Conference on Advances in Social Networks Analysis and Mining*. Beijing, China: IEEE. 10.1109/ASONAM.2014.6921628

Toole, J. L., Eagle, N., & Plotkin, J. B. (2011). Spatiotemporal correlations in criminal offense records. *ACM Transactions on Intelligent Systems and Technology*, *2*(4), 38. doi:10.1145/1989734.1989742

Traunmueller, M., Quattrone, G., & Capra, L. (2014). Mining Mobile Phone Data to Investigate Urban Crime Theories at Scale. In *Proceedings of International Conference on Social Informatics*. Berlin, Germany: Springer. 10.1007/978-3-319-13734-6_29

Uittenbogaard, A. C., & Ceccato, V. (2012). Space-time clusters of crime in Stockholm, Sweden. *Review of European Studies*, *4*(5), 148–156. doi:10.5539/res.v4n5p148

<unknown>Aghababaei, S., & Makrehchi, M. (2016). Mining Social Media Content for Crime Prediction. In Proceedings of 2016 IEEE/WIC/ACM International Conference on Web Intelligence (WI). Omaha, NE: IEEE.</conf>

Wang, H., Kifer, D., Graif, C., & Li, Z. (2016). Crime rate inference with big data. In *Proceedings of the 22nd ACM SIGKDD international conference on knowledge discovery and data mining*. San Francisco, CA: ACM Press. 10.1145/2939672.2939736

Wang, T., Rudin, C., Wagner, D., & Sevieri, R. (2013). Learning to detect patterns of crime. In *Proceedings of European conference on machine learning and knowledge discovery in databases*. Heidelberg, Germany: Springer.

Wang, X., Gerber, M. S., & Brown, D. E. (2012). Automatic Crime Prediction Using Events Extracted from Twitter Posts. In *Proceedings of Social Computing, Behavioral - Cultural Modeling and Prediction*. New York, NY: Springer. doi:10.1007/978-3-642-29047-3_28

Weisburd, D., Green, L., Gajewski, F., & Bellucci, C. (1994). Defining the Street Level Drug Market. In D. L. MacKenzie, & C. D. Uchida (Eds.), *Drugs and Crime: Evaluating Public Policy Initiatives*. Thousand Oaks, CA: Sage.

Xuemei, L. (2015). Surveying the crime analysis in U. S. prediction policing from big data. *Journal of Intelligence*, *34*(12), 16–20.

Yu, C. H., Ward, M. W., Morabito, M., & Ding, W. (2011). Crime Forecasting Using Data Mining Techniques. In *Proceedings of 2011 IEEE 11th International Conference on Data Mining Workshops*. Washington, DC: IEEE. 10.1109/ICDMW.2011.56

ADDITIONAL READING

Alsaedi, N., Burnap, P., & Rana, O. (2017). Can we predict a riot? disruptive event detection using twitter. *ACM Transactions on Internet Technology*, *17*(2), 1–26. doi:10.1145/2996183

G. Antoniou, M. Grobelnik, E. Simperl, B. Parsia, D. Plexousakis, P. De Leenheer, & J. Z. Pan (Eds.). (2011). In *Part II of Proceedings of* The Semantic Web: Research and Applications*: 8th Extended Semantic Web Conference, ESWC 2011*, Heraklion, Crete, Greece. Berlin, Germany: Springer.

Burnap, P., Colombo, W., & Scourfield, J. (2015). Machine classification and analysis of suicide-related communication on twitter. In *Proceedings of the 26th ACM conference on hypertext & social media*. New York, NY: ACM Press.

Corcoran, J., Wilson, I. D., Lewis, O. M., & Ware, J. A. (2001). Data clustering and rule abduction to facilitate crime hot spot prediction. In *Proceedings of International Conference on Computational Intelligence*. Berlin, Germany: Springer. 10.1007/3-540-45493-4_80

Kang, H. W., & Kang, H. B. (2017). Prediction of crime occurrence from multi-modal data using deep learning. *PLoS One, 12*(4). doi:10.1371/journal.pone.0176244 PMID:28437486

Malathi, A., & Santhosh Baboo, S. (2011). An enhanced algorithm to predict a future crime using data mining. *International Journal of Computers and Applications, 21*(1), 1–6. doi:10.5120/2478-3335

Ohmura, M., Kakusho, K., & Okadome, T. (2014). Social mood extraction from Twitter posts with document topic model. In *Proceedings of 2014 International Conference on Information Science & Applications*. Seoul, South Korea: IEEE. 10.1109/ICISA.2014.6847465

Saravanan, M., Thayyil, R., & Narayanan, S. (2013). Enabling Real Time Crime Intelligence Using Mobile GIS and Prediction Methods. In *Proceedings of European Intelligence & Security Informatics Conference*. Uppsala, Sweden: IEEE. doi:10.1109/EISIC.2013.27

KEY TERMS AND DEFINITIONS

Ada Boost: An iterative algorithm whose core idea is to train different classifiers (weak classifiers) according to the same training set, and then assemble these weak classifiers to form a stronger final classifier (strong classifier).

Bagging: By constructing a series of predictive functions and combining them into a predictive function in a certain way. Bagging requires a classification method of "instability" (instability is a classification method in which small changes in the index data set can result in significant changes in the classification results).

Decision Tree: A decision analysis method that, on the basis of the known probability of various situations, calculates the probability of the expected value of the net present value being greater than or equal to zero by constructing the decision tree, evaluates the project risk, and determines its feasibility.

Extra Trees: The algorithm is very similar to the random forest, but compared with random forest, all the samples used in this algorithm are only randomly selected for their features. Because the splitting is random, the results obtained are better than those obtained by random forest to some extent.

Gradient Boosting: An algorithm that selects the direction of gradient descent during iteration to ensure the best results.

K Neighbors: The idea of this method is: if most of the k most similar samples in the feature space belong to a certain category, then the sample also belongs to this category.

Logistic Regression: Logistic regression analysis is mainly used in epidemiology. The most common case is to explore the risk factors of a certain disease and predict the probability of the occurrence of a certain disease according to the risk factors.

Random Forest: A classifier that USES multiple trees to train and predict samples.

SVC: Scalable Video Coding (SVC) is based on h. 264, which is extended on the syntax and toolset to support code streams with hierarchical features. H.264svc is appendix G of H. 264 standard and the new profile of H. 264.

Compilation of References

<bok>Mukolwe, E., & Korir, J. (2016). Social Media and Entrepreneurship: Tools, Benefits, and Challenges. A Case Study of Women Online Entrepreneurs on Kilimani Mums Marketplace on Facebook. International Journal of Humanities and Social Science, 6(8), 248–256.</jrn>

<conf>Leggett, J. A., & Carter, N. T. (2012). Rio+ 20: The United Nations Conference on Sustainable Development, June 2012.</conf>

<eref>BangladeshWood. (2019). Bangladesh - Rapidly Expanding Market. Retrieved from http://www.bangladeshwood.com/#

<jrn>Chen, J., Sousa, C., & He, X. (2016). The determinants of export performance: A review of the literature 2006-2014. International Marketing Review, 33(5), 626–670.</jrn>

<unknown>Aghababaei, S., & Makrehchi, M. (2016). Mining Social Media Content for Crime Prediction. In Proceedings of 2016 IEEE/WIC/ACM International Conference on Web Intelligence (WI). Omaha, NE: IEEE.</conf>

<unknown>Assaad, W., & Gómez, J. M. (2011). Social Network in marketing (social media marketing) opportunities and risks. International Journal of Managing Public Sector Information and Communication Technologies (IJMPICT), 2(1). Retrieved from http://www.seokursu.com.tr/social-network-in-marketing.pdf</eref>

<unknown>Ja-Young, Y. (2016, March 1). Korea shifting to a cashless society. Korea Times.

<unknown>Statista. (2018). Leading social media platforms used by marketers worldwide as of January 2018. Retrieved from https://www.statista.com/statistics/259379/social-media-platforms used-by-marketers-worldwide/</eref>

<unknown>Upkere, C. L., Slabbert, A. D., & Upkere, W. I. (2014). Rising Trend in Social Media Usage by Women Entrepreneurs across the Globe to Unlock Their Potentials for Business Success Mediterranean Journal of Social Sciences, 5(10), 551–559.</jrn>

Abbar, S., Amer-Yahia, S., Indyk, P., & Mahabadi, S. (2013). Real-time recommendation of diverse related articles. In *Proceedings of the 22nd international conference on World Wide Web* (pp. 1-12). ACM.

Abbas, J., Mahmood, H. K., & Hussain, F. (2015). Information security management for small and medium size enterprises. *Science International-Lahore*, *27*(3), 2393–2398.

Abbas, J., Muzaffar, A., Mahmood, H. K., Ramzan, M. A., & Rizvi, S. S. ul H. (2014). Impact of Technology on Performance of Employees (A Case Study on Allied Bank Ltd, Pakistan). *World Applied Sciences Journal*, *29*(2), 271–276.

Abbas, J., Muzaffar, A., Shoaib, M., & Mahmood, H. K. (2014). Do Business Schools Really Fulfill Industry Requirements? An Investigation of Industrial Performance of Business Graduates. *World Applied Sciences Journal*, *31*(7), 1378–1384.

Abbas, J., & Sağsan, M. (2019). Impact of knowledge management practices on green innovation and corporate sustainable development: A structural analysis. *Journal of Cleaner Production*, *229*, 611–620. doi:10.1016/j.jclepro.2019.05.024

Abdellatif, M., Farhan, M. S., & Shehata, N. S. (2017). Overcoming business process reengineering obstacles using ontology-based knowledge map methodology. *Future Computing and Informatics Journal*.

Abdelwhab Ali, A., Panneer Selvam, D. D. D., Paris, L., & Gunasekaran, A. (2019). Key factors influencing knowledge sharing practices and its relationship with organizational performance within the oil and gas industry. *Journal of Knowledge Management*. doi:10.1108/JKM-06-2018-0394

Abdollahi, H., Razm, K., & Tan, H. (2014). TQM and market orientation's impact on SMEs & quot; performance. *Management Science Letters*, *4*(5), 887–892.

Abel, M.-H. (2015). Knowledge map-based web platform to facilitate organizational learning return of experiences. *Computers in Human Behavior, 51,* 960-966.

Abeykoon, M., & de Alwis, A. (2016). The Impact of Total Quality Management Practices on Export Performance of Apparel Exporters of Sri Lanka. *Kelaniya Journal of Human Resource Management, 10*(1).

Abiodun, S. T., & Mahmood, R. (2015). Fostering Export Performance in SMEs: The Roles of Export Market Orientation and Learning Orientation in Turbulent Environment. *International Journal of Economic Perspectives*, *9*(2), 28.

Abrahamsson, A. (2007). Sustainopreneurship-business with a cause: conceptualizing entrepreneurship for sustainability.

Academic Impact UN Report. (2019, April 18). *UNAI to Host Conference on Unlocking Your Emotions to Achieve the SDGs*. Retrieved from https://academicimpact.un.org/content/unai-host-conference-unlocking-your-emotions-achieve-sdgs

Achtenhagen, L. (2008). Understanding entrepreneurship in traditional media. *Journal of Media Business Studies*, *5*(1), 123–142. doi:10.1080/16522354.2008.11073463

Acquavita, S. P., Krummel, D. A., Talks, A., Cobb, A., & McClure, E. (2019). Assessing the digital divide among low-income perinatal women: Opportunities for provision of health information and counseling. *Telemedicine Journal and e-Health*, *25*(1), 48–54. doi:10.1089/tmj.2017.0292 PMID:29708865

ADB. (2010). Pakistan Forestry Sector Project. Retrieved from https://www.adb.org/sites/default/files/evaluation-document/35571/files/in343-10.pdf

Adler, P. A., & Adler, P. (1998). *Peer power: Pre- adolescent culture and identity*. New Brunswick, NJ: Rutgers University Press.

Adner, R., & Levinthal, D. (2001). Demand heterogeneity and technology evolution: Implications for product and process innovation. *Management Science*, *47*(5), 611–628. doi:10.1287/mnsc.47.5.611.10482

Afroze, T., Alam, K., Akther, E., & Jui, S. N. (2014). Women Entrepreneurs in Bangladesh-Challenges and Determining Factors. *Journal of Business and Technology (Dhaka)*, *9*(2), 27–41. doi:10.3329/jbt.v9i2.26194

Agarwal, R., & Prasad, J. (1998). A conceptual and operational of personal innovativeness in the domain of information technology. *Information Systems Research*, *9*(2), 204–215. doi:10.1287/isre.9.2.204

AGCXL. (2019). Product. Retrieved from http://www.agcxl.com

Agrawal, T. (2012). Role Of Rfid To Minimize Reverse Logistics: A Case Study Perspective. *IOSR Journal of Business and Management.*, *1*(4), 44–51. doi:10.9790/487X-0144451

Ahammad, I. & *Moudud-Ul- Huq, S.* (2013). Women Entrepreneurship Development in Bangladesh Challenges And Prospects. *International Journal of Innovative Research and Development*, (pp. 41–48). Retrieved from http://www.ijird.com/index.php/ijird/article/view/36096/29238

Ahmed, I., Nawaz, M. M., Ahmad, Z., Shaukat, M. Z., Usman, A., Rehman, W.-u., & Ahmed, N. (2010). Determinants of students' entrepreneurial career intentions: Evidence from business graduates. *European Journal of Soil Science, 15*(2), 14–22.

Ahmed, N. (2018). Workers' Welfare: A Comparative Study Between Public and Private Industries in Bareilly, Uttar Pradesh. *Management and Labour Studies, 43*(3), 192–204. doi:10.1177/0258042X18768307

Ajayi, B. (2016). The Impact of Entrepreneurial Orientation and Networking Capabilities on the Export Performance of Nigerian Agricultural SMEs. *Journal of Entrepreneurship and Innovation in Emerging Economies.*

Ajzen, I. (1985). From intentions to actions: A theory of planned behavior. In J. Kuhl, & J. Beckmann (Eds.), *Action Control: From cognition to behavior* (pp. 11–39). Berlin, Germany: Springer-Verlag; doi:10.1007/978-3-642-69746-3_2

Ajzen, I. (1991). The theory of planned behavior. *Organizational Behavior and Human Decision Processes, 50*(2), 179–211. doi:10.1016/0749-5978(91)90020-T

Akinseinde, S. I., & Adomi, E. E. (2004). E-Mail Usage by Technical Education Students in Nigerian Universities. *Library Hi Tech News, 21*(9), 15–20. doi:10.1108/07419050410572762

Akram, S. (2017). *Pakistan's Development Dilemma.* Retrieved from Pakistan: In U. Balakrishnan, T. Duvall, & P. Primeaux, (2003) Rewriting the bases of capitalism: reflexive modernity and ecological sustainability as the foundations of a new normative framework. *Journal of Business Ethics, 47*(4), 299–314.

Akrimi, Y., & Khemakhem, R. (2012). What Drive Consumers to Spread the Word in Social Media? *Journal of Marketing Research & Case Studies*, 1–14. doi:10.5171/2012.969979

Akter, S., Wamba, S. F., Gunasekaran, A., Dubey, R., & Childe, S. J. (2016). How to improve firm performance using big data analytics capability and business strategy alignment? *International Journal of Production Economics, 182*, 113–131. doi:10.1016/j.ijpe.2016.08.018

Al Kailani, M., & Kumar, R. (2011). Investigating Uncertainty Avoidance and Perceived Risk for Impacting Internet Buying: A Study in Three National Cultures. *International Journal of Business and Management, 6*(5), 76–92. doi:10.5539/ijbm.v6n5p76

Alavi, M., & Leidner, D. (2001). Knowledge management and knowledge management systems: Conceptual foundations and research issues. *Management Information Systems Quarterly, 25*(1), 107–136. doi:10.2307/3250961

Albert, L. (2019, May 7). *Delivering Exceptional Employee Experiences Requires Identifying and Closing Experience Gaps.* Academic Press.

Albert. H. (2018). The Infinite information technology, *Digital Transformation.* Retrieved from http://www.infiniteinformationtechnology.com/digital-transformation-what-is-digital-economy

Al-Busaidi, K. A., & Al-Shihi, H. (2012). Key factors to instructors' satisfaction of learning management systems in blended learning. *Journal of Computing in Higher Education, 24*(1), 18–39. doi:10.100712528-011-9051-x

Allur, E., Heras-Saizarbitoria, I., Boiral, O., & Testa, F. (2018). Quality and Environmental Management Linkage: A Review of the Literature. *Sustainability, 10*, 1–15. PMID:30607262

Almekhlafi, A. G., & Abulibdeh, E. S. A. (2018). K-12 teachers' perceptions of Web 2.0 applications in the United Arab Emirates? *Interactive Technology and Smart Education, 15*(3), 238–261. doi:10.1108/ITSE-11-2017-0060

Almobarraz, A. (2018). Utilization of YouTube as an information resource to support university courses. *The Electronic Library, 36*(1), 71–81. doi:10.1108/EL-04-2016-0087

Al-Qahtani, N. D., Alshehri, S. S., & Aziz, D. A. A. (2015). The impact of Total Quality Management on organizational performance. *European Journal of Business and Management, 17*(36), 1–10.

Amburgey, T. L., Kelly, D., & Barnett, W. P. (1990, August). Resetting the clock: The dynamics of organizational change and failure. In Academy of Management Proceedings (Vol. 1990, No. 1, pp. 160-164). Briarcliff Manor, NY 10510: Academy of Management.

American Psychological Association. (2010). *Publication manual of the American Psychological Association* (6th ed.). Washington, DC: American Psychological Association.

Amrit, C., Paauw, T., Aly, R., & Lavric, M. (2017). Identifying child abuse through text mining and machine learning. *Expert Systems with Applications, 88*, 402–418. doi:10.1016/j.eswa.2017.06.035

Anees-ur-Rehman, M., Saraniemi, S., Ulkuniemi, P., & Hurmelinna-laukkanen, P. (2017). The strategic hybrid orientation and brand performance of B2B SMEs. *Journal of Small Business and Enterprise Development, 24*(3), 585–606.

Angeleanu, A. (2015). New technology trends and their transformative impact on logistics and supply chain processes. *International Journal of Economic Practices and Theories, 5*(5), 413–419.

Antonio, A., & Tuffley, D. (2014). The gender digital divide in developing countries. *Future Internet, 6*(4), 673–687. doi:10.3390/fi6040673

Antony, J., Fergusson, C., Warwood, S., & Hing Yee Tsang, J. (2004). Comparing total quality management success factors in UK manufacturing and service industries: Some key findings from a survey. *Journal of Advances in Management Research, 1*(2), 32–45.

An, Y.-J., Aworuwa, B., Ballard, G., & Williams, K. (2009). Teaching with Web 2.0 Technologies: Benefits, Barriers and Best Practices. In M. Simonson (Ed.), *Proceedings of Annual Convention of the Association for Educational Communications and Technology.* Florida: Nova Southeastern University. Retrieved from https://members.aect.org/pdf/Proceedings/proceedings09/2009/09_1.pdf

App, S., & Büttgen, M. (2016). Lasting footprints of the employer brand: can sustainable HRM lead to brand commitment?, *Employee Relations, 38*(5), 703-723.

App, S., Merk, J., & Buttgen, M. (2012). Employer Branding: Sustainable HRM as a Competitive Advantage in the Market for High-Quality Employees. *Management Review, 23*(2), 262–278.

April, K., & Blass, E. (2010). Measuring Diversity Practice and Developing Inclusion. *Dimensions, 1*(1), 59–66.

Araújo, C. A. S., Tavares, E., Raupp de Vargas, E., & Rocha, E. (2015). Developing learning capabilities through a quality management program. *Service Industries Journal, 35*(9), 483–498.

Ardichvili, A., Page, V., & Wentling, T. (2003). Motivation and barriers to participation in virtual knowledge-sharing communities of practice. *Journal of Knowledge Management, 7*(1), 64–77. doi:10.1108/13673270310463626

Arons, L., May, M. A., & Brodbeck, A. J. (1963). *Television and human behavior: Tomorrow's research in mass communication.* New York, NY: Appleton-Century-Crofts.

Arthur, W. B. (1996). Increasing Returns and the New World of Business. *Harvard Business Review, 74*(4). PMID:10158472

Ashforth, B. E., & Humphrey, R. H. (1993). Emotional Labour in service roles: The influence of identity. *Academy of Management Review*, *18*(1), 88–115. doi:10.5465/amr.1993.3997508

Ashton, C., & Morton, L. (2005). Managing talent for competitive advantage: Taking a systemic approach to talent management. *Strategic HR review*, *4*(5), 28-31.

Asian Development Bank. (2018, February). Understanding the Digital Economy: What Is It and How Can It Transform Asia? Event. Asian Development Bank. Retrieved from https://www.adb.org/news/events/understanding-digital-economy-what-it-and-how-can-it-transform-asia

Aswathappa, K. (2010). *Human resources management*. New Delhi, India: Tata McGraw Hill Publishing Company Limited.

Bachner, J. (2013). *Predictive policing: preventing crime with data and analytics*. Washington, DC: IBM Center for the Business of Government.

Backhaus, K. (2004). An exploration of corporate recruitment descriptions on monster.com. *Journal of Business Communication*, *41*(2), 115–120. doi:10.1177/0021943603259585

Bader, H., Hanna, C., Douglas, C., & Fox, J. D. (2013). Illegal Timber Exploitation and Counterinsurgency Operations in Kunar Province of Afghanistan: A Case Study Describing the Nexus Among Insurgents, Criminal Cartels, and Communities Within the Forest Sector. *Journal of Sustainable Forestry*, *32*(4), 329–353. doi:10.1080/10549811.2013.767913

Baghdadi, Y. (2013). From E-commerce to social commerce: A framework to guide enabling cloud computing. *Journal of Theoretical and Applied Electronic Commerce Research*, *8*(3), 12–38. doi:10.4067/S0718-18762013000300003

Bailey, N., Breen, J., & Ward, M. (2010). *Community Education: More than Just a Course. Exploring the Outcomes and Impact of Department of Education and Skills Funded Community Education*. AONTAS The National Adult Learning Organisation. 2nd Floor 83-87 Main Street [Ireland.]. *Ranelagh, Dublin*, *6*, D6.

Baines, A. (1992). Electronic mail. *Work Study*, *41*(5), 24–25. doi:10.1108/EUM0000000002678

Bajaj, R. (2017). Top 10 Most Popular Social Sites and Apps in 2017. Retrieved from https://www.linkedin.com/pulse/top-10-most-popular-social-networking-sites-apps-2017-rajiv-bajaj

Baker, J. E. (2018). Preservation of cut flowers. In Plant growth regulating chemicals (pp. 177–191). Boca Raton, FL: CRC Press. doi:10.1201/9781351075749-10

Baker, L., Wagner, T. H., Singer, S., & Bundorf, M. K. (2003). Use of the Internet and e-mail for health care information: Results from a national survey. *Journal of the American Medical Association*, *289*(18), 2400–2406. doi:10.1001/jama.289.18.2400 PMID:12746364

Bakken, S., Marden, S., Arteaga, S. S., Grossman, L., Keselman, A., Le, P. T., ... Das, R. (2019). Behavioral Interventions Using Consumer Information Technology as Tools to Advance Health Equity. *American Journal of Public Health*, *109*(S1), S79–S85. doi:10.2105/AJPH.2018.304646 PMID:30699018

Ballesteros, G., & Roberts, L. (2018). A Closer Look at the Peruvian Healthcare System and the Patient Medical Treatment Experience.

Bal, P. M., DeCooman, R., & Mol, S. T. (2013). Dynamics of psychological contracts with work engagementand turnover intention: The influence of organizational tenure. *European Journal of Work and Organizational Psychology*, *22*(1), 107–122. doi:10.1080/1359432X.2011.626198

Balubaid, M. A. (2013). Using Web 2.0 Technology to Enhance Knowledge Sharing in an Academic Department. *Procedia: Social and Behavioral Sciences*, *102*, 406–420. doi:10.1016/j.sbspro.2013.10.756

Barhoumi, C. (2015). The Effectiveness of WhatsApp Mobile Learning Activities Guided by Activity Theory on Students' Knowledge Management. *Contemporary Educational Technology, 6*. Retrieved from https://files.eric.ed.gov/fulltext/EJ1105764.pdf

Barney, J. (1991). Firm resources and sustained competitive advantage. *Journal of Management, 17*(1), 99–120.

Bar-On, R. (1997). *The Emotional Quotient Inventory (EQ-i): Technical manual*. Toronto, Canada: Multi-Health Systems.

Barrena-Martínez, J., López-Fernández, M., & Romero-Fernández, P. M. (2017). Towards a configuration of socially responsible human resource management policies and practices: Findings from an academic consensus. *International Journal of Human Resource Management*, 1–37. doi:10.1080/09585192.2017.1332669

Barsade, S. G. (2002). The Ripple Effect: Emotional Contagion and its Influence on Group Behavior. *Administrative Science Quarterly, 47*(4), 644–675. doi:10.2307/3094912

Barsade, S., & O'Neill, O. A. (2014). What's Love got to do with it? A longitudinal study of the culture of compassionate love and employee and client outcomes in the long-term care setting. *Administrative Science Quarterly, 59*(4), 551–598. doi:10.1177/0001839214538636

Barsade, S., & O'Neil, O. A. (2016, January-February). Managing your emotional culture. *Harvard Business Review*, 58–66.

Barthes, R. (1967). *Elements of semiology* (A. Lavers, & C. Smith, Trans.). London, UK: Jonathan Cape.

Barthwal and Som. (2012). Emotional Intelligence as a measure of an employee's overall effectiveness, *Drishtikon: A Management Journal, 3*(2).

Barton, D., & Court, D. (2012). Making advanced analytics work for you. *Harvard Business Review, 90*(10), 78–83. PMID:23074867

Bartuševičienė, I., & Šakalytė, E. (2013). Organizational assessment: Effectiveness vs. efficiency. *Social Transformations in Contemporary Society, 1*(1), 45–53.

Bass, S., & Dalal-Clayton, B. (2012). *Sustainable development strategies: a resource book*. Abingdon, UK: Routledge. doi:10.4324/9781849772761

Bateman, D., & Willems, J. (2012). Facing off: Facebook and Higher Education. In D. Bateman, & J. Willems (Eds.), *Misbehavior Online in Higher Education (Cutting-edge Technologies in Higher Education* (Vol. 5, pp. 53–79). Emerald Group Publishing Limited; doi:10.1108/S2044-9968(2012)0000005007

BDCOMEX. (2018). Introducing the First Commodity Exchange in Bangladesh. Retrieved from http://bdcomex.blogspot.com/p/about-us-bangladesh-jute-commodity.html</eref>

Beasley, M., Leshuk, J. A., Yasar, M., & Cywilko, M. E. (2019). *U.S. Patent Application No. 16/077,090.*

Beazley, M., & Smith, M. (1999). Record of the DTI/social exclusion policy action team 15 visit to the Sparkbrook, Sparkhill and Tyseley Area Regeneration Initiative (SSTARI) Birmingham. Information technology and black and ethnic minority communities. School of Public Policy, University of Birmingham, in conjunction with Sparbrook, Sparkhill and Tyseley Area Regeneration Initiative, Birmingham.

Beer, M. (1997). The transformation of the human resource function: Resolving the tension between a traditional administrative and a new strategic role. *Human Resource Management: Published in Cooperation with the School of Business Administration, The University of Michigan and in alliance with the Society of Human Resources Management, 36*(1), 49-56.

Belk, R. (1975). Situational variables and consumer behavior. *Journal of Consumer Behaviour, 2*(3), 157–164. doi:10.1086/208627

Belk, R. W. (1988). Possessions and the extended self. *The Journal of Consumer Research, 15*(2), 139–168. doi:10.1086/209154

Bell, J., & Stellingwerf, J. (2012). Sustainable entrepreneurship: The motivations and challenges of sustainable entrepreneurs in the renewable energy industry.

Bell, D. (1979). *Cultural Contradictions of Capitalism*. London, UK: Heinemann.

Bell, D. (2001). *An Introduction to Cyberspace*. London, UK: Routledge.

Belz, F. M., & Binder, J. K. (2017). Sustainable entrepreneurship: A convergent process model. *Business Strategy and the Environment, 26*(1), 1–17. doi:10.1002/bse.1887

Ben-Daya, M., Hassini, E., & Bahroun, Z. (2017). Internet of things and supply chain management: A literature review. *International Journal of Production Research*, 1–24. doi:10.1080/00207543.2017.1402140

Bennett, S., Bishop, A., Dalgarno, B., Waycott, J., & Kennedy, G. (2012). Implementing Web 2.0 technologies in higher education: A collective case study. *Computers & Education, 59*(2), 524–534. doi:10.1016/j.compedu.2011.12.022

Benson, A. M., & Blackman, D. (2011). To distribute leadership or not? A lesson from the islands. *Tourism Management, 32*(5), 1141–1149. doi:10.1016/j.tourman.2010.10.002

Berawi, M. A. (2018). Utilizing big data in industry 4.0: Managing competitive advantages and business ethics. *International Journal of Technology, 3*(1), 430–433. doi:10.14716/ijtech.v9i3.1948

Bergh, J. V. den, & Behrer, M. (2016). *How cool brands stay hot: Branding to generations Y and Z*. London, UK: Kogan Page.

Berry, T. H. (1991). Managing the total quality transformation. New York, NY: McGraw-Hill.

Bersin, J. (2016). *Predictions for 2017: Everything Is Becoming Digital*. Deloitte Development LLC.

Bezrukova, K., Thatcher, S., Jehn, K. A., & Spell, C. S. (2012). The effects of alignments: Examining group faultlines, organizational cultures, and performance. *The Journal of Applied Psychology, 97*(1), 77–92. doi:10.1037/a0023684 PMID:21744943

Bhandari, R. (2014). *Impact of technology on logistics and supply chain management. IOSR Journal of Business and Management*.

Bhardwaj, M. (2019). Growth And Performance Of Organic Farming In India: What Could Be The Future Prospects? *Growth, 20*(01).

Bhattachargya, C. B., & Sen, S. (2003). Consumer –company identification: A framework for under- standing consumers' relationship with companies. *Journal of Marketing, 67*(2), 76–88. doi:10.1509/jmkg.67.2.76.18609

Bhattacharya, S., Trehan, G., & Kaur, K. (2018). Factors Determining Psychological Contract of IT Employees in India. *International Journal of Human Capital and Information Technology, 9*(1), 37–52.

Bhattacharyya, S. S., & Shrey, A. (2019). Supply Chain Startups in India: A Cross Case Comparative Analysis. In Handbook of Research on Corporate Restructuring and Globalization (pp. 320-344). Hershey, PA: IGI Global. doi:10.4018/978-1-5225-8906-8.ch016

Bhattacharyya, S. S. (2011). Reflections on strategic insights for winning in the complex emerging market of India. *International Journal of Business Excellence, 4*(1), 15–43. doi:10.1504/IJBEX.2011.037247

Bhattacharyya, S. S., & Jha, S. (2015). Mapping micro small and medium enterprises from the resource-based view and dynamic capability theory perspectives and innovation classification. *International Journal of Entrepreneurship and Small Business*, *25*(3), 331–350. doi:10.1504/IJESB.2015.069700

Bibeth, S., & Ajeya, J. (2015). Case Study: Need assessment, Pricing and Availability perceptions: A study of Cymbidium Orchids of Sikkim Himalayas. *Advances in Management*, *8*(1), 26.

Bila, B., & Egrot, M. (2009). Gender asymmetry in healthcare-facility attendance of people living with HIV/AIDS in Burkina Faso. *Social Science & Medicine*, *69*(6), 854–861. doi:10.1016/j.socscimed.2009.05.035 PMID:19539415

Bilge, H., & Bal, V. (2012). Entrepreneurship Aptitude: An Empirical Study On Undergraduate And Vocational High Scholl Students In Celal Bayar University. *Journal of Süleyman Demirel University Institute of Social Sciences*, *16*, 131–148.

Biss, J. L., & DuFrene, D. D. (2006). An Examination of Reverse Mentoring in the Workplace. *Business Education Digest*, (15).

Bista, K. (2015). Is Twitter an effective pedagogical tool in higher education? Perspectives of education graduate students. *The Journal of Scholarship of Teaching and Learning*, *15*(2), 83–102. doi:10.14434/josotl.v15i2.12825

Black, J. (1999). *Information rich, information poor: bridging the digital divide.* International Institute for Communication and Development.

Blecker, T. (Ed.). (2014). Innovative methods in logistics and supply chain management: current issues and emerging practices, (Vol. 19), epubli.

Bloomfield, G. T. (2017). The world automotive industry in transition Restructuring the global automobile industry (pp. 19–60). Abingdon-on-Thames, UK: Routledge.

Bluethmann, S. M., Coa, K. I., Alfano, C. M., & Hesse, B. W. (2018). Electronic health information exchange opportunities for self-management of care: Responses from older adults with and without cancer history in the United States. *Current Oncology Reports*, *20*(4), 30. doi:10.100711912-018-0674-1 PMID:29572671

Bogomolov, A., Lepri, B., Staiano, J., Oliver, N., Pianesi, F., & Pentland, A. (2014). Once upon a crime: towards crime prediction from demographics and mobile data. In *Proceedings of the 16th international conference on multimodal interaction*. Istanbul, Turkey: ACM Press. 10.1145/2663204.2663254

Bol, N., Helberger, N., & Weert, J. C. (2018). Differences in mobile health app use: A source of new digital inequalities? *The Information Society*, *34*(3), 183–193. doi:10.1080/01972243.2018.1438550

Bolo, M. (2016). Innovation Systems and Capability Building Among Smallholders: Lessons And Insights From Kenya's Flower Farmers. *Innovation Systems: Towards Effective Strategies in support of Smallholder Farmers*, 74.

Bommer, W. H., Rich, G. A., & Rubin, R. S. (2005). Changing attitudes about change: Longitudinal effects of transformational leader behavior on employee cynicism about organizational change. *Journal of Organizational Behavior: The International Journal of Industrial, Occupational and Organizational Psychology and Behavior*, *26*(7), 733–753. doi:10.1002/job.342

Bornstein, T. B. (2013). What does it mean to be cool? *Philosophy Now*, *80*, 1–3.

Boruah, R., Borua, S., Deka, C. R., & Borah, D. (2016). Entrepreneurial behavior of tribal winter vegetable growers in Jorhat district of Assam. *Indian Research Journal of Extension Education*, *15*(1), 65–69.

Boschee, J., & McClurg, J. (2003). Toward a better understanding of social entrepreneurship: Some important distinctions.

Boso, N., Oghazi, P., Cadogan, J. W., & Story, V. M. (2016). Entrepreneurial and market-oriented activities, financial capital, environment turbulence, and export performance in an emerging economy. *Journal of Small Business Strategy*, *26*(1), 1.

Botchkovar, E., & Broidy, L. (2013). Accumulated strain, negative emotions, and crime: A test of general strain theory in Russia. *Crime and Delinquency*, *59*(6), 837–860. doi:10.1177/0011128710382346

Bouhnik, D., & Deshen, M. (2014). WhatsApp Goes to School: Mobile Instant Messaging between Teachers and Students. *Journal of Information Technology Education: Research*, *13*, 217–231. doi:10.28945/2051

Bou-Llusar, J. C., Escrig-Tena, A. B., Roca-Puig, V., & Beltrán-Martín, I. (2009). An empirical assessment of the EFQM Excellence Model: Evaluation as a TQM framework relative to the MBNQA Model. *Journal of Operations Management*, *27*(1), 1–22.

Bourke, J., & Crowley, F. (2015). The role of HRM and ICT complementarities in firm innovation: Evidence from transition economies. *International Journal of Innovation Management*, *19*(05). doi:10.1142/S1363919615500541

Boutin, P. (2006). *Web 2.0: The new Internet doesn't live up to its name.* Retrieved from https://slate.com/technology/2006/03/web-2-0-doesn-t-live-up-to-its-name.html

Brantingham, P. J., & Brantingham, P. L. (1984). *Patterns in crime*. New York, NY: Macmillan.

Brantingham, P. L., & Brantingham, P. J. (1999). A theoretical model of crime hot spot generation. *Studies on Crime & Crime Prevention*, *8*(1), 7–26.

Brar, S., Rabbat, R., Raithatha, V., Runcie, G., & Yu, A. (2015). *Drones for Deliveries*. Berkeley, CA: University of California Berkeley, Sutardja Center for Entrepreneurship & Technology.

Brealey, R., Stewart, M., & Allen, F. (2011). Principles of Corporate Finance, New York, NY: McGraw-Hill Irwin, 191

Breiman, L. (1997). *Arcing the edge (Technical Report 486)*. Berkeley, CA: Statistics Department, University of California. Retrieved from http://www.stat.Berkeley.EDU/users/breiman/

Brengman, M., & Karimov, F. P. (2012). The effect of web communities on consumers' initial trust in B2C e-commerce websites. *Management Research Review*, *35*(9), 791–817. doi:10.1108/01409171211256569

Broadband Connectivity in South East Asia. (2018). World Telecommunication/ICT Indicators database. Retrieved from http://www.itu.int/en/ITU-D/Statistics/Pages/publications/wtid.aspx

Brodie, M., Flournoy, R. E., Altman, D. E., Blendon, R. J., Benson, J. M., & Rosenbaum, M. D. (2000). Health Information, The Internet, And The Digital Divide: Despite recent improvements, Americans' access to the Internet—and to the growing body of health information there—remains uneven. *Health Affairs*, *19*(6), 255–265. doi:10.1377/hlthaff.19.6.255 PMID:11192412

Brotheridge, C., & Lee, R. T. (2002). Testing a conservation of resources model of the dynamics of emotional labour. *Journal of Occupational Health Psychology*, *7*(1), 57–67. doi:10.1037/1076-8998.7.1.57 PMID:11827234

Brown, K. G., & Sitzmann, T. (2011). Training and employee development for improved performance.

Brownlie, D., Hewer, P., & Treanor, S. (2007). Social- ity in motion: Exploring logics of tribal consumption among cruisers. In B. Cova, R. V. Kozinets, & A. Shankar (Eds.), *Consumer Tribes* (pp. 109–128). Oxford, UK: Elsevier.

Brundtland Commission. (1987). *Our common future*. Oxford, UK: Oxford University Press.

BSE. (2018) Stock Prices. Retrieved from https://www.bseindia.com/markets/equity/EQReports/StockPrcHistori.aspx?scripcode=512289&flag=sp&Submit=G

Bucy, E. P., & Newhagen, J. E. (Eds.). (2004). Media access: Social and psychological dimensions of new technology use. London, UK: Psychology Press.

Bughin, J., Hazan, E., Lund, S., Dahlström, P., Wiesinger, A., & Subramaniam, A. (2018, May). *Skill shift: Automation and the future of the workforce*. McKinsey Global Institute. McKinsey & Company.

Burrow, S., Goldberg, L., Searle, J., & Aston, M. (2018). Vulnerability, harm, and compromised ethics revealed by the experiences of queer birthing women in rural healthcare. *Journal of Bioethical Inquiry*, *15*(4), 511–524. doi:10.100711673-018-9882-5 PMID:30402734

Burton, P. F. (1994). Electronic Mail as an Academic Discussion Forum. *The Journal of Documentation*, *50*(2), 99–110. doi:10.1108/eb026926

Burt, R. S. (1992). The social structure of competition. In N. Nohria, & R. Eccles (Eds.), *Networks and Organizations: Structure, Form and Action* (pp. 57–91). Boston, MA: Harvard Business School Press.

Büyüközkan, G., & Göçer, F. (2018). Digital supply chain: Literature review and a proposed framework for future research. *Computers in Industry*, *97*, 157–177. doi:10.1016/j.compind.2018.02.010

Byrd, T. A., Lewis, B. R., & Bryan, R. W. (2006). The leveraging influence of strategic alignment on IT investment: An empirical examination. *Information & Management*, *43*(3), 308–321. doi:10.1016/j.im.2005.07.002

Cabrera, E. F., & Cabrera, A. (2005). Fostering knowledge sharing through people management practices. *International Journal of Human Resource Management*, *16*(5), 720–735. doi:10.1080/09585190500083020

Cacioppo, J. T., Petty, R. E., Feinstein, J. A., & Jarvis, W. B. G. (1996). Dispositional differences in cognitive motivation: The life and times of individuals varying in need for cognition. *Psychological Bulletin*, *119*(2), 197–253. doi:10.1037/0033-2909.119.2.197

Cadogan, J. W., Boso, N., Story, V. M., & Adeola, O. (2016). Export strategic orientation–performance relationship: Examination of its enabling and disenabling boundary conditions. *Journal of Business Research*.

Cadogan, J. W., Diamantopoulos, A., & Mortanges, D. (1999). A measure of export market orientation: Scale development and cross-cultural validation. *Journal of International Business Studies*, *30*(4), 689–707.

Cadogan, J. W., Kuivalainen, O., & Sundqvist, S. (2009). Export market-oriented behavior and export performance: Quadratic and moderating effects under differing degrees of market dynamism and internationalization. *Journal of International Marketing*, *17*(4), 71–89.

Cadogan, J. W., Paul, N. J., Salminen, R. T., Puumalainen, K., & Sundqvist, S. (2001). Key antecedents to "export" market-oriented behaviors: A cross-national empirical examination. *International Journal of Research in Marketing*, *18*(3), 261–282.

Calvano, M., & Needham, G. (1996). Public empowerment through accessible health information. *Bulletin of the Medical Library Association*, *84*(2), 253. PMID:8826633

Campos, Y., & Villa, J. L. (2018, November). Technologies applied in the monitoring and control of the temperature in the Cold Chain. In *2018 IEEE 2nd Colombian Conference on Robotics and Automation (CCRA)* (pp. 1-6). IEEE. 10.1109/CCRA.2018.8588118

Caplan, J. M., Kennedy, L. W., Barnum, J. D., & Piza, E. L. (2015). Risk terrain modeling for spatial risk assessment. *Cityscape (Washington, D.C.)*, *17*(1), 7–16.

Carleton, T. A., & Hsiang, S. M. (2016). Social and economic impacts of climate. *Science*, *353*(6304), 9837. doi:10.1126cience.aad9837 PMID:27609899

Carley, M., & Christie, I. (2017). *Managing sustainable development*. Abingdon, UK: Routledge. doi:10.4324/9781315091525

Carlisle, R. M., Hays, D. G., Pribesh, S. L., & Wood, C. T. (2017). Educational Technology and Distance Supervision in Counselor Education. *Counselor Education and Supervision*, *56*(1), 33–49. doi:10.1002/ceas.12058

Carr, D. (2007). The global digital divide. *contexts*, *6*(3), 58.

Carr, K. (2002). Building bridges and crossing borders: Using service learning to overcome cultural barriers to collaboration between science and education departments. *School Science and Mathematics*, *102*(6), 285–298. doi:10.1111/j.1949-8594.2002.tb17886.x

Carson, J. B., Tesluk, P. E., & Marrone, J. A. (2007). Shared leadership in Tams: An investigation of antecedent conditions and performance. *Academy of Management Journal*, *50*(5), 1217–1234.

Caspary, G., & O'Connor, D. (2003). Providing Low-Cost Information Technology Access to Rural Communities in Developing Countries: What Works? What Pays? OECD Development Centre Working Paper No. 229 (Formerly Webdoc No. 17). *OECD Publishing (NJ1)*.

Cavusgil, S. T., & Zou, S. (1994). Marketing strategy-performance relationship: An investigation of the empirical link in export market ventures. *Journal of Marketing*, 1–21.

Cem, B. (2009). Long Live, YouTube: L2 Stories about YouTube in Language Learning. In *E-proceedings of the International Online Language Conference*.

Cen, L., Si, L., Li, N., & Jin, H. (2014). User Comment Analysis for Android apps and CSPI Detection with Comment Expansion. In *PIR* (pp. 25–30). SIGIR.

Central Bank of Kenya. (2015). Bank supervision annual report of 2015. Retrieved from https://www.centralbank.go.ke/uploads/399346751_2015%20Annual%20Report.pdf

Centuryply. (2018) Annual Report 2017. Retrieved from https://www.centuryply.com/files/download/43a5eb8b870d3cc

Certo, S. T., & Miller, T. (2008). Social entrepreneurship: Key issues and concepts. *Business Horizons*, *51*(4), 267–271. doi:10.1016/j.bushor.2008.02.009

Cesaroni, F. M., Demartini, P., & Paoloni, P. (2017). Women in business and social media: Implications for female entrepreneurship in emerging countries. *African Journal of Business Management*, *11*(14), 316–326. doi:10.5897/AJBM2017.8281

Cetindere, A., Duran, C., & Yetisen, M. S. (2015). The effects of total quality management on the business performance: An application in the province of Kütahya. *Procedia Economics and Finance*, *23*, 1376–1382.

CFX. (2019). Products. Retrieved from www.cfxnepal.com

Chaba, A. A., & Jagga, R. (2016). Agro-forestry: Poplar's popularity dip, no takers for PM Modi's timber farming call. Retrieved from http://indianexpress.com/article/india/india-news-india/agro-forestry-poplars-popularity-dip-no-takers-for-pm-modis-timber-farming-call-2819456/

Chainey, S., & Ratcliffe, J. (2005). *GIS and Crime Mapping*. Chichester, UK: John Wiley & Sons. doi:10.1002/9781118685181

Chakroborty, S., & Digal, S. K. (2013). Analysis of Investment Pattern of Mutual Funds Investors – An Empirical Study in Orissa, *GITAM. Journal of Management*, *11*(2), 192–207.

Chamberlain, A. (2015). *Why Is Hiring Taking Longer? New Insights from Glassdoor Data.* Mill Valley, CA: Glassdoor.

Chamberlin, L., & Lehmann, K. (2011). Twitter in higher education. In C. Wankel (Ed.), Educating Educators with Social Media (Cutting-edge Technologies in Higher Education, Volume 1) (pp. 375–391). Emerald Group Publishing Limited. doi:10.1108/S2044-9968(2011)0000001021

Chanda, N. (2000). Asian innovation awards: The digital divide. *Far Eastern Economic Review, 163*(42), 50-53.

Chang, M., & Poon, C. K. (2009). Using phrases as features in e-mail classification. *Journal of Systems and Software*, *82*(6), 1036–1945. doi:10.1016/j.jss.2009.01.013

Chani, M. I., & Shahid, M. (2012). Human capital formation and economic development in Pakistan: an empirical analysis. *Актуальні проблеми економіки*(6), 486-495.

Chan, R. C. H., Chu, S. K. W., Lee, C. W. Y., Chan, B. K. T., & Leung, C. K. (2013). Knowledge management using social media: A comparative study between blogs and Facebook. *Proceedings of the American Society for Information Science and Technology*, *50*(1), 1–9. doi:10.1002/meet.14505001069

Chaplin, N., & John, R. E. (2005). The development of self-brand connections in children and adolescents. *Journal of Consumer Behaviour*, *32*(1), 119–129. doi:10.1086/426622

Charband, Y., & Jafari Navimipour, N. (2018). Knowledge sharing mechanisms in the education. *Kybernetes*, *47*(7), 1456–1490. doi:10.1108/K-06-2017-0227

Chase, C. W. Jr. (2014). Innovations in business forecasting: Predictive analytics. *The Journal of Business Forecasting*, *33*(2), 26.

Chaston, I. (2017). Technological Entrepreneurship: Technology-driven Vs Market-driven Innovation. Berlin, Germany: Springer.

Chatterjee, A., Gupta, U., Chinnakotla, M. K., Srikanth, R., Galley, M., & Agrawal, P. (2019). Understanding Emotions in Text Using Deep Learning and Big Data. *Computers in Human Behavior*, *93*, 309–317. doi:10.1016/j.chb.2018.12.029

Chen, M., Tong, M., & Liu, C. (2017). *Recommendation of Learning Path Using an Improved ACO Based on Novel Coordinate System.* Paper presented at the Iiai International Congress on Advanced Applied Informatics. 10.1109/IIAI-AAI.2017.90

Chen. (2013). Integrated analysis of the performance of TQM tools and techniques: a case study in the Taiwanese motor industry. *International Journal of Production Research, 51*(4), 1072-1083.

Chen, F. H., Hsu, T. S., & Tzeng, G. H. (2011). A balanced scorecard approach to establish a performance evaluation and relationship model for hot spring hotels based on a hybrid MCDM model combining DEMATEL and ANP. *International Journal of Hospitality Management*, *30*(4), 908–932. doi:10.1016/j.ijhm.2011.02.001

Chen, K., Luo, P., & Wang, H. (2017). An influence framework on product word-of-mouth (WoM) measurement. *Information & Management*, *54*(2), 228–240. doi:10.1016/j.im.2016.06.010

Chen, Y. C. (2013). Effect of reverse mentoring on traditional mentoring functions. *Leadership and Management in Engineering*, *13*(3), 199–208. doi:10.1061/(ASCE)LM.1943-5630.0000227

Chen, Y., Lawell, C.-Y. L., & Wang, Y. (2017). *The Chinese automobile industry and government policy*. Retrieved from Collis, D. J. (1994). Research note: How valuable are organizational capabilities? *Strategic Management Journal*, *15*(S1), 143–152.

Chhetri, K., & Ramswamy, R. R. (2018). Socio-economic Determinants of Agripreneurship: Daramdin Floriculture Cluster In Sikkim. *Small Enterprises Development, Management & Extension (Sedme). Journal*, *45*(1).

Chiang, A. C., & Wainwright, K. (2005). *Fundamental Methods of Mathematical Economics* (pp. 645–646). Boston, MA: McGraw-Hill.

Chinn, M. D., & Fairlie, R. W. (2004). The determinants of the global digital divide. *Choice (Chicago, Ill.)*, *42*, 7–8.

Chitrakorn, K. (2016, March 31). *You feel me? Why Emotional Culture matters at work*. Retrieved from https://www.businessoffashion.com/articles/careers/you-feel-me-why-emotional-culture-matters-at-work

Choi, B., & Lee, H. (2002). Knowledge management strategy and its link to knowledge creation process. *Expert Systems with Applications*, *23*(3), 173–187. doi:10.1016/S0957-4174(02)00038-6

Choi, D. Y., & Gray, E. R. (2008). The venture development processes of "sustainable" entrepreneurs. *Management Research News*, *31*(8), 558–569. doi:10.1108/01409170810892127

Choi, M. (2011). Employees' attitudes toward organizational change: A literature review. *Human Resource Management*, *50*(4), 479–500. doi:10.1002/hrm.20434

Chopra, S., & Meindl, P. (2016), Supply chain management: Strategy, planning, and operation.

Chou, D. C., Tan, X., & Yen, D. C. (2004). Web technology and supply chain management. *Information Management & Computer Security*, *12*(4), 338–349. doi:10.1108/09685220410553550

Choudary, S. (2017). Employee welfare: A scheme of a wise investment. *International Journal of Advanced Education and Research*, *2*(1), 01-06.

Choudhury, J. K., & Hossain, A. A. (2011). Bangladesh Forestry Outlook Study, *APFSOS Working Paper*. Retrieved from http://www.fao.org/3/a-am628e.pdf

Chowdhery, H. J. (1998). *Orchid of Arunachal Pradesh* (pp. 1–824). India.

Christensen Hughes, J., & Rog, E. (2008). Talent management: A strategy for improving employee recruitment, retention and engagement within hospitality organizations. *International Journal of Contemporary Hospitality Management*, *20*(7), 743–757. doi:10.1108/09596110810899086

Christensen, C. M. (2013). The innovator's dilemma: when new technologies cause great firms to fail. Brighton, MA: Harvard Business Review Press.

Chungho, S. (2017). Designing and Developing a Novel Hybrid Adaptive Learning Path Recommendation System (ALPRS) for Gamification Mathematics Geometry Course. *Eurasia Journal of Mathematics, Science and Technology Education*, *13*(6), 2275–2298.

Chung-Wei, L., & Gwo-Hshiung, T. (2009). Identification of a threshold value for the DEMATEL method using the maximum mean deentropy algorithm to find critical services provided by a semiconductor intellectual property mall. *Expert Systems with Applications*, *8*(1), 9891–9989.

Clarke, J. (2010). Revitalizing Entrepreneurship: How Visual Symbols are Used in Entrepreneurial Performances. *Journal of Management Studies*, *48*(6), 1365–1391. doi:10.1111/j.1467-6486.2010.01002.x

Clauss, K., Wakahiu, J., & Salvaterra, M. (2013). Using technology to educate women religious in Africa. *Am. Int. J. Soc. Sci, 2*, 29–37.

Cohen, A., & Duchan, G. (2012). The usage characteristics of Twitter in the learning process. *Interdisciplinary Journal of E-Learning and Learning Objects, 8*. Retrieved from http://www.openu.ac.il/research_center/chais2011/papers.html

Cohen, B., & Winn, M. I. (2007). Market imperfections, opportunity and sustainable entrepreneurship. *Journal of Business Venturing, 22*(1), 29–49. doi:10.1016/j.jbusvent.2004.12.001

Cohen, E. L., & Lancaster, A. L. (2014). Individual differences in in-person and social media television coviewing: The role of emotional contagion, need to belong, and coviewing orientation. *Cyberpsychology & Behavior, 17*(8). PMID:24950260

Coleman, J. S. (1988). Social Capital in the Creation of Human Capital. *American Journal of Sociology, 94,* Supplement: Organizations and Institutions: Sociological and Economic Approaches to the Analysis of Social Structure, S95-S120.

Commodities Trading Company. (2019). Commodities Trading in Bhutan. Retrieved from http://www.commoditiestradingcompany.com/commodity_trading_in_bhutan.html

Comstock, G. (1989). *The evolution of American television.* Sage.

Concha-Ferreira, I. (2014). Brands, welfare and 'welfare-cool'. *Ephemera: theory and politics in organization, 14*(1), 109-117.

Connor, M. K. (1995). *What is cool? Understanding black manhood in America.* New York, NY: Crown.

Contractor, N. S., DeChurch, L. A., Carson, J., Carter, D. R., & Keegan, B. (2012). The topology of collective leadership. *The Leadership Quarterly, 23*(6), 994–1011. doi:10.1016/j.leaqua.2012.10.010

Cool Brands Council. (2013). *Cool brands-an insight into some of Britain's coolest brands.* Retrieved from http://s3.coolbrands.uk.com/files/2012/09/CB2012- 13-Selection-Process-mRb8d2.pdf

Cordova, D. (2013). Creating the environment for entrepreneurial success. *Center for International Private Enterprise.* Retrieved from http://www.cipe.org/creating-environment-entrepreneurial-success

Cormier, S. M. (2003). *Business incubation in inner-city emerging markets as an economic development tool.*

Corredor, P., & Goñi, S. (2011). TQM and performance: Is the relationship so obvious? *Journal of Business Research, 64*(8), 830–838.

Costa, D., Kehoe, T. J., & Ravindranathan, G. (2016). The Stages of Economic Growth Revisited, *Economic Policy Paper* 16-5. *Federal Reserve Bank of Minneapolis.* Retrieved from https://www.minneapolisfed.org/~/media/files/pubs/eppapers/16-5/epp-16-5-stages-of-economic-growth-revisted-part1

Costa, C., Alvelos, H., & Teixeira, L. (2019). Investigating the Use and Acceptance of Technologies by Professors in a Higher Education Institution. *International Journal of Online Pedagogy and Course Design, 9*(2), 1–20. doi:10.4018/IJOPCD.2019040101

Cotten, S. R. (2001). Implications of Internet technology for medical sociology in the new millennium. *Sociological Spectrum, 21*(3), 319–340. doi:10.1080/027321701300202019

Cotten, S. R., & Gupta, S. S. (2004). Characteristics of online and offline health information seekers and factors that discriminate between them. *Social Science & Medicine, 59*(9), 1795–1806. doi:10.1016/j.socscimed.2004.02.020 PMID:15312915

Cotton, A. P., Sohail, M., & Scott, R. E. (2005). Towards improved labour standards for construction of minor works in low income countries. *Engineering, Construction, and Architectural Management*, *12*(6), 617–632. doi:10.1108/09699980510634164

Cotugna, N., & Vickery, C. (1998). Reverse mentoring: A twist to teaching technology. *Journal of the American Dietetic Association*, *98*(10), 1166–1168. doi:10.1016/S0002-8223(98)00270-3 PMID:9787725

Council, N. E. (2014). *Pakistan Vision 2025*. Pakistan.

Coventry, W. F., & Barker, J. K. (1988). *Management* (International Edition). Portsmouth, NH: Heinemann Professional Publishing.

Covin, J. G., & Miles, M. P. (1999). Corporate entrepreneurship and the pursuit of competitive advantage. *Entrepreneurship Theory and Practice*, *23*(3), 47–47.

Covin, J. G., & Slevin, D. P. (1989). Strategic management of small firms in hostile and benign environments. *Strategic Management Journal*, *10*(1), 75–87.

Coyle Shapiro, J., & Kessler, I. (1998). The pshycological contract in the UK public sector. In S. J. Havlovic (Ed.), Employee and employer obligations and contract fulfillment (pp. 1-7). NPS.

Creswell, J. W., & Poth, C. N. (2017). *Qualitative inquiry and research design: Choosing among five approaches*. New York, NY: Sage.

Crick, D., & Dana, L.-P. (2004). 14. A comparative, exploratory investigation into the perceptions of internationalizing firms in Singapore and the UK. *Emerging Paradigms in International Entrepreneurship*, 319.

Cropanzano, R., & Mitchell, M. (2005). Social Exchange Theory:An Interdisciplinary Review. *Southern Management Association*, *31*(6), 874–900.

Croxton, K. L., Garcia-Dastugue, S. J., Lambert, D. M., & Rogers, D. S. (2001). The supply chain management processes. *International Journal of Logistics Management*, *12*(2), 13–36. doi:10.1108/09574090110806271

Cullen, J. B., & Levitt, S. D. (1999). Crime, urban flight, and the consequences for cities. *The Review of Economics and Statistics*, *81*(2), 159–169. doi:10.1162/003465399558030

Culpepper, W. L. (2009). *Incentive pay eligibility varies by job and location*. Retrieved from http://www.shrm.org/hrdisciplines/compensation/Articles/Pages/IncentiveEligibility.aspx

Cvitani'c, J., & Fernando, Z. (2004). *Economics and Mathematics of Financial Markets* (pp. 17–159). Cambridge, MA: MIT Press.

D'alessandro, D. M., & Dosa, N. P. (2001). Empowering children and families with information technology. *Archives of Pediatrics & Adolescent Medicine*, *155*(10), 1131–1136. doi:10.1001/archpedi.155.10.1131 PMID:11576008

D'souza, D. J., & Joshi, H. G. (2019). E-Commerce Framework for Strategic Marketing of Udupi Jasmine. *AGRIS On-Line Papers in Economics and Informatics*, *11*(1), 17–26. doi:10.7160/aol.2019.110102

Daghfous, A., & Barkhi, R. (2009). The strategic management of information technology in UAE hotels: An exploratory study of TQM, SCM, and CRM implementations. *Technovation*, *29*(9), 588–595. doi:10.1016/j.technovation.2009.05.007

Dahlgaard-Park, S. M. (2012). Core values–the entrance to human satisfaction and commitment. *Total Quality Management & Business Excellence*, *23*(2), 125–140.

Daichao, L. I., & Sheng, W. U. (2014). Theme-Oriented Visual Analysis of Crime with Big Data. *Journal of Geo-Information Science*, *220*(1), 96–100.

Dajian, Z., & Plan, B. (2003). Rescuing a planet under stress and a civilization in trouble. *Chinese Journal of Population, Resources and Environment*(6), 4.

Dalkir, K. (2011). Knowledge Management in Theory and Practice (2nded.). Cambridge, MA: The MIT Press.

Dana, L. P. (2008). Handbook of research on European business and entrepreneurship: Towards a theory of internationalization. Cheltenham, UK: Edward Elgar Publishing.

Dana, L. P., Dana, T. E., & Spence, M. (2004). Public policy and international expansion of high-technology SMEs: A research agenda. *International Journal of Entrepreneurship and Innovation Management*, *4*(2-3), 116–123.

Dana, L. P., Grimwood, S., & William, G. (2009). Export incentives and international entrepreneurship in New Zealand firms. *Journal for International Business and Entrepreneurship Development*, *4*(1-2), 1–21.

Dana, L. P., Hamilton, R. T., & Pauwels, B. (2007). Evaluating offshore and domestic production in the apparel industry: The small firm's perspective. *Journal of International Entrepreneurship*, *5*(3-4), 47.

Dana, L. P., & Paulin, C. (2008). Internationalisation of the New Zealand nutraceutical industry. *Journal for International Business and Entrepreneurship Development*, *3*(3-4), 171–187.

Danesi, M. (1994). *Cool: The signs and meanings of adolescence*. Toronto, Canada: University of Toronto. doi:10.3138/9781442673472

Dang, Y., Zhang, Y., Chen, H., & Larson, C. A. (2011). Knowledge Mapping for Bioterrorism-Related Literature. In C. Castillo-Chavez, H. Chen, W. Lober, M. Thurmond, & D. Zeng (Eds.), *Infectious Disease Informatics and Biosurveillance. Integrated Series in Information Systems* , 27, 311–338. Boston, MA: Springer.

Dar, M. A. (2015). *Impact of Floriculture Development Programme on Registered Flower Growers of Central Kashmir* (Doctoral dissertation, SKUAST Kashmir).

Darnell, M. J. (2015). The interactive TV experience: Where we came from and where we are going. In R. Nakatsu, M. Rauterberg, & P. Ciancarini (Eds.), *Handbook of digital games and entertainment technologies*. New York, NY: Springer. doi:10.1007/978-981-4560-52-8_52-1

Dar-Nimrod, I., Hansen, I. G., Proulx, T., Lehman, D. R., Chapman, B. P., & Duberstein, P. R. (2012). Coolness: An empirical investigation. *Journal of Individual Differences*, *33*(3), 175–185. doi:10.1027/1614-0001/a000088

Dash, S. (2019). Google classroom as a learning management system to teach biochemistry in a medical school. *Biochemistry and Molecular Biology Education*, bmb.21246. doi:10.1002/bmb.21246

Davenport, T. H. (1993). Process innovation: reengineering work through information technology. Brighton, MA: Harvard Business Press.

Davenport, T. H., & Prusak, L. (1998). *Working Knowledge*. Boston, MA: Harvard Business School Press.

Davenport, T., & Beck, J. (2002). *The attention economy: Understanding the new currency of business*. Boston: Harvard Business School Press.

Davidson, S. L., & Butcher, J. (2019). Rural Superintendents' Experiences with Empowerment and Alignment to Vision in the Application of Principle-Centered Leadership. *Rural Educator*, *40*(1).

Davila, T., Epstein, M., & Shelton, R. (2012). Making innovation work: How to manage it, measure it, and profit from it. Upper Saddle River, NJ: FT Press.

Davis, J., & Goadrich, M. (2006). The relationship between Precision-Recall and ROC curves. In *Proceedings of the International Conference on Machine Learning*. New York, NY: ACM Press. 10.1145/1143844.1143874

Dawn. (2016). Pakistan's development challenges. *Dawn.*

Daya, P. (2014). Diversity and inclusion in an emerging market context, *Equality. Diversity & Inclusion*, *33*(3), 293–308. doi:10.1108/EDI-10-2012-0087

DCXNepal. (2019). Products. Retrieved from www.dcxnepal.com

De Carolis, D. M., Litzky, B. E., & Eddleston, K. A. (2009). Why Networks Enhance the Progress of New Venture Creation: The Influence of Social Capital and Cognition. *Entrepreneurship Theory and Practice*, *33*(2), 527–545. doi:10.1111/j.1540-6520.2009.00302.x

De Clercq, D., Haq, I. U., Azeem, M. U., & Ahmad, H. N. (2019). The relationship between workplace incivility and helping behavior: Roles of job dissatisfaction and political skill. *The Journal of Psychology*, 1–21. PMID:30696391

De Clercq, D., & Zhou, L. (2014). Entrepreneurial strategic posture and performance in foreign markets: The critical role of international learning effort. *Journal of International Marketing*, *22*(2), 47–67.

De Jong, J. P., & Hartog, D. N. D. (2007). How leaders influence employees' innovative behaviour. *European Journal of Innovation Management*, *10*(1), 41–64. doi:10.1108/14601060710720546

Deery, M. (2008). Talent management, work-life balance and retention strategies. *International Journal of Contemporary Hospitality Management*, *20*(7), 792–806. doi:10.1108/09596110810897619

del Mar García-Calvente, M., Hidalgo-Ruzzante, N., del Río-Lozano, M., Marcos-Marcos, J., Martínez-Morante, E., Maroto-Navarro, G., ... Gil-García, E. (2012). Exhausted women, tough men: A qualitative study on gender differences in health, vulnerability and coping with illness in Spain. *Sociology of Health & Illness*, *34*(6), 911–926. doi:10.1111/j.1467-9566.2011.01440.x PMID:22443288

Delaney, R., & D'Agostino, R. (2015). The Challenges of Integrating New Technology into an Organization.

deLeede, J., & Looise, J. K. (2005). Innovation and HRM: Towards an integrated framework. *Creativity and Innovation Management*, *14*(2), 108–117. doi:10.1111/j.1467-8691.2005.00331.x

Deloitte Global Human Capital Trends- Rewriting the rules for the digital age. (2017). *The employee experience - Culture, engagement, and beyond.* Deloitte University Press.

Deloitte. (2012). *Waiter, is that inclusion in my soup? A new recipe to improve business performance*, Deloitte Research Report, Australia.

Deloitte. (2017). Role of digital in Asia's economic growth | Deloitte Insights, 2017. Retrieved from https://www2.deloitte.com/insights/us/en/economy/voice-of-asia/may-2017/digital-role-economic-growth.html

Deloitte. (2017). What is the digital economy? Deloitte Malta, 2017. Retrieved from https://www2.deloitte.com/mt/en/pages/technology/articles/mt-what-is-digital-economy.html

DelVecchio, G. (1997). Creating ever cool: A marketer's guide to a kid's heart. Gretna, LA: Pelican.

Denise, L. Y. (2018, June 2). *2018 Will Be the Year of Employee Experience*. Retrieved 06 03, 2019, from Forbes: https://www.forbes.com/sites/deniselyohn/2018/01/02/2018-will-be-the-year-of-employee-experience/#1af202ae1c8f

DevelopmentsS. M. E. in ASEAN. (2017). Retrieved from http://asean.org/asean-economic-community/sectoral-bodies-underthe-purview-of-aem/micro-small-and-medium-enterprises/overview/

Dhamodharan, V., Daniel, B. J. C., & Ambuli, T. V. (2010). An empirical study on assessing trainees' expectations and their perceptions. *International Business Research*, *3*(2), 174. doi:10.5539/ibr.v3n2p174

Dhingra, N., Emmett, J., & Samadani, M. (2018, March 12). *Employee experience: Essential to compete*. McKinsey & Company.

Di Fabio, A., & Kenny, M. E. (2018). Connectedness to Nature, Personality Traits and Empathy from a Sustainability Perspective. *Current Psychology (New Brunswick, N.J.)*. doi:10.100712144-018-0031-4

Díaz-Casero, J. C., Fernández-Portillo, A., Sánchez-Escobedo, M.-C., & Hernández-Mogollón, R. (2017). The Influence of University Context on Entrepreneurial Intentions Entrepreneurial Universities (pp. 65–81). Berlin, Germany: Springer.

Dick, P., & David, R. (2000). *Cool rules: Anatomy of an attitude*. London, UK: Reaktion Books.

Diebold, F. X., & Mariano, R. S. (1995). Comparing predictive accuracy. *Journal of Business & Economic Statistics*, *13*(3), 253–263.

Digital Efficiency Report, Cabinet Office. (2012, November). Retrieved from https://www.gov.uk/government/publications/digital-efficiency-report

Digital Societies Report. (2016). Australia, Japan, and Singapore were selected as representative countries.

Digital Societies Report. (2016). Indonesia and Thailand.

Dittmann, P. (2017). New Supply Chain Technology Best Practices. Available at https://haslam.utk.edu/sites/default/files/GSCI%20InnovationPaper%20FIN4-4-17.pdf

Diwan, M. (2016, March). Internet of Things In Logistics: Towards Autonomous Logistics & Smart Logistics Entities. Paper presented at International Maritime transport & Logistic Conference. Available at https://marlog.aast.edu/archive/2016/pdf/Papers/s06p02.pdf

Dixon, S. E., & Clifford, A. (2007). Ecopreneurship–a new approach to managing the triple bottom line. *Journal of Organizational Change Management*, *20*(3), 326–345. doi:10.1108/09534810710740164

do Nascimento Simões, A., Diniz, N. B., da Silva Vieira, M. R., Ferreira-Silva, S. L., da Silva, M. B., Minatel, I. O., & Lima, G. P. P. (2018). Impact of GA3 and spermine on postharvest quality of anthurium cut flowers (Anthurium andraeanum) cv. Arizona. *Scientia Horticulturae*, *241*, 178–186. doi:10.1016/j.scienta.2018.06.095

Dobusch, L. (2014). How exclusive are inclusive organisations?. *Equality, Diversity and Inclusion: An International Journal, 33*(3), 220-234.

Docherty, P. (2016). Excel to derive the efficient frontier. Retrieved from https://www.youtube.com/watch?v=qV1lDqbUwPo

Dolan, P., Fujiwara, D., & Metcalfe, R. (2012). Review and update of research into the wider benefits of adult learning. *BIS Research Paper,(90)*, 1-47.

Dong, Y. (2015). On the Data Management and Data Mining of Film and TV Drama Played On-Line. *In Conference on Engineering And Technology Management* (p. 69). GCETM.

Doughty, M., Rowland, D., & Lawson, S. (2011). Co-viewing live TV with digital backchannel streams. In *Proceedings of the 9th International Interactive Conference on Interactive Television* (pp. 141-144). New York, NY: Association for Computing Machinery. 10.1145/2000119.2000147

Douglas, M., & Isherwood, B. (1996). *The world of good: Towards an anthropology of consumption*. London, UK: Routledge.

Drescher, S. (2004). *The mighty experiment: Free labor versus slavery in British emancipation*. Oxford, UK: Oxford University Press.

Driver, S. (2018, Oct. 15). Social Media for Business: A Marketer's Guide. Business News Daily. Retrieved from https://www.businessnewsdaily.com/7832-social-media-for-business.html

Drucker, P. F. (1996). Your leadership is unique. *Leadership*, *17*(4), 54.

Du Pisani, J. A. (2006). Sustainable development–historical roots of the concept. *Environmental Sciences*, *3*(2), 83–96. doi:10.1080/15693430600688831

Du Puy, D., & Cribb, P. (1988). *The genus Cym-bidiwn*. Portland, OR: Timber Press.

Dubbah, M. M. (2010). *International and Comparative Competition Law* (p. 115). Cambridge, UK: Cambridge University Press. doi:10.1017/CBO9780511777745

Dubey, P. (2009). Prospects and Challenges for the Emerging Timber Import Market in India. *Journal of Forestry*, *107*(1), 23–28.

Duggal, E., & Verma, H. V. (2019). Cool perspectives, Indian cool and branding. *South Asian Journal of Business Studies*, *2*(8), 130–145. doi:10.1108/SAJBS-07-2018-0083

Dukes, E. (2019). *5 Big Employee Experience Challenges to Overcome in 2019*. Houston, TX: Office.

Dumbill, E. (2013). Big data and thought crime: an interview with Jim Adler. *Big Data*, *1*(1), 10-13.

Durand, G., Belacel, N., & Laplante, F. (2013). Graph theory based model for learning path recommendation. *Information Sciences*, *251*(4), 10–21. doi:10.1016/j.ins.2013.04.017

Durga Prasad, K. G., Venkata Subbaiah, K., & Narayana Rao, K. (2012). Aligning the competitive strategy with supply chain strategy through QFD. *Journal of Advances in Management Research*, *9*(2), 189–198. doi:10.1108/09727981211271931

Dwivedi, P., Kant, V., & Bharadwaj, K. K. (2017). Learning path recommendation based on modified variable length genetic algorithm. *Education and Information Technologies*.

Dyer, J. H., & Chu, W. (2003). The Role of Trustworthiness in Reducing Transaction Costs and Improving Performance: Empirical Evidence from the United States, Japan, and Korea. *Organization Science*, *14*(1), 57–68. doi:10.1287/orsc.14.1.57.12806

Easterday, A., Driscoll, D., & Ramaswamy, S. (2019). Rural homelessness: Its effect on healthcare access, healthcare outcomes, mobility, and perspectives of novel technologies. *Journal of Social Distress and the Homeless*, *28*(1), 56–64. doi:10.1080/10530789.2019.1567978

Eby, L. T., & Allen, T. D. (2002). Further investigation of proteges' negative mentoring experiences: Patterns and outcomes. *Group & Organization Management*, *27*, 456–479. doi:10.1177/1059601102238357

Eby, L. T., Butts, M. M., Lockwood, A., & Simon, S. A. (2004). Proteges' negative mentoring experiences: Construct development and nomological validation. *Personnel Psychology*, *57*(2), 411–447. doi:10.1111/j.1744-6570.2004.tb02496.x

Echeng, R., Usoro, A., Echeng, R., & Usoro, A. (2016). Enhancing the use of Web 2.0 Technologies in Higher Education: Students' and Lectures' Views. *Journal of International Technology and Information Management*, *25*(1), 89–106. Retrieved from http://scholarworks.lib.csusb.edu/jitimhttp://scholarworks.lib.csusb.edu/jitim/vol25/iss1/6

Edvardsson, I. R. (2008). HRM and knowledge management. *Employee Relations*, *30*(5), 553–561. doi:10.1108/01425450810888303

Ehnert, I. (2009). *Sustainable Human Resource Management- A conceptual and exploratory analysis from a paradox perspective* (1st ed.). Physica-Verlag Heidelberg.

Ehnert, I. (2009). *Sustainable Human Resource Management: A Conceptual and Exploratory Analysis from a Paradox Perspective*. Heidelberg, Germany: Physica-Verlag. doi:10.1007/978-3-7908-2188-8

Einsfeld, K., Ebert, A., Kerren, A., & Deller, M. (2009). Knowledge generation through human-centered information visualization. *Information Visualization*, *8*(3), 180–196. doi:10.1057/ivs.2009.15

Elliott, R. (1999). Symbolic meaning and post- mod- ern consumer culture. In D. Brownlie, M. Saren, R. Wensley, & R. Whittington (Eds.), *Rethinking marketing*. London, UK: Sage.

Elliott, R., & Wattansuwan, K. (1998). Brands as symbolic resources for the construction of identity. *International Advertising*, *17*(2), 135–145.

Elson Anderson, K., & Still, J. M. (2011). An introduction to Google Plus. *Library Hi Tech News*, *28*(8), 7–10. doi:10.1108/07419051111187842

Engenke, L., & Yuen, J. C. (2008). Types of Commodity Investments. In F. Fabbozi, R. Fuss, & D. G. Kaiser (Eds.), *The Handbook of Commodity Investing* (pp. 549–569). Hoboken, NJ: John Wiley and Sons.

England, M. I., Stringer, L. C., Dougill, A. J., & Afionis, S. (2018). How do sectoral policies support climate compatible development? An empirical analysis focusing on southern Africa. *Environmental Science & Policy*, *79*, 9–15. doi:10.1016/j.envsci.2017.10.009

Epitropaki, O., & Martin, R. (2013). Transformational–transactional leadership and upward influence: The role of Relative Leader–Member Exchanges (RLMX) and Perceived Organizational Support (POS). *The Leadership Quarterly*, *24*(2), 299–315. doi:10.1016/j.leaqua.2012.11.007

Eppler, M. J., & Simon, H. A. (2008). A process-based classification of knowledge maps and application examples. *Knowledge and Process Management*, *15*(1), 59–71. doi:10.1002/kpm.299

Erasmus, B., & Schenk, H. (2008). *South African human resource management: Theory & practice*. Cape Town, South Africa: Juta and Company Ltd.

Ernest III, J., Wilson, E. J., & Wilson III III, E. J. (2004). *The information revolution and developing countries*. Cambridge, MA: MIT Press.

Escalas, J. E. (2004). Narrative processing: Building consumer connections to brands. *Journal of Consumer Psychology*, *12*(1-2), 168–179. doi:10.120715327663jcp1401&2_19

Escoffery, C., Miner, K. R., Adame, D. D., Butler, S., McCormick, L., & Mendell, E. (2005). Internet use for health information among college students. *Journal of American College Health*, *53*(4), 183–188. doi:10.3200/JACH.53.4.183-188 PMID:15663067

Etemad, H., Wilkinson, I., & Dana, L. P. (2010). Internetization as the necessary condition for internationalization in the newly emerging economy. *Journal of International Entrepreneurship*, *8*(4), 319–342.

ETF.com. (2016). CUT Fund Report. Retrieved from http://www.etf.com/pdf-version/CUT

ETFdb. (2018). Invesco MSCI Global Timber ETF. Retrieved from http://etfdb.com/etf/CUT/

European Foreign Institute. (2017). The Vietnam-EU Voluntary Partnership Agreement. Retrieved from http://www.euflegt.efi.int/publications/the-vietnam-eu-voluntary-partnership-agreement

European Foundation for Quality Management. (2013). EFQM Model 2013. Retrieved from http://www.efqm.org/index.php/efqm-model-2013/

Facility, E. U. F. L. E. G. T. (2017). What is FLEGT. Retrieved from http://www.euflegt.efi.int/publications/cross-border-timber-trade-in-the-saarc-area

Fairs, M. (2019). China is fast becoming the world's creative superpower. Retrieved from https://www.dezeen.com/2019/04/17/china-design-power-opinion-marcus-fairs/

Falatoonitoosi, E., Leman, Z., Sorooshian, S., & Salimi, M. (2013). Decision-Making Trial and Evaluation Laboratory, Research Journal of Applied Sciences. *Engineering and Technology*, *5*(13), 3476–3480.

Fang, M. L., Canham, S. L., Battersby, L., Sixsmith, J., Wada, M., & Sixsmith, A. (2018). Exploring privilege in the digital divide: Implications for theory, policy, and practice. *The Gerontologist*, *59*(1), e1–e15. PMID:29750241

FAO(2019b). Chapter 23: South Asia. Retrieved from http://www.fao.org/docrep/004/y1997e/y1997e0s.htm

FAO. (2019a). The State of Forestry in Napal. Retrieved from http://www.fao.org/docrep/w7719e/w7719e04.htm

Farhin, N. (2018, Jan. 9). How small businesses use Facebook to promote products and services. Dhaka Tribune. Retrieved from https://www.dhakatribune.com/ business/2018/01/09/small-businesses-use-facebook-promote-products-services

Farsi, J. Y., & Toghraee, M. T. (2014). Identification the main challenges of small and medium sized enterprises in exploiting of innovative opportunities (Case study: Iran SMEs). *Journal of Global Entrepreneurship Research*, *4*(1), 4. doi:10.1186/2251-7316-2-4

Fathy, S., Said, E., & Fattah, A. (2015). The Effectiveness of Using WhatsApp Messenger as One of Mobile Learning Techniques to Develop Students' Writing Skills. *Journal of Education and Practice*, *6*(32), 115–127.

Fattah, H. (2000). Politics or real problem? *Technology Marketing Intelligence*, *20*(9), 83–87.

Feifei, S., Zhuo, C., & Xiaolei, X. (2014). Application of an improved random forest based classifier in crime prediction domain. *Journal of Intelligence*, *33*(10), 148–152.

Feinstein*, L., & Hammond, C. (2004). The contribution of adult learning to health and social capital. *Oxford Review of Education, 30*(2), 199-221.

Feizabadi, J., Singh, M., & Motlagh, S. A. (2014). Contribution of supply chain to corporate strategy: A case study in agriculture machinery industry. *International Journal of Logistics Systems and Management*, *18*(4), 473–499. doi:10.1504/IJLSM.2014.063981

Feldman, R., & Dagan, I. (1995). *KDT-Knowledge Discovery in Texts.* Paper presented at the First International Conference on Knowledge Discovery (KDD).

Felea, M., & Albăstroiu, I. (2013). Defining the concept of supply chain management and its relevance to Romanian academics and practitioners. *Amfiteatru Economic Journal*, *15*(33), 74–88.

Ferguson, S. (2011). A global culture of cool? Generation Y and their perception coolness. *Young Consumers*, *12*(3), 265–275. doi:10.1108/17473611111163313

Fernandes Bolina, A., Rodrigues, R. A. P., Tavares, D. M. D. S., & Haas, V. J. (2019). Factors associated with the social, individual and programmatic vulnerability of older adults living at home. *Revista da Escola de Enfermagem da U S P.*, 53. PMID:30810627

Field, J. (2009). *Well-being and happiness*. National Institute of Adult Continuing Education.

Fischer, E., & Reuber, A. R. (2011). Social interaction via new social media: (How) can interactions on Twitter affect effectual thinking and behavior? *Journal of Business Venturing*, *26*(1), 1–18. doi:10.1016/j.jbusvent.2010.09.002

Fish, L. A. (2011). Supply chain quality management. In Supply Chain Management-Pathways for Research and Practice. *IntechOpen*.

Fishbein, M., & Ajzen, I. (1975). *Belief, attitude, intention, and behavior: An introduction to theory and research*. Reading, MA: Addison-Wesley.

Fisher, C. (2019). *Emotions in Organisations, Oxford Research Encyclopaedia of Business and Management*. USA: Oxford University Press.

Fisher, I. (1930). *The Theory of Interest as Determined by Impatience to Spend Income and Opportunity to Invest it*. London, UK: Macmillan.

Flomenbaum, A. (2015). Accenture Report: 87% of consumers use second screen device while watching TV. *Adweek*. Retrieved from http://www.adweek.com/lostremote/accenture-report-87-of-consumers-use-second-screen-device-while-watching-tv/51698

Foresee. (2014). *Measuring employee experience to drive positive employee engagement*. A Foresee White Paper.

Forest Survey of India. (2009). State of Forest Report. Dehra Dun, India Forest Survey of India, Ministry of Environment and Forests, Government of India.

Forestry Department. (2010). Global Forest Resources Assessment Country Reports Maldives. Retrieved from http://www.fao.org/docrep/013/al559E/al559E.pdf

Forman, G., & Kirshenbaum, E. (2008). *Extremely fast text feature extraction for classification and indexing*. Paper presented at the 17th ACM Conference on Information and Knowledge Management, California, USA. 10.1145/1458082.1458243

Foster, H. (1983). *Postmodernism: A preface. The anti-aesthetic: Essays on postmodern culture*. Port Townsend, WA: Bay Press.

Fournier, S. M. (1998). Consumer and their brands: Developing relationship theory in consumer research. *The Journal of Consumer Research*, *24*(4), 343–373. doi:10.1086/209515

Fox, S. (2006). Online Health Search 2006. Pew Internet and American Life Project. Retrieved from http://www. pewInternet. org/pdfs/PIP_Online_Health_2006. pdf

Fox, G., & Connolly, R. (2018). Mobile health technology adoption across generations: Narrowing the digital divide. *Information Systems Journal*, *28*(6), 995–1019. doi:10.1111/isj.12179

Francis, J., Ball, C., Kadylak, T., & Cotten, S. R. (2019). Aging in the Digital Age: Conceptualizing Technology Adoption and Digital Inequalities. In *Ageing and Digital Technology* (pp. 35–49). Singapore: Springer. doi:10.1007/978-981-13-3693-5_3

Frank, H., Kessler, A., & Fink, M. (2010). Entrepreneurial orientation and business performance-a replication study. *Schmalenbach Business Review*, *62*, 175–198.

Frank, T. (1977). *Jazz: A history*. New York, NY: W.W. Norton & Co.

Fredrickson, B. L. (2001). The role of positive emotions in positive psychology. *The American Psychologist*, *56*(3), 218–226. doi:10.1037/0003-066X.56.3.218 PMID:11315248

Fredrickson, B. L., & Branigan, C. (2005). Positive emotions broaden the scope of attention and thought-action repertoires. *Cognition and Emotion*, *19*(3), 313–333. doi:10.1080/02699930441000238 PMID:21852891

Freeman, E. A., & Moisen, G. G. (2008). A comparison of the performance of threshold criteria for binary classification in terms of predicted prevalence and kappa. *Ecological Modelling*, *217*(1-2), 48–58. doi:10.1016/j.ecolmodel.2008.05.015

Friedman, J. H. (2001). Greedy function approximation: A gradient boosting machine. *Annals of Statistics*, *29*(5), 1189–1232. doi:10.1214/aos/1013203451

Fries, D., Rossegger, A., Endrass, J., & Singh, J. P. (2013). The prediction of criminal recidivism using routinely available file information. *International Journal of Psychological Research*, *6*(2), 8–14. doi:10.21500/20112084.671 PMID:25374652

Fruhling, A. L., & Digman, L. A. (2000). The Impact of Electronic Commerce on Business-Level Strategies. *Journal of Electronic Commerce Research*, *1*(1), 13–22.

Fuchs, C. (2008). *Internet and Society: Social theory in the information age*. New York, NY: Routledge.

Gallup state of the American workplace: Employee engagement insights for U.S. business leaders. (2013). Retrieved from http://www.michaeljbeck.com/documents/State

Gallup. (2006). Gallup study: Engaged employees inspire company innovation: national survey finds that passionate workers are most likely to drive organisations forward. *The Gallup Management Journal*.

Ganesan, K., & Zhai, C. (2012). Opinion-based entity ranking. *Information Retrieval*, *15*(2), 116–150. doi:10.100710791-011-9174-8

Ganguly, P. (2016, January). Selecting the right IoT cloud platform. In *2016 International Conference on Internet of Things and Applications (IOTA)* (pp. 316-320), IEEE. 10.1109/IOTA.2016.7562744

Gann, R. (1986). The health information handbook: resources for self care. Abingdon-on-Thames, UK: Gower Pub.

Gann, B., & Needham, G. (Eds.). (1992). *Promoting Choice: Consumer Health Information in the 1990's*. Consumer Health Information Consortium.

Gardner, M. (1985). Mood states and consumer behavior:A critical review. *The Journal of Consumer Research*, *12*(3), 281–300. doi:10.1086/208516

Garrison, G., Wakefield, R. L., & Kim, S. (2015). The effects of IT capabilities and delivery model on cloud computing success and firm performance for cloud supported processes and operations. [</unknown>]. *International Journal of Information Management*, *35*(4), 377–393.

Gawde, P. (2018). Effect of packaging materials on vase life of tuberose (Polianthes tuberose) cv. Shringar. *IJCS*, *6*(5), 2011–2014.

Gebreeyesus, M., & Sonobe, T. (2009). Governance of global value chain and firms' capability in African floriculture, 1st Draft.

Gebreeyesus, M. (2015). Firm adoption of international standards: Evidence from the Ethiopian floriculture sector. *Agricultural Economics*, *46*(S1), 139–155. doi:10.1111/agec.12203

Genç, M., & Oksüz, B. (2015). A fact or an Illusion: Effective social media usage of female entrepreneurs. *Procedia: Social and Behavioral Sciences*, *195*, 293–300. doi:10.1016/j.sbspro.2015.06.345

George, S., Daniels, K., & Fioratou, E. (2018). A qualitative study into the perceived barriers of accessing healthcare among a vulnerable population involved with a community centre in Romania. *International Journal for Equity in Health*, *17*(1), 41. doi:10.118612939-018-0753-9 PMID:29615036

Gerguri, S., & Ramadani, V. (2010). The impact of innovation into the economic growth.

Gerpott, F. H., Lehmann-Willenbrock, N., & Voelpel, S. C. (2017). A phase model of intergenerational learning in organizations. *Academy of Management Learning & Education*, *16*(2), 193–216. doi:10.5465/amle.2015.0185

Gershon, R. A. (2019). Transnational Media and the Economics of Global Competition. *Global Communication. Multicultural Perspectives*, 37.

Ghosh, S., & Bhowmick, B. (2014, October). Technological uncertainty: Exploring factors in indian start-ups. In *IEEE Global Humanitarian Technology Conference (GHTC 2014)* (pp. 425-432), IEEE. 10.1109/GHTC.2014.6970317

Gibson, E. C., & Matthews, D. (2013, July). The impact of rapidly changing technology on the supply chain. In *2013 Proceedings of PICMET'13: Technology Management in the IT-Driven Services (PICMET)* (pp. 1944-1950), IEEE.

Giddens, D., & Phillips, R. (2009, September). Reverse Mentoring: Finding a new way of working while discovering Web 2.0, ALIA National Library and Information Technicians Conference, Adelaide Convention Centre, Adelaide, SA, Australia.

Giddens, A. (1991). *Modernity and self-identity*. Cambridge, UK: Polity Press.

Gilchrist, K. (2018, July 17). Facebook and 3 millennials are changing the start-up scene in Bangladesh. CNBC. Retrieved from https://www.cnbc.com/2018/07/17/shopup-bangladesh-start-up-uses-facebook-to-help-micro-entrepreneurs.html

Gill, K., Brooks, K., McDougall, I., & Patel, P. K., (2010) A. Bridging the Gender Divide: How Technology Can Advance Women Economically; International Centre for Research on Women: Washington, DC.

Gladwell, M. (1997). Annals of style: The cool hunt. *New Yorker (New York, N.Y.)*, (March): 17.

Glas, A. H., & Kleemann, F. C. (2016). The impact of industry 4.0 on procurement and supply management: A conceptual and qualitative analysis. *International Journal of Business and Management Invention*, *5*(6), 55–66.

Global Canopy Programme. (2009). The Little Climate Finance Book. Retrieved from https://www.ecbi.org/sites/default/files/Little_Climate_Finance_Book_-_English-libre_pdf

Gob´e, M., & Zyman. (2001). *Emotional branding: The new paradigm for connecting brands to people*. New York: Allworth Press.

Godbole, S., & Sarawagi, S. (2004). Discriminative methods for multi-labeled classification. *Lecture Notes in Computer Science*, *3056*, 22–30. doi:10.1007/978-3-540-24775-3_5

Go, F. M., & Govers, R. (2000). Integrated quality management for tourist destinations: A European perspective on achieving competitiveness. *Tourism Management*, *21*(1), 79–88.

Gold, H. A., Malhotra, A., & Segars, H. A. (2001). Knowledge Management: An organizational Capabilities Perspective. *Journal of Management Information Systems*, *18*(1), 185–214. doi:10.1080/07421222.2001.11045669

Goldsmith, J. (2000). How will the Internet change our health system? *Health Affairs*, *19*(1), 148–156. doi:10.1377/hlthaff.19.1.148 PMID:10645081

Goleman, D. (1998). *Working with emotional intelligence*. New York, NY: Bantam.

Gomez-Mejia, L. R., Balkin, D. B., & Cardy, R. L. (2001). *Managing Human Resources* (3rd ed.). Upper Saddle River, NJ: Prentice Hall.

Gonzalez, K. M., Shaughnessy, M. J., Kabigting, E. N. R., Tomasulo West, D., Callari Robinson, J. F., Chen, Q., & Stewart Fahs, P. (2018). A Systematic Review of the Health of Vulnerable Populations within US Rural Societies. *Online Journal of Rural Nursing and Health Care: the Official Journal of the Rural Nurse Organization, 18*(1), 112–147. doi:10.14574/ojrnhc.v18i1.507

González-Ramírez, R., Gascó, J. L., & Llopis Taverner, J. (2015). Facebook in teaching: Strengths and weaknesses. *International Journal of Information and Learning Technology, 32*(1), 65–78. doi:10.1108/IJILT-09-2014-0021

Gordon, N. P., & Hornbrook, M. C. (2018). Older adults' readiness to engage with eHealth patient education and self-care resources: A cross-sectional survey. *BMC Health Services Research, 18*(1), 220. doi:10.118612913-018-2986-0 PMID:29587721

Gosserand, R. H., & Diefendorff, J. M. (2005). Emotional Display Rules and Emotional Labor: The Moderating Role of Commitment. *The Journal of Applied Psychology, 90*(6), 1256–1264. doi:10.1037/0021-9010.90.6.1256 PMID:16316278

Gottdiener, M. (2001). *The theming of America – American dreams, Media Fan*. Boulder, CO: Westview Press.

Government of India. (1969). *Report of the Committee on Labour Welfare*. India: Ministry of Labour& Employment.

Govindan, K., Kaliyan, M., Kannan, D., & Haq, A. N. (2014). Barriers analysis for green supply chain management implementation in Indian industries using analytic hierarchy process. *International Journal of Production Economics, 147*, 555–568. doi:10.1016/j.ijpe.2013.08.018

Govindan, K., Soleimani, H., & Kannan, D. (2015). Reverse logistics and closed-loop supply chain: A comprehensive review to explore the future. *European Journal of Operational Research, 240*(3), 603–626. doi:10.1016/j.ejor.2014.07.012

Grandey, A. A. (2000). Emotion Regulation in the Workplace: A New Way to Conceptualize Emotional Labor. *Journal of Occupational Health Psychology, 5*(1), 95–110. doi:10.1037/1076-8998.5.1.95 PMID:10658889

Grant, R. M. (1996b). Toward a knowledge-based theory of the firm. *Strategic Management Journal, 17*(S2), 109–111. doi:10.1002mj.4250171110

Graver, M. (2009). *Stoicism and emotion*. Chicago, IL: University of Chicago Press.

Greblikaite, J. (2012). Development of Social Entrepreneurship: Challenge for Lithuanian Researchers. *European Integration Studies*(6), 210-215.

Greenberg, A. J., Haney, D., Blake, K. D., Moser, R. P., & Hesse, B. W. (2018). Differences in access to and use of electronic personal health information between rural and urban residents in the United States. *The Journal of Rural Health, 34*, s30–s38. doi:10.1111/jrh.12228 PMID:28075508

Greengard, S. (2002). Moving forward with reverse mentoring. *Workforce, 81*(1), 15.

Greenply. (2018). Annual Report 2018. Retrieved from http://www.greenply.com/images/pdf/Annual-Report-2017-18.pdf

Green, T. J. (2012). TQM and organisational culture: How do they link? *Total Quality Management & Business Excellence, 23*(2), 141–157.

Greve, A., & Salaff, J. W. (2003). *Social Networks and Entrepreneurship, Entrepreneurship Theory and Practice*, (Fall): 1–22. doi:10.1111/1540-8520.00029

Grieger, L., & Hosser, D. (2012). Attention deficit hyperactivity disorder does not predict criminal recidivism in young adult offenders: Results from a prospective study. *International Journal of Law and Psychiatry*, *35*(1), 27–34. doi:10.1016/j.ijlp.2011.11.005 PMID:22142896

Grimes, S., Phillips, T., Hahn, V., Capezzone, F., & Graeff-Hönninger, S. (2018). Growth, Yield Performance and Quality Parameters of Three Early Flowering Chia (Salvia hispanica L.) Genotypes Cultivated in Southwestern Germany. *Agriculture*, *8*(10), 154. doi:10.3390/agriculture8100154

Grossman, L. (2003). The quest for cool. *Time*, *162*, 48.

GSMA. (2014). Mobile Economy Report Sub-Saharan Africa. Available at http://www.gsma.com/mobileeconomy/archive/GSMA_ME_SubSaharanAfrica_2014.pdf, page 29.

GSMA. (2014). State of the Industry: Mobile Financial Services for the Unbanked. Retrieved from https://www.gsma.com/mobilefordevelopment/wp-content/uploads/2015/03/SOTIR_2014.pdf

GSMA. (2015). Competition Policy in the Digital Age. A practical handbook. Available at http://www.gsma.com/competition-policy-handbook

GSMA. (2016). Advancing Digital Societies in Asia. Available at http://www.gsma.com/newsroom/press-release/new-gsma-study-tracks-digitalsociety-progress-asia/, page 20 44.

Guillén, M. F., & Suárez, S. L. (2005). Explaining the global digital divide: Economic, political and sociological drivers of cross-national Internet use. *Social Forces*, *84*(2), 681–708. doi:10.1353of.2006.0015

Gunasekaran, A., Lai, K. H., & Cheng, T. E. (2008). Responsive supply chain: A competitive strategy in a networked economy. *Omega*, *36*(4), 549–564. doi:10.1016/j.omega.2006.12.002

Gupta, M., & Kohli, A. (2006). Enterprise resource planning systems and its implications for operations function. *Technovation*, *26*(5-6), 687–696. doi:10.1016/j.technovation.2004.10.005

Habersang, S., Küberling-Jost, J., Reihlen, M., & Seckler, C. (2019). A Process Perspective on Organizational Failure: A Qualitative Meta-Analysis. *Journal of Management Studies*, *56*(1), 19–56. doi:10.1111/joms.12341

Habib, M., Abbas, J., & Noman, R. (2019). Are human capital, intellectual property rights, and research and development expenditures really important for total factor productivity? An empirical analysis. *International Journal of Social Economics*. doi:10.1108/IJSE-09-2018-0472

Hafkin, N., & Taggart, N. (2001). Gender, information technology, and developing countries: An analytic study (pp. 42-48). Office of Women in Development, Bureau for Global Programs, Field Support and Research, United States Agency for International Development.

Hafkin, N., & Huyer, S. (2007). Women and gender in ICT statistics and indicators for development. *Information Technologies and International Development*, *4*(2), 25–41. doi:10.1162/itid.2008.00006

Hagel, J., & Armstrong, A. G. (1997). *Net gain: Expanding markets through virtual communities*. Boston, MA: Harvard Business School Press.

Hahn, J., & Subramani, M. R. (2000). A framework of knowledge management systems: issues and challenges for theory and practice. *Proceedings of the International Conference on Information Systems, ICIS'2000*, 302-12.

Hall, B. H., & Khan, B. (2003). Adoption of new technology (No. w9730). *National bureau of economic research*.

Hall, A., Carberry, P., Djikeng, A., Roy-Macauley, H., Pengelly, B., Njoya, A., ... Keating, B. (2016). The Journey to R4D: An institutional history of an Australian initiative on food security in Africa. In J. Francis, L. Mytelka, A. van Huis, & N. Röling (Eds.), *Innovation Systems: Towards Effective Strategies in Support of Smallholder Farmers. Technical Centre for Agricultural and Rural Cooperation (CTA) and Wageningen University and Research (WUR) Convergence of Sciences-Strengthening Innovation Systems (CoS-SIS) programme* (pp. 183–201). Wageningen, The Netherlands.

Hall, J. K., Daneke, G. A., & Lenox, M. J. (2010). Sustainable development and entrepreneurship: Past contributions and future directions. *Journal of Business Venturing*, *25*(5), 439–448. doi:10.1016/j.jbusvent.2010.01.002

Hamermesh, D. S. (2011). *Beauty pays: Why at- tractive people are more successful*. Princeton, NJ: Princeton University Press.

Hammond, C., & Feinstein, L. (2006). *Are those who flourished at school healthier adults? What role for adult education?* (Wider Benefits of Learning Research Report No. 17). Centre for Research on the Wider Benefits of Learning.

Handzic, M. (2011). Integrated socio-technical knowledge management model: An empirical evaluation. *Journal of Knowledge Management*, *15*(2), 198–211. doi:10.1108/13673271111119655

Hang, M., & Van Weezle, A. (2007). Media and entrepreneurship: A survey of the literature relating both concepts. *Journal of Media Business Studies*, *4*(1), 51–70. doi:10.1080/16522354.2007.11073446

Hanna, R., Rohm, A., & Crittenden, V. L. (2011). We're all connected: The power of the social media ecosystem. *Business Horizons*, *54*(3), 265–273. doi:10.1016/j.bushor.2011.01.007

Harboe, G., Metchalf, C. J., Bentley, F., & Tullio, J., Massey, N., & Romano, G. (2008). Ambient social TV: Drawing people into a shared experience. In M. Czerwinski, & A. Lund (Eds.), *Proceedings of the SIGCHI Conference on Human Factors in Computing Systems* (pp. 1-10). New York, NY: ACM. 10.1145/1357054.1357056

Hardey, M. (1999). Doctor in the house: The Internet as a source of lay health knowledge and the challenge to expertise. *Sociology of Health & Illness*, *21*(6), 820–835. doi:10.1111/1467-9566.00185

Hargadon, S. (2008). *Web 2.0 Is the Future of Education.* Retrieved from http://www.stevehargadon.com/2008/03/web-20-is-future-of-education.html

Harrington, B., & Ladge, J. (2009). Work–Life Integration:Present Dynamics andFuture Directions for Organizations. *Organizational Dynamics*, *38*(2), 148–157. doi:10.1016/j.orgdyn.2009.02.003

Harris, P. (2007). We the people: The importance of employees in the process of building customer experience. *Brand Management*, *15*(2), 102–114. doi:10.1057/palgrave.bm.2550123

Hart, S. L. (2005). Capitalism at the crossroads: The unlimited business opportunities in solving the world's most difficult problems. London, UK: Pearson Education.

Hartman, L. P., Shaw, B., & Stevenson, R. (2000). Human Resources Opportunities to Balance Ethics and Neoclassical Economics in Global Labor Standards. *Business & Professional Ethics Journal*, *19*(3/4), 73–116. doi:10.5840/bpej2000193/42

Harwood, J. (2007). *Understanding communication and aging: Developing knowledge and awareness*. Thousand Oaks, CA: Sage.

Hasan, I., Khan, A. N. M. S., Karim, M. A., Khan, S. R., Alam, S., & Sanjana, B. (2018). Health and safety compliance in the readymade garment sector of Bangladesh: Practices and Observations. *Independent Business Review*, *11*(1-2), 25–32.

Hays-Thomas, R., Bowen, A., & Boudreaux, M. (2012). Skills for diversity and inclusion in organizations: A review and preliminary investigation. *The Psychologist Manager Journal*, *15*(2), 128–141. doi:10.1080/10887156.2012.676861

Head, J. H. ACC, & Freedman, J. (2014, Jan. 2). *Inspiring Employee Engagement through Emotional Intelligence.* Retrieved from https://www.6seconds.org/2014/01/02/employee-engagement-emotional-intelligence/

Head, J. H., & Freedman, J. (2014, Jan 2). *Inspiring Employee Engagement through Emotional Intelligence.* Retrieved June 01, 2018, from EQ Business: https://www.6seconds.org/2014/01/02/employee-engagement-emotional-intelligence/

Heisig, P. (2009). Harmonisation of knowledge management–comparing 160 KM frameworks around the globe. *Journal of Knowledge Management*, *13*(4), 4–31. doi:10.1108/13673270910971798

Hellström, T., & Husted, K. (2004). Mapping knowledge and intellectual capital in academic environments: A focus group study. *Journal of Intellectual Capital*, *5*(1), 165–180. doi:10.1108/14691930410512987

Helmedag, F. (2008). The Optimal Rotation Period of Renewable Resources: Theoretical Evidence from the Timber Sector. In F. Fabbozi, R. Fuss, & D. G. Kaiser (Eds.), *The Handbook of Commodity Investing* (pp. 145–166). Hoboken, NJ: John Wiley and Sons. doi:10.1002/9781118267004.ch6

Henderson, J. C., & Venkatraman, H. (1999). Strategic alignment: Leveraging information technology for transforming organizations. *IBM systems journal, 38*(2.3), 472-484.

Hennart, J.-F., Majocchi, A., & Forlani, E. (2017). The myth of the stay-at-home family firm: How family-managed SMEs can overcome their internationalization limitations. *Journal of International Business Studies*, 1–25.

Hennessy, C. M., Kirkpatrick, E., Smith, C. F., & Border, S. (2016). Social media and anatomy education: Using twitter to enhance the student learning experience in anatomy. *Anatomical Sciences Education*, *9*(6), 505–515. doi:10.1002/ase.1610 PMID:27059811

Henning, B., & Vorderer, P. (2001). Psychological escapism: Predicting the amount of television viewing by need for cognition. *Journal of Communication*, *51*(1), 100–120. doi:10.1111/j.1460-2466.2001.tb02874.x

Henrekson, M., & Sanandaji, T. (2014). Small business activity does not measure entrepreneurship. *Proceedings of the National Academy of Sciences of the United States of America*, *111*(5), 1760–1765. doi:10.1073/pnas.1307204111 PMID:24449873

Henwood, F., Wyatt, S., Hart, A., & Smith, J. (2003). 'Ignorance is bliss sometimes': Constraints on the emergence of the 'informed patient'in the changing landscapes of health information. *Sociology of Health & Illness*, *25*(6), 589–607. doi:10.1111/1467-9566.00360 PMID:12919447

Hernandez-Perlines, F. (2018). Moderating effect of absorptive capacity on the entrepreneurial orientation of international performance of family businesses. *Journal of Family Business Management.*

Hern, M. J., Weitkamp, T., Hillard, P. J. A., Trigg, J., & Guard, R. (1998). Promoting women's health via the World Wide Web. *Journal of Obstetric, Gynecologic, and Neonatal Nursing*, *27*(6), 606–610. doi:10.1111/j.1552-6909.1998.tb02629.x PMID:9836154

Herscovitch, L., & Meyer, J. P. (2002). Commitment to organizational change: Extension of a three-component model. *The Journal of Applied Psychology*, *87*(3), 474–487. doi:10.1037/0021-9010.87.3.474 PMID:12090605

Hersey, R. B. (1932). *Workers' Emotions in Shop and Home: A Study of Individual Workers from the Psychological and Physiological Standpoint.* Philadelphia, PA: University of Pennsylvania Press.

Hervani, A. A., Helms, M. M., & Sarkis, J. (2005). Performance measurement for green supply chain management. *Benchmarking International Journal (Toronto, Ont.)*, *12*(4), 330–353.

Heutger, M., & Kückelhaus, M. (2015). Unmanned Aerial Vehicle in Logistics. Available at http://www.dhl.com/content/dam/downloads/g0/about_us/logistics_insights/DHL_TrendReport_UAV.pdf

Hilbert, M. (2011 a). Digital gender divide or technologically empowered women in developing countries? A typical case of lies, damned lies and statistics. *Women's Studies International Forum*, *34*(6), 479–489. doi:10.1016/j.wsif.2011.07.001

Hilbert, M. (2011). The end justifies the definition: The manifold outlooks on the digital divide and their practical usefulness for policy-making. *Telecommunications Policy*, *35*(8), 715–736. doi:10.1016/j.telpol.2011.06.012

Hinings, C. R., Thibault, L., Slack, T., & Kikulis, L. M. (1996). Values and organizational structure. *Human Relations*, *47*(7), 885–916. doi:10.1177/001872679604900702

Hirschman, E. C., & Holbrook, M. B. (1982). Hedonic consumption: Emerging concepts, methods, and propositions. *Journal of Marketing*, *46*(3), 92–101. doi:10.1177/002224298204600314

Hoang, D. T., Igel, B., & Laosirihongthong, T. (2010). Total quality management (TQM) strategy and organisational characteristics: Evidence from a recent WTO member. *Total Quality Management*, *21*(9), 931–951.

Hoberg, K., & Alicke, K. (2016). How SC4. 0 will Enhance the Customer Experience. *Supply Chain Management Review*, (September/October), 28-37.

Hobman, E. V., Bordia, P., & Gallois, C. (2003). Consequences of feeling dissimilar from others in a work teams. *Journal of Business and Psychology*, *17*(3), 301–304. doi:10.1023/A:1022837207241

Hochschild, A. (1983). *The managed heart*. Berkeley, CA: University of California Press.

Hochschild, A. R. (2003). *The Managed Heart: Commercialization of Human Feeling*. Los Angeles, CA: University of California Press.

Hoekman, B., & Shingal, A. (2017). Aid for trade and international transactions in goods and services.

Hoffman, D. L., & Novak, T. P. (1998). Bridging the Digital Divide: The Impact of Race on Computer Access and Internet Use.

Höflinger, P. J., Nagel, C., & Sandner, P. (2018). Reputation for technological innovation: Does it actually cohere with innovative activity? *Journal of Innovation & Knowledge*, *3*(1), 26–39. doi:10.1016/j.jik.2017.08.002

Hofmann, E. (2010). Linking corporate strategy and supply chain management. *International Journal of Physical Distribution & Logistics Management*, *40*(4), 256–276. doi:10.1108/09600031011045299

Holmes, K., & Prieto-Rodriguez, E. (2018). Student and Staff Perceptions of a Learning Management System for Blended Learning in Teacher Education. *Australian Journal of Teacher Education*, *43*(3), 21–34. doi:10.14221/ajte.2018v43n3.2

Holt, D. B. (2005). How societies desire brands: Using cultural theory to explain brand symbolism. In S. Ratneshwar, & D. G. Mick (Eds.), *Inside consumption* (pp. 273–291). London, UK: Routledge.

Holvino, E., Ferdman, B. M., & Merrill-Sands, D. (2004). Creating and sustaining diversity and inclusion in organizations: Strategies and approaches. In M. S. Stockdale, & F. J. Crosby (Eds.), *The psychology and management of workplace diversity* (pp. 245–276). Malden, MA: Blackwell Publishing.

Honneth. (1994). The social dynamics of disrespect: on the location of critical theory today, *Constellations* 1(1), 255-269.

Hopwood, B., Mellor, M., & O'Brien, G. (2005). Sustainable development: Mapping different approaches. *Sustainable Development, 13*(1), 38–52. doi:10.1002d.244

Hora, M. T. (2019). *Beyond the skills gap: Preparing college students for life and work.* Cambridge, MA: Harvard Education Press.

Horton, M., Read, J. C., Fitton, D., Little, L., & Toth, N. (2012). Too Cool at School - Understanding Cool Teenagers. *PsychNology Journal, 10*(2), 73–91.

Hossain, S. F. A., Ying, Y., & Saha, S. K. (2019). Systematic Mobile Device Usage Behavior and Successful Implementation of TPACK Based on University Students Need. In *Science and Information Conference* (pp. 729-746). Cham, Switzerland: Springer.

Hossain, S. F. A. (2019). Social Networking and Its Role in Media Entrepreneurship: Evaluating the Use of Mobile Phones in the Context of Online Shopping–A Review. [JMME]. *Journal of Media Management and Entrepreneurship, 1*(1), 73–86. doi:10.4018/JMME.2019010105

Hossain, S. F. A., Nurunnabi, M., Hussain, K., & Saha, S. K. (2019). Effects of variety-seeking intention by mobile phone usage on university students' academic performance. *Cogent Education, 6*(1). doi:10.1080/2331186X.2019.1574692

Hover-Dixon, T. (2006). The Upside of Down: Catastrophe, Creativity and the Renewal of Civilisation. London, UK: Souvenir Press.

Howell, K. E., & Annansingh, F. (2013). Knowledge generation and sharing in UK universities: A tale of two cultures? *International Journal of Information Management, 33*(1), 32–39. doi:10.1016/j.ijinfomgt.2012.05.003

Hsieh, H. F., & Shannon, S. E. (2005). Three approaches to qualitative content analysis. *Qualitative Health Research, 15*(9), 1277–1288. doi:10.1177/1049732305276687 PMID:16204405

Huang, C. Y., Shyu, J. Z., & Tzeng, G. H. (2007). Reconfiguring the innovation policy portfolios for Taiwan's SIP Mall industry. *Technovation, 27*(12), 744–765. doi:10.1016/j.technovation.2007.04.002

Huang, H.-M., & Liaw, S.-S. (2004). The Framework of Knowledge Creation for Online Learning Environments. *Canadian Journal of Learning and Technology, 30*(1). doi:10.21432/T26G6Z

Huang, Y. Y., Li, C. T., & Jeng, S. K. (2015). Mining location-based social networks for criminal activity prediction. In *Proceedings of Wireless & Optical Communication Conference*. Taipei, China: IEEE. doi:10.1109/WOCC.2015.7346202

Huber, N., Michael, K., & McCathie, L. (2007, September). Barriers to RFID adoption in the supply chain. In *2007 1st Annual RFID Eurasia* (pp. 1–6). IEEE. doi:10.1109/RFIDEURASIA.2007.4368128

Huda, K. N. (2016). Towards Sustainopreneurship Development at the Tertiary Level Education: A Case study on Southern University Bangladesh. *Journal on Innovation and Sustainability, 7*(2), 3–16.

Hull, K., & Dodd, J. E. (2017). Faculty use of Twitter in higher education teaching. *Journal of Applied Research in Higher Education, 9*(1), 91–104. doi:10.1108/JARHE-05-2015-0038

Hung, C. L. (2004). The political economy of the digital divide in Taiwan.

Hung, Y. H., Chou, S. C., & Tzeng, G. H. (2011). Knowledge management adoption and assessment for SMEs by a novel MCDM approach. *Decision Support Systems, 51*(2), 270–291. doi:10.1016/j.dss.2010.11.021

Hunt, T. (2014). Has social media changed television? The Internet and social media have enhanced TV viewing. *Social media Explorer*. Retrieved from http://www.socialmediaexplorer.com/social-media-marketing/has-social-media-changed-television/

Huo, B., Gu, M., & Jiang, B. (2018). China-related POM research: Literature review and suggestions for future research. *International Journal of Production Economics*.

Hurley, R. F., & Hult, G. T. M. (1998). Innovation, market orientation, and organizational learning: An integration and empirical examination. *Journal of Marketing*, 42–54.

Hu, T., Zhu, X., Duan, L., Guo, W., & Arcaute, E. (2018). Urban crime prediction based on spatio-temporal bayesian model. *PLoS One*, *13*(10). doi:10.1371/journal.pone.0206215 PMID:30379897

Hutcheon, L. (1988). *A poetics of postmodernism: History, theory, fiction*. New York, NY: Routledge. doi:10.4324/9780203358856

IBM & Globoforce. (2016). *The Employee Experience Index- A new global measure of a human workplace and its impact*. Software Group.

ICEX. (2019). Products. Retrieved from https://www.icexindia.com/static/products.aspx

ICIMOD. (2016) Building Timber Value Chains for REDD+ in Nepal. *ICIMOD Working Paper 2016/9*. Retrieved from http://lib.icimod.org/record/32385/files/WP%202016_9_Value%20chain.pdf

Iizuka, M., & Gebreeyesus, M. (2017). Using Functions of Innovation Systems to Understand the Successful Emergence of Non-traditional Agricultural Export Industries in Developing Countries: Cases from Ethiopia and Chile. *European Journal of Development Research*, *29*(2), 384–403. doi:10.105741287-016-0004-0

Iles, P., Chuai, X., & Preece, D. (2010). Talent management and HRM in multinational companies in Beijing: Definitions, differences and drivers. *Journal of World Business*, *45*(2), 179–189. doi:10.1016/j.jwb.2009.09.014

Imran, M., Aziz, A., & Hamid, S. (2017). Determinants of SME export performance. *International Journal of Data and Network Science*, *1*(2), 39–58.

Imran, M., Aziz, A., & Hamid, S. (2017). Total Quality Management, Export Market Orientation and Firm Export Performance: A Conceptual Framework. *International Journal of Academic Research in Business and Social Sciences*, *7*(9), 591–601. doi:10.6007/IJARBSS/v7-i9/3382

Imran, M., Aziz, A., Hamid, S., Shabbir, M., Salman, R., & Jian, Z. (2018). The mediating role of total quality management between entrepreneurial orientation and SMEs export performance. *Management Science Letters*, *8*(6), 519–532.

Imran, M., Hamid, S., & Aziz, A. (2018). The influence of TQM on export performance of SMEs: Empirical evidence from manufacturing sector in Pakistan using PLS-SEM. *Management Science Letters*, *8*(5), 483–496.

Inbaek, S., Kang, B. K., Doo, I. C., Mee, Y. E. S., & Park, H. (2017). Implementation of crime prevention system using public big data. *Advanced Science Letters*, *23*(10), 9574–9578. doi:10.1166/asl.2017.9750

India. Planning Commission. (2008). Sikkim Development Report. Academic Foundation.

India's growth rate set to surpass China this year: World Bank. The Economic Times. Retrieved September 11, 2015.

Ingelsson, P., Eriksson, M., & Lilja, J. (2012). Can selecting the right values help TQM implementation? A case study about organisational homogeneity at the Walt Disney Company. *Total Quality Management & Business Excellence*, *23*(1), 1–11.

International Labour Organisation. (1963). *Asian Regional Conference*. Labour laws and legislation, 5-6.

Investing.com. (2018). India 10-year Bond Yield. Retrieved from https://in.investing.com/rates-bonds/india-10-year-bond-yield-historical-data

Investment Company Institute. (2015). 2015 Investment Company Fact Book. Retrieved from http://www.icifactbook.org/fb_ch3.html

Investopedia. (2016). Understanding Sharpe Ratio. Retrieved from http://www.investopedia.com/articles/07/sharpe_ratio.asp

İpek, İ. (2017). The Resource-Based View within the Export Context: An Integrative Review of Empirical Studies. *Journal of Global Marketing*, 1–23.

Isen, A. M. (1999). Positive affect. In T. Dalgleish, & M. J. Powers (Eds.), *Handbook of Cognition and Emotion* (pp. 25–521). Hoboken, NJ: Wiley.

Isenberg, D. J. (2010). How to start an entrepreneurial revolution. *Harvard Business Review*, *88*(6), 40–50.

Ishares. (2018). Prospectus. Retrieved from https://www.ishares.com/us/library/stream-document?stream=reg&product=I-TIMBR&shareClass=NA&documentId=925986~925792~926067~925559~925569&iframeUrlOverride=%2Fus%2Fliterature%2Fprospectus%2Fp-ishares-global-timber-and-forestry-etf-3-31.pdf

Islam, N., & Ahmed, R. (2016). Factors Influencing the Development of Women Entrepreneurship in Bangladesh. doi:. doi:10.2139srn.2851786

Islam, M. Z., Jasimuddin, S. M., & Hasan, I. (2018). Determinants that influence knowledge sharing: An integrated literature review. *International Journal of Knowledge Management Studies*, *9*(4), 363–380. doi:10.1504/IJKMS.2018.096318

ITTO. (2009). Encouraging Industrial Forest Plantations in the Tropics. Retrieved from http://www.itto.int/direct/topics/topics_pdf_download/topics_id=2165&no=0&disp=inline

Ivanov, D., & Sokolov, B. (2010). Conceptual Frameworks for Supply Chain Management. *Adaptive Supply Chain Management*, 19-33.

Ivanov, A., & Cyr, D. (2006). The concept plot: A concept mapping visualization tool for asynchronous web-based brainstorming sessions. *Information Visualization*, *5*(3), 185–191. doi:10.1057/palgrave.ivs.9500130

İyigün, N. Ö. (2015). What could entrepreneurship do for sustainable development? A corporate social responsibility-based approach. *Procedia: Social and Behavioral Sciences*, *195*, 1226–1231. doi:10.1016/j.sbspro.2015.06.253

Jabbour, C. J. C., & Santos, F. C. A. (2008). The central role of human resource management in the search for sustainable organizations. *International Journal of Human Resource Management*, *19*(12), 2133–2154. doi:10.1080/09585190802479389

Jabbour, C., & Santos, F. (2008). Relationships between human resource dimensions and environmental management in companies: Proposal of a model. *Journal of Cleaner Production*, *16*(1), 51–58. doi:10.1016/j.jclepro.2006.07.025

Jabeen, R., & Mahmood, R. (2015). The Effects of Total Quality Management and Market Orientation on Business Performance of Small and Medium Enterprises in Pakistan. *British Journal of Economics. Management & Trade*, *5*(4), 408–418. doi:10.9734/BJEMT/2015/14226

Jackson, J. (1966). A conceptual and measurement model for norms and roles. *Pacific Sociological Review*, *9*(1), 35–47. doi:10.2307/1388305

Jacob, M., & Goldsmith, M. (2017). *The Employee Experience Advantage: How to Win the War for Talent by Giving Employees the Workspaces they Want, the Tools they Need, and a Culture They Can Celebrate*. New York: John Wiley & Sons.

Jagongo, A., & Kinyua, C. (2013). The Social Media and Entrepreneurship Growth (A New Business Communication Paradigm among SMEs in Nairobi). *International Journal of Humanities and Social Science*, *3*(10), 213–227.

Jain, S. (2016). Study of employee welfare & benefit practices at Indian oil corporation limited (Lubes Plant, Vashi, Navi Mumbai). *NCRD's Business Review: e-Journal, 2*(2), 1-10.

Jain, S., & Sinha, A. (2018). Social Network Analysis: Tools, Techniques, and Technologies. In Social Network Analytics for Contemporary Business Organizations (pp. 1-18). Hershey, PA: IGI Global.

Jain, A. K. (2010). Data clustering: 50 years beyond k-means. *Pattern Recognition, 31*(8), 651–666. doi:10.1016/j.patrec.2009.09.011

Jain, J., Dangayach, G. S., Agarwal, G., & Banerjee, S. (2010). Supply chain management: Literature review and some issues. *Journal of Studies on Manufacturing, 1*(1).

Janakiram, T., Jain, R., Vyas, A. K., & Srilekha, R. (2018). Florienterpreneurship in India: New Avenue of Economic Empowerment of Women. *ASCI Journal of Management, 47.*

Jang, Y., Park, N. S., Yoon, H., Huang, Y. C., Rhee, M. K., Chiriboga, D. A., & Kim, M. T. (2018). The risk typology of healthcare access and its association with unmet healthcare needs in Asian Americans. *Health & Social Care in the Community, 26*(1), 72–79. doi:10.1111/hsc.12463 PMID:28620950

Jaskiewicz, W., & Tulenko, K. (2012). Increasing community health worker productivity and effectiveness: A review of the influence of the work environment. *Human Resources for Health, 10*(1), 38. doi:10.1186/1478-4491-10-38 PMID:23017131

Jensen, J. F. (2008). The concept of interactivity - revisited: Four new typologies for a new media landscape. In *UXTV '08 Proceedings of the 1st International Conference on Designing Interactive User Experiences for TV and Video* (pp. 129-132).

Jeong, K. S., Moon, T. H., & Jeong, J. H. (2010). Hotspot analysis of urban crime using space-time scan statistics. *Journal of the Korean Association of Geographic Information Studies, 13*(3), 14–28.

Jha, A., Sharma, B., & Kumar, J. (2018). Life-Style Distinction Between Customers and Non-Customers of Sikkim Cymbidium. In Management Strategies and Technology Fluidity in the Asian Business Sector (pp. 14–26). Hershey, PA: IGI Global. doi:10.4018/978-1-5225-4056-4.ch002

Jiang, J., & Xia, R. (2016). Microblog sentiment classification via combining rule-based and machine learning methods.

Jiann, L. Y., & Gwo-Hshiung, T. (2011). An integrated MCDM technique combined with DEMATEL for a novel cluster-weighted with ANP method. *Expert Systems with Applications, 7*(2), 1417–1424.

Jin, B., & Jung, S. (2016). Toward a deeper understanding of the roles of personal and business networks and market knowledge in SMEs' international performance. *Journal of Small Business and Enterprise Development, 23*(3), 812–830.

Jing, Z., & Zhu, M. (2016). Market orientation, product innovation and export performance: Evidence from Chinese manufacturers. *Journal of Strategic Marketing, 25*(5), 377–397.

Joia, L. A. (2000). Using intellectual capital to evaluate educational technology projects. *Journal of Intellectual Capital, 1*(4), 341–356. doi:10.1108/14691930010359243

Jones, A. (2011). How Twitter Saved My Literature Class: A Case Study with Discussion. In C. Wankel (Ed.), Teaching Arts and Science with the New Social Media (Cutting-edge Technologies in Higher Education, Volume 3) (pp. 91–105). Emerald Group Publishing Limited. doi:10.1108/S2044-9968(2011)0000003008

Jones, M. C. (2012). Cool for kids. *The Marketing Week*, July, 1.

Joseph, B., Joseph, I., & Varghese, R. (2009). Labour Welfare in India. *Journal of Workplace Behavioral Health, 24*(1&2), 221–242. doi:10.1080/15555240902849131

Jo, T. (2010). NTC (Neural Text Categorizer): Neural network for text categorization. *International Journal of Information Science, 2*(2), 83–96.

Judy, R. W., & d'Amico, C. (1997). *Workforce 2020: Work and workers in the 21st century*. Hudson Institute, Herman Kahn Center, PO Box 26-919, Indianapolis, IN 46226; tele.

Junker, M., Hoch, R., & Dengel, A. (1999). On the Evaluation of Document Analysis Components by Recall, Precision, and Accuracy. In *Proceedings of International Conference on Document Analysis & Recognition*. Bangalore, India: IEEE. 10.1109/ICDAR.1999.791887

Kadam, A., & Ayarekar, S. (2014). Impact of Social Media on Entrepreneurship and Entrepreneurial Performance: Special Reference to Small and Medium Scale Enterprises. *SIES Journal of Management, 10*(1), 3–11.

Kahn, W. A. (1990). Psychological Conditions of Personal Engagement and Disengagement at Work. *Academy of Management Journal, 33*(4), 692–724.

Kalantzis-Cope, P. (2011). Properties of Technology. In V. Kalantzis-Cope, & K. Gherab-Martin (Eds.), *Emerging Digital Spaces in Contemporary Society: Properties of Technology* (pp. 3–9). London, UK: Palgrave Macmillan.

Kalema, B. M. M., Motsi, L., & Motjolopane, I. M. (2017). Utilizing IT to Enhance Knowledge Sharing for School Educators in Developing Countries. *The Electronic Journal on Information Systems in Developing Countries, 73*(1), 1–22. doi:10.1002/j.1681-4835.2016.tb00533.x

Kaličanin, Đ., Veljković, S., & Bogetić, Z. (2015). Brand orientation and financial performance nexus. *Industrija, 43*(1), 155–173.

Kalichman, S. C., Weinhardt, L., Benotsch, E., & Cherry, C. (2002). Closing the digital divide in HIV/AIDS care: Development of a theory-based intervention to increase Internet access. *AIDS Care, 14*(4), 523–537. doi:10.1080/09540120208629670 PMID:12204154

Kals, E., & Maes, J. (2002). *Sustainable Development and Emotions*. doi:. doi:10.1007/978-1-4615-0995-0_6

Karaa, D., Uysalb, M., Sirgyc, M. J., & Leed, G. (2013). The effects of leadership style on employee well-being in hospitality. *International Journal of Hospitality Management, 34*, 9–18. doi:10.1016/j.ijhm.2013.02.001

Karan, K. K. (1989). *George Bernard Shaw and the concept of Superman*. New Delhi, India: Vani- Prakashan.

Karim, A. S., Zaki, A. R., & Mubeen, M. H. (2019). Managing Workforce Diversity in Multicultural Organizations. *Journal of European Studies, 35*(1), 79–91.

Karlaftis, M. G. (2004). A DEA approach for evaluating the efficiency and effectiveness of urban transit systems. *European Journal of Operational Research, 152*(2), 354–364. doi:10.1016/S0377-2217(03)00029-8

Kaše, R., Saksida, T., & Mihelič, K. K. (2019). Skill development in reverse mentoring: Motivational processes of mentors and learners. *Human Resource Management, 58*(1), 57–69. doi:10.1002/hrm.21932

Kavitha, V., & Punithavalli, M. (2010). Clustering time series data stream - A literature survey. *International Journal of Computer Science and Information Security, 8*(1), 289–294.

Kay, C. (2018). Rural Latin America: exclusionary and uneven agricultural development. In Capital, Power, and Inequality in Latin America (pp. 21-52). Abingdon-on-Thames, UK: Routledge. doi:10.4324/9780429501869-2

Kearney, E., Gebert, D., & Voelpel, S. C. (2009). When and how diversity benefits teams: The importance of team members' need for cognition. *Academy of Management Journal*, *52*(3), 581–598. doi:10.5465/amj.2009.41331431

Kerr, M., & Clegg, C. (2007). Sharing knowledge: Contextualizing socio–technical thinking and practice. *The Learning Organization*, *14*(5), 423–435. doi:10.1108/09696470710762646

Khajeheian, D. (2013). New Venture Creation in Social Media Platform; Towards a Framework for Media Entrepreneurship. Handbook of Social Media Management Value Chain and Business Models in Changing Media Markets, 125-142. doi:10.1007/978-3-642-28897-5_8

Khalessizadeh, S. M., Zaefarian, R., Nasseri, S. H., & Ardil, E. (2006). Genetic mining: Using genetic algorithm for topic based on concept distribution. *Journal of Word Academy of Science. Engineering and Technology*, *13*(2), 144–147.

Khandave, S., Prajapati, M. R., & Patel, V. T. (2017). Entrepreneurial Attributes Of Nursery Growers. *Gujarat Journal of Extension Education*, 418.

Khan, M. S., Khan, M. M. I., & Haleem, A. (2018, August). Towards effective management of cold chain: A DEMATEL approach. [IOP Publishing]. *IOP Conference Series. Materials Science and Engineering*, *404*(1). doi:10.1088/1757-899X/404/1/012019

Kim, K., Jeong, B., & Jung, H. (2014). Supply chain surplus: Comparing conventional and sustainable supply chains. *Flexible Services and Manufacturing Journal*, *26*(1-2), 5–23. doi:10.100710696-012-9163-2

Kim, S., Suh, E., & Hwang, H. (2003). Building the knowledge map: An industrial case study. *Journal of Knowledge Management*, *7*(2), 34–45. doi:10.1108/13673270310477270

Kim, T. T., & Lee, G. (2013). Hospitality employee knowledge-sharing behaviors in the relationship between goal orientations and service innovative behavior. *International Journal of Hospitality Management*, *34*, 324–337. doi:10.1016/j.ijhm.2013.04.009

Kim, T., & Lee, G. (2012). A modified and extended Triandis model for the enablers–process–outcomes relationship in hotel employees' knowledge sharing. *Service Industries Journal*, *32*(13), 2059–2090. doi:10.1080/02642069.2011.574276

Kim, W. C., & Mauborgne, R. (1999). Creating new market space. *Harvard Business Review*, *77*(1), 83–93. PMID:10345394

Kim, Y., & Ployhart, R. E. (2014). The effects of staffing and training on firm productivity and profit growth before, during, and after the Great Recession. *The Journal of Applied Psychology*, *99*(3), 361–389. doi:10.1037/a0035408 PMID:24377393

Kinnunen, T., & Li, H. (2010). An overview of text-independent speaker recognition: From features to supervectors. *Speech Communication*, *52*(1), 12–40. doi:10.1016/j.specom.2009.08.009

Kiplinger. (2012). *Cool its earliest slang meaning*. Retrieved from http://kiplinger.tumblr.com/ post/25889435298/cool-its-earliest-slang-meaning-dates-to-1728

Kirzner, I. M. (1997). Entrepreneurial discovery and the competitive market process: An Austrian approach. *Journal of Economic Literature*, *35*(1), 60–85.

Kjeldgaard, D., & Askegaard, S. (2006). The globalization of youth culture: The global youth segment as structures of common difference. *The Journal of Consumer Research*, *33*(2), 231–247. doi:10.1086/506304

Klein, N. (2000). No logo. London, UK: Hammersmith.

Klein, A. Z., Junior, J. C., Mattiello da Silva, J. V. V. M., Barbosa, J. L. V., & Baldasso, L. (2018). The Educational Affordances of Mobile Instant Messaging (MIM). *International Journal of Distance Education Technologies*, *16*(2), 51–64. doi:10.4018/IJDET.2018040104

Klomp, J., & Hoogezand, B. (2018). Natural disasters and agricultural protection: A panel data analysis. *World Development*, *104*, 404–417. doi:10.1016/j.worlddev.2017.11.013

Klugman, J., Rodríguez, F., & Choi, H.-J. (2011). The HDI 2010: New controversies, old critiques. *The Journal of Economic Inequality*, *9*(2), 249–288. doi:10.100710888-011-9178-z

Knoke, D. (2018). Changing organizations: Business networks in the new political economy. Abingdon-on-Thames, UK: Routledge.

Kohlenberger, J. (2016). The New Formula for Cool: Science, Technology and The Popular in the American Imagination. New York, NY: Columbia University Press.

Kohli, A. K., & Jaworski, B. J. (1990). Market orientation: The construct, research propositions, and managerial implications. *Journal of Marketing*, 1–18.

Kohli, A. K., Jaworski, B. J., & Kumar, A. (1993). MARKOR: A measure of market orientation. *JMR, Journal of Marketing Research*, 467–477.

Koh, N. S., Hu, N., & Clemons, E. K. (2010). Do online reviews reflect a product's true perceived quality? An investigation of online movie reviews across cultures. *Electronic Commerce Research and Applications*, *9*(5), 374–385. doi:10.1016/j.elerap.2010.04.001

Kongklom, N., Chuensangjun, C., Chisti, Y., & Sirisansaneeyakul, S. (2018). Improved keeping quality of Dendrobium "Bom" orchids using nutrients entrapped in a biodegradable hydrogel. *Scientia Horticulturae*, *234*, 184–192. doi:10.1016/j.scienta.2018.02.031

Korosec, R. L., & Berman, E. M. (2006). Municipal support for social entrepreneurship. *Public Administration Review*, *66*(3), 448–462. doi:10.1111/j.1540-6210.2006.00601.x

Kosik, A. (2007). The implications of Facebook, sharing the Commonwealth: Critical issues in higher education. *Critical Issues in Higher Education*, 9–10.

Kram, K. E. (1983). Phases of the mentor relationship. *Academy of Management Journal*, *26*(4), 608–625.

Kreps, G. L. (2001). The evolution and advancement of health communication inquiry. *Annals of the International Communication Association*, *24*(1), 231–253. doi:10.1080/23808985.2001.11678988

Kulakli, A., & Mahony, S. (2014). Knowledge Creation and Sharing with Web 2.0 Tools for Teaching and Learning Roles in So-called University 2.0. *Procedia: Social and Behavioral Sciences*, *150*, 648–657. doi:10.1016/j.sbspro.2014.09.084

Kulkarni, N. P., & Jahagirdar, K. A. (2019). Entrepreneurial Behavior and Constraints Faced by the Rose Growers. *Asian Journal of Agricultural Extension. Economia e Sociologia (Evora, Portugal)*, 1–8.

Kumar, R., & Nagpal, B. (2018). Analysis and prediction of crime patterns using big data. *International Journal of Information Technology*, (4), 1-7.

Kumar, S., & Gulati, R. (2010). Assessing the Effect of Ownership on the Efficiency of Indian Domestic Banks. *IUP Journal of Bank Management, 9*(3).

Kumar, G. S. A., & Kumar, K. A. (2018). A Study on Labour Welfare Measures in Singareni Collieries Company Limited. *International Journal of Engineering Technology Science and Research*, *5*(3), 1376–1382.

Kumari, M., Mehta, P., & Raina, K. K. (2016). Farmers' Perceptions towards Marketing Problems and Challenges in Floriculture in Solan District of Himachal Pradesh, India. *International Journal of Economic Plants*, *3*(4), 143–149.

Kumar, P., & Rai, S. C. (2018). Agricultural Diversities and Its Sustainability in Sikkim Himalaya: An Analysis. *Political Economy Journal of India*, *27*(1-2), 91.

Kumar, S., & Yadav, S. S. (2002). Satisfaction level from labour welfare schemes in sugar factories of Gorakhpur division. *Indian Journal of Economics*, *33*(329), 171–188.

Kumar, V. (2012). *101 design methods: A structured approach for driving innovation in your organization*. Hoboken, NJ: John Wiley & Sons.

Kuteesa, H. (2018, Nov. 3). Hangzhou: Inside one of the world's leading cashless cities. The New Times, Rwanda.

Laborde, Z. B., Burbano, K. B., Reinoso, V. G., Bangeppagari, M., Mulla, S. I., & Selvanayagam, M. (2019). Emotional Intelligence Models as Generators of Business Management Change in the Human Talent Area. *Journal of Artificial Intelligence*, *12*(1), 1–10. doi:10.3923/jai.2019.1.10

Lach, J. (2000). Crossing the digital divide. *American Demographics*, *22*(6), 9–12.

Lages, L. F., Silva, G., & Styles, C. (2009). Relationship capabilities, quality, and innovation as determinants of export performance. *Journal of International Marketing*, *17*(4), 47–70.

Laroche, M., Habibi, M. R., Richard, M. O., & Sankaranarayanan, R. (2012). The effects of social media based brand communities on brand community markers, value creation practices, brand trust and brand loyalty. *Computers in Human Behavior*, *28*(5), 1755–1767. doi:10.1016/j.chb.2012.04.016

Laukkanen, T., Nagy, G., Hirvonen, S., Reijonen, H., & Pasanen, M. (2013). The effect of strategic orientations on business performance in SMEs: A multigroup analysis comparing Hungary and Finland. *International Marketing Review*, *30*(6), 510–535.

LEAD. (2016). *Sustainable Development Goals (SDGs) (2015-2030)* Knowledge Hub on SDGs. Briefing Notes. Leadership for Environment and Development (LEAD), Pakistan. Islamabad, Pakistan. Retrieved from http://www.lead.org.pk

Leavengood, S., Anderson, T. R., & Daim, T. U. (2014). Exploring linkage of quality management to innovation. *Total Quality Management & Business Excellence*, *25*(9-10), 1126–1140.

Lee, J. H., & Segev, A. (2012). Knowledge maps for e-learning. *Computers & Education*, *59*(2), 0-364.

Lee, H. L. (2002). Aligning supply chain strategies with product uncertainties. *California Management Review*, *44*(3), 105–119. doi:10.2307/41166135

Lee, H., & Choi, B. (2003). Knowledge management enablers, process, and organizational performance: An integrative view and empirical examination. *Journal of Management Information Systems*, *20*(1), 179–228. doi:10.1080/07421222.2003.11045756

Leidner, R. (1999). Emotional Labor in service work, *THE. The Annals of the American Academy of Political and Social Science*, *561*(1), 81–95. doi:10.1177/000271629956100106

Leonardo, D. (2012). Google Books: Primary sources in the public domain. *Collection Building*, *31*(3), 103–107. doi:10.1108/01604951211243498

Levine, A., & Alexander, B. (2008). Web 2.0 Storytelling: Emergence of a New Genre. *EDUCAUSE Review*, *43*(6). Retrieved from https://er.educause.edu/articles/2008/10/web-20-storytelling-emergence-of-a-new-genre

Levy, H., Janke, A. T., & Langa, K. M. (2015). Health literacy and the digital divide among older Americans. *Journal of General Internal Medicine*, *30*(3), 284–289. doi:10.100711606-014-3069-5 PMID:25387437

Levy, S. (1959). Symbols for sale. *Harvard Business Review*, *37*(4), 117–124.

Lewis, L. (2019). *Organizational change: Creating change through strategic communication.* Hoboken, NJ: Wiley-Blackwell. doi:10.1002/9781119431503

Lewis, R. E., & Heckman, R. J. (2006). Talent management: A critical review. *Human Resource Management Review*, *16*(2), 139–154. doi:10.1016/j.hrmr.2006.03.001

Li, X. (2009). Dis/locating audience: transnational media flows and the online circulation of East Asian television drama (*Doctoral dissertation, Massachusetts Institute of Technology*).

Liley, M., Feliciano, P., & Laurs, A. (2017). *Employee Experience Reimagined.* Accenture.

Lin, C. H., & Sanders, K. (2017). HRM and innovation: A multi-level organisational learning perspective. *Human Resource Management Journal*, *27*(2), 300–317. doi:10.1111/1748-8583.12127

Lines, R. (2004). Influence of participation in strategic change: Resistance, organizational commitment and change goal achievement. *Journal of Change Management*, *4*(3), 193–215. doi:10.1080/1469701042000221696

Lin, J. (2011). Technological adaptation, cities, and new work. *The Review of Economics and Statistics*, *93*(2), 554–574. doi:10.1162/REST_a_00079

Lin, J., Luo, Z., Cheng, X., & Li, L. (2019). Understanding the interplay of social commerce affordances and swift guanxi: An empirical study. *Information & Management*, *56*(2), 213–224. doi:10.1016/j.im.2018.05.009

Linnenluecke, M., & Griffiths, A. (2010). Beyond adaptation: Resilience for business in light of climate change and weather extremes. *Business & Society*, *49*(3), 477–511. doi:10.1177/0007650310368814

Liou, J. J. H., Tzeng, G. H., & Chang, H. C. (2007). Airline safety measurement using a hybrid model. *Journal of Air Transport Management*, *13*(4), 243–249. doi:10.1016/j.jairtraman.2007.04.008

Lirio, P., Lee, M. D., Williams, M. L., Haugen, L. K., & Kossek, E. E. (2008). The inclusion challenge with reduced load professionals: The role of the manager. *Human Resource Management*, *47*(3), 443–461. doi:10.1002/hrm.20226

Li, S., Ragu-Nathan, B., Ragu-Nathan, T. S., & Rao, S. S. (2006). The impact of supply chain management practices on competitive advantage and organizational performance. *Omega*, *34*(2), 107–124. doi:10.1016/j.omega.2004.08.002 PMID:17876965

Liu, Z., & Liu, L. (2012). Empirical study of sentiment classification for Chinese microblog based on machine learning. Jisuanji Gongcheng yu Yingyong(*Computer Engineering and Applications*), *48*(1), 1-4.

Liu, C. H. G. H., Tzeng, G.-H., & Lee, M.-H. (2012). Improving tourism policy implementation— The use of hybrid MCDM models. *Tourism Management*, *33*(2), 239–488. doi:10.1016/j.tourman.2011.05.002

Li, W. (2018). Research on the Innovative Development Mode of Quality Education of College Students Based on the Perspective of Human Resource Management. *Educational Sciences: Theory and Practice*, *18*(5), 2447–2454.

Li, X. (2014). Operations management of logistics and supply chain: Issues and directions. *Discrete Dynamics in Nature and Society*.

Loader, B., & Keeble, L. (2004). Challenging the digital divide?: a literature review of community informatics initiatives. York, UK: Joseph Rowntree Foundation.

Logasakthi, K., & Rajagopal, K. (2013). A study on employee health, safety and welfare measures of chemical industry in the view of Sleam region, Tamil Nadu, India. *International Journal of Research in Business Management, 1*(1), 1–10.

Lombard, M., Snyder-Duch, J., & Bracken, C. C. (2002). Content analysis in mass communication: Assessment and reporting of intercoder reliability. *Human Communication Research, 28*(4), 587–604. doi:10.1111/j.1468-2958.2002.tb00826.x

London, M. (2012). Generative team learning in Web 2.0 environments. *Journal of Management Development, 32*(1), 73–95. doi:10.1108/02621711311287035

López, L., Green, A. R., Tan-McGrory, A., King, R. S., & Betancourt, J. R. (2011). Bridging the digital divide in health care: The role of health information technology in addressing racial and ethnic disparities. *Joint Commission Journal on Quality and Patient Safety, 37*(10), 437–445. doi:10.1016/S1553-7250(11)37055-9 PMID:22013816

López, S. P., Peón, J. M. M., & Ordás, C. J. V. (2009). *Information Technology as an Enabler Of Knowledge Management: An Empirical Analysis* (pp. 111–129). Boston, MA: Springer; doi:10.1007/978-1-4419-0011-1_8

Losh, S. C. (2004). Gender, educational, and occupational digital gaps 1983-2002. *Social Science Computer Review, 22*(2), 152–166. doi:10.1177/0894439303262557

Love, L. F., & Singh, P. (2011). Workplace branding: Leveraging human resources management practices for competitive advantage through "Best Employer" surveys. *Journal of Business and Psychology, 26*(2), 175–181. doi:10.100710869-011-9226-5

Low, J. (2000). The value creation index. *Journal of Intellectual Capital, 1*(3), 252–262. doi:10.1108/14691930010377919

Lubis, M. A., Yunus, M. M., Lampoh, A. A., & Ishak, N. M. (2010). The use of ICT in teaching islamic subjects in Brunei Darussalam. In *International Conference on Education and Educational Technologies* (pp. 212–217). Iwate. Retrieved from https://ukm.pure.elsevier.com/en/publications/the-use-of-ict-in-teaching-islamic-subjects-in-brunei-darussalam

Lucksom, S. (2007). *The Orchids of Sikkim and North East Himalaya,* (pp. 1–965). Open Library: S. Z. Lucksom.

Ludu. (2013). Burglary Crime Analysis Using Logistic Regression. In S. Yamamoto (Ed.), *Human Interface and the Management of Information. Information and Interaction for Learning, Culture, Collaboration and Business*. Berlin, Germany: Springer.

Luke, T. W. (2013). Corporate social responsibility: An uneasy merger of sustainability and development. *Sustainable Development, 21*(2), 83–91. doi:10.1002d.1558

Lumpkin, G. T., & Dess, G. G. (1996). Clarifying the entrepreneurial orientation construct and linking it to performance. *Academy of Management Review, 21*(1), 135–172.

Lundrigan, M., Tangsuvanich, V. L., Wu, S., & Mujtaba, B. (2012). Coaching a diverse workforce: The impact of changing demographics for modern leaders. *International Journal of Humanities and Social Science, 2*(3), 40–48.

Maak, Th., & Pless, N. M. (2006). Responsible Leadership: A Relational Approach. In Th. Maak, & N. M. Pless (Eds.), *Responsible Leadership*. London, UK: Routledge. doi:10.4324/9780203002247

Macaskill, W., & Owen, D. (2006). Web 2.0 to go. In *Proceedings of LIANZA Conference*. Wellington, New Zealand.

Macey, W. H., & Schneider, B. (2008). The meaning of employee engagement. *Industrial and Organizational Psychology: Perspectives on Science and Practice, 1*(1), 3–30. doi:10.1111/j.1754-9434.2007.0002.x

Madden, A., Ruthven, I., & McMenemy, D. (2013). A classification scheme for content analyses of YouTube video comments. *The Journal of Documentation, 69*(5), 693–714. doi:10.1108/JD-06-2012-0078

Mael, F., & Ashforth, B. E. (1992). Alumni and alma mater: A partial test of the reformulated model of organizational identification. *Journal of Organizational Behavior*, *12*(2), 103–123. doi:10.1002/job.4030130202

Mahbub Ul Haq Research Center. (2018). History. Retrieved from https://mhrc.lums.edu.pk/history-3

Mahmood, H. K., Hashmi, M. S., Shoaib, D. M., Danish, R., & Abbas, J. (2014). Impact of TQM Practices on Motivation of Teachers in Secondary Schools Empirical Evidence from Pakistan. *Journal of Basic and Applied Scientific Research*, *4*(6), 1–8.

Mahmood, K., Mahmood Ahmad Qureshi, I., & Nisar, A. (2014). An empirical study on measurement of performance through TQM in Pakistani aviation manufacturing industry. *International Journal of Quality & Reliability Management*, *31*(6), 665–680.

Mahmoud, Y., Makoond, A., & Naik, A. (2017). *Entrepreneurship for Sustaining Peace*. International Peace Institute Issue.

Maina, A. (2018). 20 Popular Social Media Sites Right Now. Retrieved from https://smallbiztrends.com/2016/05/popular-social-media-sites.html

Majid, I. A., & Koe, W.-L. (2012). Sustainable entrepreneurship (SE): A revised model based on triple bottom line (TBL). *International Journal of Academic Research in Business and Social Sciences*, *2*(6), 293.

Majors, R., & Billson, J. M. (1992). *Cool pose: The dilemmas of black manhood in America*. New York, NY: Lexington Books.

Malik, I. (2003). *An Entrepreneurial Vacuum*. Pakistan.

Mallahi, T., Ramezanian, A., Saharkhiz, M. J., Javanmardi, J., & Iraji, A. (2018). Antimicrobial activities of Asafoetida and Shirazi thyme essential oils improve the vase life of gerbera cut flowers. *Acta Ecologica Sinica*, *38*(3), 228–233. doi:10.1016/j.chnaes.2017.08.009

Mangalam Timber. (2018). Annual Report 2017. Retrieved from http://www.mangalamtimber.com/images/investors_desk_2016-17.pdf

Mangold, W. G., & Faulds, D. J. (2009). Social media: The new hybrid element of the promotion mix. *Business Horizons*, *52*(4), 357–365. doi:10.1016/j.bushor.2009.03.002

Manninen, J., Sgier, I., Fleige, M., Thöne-Geyer, B., Kil, M., Možina, E., & Diez, J. (2014). *Benefits of lifelong learning in Europe: Main results of the BeLL-project*. German Institute for Adult Education DIE.

Manzini, H., & Gwandure, C. (2011). *The Provision of Employee Assistance Programmes in South Africa Football Clubs*. Johannesburg, South Africa: University of the Witwatersrand.

Manzoor, F., Wei, L., Bányai, T., Nurunnabi, M., & Subhan, Q. A. (2019). An Examination of Sustainable HRM Practices on Job Performance: An Application of Training as a Moderator. *Sustainability*, *11*(8), 2263. doi:10.3390u11082263

Marathe, S., Sundar, S. S., Nije Bijvank, M., Van Vugt, H., & Veldhuis, J. (2007). Who are these power viewers anyway? Building a psychological profile. *Paper presented at the International Communication Association*, San Francisco, CA.

Marić, I. (2013). Stakeholder analisys of higher education institutions. *Interdisciplinary Description of Complex Systems*, *11*(2), 217–226. doi:10.7906/indecs.11.2.4

Marikyan, D., Papagiannidis, S., & Alamanos, E. (2019). A systematic review of the smart home literature: A user perspective. *Technological Forecasting and Social Change*, *138*, 139–154. doi:10.1016/j.techfore.2018.08.015

MarketingCharts. (2014). Are young people watching less TV? (Updated – Q4 2013 Data). Retrieved from http://www.marketingcharts.com/wp/television/are-young-people-watching-less-tv-24817/

Marr, B. (2015). *Big Data: Using SMART big data, analytics and metrics to make better decisions and improve performance*. Hoboken, NJ: John Wiley & Sons.

Mason, L., Baxter, J., Bartlett, P., & Frean, M. (1999). Boosting algorithms as gradient descent. In *Proceedings of International Conference on Neural Information Processing Systems*. Cambridge, MA: MIT Press.

Matthews, C. (2017, Oct. 30). *Here's why India is going to become the World's most important economy.* Retrieved from http://fortune.com/2016/04/08/india-economy/

Mayer, J., Salovey, P., Caruso, D., & Cherkasskiy, L. (2011). Emotional Intelligence. In R. Sternberg, & S. Kaufman (Eds.), The Cambridge Handbook of Intelligence (Cambridge Handbooks in Psychology, pp. 528-549). Cambridge, UK: Cambridge University Press. doi:10.1017/CBO9780511977244.027

Mayring, P. (2004). Qualitative content analysis. *A companion to qualitative research, 1*, 159-176.

Mazrui, A. (Ed.). (1977). *The warrior tradition in modern Africa*. Leiden, The Netherlands: Brill.

Mcafee, A., & Brynjolfsson, E. (2012). Big data: The management revolution. *Harvard Business Review*, *90*(10), 60–66, 68, 128. PMID:23074865

McCracken, G. (1986). Culture and consumption: A theoretical account of the structure and movement of cultural meaning of consumer goods. *The Journal of Consumer Research*, *13*(1), 71–84. doi:10.1086/209048

McKay, P., Avery, D. R., & Morris, M. (2009). A tale of two climates: Diversity climate from subordinates' and managers' perspectives and their role in store unit sales. *Personnel Psychology*, *62*(4), 767–791. doi:10.1111/j.1744-6570.2009.01157.x

McLaren, J., & Zappala, G. (2002). The'digital divide'among financially disadvantaged families in Australia. *First Monday*, *7*(11). doi:10.5210/fm.v7i11.1003

McNeill, M., Ming Diao, M., & Gosper, M. (2011). Student uses of technology in learning: Two lenses. *Interactive Technology and Smart Education*, *8*(1), 5–17. doi:10.1108/17415651111125478

MCX. (2019). Market Activity. Retrieved from https://www.mcxindia.com/home

Meadows, D. H., Meadows, D. H., Randers, J., & Behrens, W. W. III. (1972). *The limits to growth: a report to the club of Rome (1972)*. Google Scholar.

Medium. (n.d.). Internet Economy vs. Sharing Economy vs. Token Economy. Retrieved from https://medium.com/@dennis_z/internet-economy-vs-sharing-economy-vs-token-economy-13c69905946b

Mehlum, H., Moene, K., & Torvik, R. . (2005). Crime induced poverty traps. *Journal of Development Economics, 77*(2), 0-340.

Mehmood, Z. U. I., & Arif, M. I. (2011). Leadership and HRM: Evaluating new leadership styles for effective human resource management. *International Journal of Business and Social Science*, *2*(15).

Mehra, G. (2017). 105 Social Networks Worldwide. Retrieved from https://www.practicalecommerce.com/105-leading-social-networks-worldwide

Mehralian, G., Nazari, J. A., Zarei, L., & Rasekh, H. R. (2016). The effects of corporate social responsibility on organizational performance in the Iranian pharmaceutical industry: The mediating role of TQM. *Journal of Cleaner Production, 135*, 689–698.

Melhem, S., Morell, C., & Tandon, N. (2009). Information and communication technologies for women's socio-economic empowerment. Washington, DC: The World Bank. doi:10.1596/978-0-8213-8133-5

Melissa, E., Hamidati, A., & Saraswati, M. S. (2013). Social Media Empowerment: How Social Media Helps to Boost Women Entrepreneurship in Indonesian Urban Areas. *The IAFOR Journal of Media*, Communication and Film, 1(1), pp. 77-90.

Men, L. R., & Yue, C. A. (2019). Creating a positive emotional culture: Effect of internal communication and impact on employee supportive behaviours. *Public Relations Review*, *45*(3). doi:10.1016/j.pubrev.2019.03.001

Menon, R. (2013). Taking to the Trees. Retrieved from http://indiatoday.intoday.in/story/gujarat-farmers-switch-to-highly-profitable-cultivation-of-eucalyptus-trees/1/391831.html

Mercantile Exchange of Nepal Limited. (2019). Products. Retrieved from http://www.mexnepal.com/webpages/products_mex.html

Merinews. (2019). Stages of Economic Growth: An Analysis. Retrieved from http://www.merinews.com/article/stages-of-economic-growth-an-analysis/15811108.shtml

Merlino, M., & Sproģe, I. (2017). The augmented supply chain. *Procedia Engineering*, *178*, 308–318. doi:10.1016/j.proeng.2017.01.053

Merrilees, B., Rundle-Thiele, S., & Lye, A. (2011). Marketing capabilities: Antecedents and implications for B2B SME performance. *Industrial Marketing Management*, *40*(3), 368–375.

Microsoft. (2015, July 23). Reverse mentoring: How millennials are becoming the new mentors (Blog post). Retrieved from https://news.microsoft. com/Europe/features/reverse-mentoring-how-millennials-are-becoming-the-new-mentors

Mikulincer, M., & Shaver, P. R. (2007). *Attachment to adulthood: Structure, dynamics and change*. New York, NY: The Guilford Press.

Milfelner, B., Potočnik, A., & Žižek, S. Š. (2015). Social responsibility, human resource management and organizational performance. *Systems Research and Behavioral Science*, *32*(2), 221–229. doi:10.1002res.2263

Miller, D. (1983). The correlates of entrepreneurship in three types of firms. *Management Science*, *29*(7), 770–791.

Miller, D., & Friesen, P. H. (1978). Archetypes of strategy formulation. *Management Science*, *24*(9), 921–933.

Miller, J. D. (2001). Who is using the web for science and health information? *Science Communication*, *22*(3), 256–273. doi:10.1177/1075547001022003003

Mittermeier, R. A., Gils, P. R., Hoffman, M., Pilgrim, J., Brooks, T., Mittermeier, C. G., ... Da-Fonseca, G. A. B. (Eds.). (2004). *Hotspots Revisited. Earth's Biologically Richest and Most Endangered Terrestrial Ecoregions*. Arlington, VA: Conservation International.

Mohiuddin, K. G. B., Gordon, R., Magee, C., & Lee, J. K. (2016). A conceptual framework of cool for social marketing. *Journal of Social Marketing*, *6*(2), 121–143. doi:10.1108/JSOCM-07-2015-0046

Momaya, K., & Ajitabh, A. (2005). Technology management and competitiveness: Is there any relationship? *International Journal of Technology Transfer and Commercialisation*, *4*(4), 518–524. doi:10.1504/IJTTC.2005.006702

Monappa, A. (1995). *Industrial Relations (Ninth print). Quality Control. Tool for Worker Development*. New Delhi, India: Tata Mc-Graw Hill Publishing Company Ltd.

Mooijman, M., van Dijk, W. W., van Dijk, E., & Ellemers, N. (2019). Leader power, power stability, and interpersonal trust. *Organizational Behavior and Human Decision Processes*, *152*, 1–10. doi:10.1016/j.obhdp.2019.03.009

Moolman, J., Primo, N., & Shackleton, S. (2007). Taking a byte of technology: Women and ICTs. *Empower. Women Gend. Equity*, *21*, 4–14.

Moorthy, M. V. (1968). *Principles of Labour Welfare* (1st ed.). Visakhapatnam, India: Gupta Brothers Books.

Morais, U. P., Jacqueline, P., Kevin, S., Lucien, S., Roiner, R., & Yesenia Rivera, M. B. (2014). Managing Diverse Employees at Starbucks: Focusing on Ethics and Inclusion. *International Journal of Learning & Development*, *4*(3), 35–50. doi:10.5296/ijld.v4i3.5994

Morgan, J. (2018). The technological environment in employee experience. *Leader to Leader*, *2018*(87), 28–35. doi:10.1002/ltl.20340

Morley, D. (1986). *Family television: Cultural power and domestic leisure*. London, UK: Comedia.

Morley, D. (1988). Domestic relations: The framework of family viewing in Great Britain. In J. Lull (Ed.), *World families watch television* (pp. 22–48). Newbury Park, CA: Sage.

Morris, W. (Ed.). (1969). *The American heritage dictionary*. New York, NY: American Heritage Publishing Co., and Houghton Mifflin.

Mossberger, K., Tolbert, C. J., & Gilbert, M. (2006). Race, place, and information technology. *Urban Affairs Review*, *41*(5), 583–620. doi:10.1177/1078087405283511

Moussa, M., & Somjai, K. (2019). *Job dissatisfaction and employee turnover: A qualitative case study in Thailand*. Sage Publications: Sage Business Cases Originals.

Mozaffarian, D., Angell, S. Y., Lang, T., & Rivera, J. A. (2018). Role of government policy in nutrition—Barriers to and opportunities for healthier eating. *BMJ (Clinical Research Ed.)*, *361*, k2426. doi:10.1136/bmj.k2426 PMID:29898890

Mullen, T., & Collier, N. (2004). Sentiment analysis using support vector machines with diverse information sources. *In Proceedings of the 2004 conference on empirical methods in natural language processing*.

Müller, F., & Hermes, J. (2010). The performance of cultural citizenship: Audiences and the politics of multicultural television drama. *Critical Studies in Media Communication*, *27*(2), 193–208. doi:10.1080/15295030903550993

Munizu, M. (2013). The Impact of total quality management practices towards competitive advantage and organizational performance: Case of fishery industry in South Sulawesi Province of Indonesia.

Muñoz, P., & Dimov, D. (2015). The call of the whole in understanding the development of sustainable ventures. *Journal of Business Venturing*, *30*(4), 632–654. doi:10.1016/j.jbusvent.2014.07.012

Muraro, D., Negrelle, R. R., Cuquel, F. L., & Anacleto, A. (2016). Market management: The impact on the development of an ornamental plants supply chain in Curitiba, Brazil. *Ciencia e Investigación Agraria*, *42*(3), 453–460.

Musau, S. M. (2017). *The Role of Strategic Management Practices on Competitiveness of Floriculture Industry in Kenya: A Case of Kiambu County* (Doctoral dissertation, United States International University-Africa).

Mutlu, H. M., & Aksoy, H. (2014). Strategic Orientations, Firm Capabilities, and Export Performance: An Empirical Analysis in Turkey. *International Journal of Economic Practices and Theories*, *4*(2), 214–221.

Mwesige, P. (2018). *Value Chain governance and standards compliance in fresh fruit and vegetable (FFV) export sector in Uganda* (Doctoral dissertation, Makerere university business school Institutional repository).

Myrberg, C., & Wiberg, N. (2015). Screen vs. paper: What is the difference for reading and learning? *Insights the UKSG Journal*, *28*(2), 49–54. doi:10.1629/uksg.236

Nabitz, U., Klazinga, N., & Walburg, J. (2000). The EFQM excellence model: European and Dutch experiences with the EFQM approach in health care. *International Journal for Quality in Health Care*, *12*(3), 191–202. PMID:10894190

Nadler, D. A., & Tushman, M. L. (1988). Organizational frame bending: Principles for managing reorientation. *The Academy of Management Perspectives*, *3*(3), 194–204. doi:10.5465/ame.1989.4274738

Naeem, M. (2019). Uncovering the role of social media and cross-platform applications as tools for knowledge sharing. *VINE Journal of Information and Knowledge Management Systems*, VJIKMS. doi:10.1108/VJIKMS-01-2019-0001

Nair, N., & Vohra, N. (2015). *Diversity and Inclusion at the Workplace: A Review of Research and Perspectives*, IIMA Working Papers WP2015-03-34, Indian Institute of Management Ahmedabad, Research and Publication Department.

Nancarrow, R., Nancarrow, P., & Page, J. (2002). An analysis of the concept of cool and its marketing implications. *Journal of Consumer Behaviour*, *1*(4), 311–322. doi:10.1002/cb.77

Nanda, N., & Panda, J. K. (2013). Challenges and effectiveness of industrial relation environment in Indian Industries study on Rourkela Steel Plant. *International Journal of Financial Services and Management Research*, *2*(6), 163–174.

Naranjo-Valencia, J. C., Calderón-Hernández, G., Jiménez-Jiménez, D., & Sanz-Valle, R. (2018). Entrepreneurship and innovation: Evidence in Colombian SMEs. In Handbook of Research on Intrapreneurship and Organizational Sustainability in SMEs (pp. 294–316). Hershey, PA: IGI Global. doi:10.4018/978-1-5225-3543-0.ch014

Narushima, M., Liu, J., & Diestelkamp, N. (2018). Lifelong learning in active ageing discourse: It's conserving effect on wellbeing, health and vulnerability. *Ageing and Society*, *38*(4), 651–675. doi:10.1017/S0144686X16001136 PMID:29551843

Narver, J. C., & Slater, S. F. (1990). The effect of a market orientation on business profitability. *Journal of Marketing*, 20–35.

Nasridinov, A., Ihm, S. Y., & Park, Y. H. (2013). A Decision Tree-Based Classification Model for Crime Prediction. In J. Park, L. Barolli, F. Xhafa, & H. Y. Jeong (Eds.), *Information Technology Convergence*. Dordrecht, The Netherlands: Springer.

Nass, C., Steuer, J., & Tauber, E. R. (1994). Computers are social actors. In *Proceedings of the SIGCHI conference on Human factors in computing systems* (pp. 72-78). ACM.

Nathanson, A. I., Perse, E. M., & Ferguson, D. A. (1997). Gender differences in television use: An exploration of the instrumental-expressive dichotomy. *Communication Research Reports*, *14*(2), 176–188. doi:10.1080/08824099709388659

Naudé, W. (2013). *Entrepreneurship and economic development: Theory, evidence and policy*. Browser Download This Paper.

Nayak, B., & Bhattacharyya, S. S. (2019). Integrating Digital Wisdom and Human Capital. *Journal for Quality and Participation*, *41*(4), 20–23.

NCDEX. (2019). Products. Retrieved from https://www.ncdex.com/index.aspx

Neetha, N. (2001). *Gender and Technology: Impact of flexible Organisation and Production on Female Labour in the Tiruppur Knitwear Industry*. Noida, India: V. V. Giri National Labour Institute, 82.

Negi, M. (2016). Indian Forestry: 10 Main Problems Faced by the Indian Forestry. Retrieved from http://www.yourarticlelibrary.com/environment/forest/indian-forestry-10-main-problems-faced-by-the-indian-forestry/13861/

Negi, S., & Anand, N. (2018). Wastage and Cold Chain Infrastructure Relationship in Indian Food Supply Chain: A Study From Farm to Retail. In Supply Chain Management Strategies and Risk Assessment in Retail Environments (pp. 247-266). Hershey, PA: IGI Global. doi:10.4018/978-1-5225-3056-5.ch014

Nelson, E. (2017). *Employee Experience-How to build an EX-centric organization*. Kennedy Fitch.

Neuman, W. R. (1982). Television and American culture: The mass medium and the pluralist audience. *Public Opinion Quarterly*, *46*(4), 471–487. doi:10.1086/268745

Neumayer, E. (2001). The human development index and sustainability—A constructive proposal. *Ecological Economics*, *39*(1), 101–114. doi:10.1016/S0921-8009(01)00201-4

Newhagen, J. E., & Rafaeli, S. (1996). Why communication researchers should study the internet: A dialogue. *Journal of Communication*, *46*(1), 4–13. doi:10.1111/j.1460-2466.1996.tb01458.x

New, S. (2010). The transparent supply chain. *Harvard Business Review*, *88*, 1–5.

Ngambi, M. T., & Nkemkiafu, A. G. (2015). The Impact of Total Quality Management on Firm's Organizational Performance. *American Journal of Management*, *15*(4), 69.

Ng, P. K., & Jee, K. S. (2012). Innovating TQM, CE and KM for productive manufacturing in a Malaysian firm. *Total Quality Management & Business Excellence*, *23*(9-10), 1089–1105.

Nie, D., Lämsä, A. M., & Pučėtaitė, R. (2018). Effects of responsible human resource management practices on female employees' turnover intentions. *Business Ethics (Oxford, England)*, *27*(1), 29–41. doi:10.1111/beer.12165

Nielsen. (2009). Online engagement deepens as social media and video sites reshape the Internet. Retrieved from http://blog.nielsen.com/nielsenwire/wp-content/uploads/2009/04/nielsen-online-global-_pr.pdf

Nielsen. (2014). Living social: How second screens are helping TV make fans. Retrieved from http://www.nielsen.com/us/en/insights/news/2014/living-social-how-second-screens-are-helping-tv-make-fans.html

Nielsen. (2015). Local watch: Where you live and its impact on your choice. Retrieved from http://go.brandlift.com/rs/vizu/images/q1-2015-local-watch-report.pdf

Nilakant, V., & Ramnarayan, S. (2006). *Change management: Altering mindsets in a global context*. Sage Publications India.

Nkhoma, C. A., Thomas, S., Nkhoma, M. Z., Sriratanaviriyakul, N., Truong, T. H., & Vo, H. X. (2018). Measuring the impact of out-of-class communication through instant messaging. *Education + Training*, *60*(4), 318–334. doi:10.1108/ET-12-2017-0196

Nohria, N., Groysberg, B., & Lee, L. E. (2008). *Motivating Employees: A Powerful New Model. Harvard Business Review*. Boston, MA: Harvard Business Publishing.

Nonaka, I., & Takeuchi, H. (1995). *The Knowledge Creation Company: How Japanese Companies Create the Dynamics of Innovation*. New York, NY: Oxford University Press.

Norbury, F., Johansson, P.-O., & Lofgren, K.-G. (1986). The Economics of Forestry and Natural Resources. *American Journal of Agricultural Economics*, *68*(3), 760. doi:10.2307/1241574

Norman, M. (1957). *The white negro*. Dissent Publishing Associates.

Noss, R. T.-T., Selwyn, N., Crook, C. B., Carr, D., Carmichael, P., & Laurillard, D. (2008). Education 2.0? Designing the web for teaching and learning: A Commentary by the Technology Enhanced Learning phase of the Teaching and Learning Research Programme. Retrieved from https://www.semanticscholar.org/paper/Education-2.0-Designing-the-web-for-teaching-and-A-Noss-Tlrp-Tel/41c6c7f9316a30c422c447a618dd777b41edbb9d

Noyes, J. M., & Garland, K. J. (2008). Computer- vs. paper-based tasks: Are they equivalent? *Ergonomics*, *51*(9), 1352–1375. doi:10.1080/00140130802170387 PMID:18802819

Nwachukwu, C., Chladkova, H., & Zufan, P. (2017). Empirical assessment of microfinance banks in nigeria using efqm excellence model. *International Journal of Qualitative Research*, *11*(2).

O'Brian, B. (2013). *Buddhism and equanimity: Why equanimity is an essential Buddhist virtue*. Retrieved from http://buddhism.about.com/od/basicbuddhistteachings/a/Buddhism-And-Equa- nimity.htm

O'Donnell, K. A., & Wardlow, D. L. (2010). Atheory on the origins of coolness. *Advances in Consumer Research. Association for Consumer Research (U. S.)*, *27*(1), 5.

O'Neill, P., Sohal, A., & Teng, C. W. (2016). Quality management approaches and their impact on firms? financial performance – An Australian study. *International Journal of Production Economics*, *171*, 381–393.

O'Reilly, T. (2007). What Is Web 2.0: Design Patterns and Business Models for the Next Generation of Software. *International Journal of Digital Economics*, *65*, 17–37. Retrieved from https://mpra.ub.uni-muenchen.de/4580/

Obara-Okeyo, P., Fujii, K., & Kako, S. (1997). Enzyme polymorphism in Cymbidium orchid cultivars and inheritance of leucine aminopeptidase. *HortScience*, *32*(7), 1267–1271. doi:10.21273/HORTSCI.32.7.1267

OECD. (2014). The digital economy, new business models and key features. In *Addressing the Tax Challenges of the Digital Economy*. Paris, France: OECD Publishing. doi:10.1787/9789264218789-7-

OECD. (2015). *Measuring the Digital Economy, A New Perspective*. OECD Publishing.

Oke, D. F. (2013) The Effect of Social Network on women entrepreneurs in Nigeria: A case study of Ado-Ekiti Small scale Enterprise. *International Journal of Education and Research, 1*(11), pp. 1-14.

Oliveira, G., Corrêa, J., Balestrassi, P., Martins, R., & Turrioni, J. (2017). Investigation of TQM implementation: Empirical study in Brazilian ISO 9001-registered SMEs. *Total Quality Management & Business Excellence*, 1–19.

Oltra, V. (2005). Knowledge management effectiveness factors: The role of HRM. *Journal of Knowledge Management*, *9*(4), 70–86. doi:10.1108/13673270510610341

Ono, H., & Zavodny, M. (2007). Digital inequality: A five country comparison using microdata. *Social Science Research*, *36*(3), 1135–1155. doi:10.1016/j.ssresearch.2006.09.001

Oskam, I., Bossink, B., & de Man, A. P. (2018). The interaction between network ties and business modeling: Case studies of sustainability-oriented innovations. *Journal of Cleaner Production*, *177*, 555–566. doi:10.1016/j.jclepro.2017.12.202

Oxford Business Group. (2011). *The report Brunei Darussalam 2011*.

Ozcelik, H., Langton, N., & Aldrich, H. (2008). Doing well and doing good: The relationship between leadership practices that facilitate a positive emotional climate and organizational performance. *Journal of Managerial Psychology*, *23*(2), 186–203. doi:10.1108/02683940810850817

Packard, T. (2009). Leadership and performance in human services organizations. Chapter 7 in The handbook of human services management, Sag5e, 143-164.

Palanivelu, P., & Dhawan, M. (2015). Sustainability and Green Logistics in Supply Chain. Available at https://www.academia.edu/28094615/Green_Logistics_Whitepaper

Panayiotou, A., Putnam, L. L., & Kassinis, G. (2019). Generating tensions: A multilevel, process analysis of organizational change. *Strategic Organization*, *17*(1), 8–37. doi:10.1177/1476127017734446

Pande, R. (Ed.). (2012). Globalization, Technology Diffusion and Gender Disparity: Social Impacts of ICTs: Social Impacts of ICTs. Hershey, PA: IGI Global. doi:10.4018/978-1-4666-0020-1

Pandey, S. K., Hart, J. J., & Tiwary, S. (2003). Women's health and the Internet: Understanding emerging trends and implications. *Social Science & Medicine*, *56*(1), 179–191. doi:10.1016/S0277-9536(02)00019-9 PMID:12435560

Panel of Experts Discuss Using Emotional Intelligence to Achieve the SDGS. (2019, May). Retrieved from https://academicimpact.un.org/content/panel-experts-discuss-using-emotional-intelligence-achieve-sdgs

Park, C. W., Bernard, J. J., & Deborah, J. (1986). Strategic brand concept-image management. *Journal of Marketing*, *50*(4), 135–145. doi:10.1177/002224298605000401

Park, J. Y., & Sung, C. S. (2017). Does Social Media Use Influence Entrepreneurial Opportunity? A Review of its Moderating Role. *Sustainability*, *9*(1593), 1–16. doi:10.3390u9091593

Parrado-Moreno, C. A., Ricardo-Hernández, R. E., Velásquez-Arredondo, H. I., Lopera-Castro, S. H., & Hasenstab, C. (2019). An Environmental Evaluation of the Cut-Flower Supply Chain (Dendranthema grandiflora) Through a Life Cycle Assessment. *Revista EIA*, *16*(31), 27–42. doi:10.24050/reia.v16i31.747

Parrish, B. D., & Foxon, T. J. (2009). Sustainability entrepreneurship and equitable transitions to a low-carbon economy. *Greener Management International* (55).

Parul, P. B., & Kumar, A. M. (2013). Provision of Welfare under Factories Act & Its Impact on Employee Satisfaction. *Journal of Business Management & Social Sciences Research*, *2*(2), 57–69.

Pascale, R., Millemann, M., & Gioja, L. (1997). Changing the way we change. *Harvard Business Review*, *75*(6), 126. PMID:10174794

Pastakia, A. (1998). Grassroots ecopreneurs: Change agents for a sustainable society. *Journal of Organizational Change Management*, *11*(2), 157–173. doi:10.1108/09534819810212142

Patel, B. H. (2012). Export Market Challenges for Indian Timber Products. Retrieved from https://www.chathamhouse.org/sites/files/chathamhouse/public/Research/Energy,%20Environment%20and%20Development/10022012Patel1.pdf

Patel, A., Gohil, A., & Shah, H. (2017). Study on Labour Welfare Measures and Social Security on Selected Engineering Unit of Ahmadabad. *IBMRD's. Journal of Management Research*, *6*(1), 19–26. doi:10.17697/ibmrd/2017/v6i1/111654

Patrick, K., & Dotsika, F. (2007). Knowledge sharing: Developing from within. *The Learning Organization*, *14*(5), 395–406. doi:10.1108/09696470710762628

Patro, C. S. (2016b). Influence of Retention Policies on Employee Efficiency and Organisation Productivity. In U. Aung, & P. Ordoñez de Pablos (Eds.), Managerial Strategies and Practice in the Asian Business Sector (pp. 124-149). Hershey, PA: Business Science Reference (IGI Global). doi:10.4018/978-1-4666-9758-4.ch008

Patro, C. S. (2017a). A Study on Adoption of Employee Welfare Schemes in Industrial and Service Organisations: In Contrast with Public and Private Sectors. In M. Khosrow-Pour (Ed.), Public Health and Welfare: Concepts, Methodologies, Tools, and Applications (pp. 809-824). Hershey PA: Business Science Reference (IGI Global). doi:10.4018/978-1-5225-1674-3.ch038

Patro, C. S. (2017b). Employee Welfare Measures in Public and Private Sectors: A Comparative Analysis. In M. Khosrow-Pour (Ed.), Public Health and Welfare: Concepts, Methodologies, Tools, and Applications (pp. 1026-1042). Hershey PA: Business Science Reference (IGI Global). doi:10.4018/978-1-5225-1674-3.ch047

Patro, C. S. (2017d). Welfare Regime: A Critical Discourse. In B. Christiansen, & C. Harish Chandan (Eds.) Handbook of Research on Human Factors in Contemporary Workforce Development (pp. 110-131). Hershey PA: Business Science Reference (IGI Global). doi:10.4018/978-1-5225-2568-4.ch005

Patro, C. S. (2019). Welfare Programs as a Strategy of Promoting Employees' Economic Growth and Work Productivity. In B. Christiansen, I. Sysoeva, A. Udovikina, & A. Ketova (Eds.), Emerging Economic Models for Global Sustainability and Social Development (pp. 291-311). Hershey PA: Business Science Reference (IGI Global). doi:10.4018/978-1-5225-5787-6.ch016

Patro, C. S., & Raghunath, K. M. K. (2018). Employee Welfare Measures: The Impact on Employees' Efficacy and Organisations Productivity. In P. Ordóñez de Pablos (Ed.), Management Strategies and Technology Fluidity in the Asian Business Sector (pp. 215-234). Hershey PA: Business Science Reference (IGI Global). doi:10.4018/978-1-5225-4056-4.ch013

Patro, C. S. (2012). Employee Welfare Activities in Private Sector and their Impact on Quality of Work Life. *International Journal of Productivity Management and Assessment Technologies*, *1*(2), 18–29. doi:10.4018/ijpmat.2012040102

Patro, C. S. (2014). A Study on the Impact of Employee Retention Policies on Organisation Productivity in Private Sector. *International Journal of Asian Business and Information Management*, *5*(3), 48–63. doi:10.4018/ijabim.2014070104

Patro, C. S. (2015). Employee Welfare Measures in Public and Private Sectors: A Comparative Analysis. *International Journal of Service Science, Management, Engineering, and Technology*, *6*(1), 22–37. doi:10.4018/ijssmet.2015010102

Patro, C. S. (2016a). A Study on Adoption of Employee Welfare Schemes in Industrial and Service Organisations: In Contrast with Public and Private Sectors. *International Journal of Service Science, Management, Engineering, and Technology*, *7*(2), 16–33. doi:10.4018/IJSSMET.2016040102

Patro, C. S. (2017c). Espousal of Welfare Schemes: A Means for Employees' Satisfied Work Life with Reference to Pharmaceutical Companies. *International Journal of Asian Business and Information Management*, *8*(3), 36–51. doi:10.4018/IJABIM.2017070103

Patro, C. S., & Raghunath, K. M. K. (2016). A Take on Employee Welfare Facilities and Employees' Efficiency. *International Journal of Asian Business and Information Management*, *7*(3), 54–70. doi:10.4018/IJABIM.2016070104

Patterson, K. A., Grimm, C. M., & Corsi, T. M. (2003). Adopting new technologies for supply chain management. *Transportation Research Part E, Logistics and Transportation Review*, *39*(2), 95–121. doi:10.1016/S1366-5545(02)00041-8

Patzelt, H., & Shepherd, D. A. (2011). Recognizing opportunities for sustainable development. *Entrepreneurship Theory and Practice*, *35*(4), 631–652. doi:10.1111/j.1540-6520.2010.00386.x

Pedro, N., Soares, F., Matos, F., & Santos, M. (2008). *The Use of Learning Management Platforms in School Context - a National Study*. Retrieved from http://repositorio.ul.pt/bitstream/10451/5305/1/The_use_of_learning_manegement_national_study_english_version_.pdf

Peizer, J. (2000). What do we mean when we say 'digital divide?'. *Retrieved December 2*, 2000.

Pelled, L. H., Ledford, G. E., & Mohrman, S. A. (1999). Demographic dissimilarity and workplace inclusion. *Journal of Management Studies*, *36*(7), 1013–1031. doi:10.1111/1467-6486.00168

Pereira, M. G., Caramelo, L., Orozco, C. V., Costa, R., & Tonini, M. (2015). Space-time clustering analysis performance of an aggregated dataset: The case of wildfires in Portugal. *Environmental Modelling & Software*, *72*, 239–249. doi:10.1016/j.envsoft.2015.05.016

Perju, A. (2015). Gender Differences in Modeling the Influence of Online Marketing Communication on Behavioral Intentions. *Procedia Economics and Finance*, *27*, 567–573. doi:10.1016/S2212-5671(15)01034-5

Petrides, K. V., & Furnham, A. (2003). Trait emotional intelligence: Behavioural validation in two studies of emotion recognition and reactivity to mood induction. *European Journal of Personality*, *17*(1), 39–57. doi:10.1002/per.466

Petty, R. E., & Cacioppo, J. T. (1984). The effects of involvement on response to argument quantity and quality: Central and peripheral routes to persuasion. *Journal of Personality and Social Psychology*, *46*(1), 69–81. doi:10.1037/0022-3514.46.1.69

Petty, R. E., Cacioppo, J. T., & Kao, C. F. (1984). The efficient assessment of need for cognition. *Journal of Personality Assessment*, *48*(3), 306–307. doi:10.120715327752jpa4803_13 PMID:16367530

Pfau, L. D. (1989). Total quality management gives companies a way to enhance position in global marketplace. *Industrial Engineering (American Institute of Industrial Engineers)*, *21*(4), 17–25.

Pflueger, M. O., Franke, I., Graf, M., & Hachtel, H. (2015). Predicting general criminal recidivism in mentally disordered offenders using a random forest approach. *BMC Psychiatry*, *15*(1), 62. doi:10.118612888-015-0447-4 PMID:25885691

Pieri, M., & Diamantini, D. (2014). An E-learning Web 2.0 Experience. *Procedia: Social and Behavioral Sciences*, *116*, 1217–1221. doi:10.1016/j.sbspro.2014.01.371

Pine, B., & Gilmore, J. (1999). *The experience economy*. Boston: Harvard Business.

Pinprayong, B., & Siengtai, S. (2012). Restructuring for organizational efficiency in the banking sector in Thailand: A case study of Siam Commercial Bank. *Far East Journal of Psychology and Business*, *8*(2), 29–42.

Plaskoff, J. (2017). Employee Experience- The new human resource management approach. *Strategic HR Review*, *16*(3), 136–141. doi:10.1108/SHR-12-2016-0108

Pless, N., & Maak, T. (2004). Building an Inclusive Diversity Culture: Principles, Processes and Practice. *Journal of Business Ethics*, *54*(2), 129–147. doi:10.100710551-004-9465-8

PMEX. (2019). Products. Retrieved from https://www.pmex.com.pk/

Polat, R. K. (2012). Digital exclusion in Turkey: A policy perspective. *Government Information Quarterly*, *29*(4), 589–596. doi:10.1016/j.giq.2012.03.002

Polkinghorne, D. (1983). *Methodology for the Human Sciences*. Albany, NY: Human Science Press.

Pollay, R. W. (1985). The subsiding sizzle: A descriptive history of print advertising, 1900-1980. *Journal of Marketing*, *49*, 24–37.

Popescu, E. (2010). Students' Acceptance of Web 2.0 Technologies in Higher Education: Findings from a Survey in a Romanian University. In 2010 Workshops on Database and Expert Systems Applications (pp. 92–96). IEEE. doi:10.1109/DEXA.2010.38

Popescu, A. M., & Etzioni, O. (2007). Extracting product features and opinions from reviews. In *Natural language processing and text mining* (pp. 9–28). London, UK: Springer. doi:10.1007/978-1-84628-754-1_2

Porter, G. W. (2013). Free choice of learning management systems. *Interactive Technology and Smart Education*, *10*(2), 84–94. doi:10.1108/ITSE-07-2012-0019

Porter, M. E. (1996). What is strategy? *Harvard Business Review*, *74*(6), 61–78. PMID:10158474

Portney, P. R. (2005). *Corporate Social Responsibility*. Washington, DC: REF Books Publishers.

Pountain, D., & Robbins, D. (2000). *Cool rules*. London, UK: Reaktion Books.

Pozo, H., Souza, C. P., & Tachizawa, T. (n.d.). Supply chain management as a competitive strategy for costs reduction: a case study in two small manufacturing companies.

Prabhu, G., Thamaraiselvi, S. P., Aruna, P., & Sudhakar, R. (2018). Evaluation of chrysanthemum (Dendranthema grandiflora Tzelev.) genotypes for loose flower production under Coimbatore conditions. *IJCS*, *6*(4), 1618–1621.

Press Information Bureau. (2018). Cabinet approves national policy on Biofuels. Retrieved from http://pib.nic.in/newsite/PrintRelease.aspx?relid=179313

Press, L., Foster, W., Wolcott, P., & McHenry, W. (2003). The Internet in India and China. *Information Technologies & International Development, 1*(1), pp-41.

Prime, J., & Salib, E. R. (2014). *Inclusive leadership: The view from six countries*. New York, NY: Catalyst.

Proctor, R. W., & Cho, Y. S. (2006). Polarity correspondence: A general principle for performance of speeded binary classification tasks. *Psychological Bulletin*, *132*(3), 416–442. doi:10.1037/0033-2909.132.3.416 PMID:16719568

Psomas, E. L., & Jaca, C. (2016). The impact of total quality management on service company performance: Evidence from Spain. *International Journal of Quality & Reliability Management*, *33*(3), 380–398.

Pudaruth, S., & Busviah, D. (2018). Developing and Testing a Pioneer Model for Online Shopping Behavior for Natural Flowers: Evidence from Mauritius. *Studies in Business and Economics*, *13*(1), 128–147. doi:10.2478be-2018-0011

Pullman & Gross. (2004). Ability of Experience Design Elements to Elicit Emotions and Loyalty Behaviors. *Decision Sciences, 35*(3), 551-578.

Pyo, S. (2005). Knowledge map for tourist destinations—needs and implications. *Tourism Management*, *26*(4), 583-594.

Qaiser Gillani, D., Asghar, N., & Farooq, F. (2014). Socio-Economic Determinants of Self-Employment: Evidence from Southern Punjab (Pakistan). *Pakistan Journal of Social Sciences*, *34*(2).

Qhogwana, X. (2017). *The use of innovative strategies by automotive component manufacturers in Gauteng*.

Quental, N., Lourenço, J. M., & Da Silva, F. N. (2011). Sustainable development policy: Goals, targets and political cycles. *Sustainable Development*, *19*(1), 15–29. doi:10.1002d.416

R. D., S. (n.d). *Introduction to sustainable development: World conservation strategy of the international union for the conservation of nature and natural resources (iucn)*. York University, Toronto, Canada.

Rajeevan, P. K., Geetha, C. K., & Rajendran, P. (2016,). Orchid-centric floriculture development in Kerala, India. In *International Symposium on Succulents and Other Ornamentals 1165* (pp. 15-26).

Ramage, M., Burridge, H., Busse-Wicher, M., Fereday, G., Reynolds, T., Shah, D. U., ... Scherman, O. (2017). The wood from the trees: The use of timber in construction. *Renewable & Sustainable Energy Reviews*, *68*(Part 1), 333–359. doi:10.1016/j.rser.2016.09.107

Ranson, M. (2014). Crime, weather, and climate change. *Journal of Environmental Economics and Management, 67*(3), 274–302. doi:10.1016/j.jeem.2013.11.008

Rao, P. V., Patro, C. S., & Raghunath, K. M. K. (2015). Employee Welfare is the Key - An Insight. *International Journal of Business and Administration Research Review, 3*(11), 40–47.

Rasool, F., Gulzar, A., & Naseer, S. (n.d.), Drivers of Entrepreneurship: Linking with Economic Growth and Employment Generation.

Rathi, N., & Rastogi, R. (2009). Assessing the Relationship between Emotional Intelligence, Occupational Self-Efficacy and Organizational Commitment. *Journal of the Indian Academy of Applied Psychology, 35*, 93–102.

Ratnajeewa, D., & Bandara, J. (2018, November). A Review of Research on Green Logistics Distribution Practices. Paper presented at the KDU International Research Conference, Sri Lanka. Available at http://ir.kdu.ac.lk/bitstream/handle/345/1552/msh-042.pdf?sequence=1&isAllowed=y

Reddy, J. S. (2005). Gaining competitive advantage through Supply chain management. *Indian Journal of Marketing, 35*(6).

Reed, A. (2002). Exploring the links between brand name and consumer identity. Knowledge@Wharton, Nov. 6 (pp. 1-4).

Rehman, S., Zahid, M., Rahman, H. U., & Habib, M. N. (2019). A Partial Least Squares Approach to the Leadership Styles, Organizational Culture, and Employees' Productivity: A Case of Pakistan Banking Industry. *International Journal of Asian Business and Information Management, 10*(1), 55–64. doi:10.4018/IJABIM.2019010104

Reichheld, F. (1996). *The loyalty effect: The hidden forces behind growth, profits,and lasting value*. Boston: Harvard Business School Press.

Reijonen, H., Hirvonen, S., Nagy, G., Laukkanen, T., & Gabrielsson, M. (2015). The impact of entrepreneurial orientation on B2B branding and business growth in emerging markets. *Industrial Marketing Management, 51*, 35–46.

Report-UNDP, H. D. (2010). The real wealth of nations: Pathways to human development: United Nations Development Program New York.

Resnick, D., Haggblade, S., Babu, S., Hendriks, S. L., & Mather, D. (2018). The Kaleidoscope Model of policy change: Applications to food security policy in Zambia. *World Development, 109*, 101–120. doi:10.1016/j.worlddev.2018.04.004

Rice, J. (2002). *What is cool? Notes on intellectualism, popular culture, and writing*. Retrieved http:// www.ctheory.net/articles.aspx?id=338

Rice, M., & Springett, M. (2015). Digital interactive television and the older generation. In R. Nakatsu, M. Rauterberg, & P. Ciancarini (Eds.), *Handbook of digital games and entertainment technologies* (pp. 1–28). Singapore: Springer. doi:10.1007/978-981-4560-52-8_38-2

Ricoy, M.-C., & Feliz, T. (2016). Twitter as a Learning Community in Higher Education. *Journal of Educational Technology & Society*. International Forum of Educational Technology & Society. doi:10.2307/jeductechsoci.19.1.237

Ridley, J., & Channing, J. (2008). *Safety at work*. Oxford, UK: Butterworth-Heinemann. doi:10.4324/9780080557137

Rifkin, J. (2001). *The Age of Access: The New Culture of Hypercapitalism Where All of Life is a Paid-for Experience*. New York, NY: TarcherPerigee/Putnam.

Road map towards the implementation of the United Nations Millennium Declaration: Report of the Secretary General . (2001). UN GAOR, 56th Sess., Annex, Agenda Item 40, UN Doc. A/56/326 55.

Robinson, A., Sparrow, P., Clegg, C., & Birdi, K. (2006). Forecasting Future Competency Requirements: A Three-Phase Methodology. *Personnel Review*, *36*(1), 65–90. doi:10.1108/00483480710716722

Robinson, J. P. (2011). IT, TV and time displacement: What Alexander Szalai anticipated but couldn't know. *Social Indicators Research*, *101*(2), 193–206. doi:10.100711205-010-9653-0 PMID:21475389

Robin, X., Turck, N., Hainard, A., Tiberti, N., Lisacek, F., Sanchez, J.-C., & Müller, M. (2011). Proc: An open-source package for R and S+ to analyze and compare ROC curves. *BMC Bioinformatics*, *12*(1), 77. doi:10.1186/1471-2105-12-77 PMID:21414208

Rocco, T. S., & Plakhotnik, M. S. (2009). Literature reviews, conceptual frameworks, and theoretical frameworks: Terms, functions, and distinctions. *Human Resource Development Review*, *8*(1), 120–130. doi:10.1177/1534484309332617

Roethlisberger, F. J., & Dickson, W. J. (1939). *Management and the Worker*. Cambridge, MA: Harvard University Press.

Rogers, E. M. (1962). *Diffusion of innovations*. New York, NY: Free Press.

Rolinson, J. (1996). *Health information for the teenage years: what do they want to know? Department of Information and Library Studies*. Loughborough University, UK.

Ropes, D. (2013). Intergenerational learning in organizations. *European Journal of Training and Development*, *37*(8), 713–727. doi:10.1108/EJTD-11-2012-0081

Rose, G. M., & Shoham, A. (2002). Export performance and market orientation: Establishing an empirical link. *Journal of Business Research*, *55*(3), 217–225.

Rosenbusch, N., Brinckmann, J., & Bausch, A. (2011). Is innovation always beneficial? A meta-analysis of the relationship between innovation and performance in SMEs. *Journal of Business Venturing*, *26*(4), 441–457. doi:10.1016/j.jbusvent.2009.12.002

Rosete, D. (2005, June). *A leader's edge – what attributes make an effective leader?* Paper presented at the Fifth Annual Emotional Intelligence Conference, The Netherlands.

Rosete, D. (2009). *A leader's edge – what attributes make an effective leader?* Manuscript in preparation.

Rowley, J. (2000). From learning organisation to knowledge entrepreneur. *Journal of Knowledge Management*, *4*(1), 7–15. doi:10.1108/13673270010315362

Rowley, J. (2012). Conducting research interviews. *Management Research Review*, *35*(3/4), 260–271. doi:10.1108/01409171211210154

Royal Government of Bhutan. (2006). Forest and Nature Conservation Rules of Bhutan. Retrieved from www.moaf.gov.bt/download/.../Foest-and-Nature-Conservation-Rules-2006.pdf

Roy, T. N., Das, K. K., & Rai, D. (2015). Floriculture in West Bengal in Augmenting Income and Export. In *Diversification of Agriculture in Eastern India* (pp. 215–223). New Delhi, India: Springer. doi:10.1007/978-81-322-1997-2_19

Rumi, S. K., Deng, K., & Salim, F. D. (2018). Crime event prediction with dynamic features. *EPJ Data Science*, *7*(1), 43. doi:10.1140/epjds13688-018-0171-7

Runyan, R. C., Noh, M., & Mosier, J. (2013). What is cool? Operationalizing the construct in an apparel context. *Journal of Fashion Marketing and Management*, *17*(3), 322–340. doi:10.1108/JFMM-01-2012-0001

Russell, B. (1972). *A history of Western philosophy*. New York, NY: Simon & Schuster, Inc.

Russell, E., Lloyd-Houldey, A., Memon, A., & Yarker, J. (2019). Factors influencing uptake and use of a new health information app for young people. *Journal of Technology in Human Services*, 1–19.

Russell, R. S., & Taylor-Iii, B. W. (2008). *Operations management along the supply chain.* Hoboken, NJ: John Wiley & Sons.

S&P Global. (2016). *S&P Global Timber and Forestry Index.* Retrieved from http://us.spindices.com/indices/equity/sp-global-timber-and-forestry-index

Saad, L. (2013). TV Is Americans' main source of news. Retrieved from http://www.gallup.com/poll/163412/americans-main-source-news.aspx

Sabbaghi, A., & Sabbaghi, N. (2004). Global supply-chain strategy and global competitiveness. *International Business and Economics Research Journal, 3*, 63–76.

Sachon, M. R. J., Zhang, D., Zhang, Y., & Castillo, C. (2016). The Chinese Automotive Industry in 2016. Spain: Universidad de Navarra.

Sachs, J. D. (2004). Stages of Economic Development. *Speech at the Chinese Academy of Arts and Sciences.* Retrieved from http://www.earth.columbia.edu/sitefiles/file/about/director/documents/china_speech061904.pdf

Sachs, J. D. (2012). From millennium development goals to sustainable development goals. *Lancet, 379*(9832), 2206–2211. doi:10.1016/S0140-6736(12)60685-0 PMID:22682467

Sadikoglu, E., & Olcay, H. (2014). The effects of total quality management practices on performance and the reasons of and the barriers to TQM practices in Turkey. *Advances in Decision Sciences*, 2014.

Sagar, A. D., & Najam, A. (1998). The human development index: A critical review1. *Ecological Economics, 25*(3), 249–264. doi:10.1016/S0921-8009(97)00168-7

Sahoo, R., & Sahoo, C. K. (2018). Drivers of Cordial Employee Relations: The Study of a State-owned Public Sector Undertaking. *Management and Labour Studies, 43*(1&2), 123–139. doi:10.1177/0258042X18754428

Sailesh, S. (2012). Role of Organisation in welfare measures for employees. *International Journal of Research in IT and Management, 2*(9), 3640.

Sajid, M., Rab, A., & Khan, I. A. (2018). The pre-harvest foliar application influenced the flower quality and vase life of chrysanthemum cultivars. *Horticult Int J, 2*(4), 145–152.

Saks, A. (2006). Antecedents and Consequences of Employee Engagement. *Journal of Managerial Psychology, 21*(7), 600–619. doi:10.1108/02683940610690169

Salem, M. I. (2014). The role of business incubators in the economic development of Saudi Arabia. *The International Business & Economics Research Journal (Online), 13*(4), 853.

Salovey, P., & Mayer, J. D. (1990). Emotional intelligence. *Imagination, Cognition and Personality, 9*(3), 185–211. doi:10.2190/DUGG-P24E-52WK-6CDG

Samsung Tomorrow. (2015). History of Samsung smart TV. Retrieved from http://global.samsungtomorrow.com/infographic-history-of-samsung-smart-tv/

Samuelson, P. (1976). Economics of Forestry in an Evolving Society. *Economic Enquiry, 14*(4), 466–492. doi:10.1111/j.1465-7295.1976.tb00437.x

Sánchez, R. A., & Hueros, A. D. (2010). Motivational factors that influence the acceptance of Moodle using TAM. *Computers in Human Behavior*, *26*(6), 1632–1640. doi:10.1016/j.chb.2010.06.011

Sandnes, F. E., Jian, H.-L., Hagen, S., & Talberg, O. (2007). Student Evaluation of the Learning Management System Fronter From an HCI Perspective. In *International Conference on Engineering Education – ICEE 2007*. Coimbra, Protugal. Retrieved from http://icee2007.dei.uc.pt/proceedings/papers/86.pdf

Santangelo, G. D., & Pini, P. (2011). New HRM practices and exploitative innovation: A shopfloor level analysis. *Industry and Innovation*, *18*(6), 611–630. doi:10.1080/13662716.2011.591977

Sarda Plywood. (2017). Annual Report 2017. Retrieved from https://www.sardaplywood.in/investors.php

Saripalle, M. (2016). Jasmine cultivation in Tamil Nadu: Market structure and pricing. *World Development Perspectives*, *1*, 12–14. doi:10.1016/j.wdp.2016.05.004

Satayanarayan, M. R., & Redhi, R. J. (2012). Labour welfare measure in cement industries in India. *International Journal of Physical and Social Sciences*, *2*(7), 257–254.

Saunders, A., & Cornett, M. M. (2008). *Financial Institutions Management*. New York, NY: McGraw-Hill Irwin, 214-220

Sayan, H. (2016). Affecting Higher Students Learning Activity by Using Whatsapp. *European Journal of Research and Reflection in Educational Sciences*, *4*(3), 88–93. Retrieved from www.idpublications.org

Schaltegger, S., & Lüdeke-Freund, F. (2013). Business cases for sustainability. Encyclopedia of corporate social responsibility, 245-252.

Schaper, M. (2002). The essence of ecopreneurship.

Schein, E. (1985). *Organizational Culture and Leadership*. San Francisco, CA: Jossey-Bass.

Schiederig, T., Tietze, F., & Herstatt, C. (2012). Green innovation in technology and innovation management–an exploratory literature review. *Research Management*, *42*(2), 180–192.

Schlange, L. E. (2009). Stakeholder Identification in Sustainability Entrepreneurship. *Greener Management International*(55).

Schmitt, B. (1999). *Experiential marketing*. New York: The Free Press.

Schmitt, B. H. (2003). *Customer Experience Management*. Hoboken, NJ: Wiley.

Schuller, T., Preston, J., Hammond, C., Brassett-Grundy, A., & Bynner, J. (2004). The benefits of learning: The impact of education on health, family life and social capital. Abingdon-on-Thames, UK: Routledge. doi:10.4324/9780203390818

Schumpeter, J. (1942). *Capitalism, socialism and democracy*. New York, NY: Harper.

Schumpeter, J. A. (1932). *The Theory of Economic Development*. London, UK: Transaction Publishers.

Schumpeter, J. A. (1934). *The schumpttr: Theory economic development*. Cambridge, MA: Harvard University Press.

Schwartz, B. (2015, August 28). Rethinking Work. *The New Times*. Retrieved from http://www.nytimes.com

Schweyer, A. (2010). *Talent management systems: Best practices in technology solutions for recruitment, retention and workforce planning*. Hoboken, NJ: John Wiley & Sons.

Seeck, H., & Diehl, M. R. (2017). A literature review on HRM and innovation–taking stock and future directions. *International Journal of Human Resource Management*, *28*(6), 913–944. doi:10.1080/09585192.2016.1143862

Sekliuckiene, J., & Kisielius, E. (2015). Development of Social Entrepreneurship Initiatives: A Theoretical Framework. *Procedia: Social and Behavioral Sciences*, *213*, 1015–1019. doi:10.1016/j.sbspro.2015.11.519

Semeraro, J., & Moore, N. S. (2016). The Use of Google Docs Technology to Support Peer Revision. In E. Ortlieb, E. H. J. Cheek, & W. Verlaan (Eds.), Writing Instruction to Support Literacy Success (Literacy Research, Practice and Evaluation, Volume 7) (pp. 203–220). Emerald Group Publishing Limited. doi:10.1108/S2048-045820160000007013

Senecal, C., Widmer, R. J., Bailey, K., Lerman, L. O., & Lerman, A. (2018). Usage of a Digital Health Workplace Intervention Based on Socioeconomic Environment and Race: Retrospective Secondary Cross-Sectional Study. *Journal of Medical Internet Research*, *20*(4), e145. doi:10.2196/jmir.8819 PMID:29685862

Senge, P. M. (2006). The Fifth Discipline: The Art and Practice of the Learning Organization. Currency, New York, NY.

Seyal, A. H. (2012). A preliminary study of school administrators' use of information and communication technologies: Bruneian perspective. *International Journal of Education and Development using Information and Communication Technology (IJEDICT)* (Vol. 8). Retrieved from www.bit.gov.bn

Seyed-Hosseini, S. M., Safaei, N., & Asgharpour, M. J. (2005). Reprioritization of failures in a system failure mode and effects analysis by decision making trial and evaluation laboratory chnique. *Reliability Engineering & System Safety*, *91*(8), 872–881. doi:10.1016/j.ress.2005.09.005

Seyoum, B. (2017). Export Controls and International Business: A Study with Special Emphasis on Dual-Use Export Controls and Their Impact on Firms in the US. *Journal of Economic Issues*, *51*(1), 45–72.

Shabbir, M. S., Ghazi, M. S., & Mehmood, A. R. (2016). Impact of Social Media Applications on Small Business Entrepreneurs. *Arabian Journal of Business and Management Review*6(203), pp. 1-3. doi:. doi:10.4172/2223-5833.1000203

Shafiq, M. (2012). Implementation of quality management systems and business excellence frameworks in Pakistani textile companies. *Journal of Quality and Technology Management*, *7*(2), 11–23.

Shafiq, M., Lasrado, F., & Hafeez, K. (2017). The effect of TQM on organisational performance: Empirical evidence from the textile sector of a developing country using SEM. *Total Quality Management & Business Excellence*, 1–22. do i:10.1080/14783363.2017.1283211

Shah, D. V., Hanna, A., Bucy, E. P., Wells, C., & Quevedo, V. (2015). The power of television images in a social media age: Linking biobehavioral and computational approaches via the second screen. *The Annals of the American Academy of Political and Social Science*, *659*(1), 225–245. doi:10.1177/0002716215569220

Shahzad, M., Ying, Q., Ur Rehman, S., Zafar, A., Ding, X., & Abbas, J. (2019). Impact of knowledge Absorptive Capacity on Corporate Sustainability with Mediating Role of CSR: Analysis from the Asian Context. *Journal of Environmental Planning and Management*, 1–27. doi:10.1080/09640568.2019.1575799

Shannon, C. E., & Weaver, W. (1949). *The mathematical theory of communication*. Urbana, IL: University of Illinois Press.

Shapiro, H. (1999). *Waiting for men*. London, UK: Helter Skelter.

Share of Population in Selected Countries Who Are Active WhatsApp Users as of 3rd quarter. (2017). Retrieved from https:// www.statista.com/statistics/291540/mobile-internet-user-whatsapp/

Sharma, S. N. (2018, Dec.). *Why Indian Railways is testing emotional intelligence of its officers*, Economic times, Retrieved from https://economictimes.indiatimes.com/industry/transportation/railways/can-emotional-intelligence-training-for-staff-make-indian-railways-safer/article

Sheer, V. C., & Fung, T. K. (2007). Can Email Communication Enhance Professor-Student Relationship and Student Evaluation of Professor?: Some Empirical Evidence. *Journal of Educational Computing Research*, *37*(3), 289–306. doi:10.2190/EC.37.3.d

Shen, J., & Benson, J. (2016). When CSR is a social norm: How socially responsible human resource management affects employee work behavior. *Journal of Management*, *42*(6), 1723–1746. doi:10.1177/0149206314522300

Shen, J., & Jiuhua Zhu, C. (2011). Effects of socially responsible human resource management on employee organizational commitment. *International Journal of Human Resource Management*, *22*(15), 3020–3035. doi:10.1080/09585192.2011.599951

Shepherd, D. A., & Patzelt, H. (2011). The new field of sustainable entrepreneurship: Studying entrepreneurial action linking "what is to be sustained" with "what is to be developed". *Entrepreneurship Theory and Practice*, *35*(1), 137–163. doi:10.1111/j.1540-6520.2010.00426.x

Sheth, J. N., Sisodia, R. S., & Sharma, A. (2000). The antecedents and consequences of customer-centric marketing. *Journal of the Academy of Marketing Science*, *28*(1), 55–66. doi:10.1177/0092070300281006

Shih, C. P., & Orochena, O. D. C. P. (2016). Analyzing the Effect of Transformational Leadership on Innovation and Organizational Performance. *International Journal of Productivity Management and Assessment Technologies*, *4*(2), 11–27. doi:10.4018/IJPMAT.2016070102

Shih-Chi, C., Sun, C. C., & Herchan, A. (2011). The DEMATEL approach applied to solar cell industry material selection process in Taiwan. *Session Interd. Manage. Sem.*, *15*(13), 253–267.

Shih, H. A., & Chiang, Y. H. (2005). Strategy alignment between HRM, KM, and corporate development. *International Journal of Manpower*, *26*(6), 582–603. doi:10.1108/01437720510625476

Shin, D.-H. (2009). An empirical investigation of a modified technology acceptance model of IPTV. *Behaviour & Information Technology*, *28*(4), 361–372. doi:10.1080/01449290701814232

Shin, D.-H. (2016). Do users experience real sociability through social TV? Analyzing parasocial behavior in relation to social TV. *Journal of Broadcasting & Electronic Media*, *60*(1), 140–159. doi:10.1080/08838151.2015.1127247

Shin, D.-H., Hwang, Y., & Choo, H. (2013). Smart TV: Are they really smart in interacting with people? Understanding the interactivity of Korean Smart TV. *Behaviour & Information Technology*, *32*(2), 156–172. doi:10.1080/0144929X.2011.603360

Shipton, H., West, M. A., Dawson, J., Birdi, K., & Patterson, M. (2006). HRM as a predictor of innovation. *Human Resource Management Journal*, *16*(1), 3–27. doi:10.1111/j.1748-8583.2006.00002.x

Shoham, A. (1998). Export performance: A conceptualization and empirical assessment. *Journal of International Marketing*, 59–81.

SHRM. (2009). *Global Diversity and Inclusion-Perceptions, Practices and Attitudes*. Society for Human Resource Management.

Shuhaiber, A., & Mashal, I. (2019). (in press). Understanding users' acceptance of smart homes. *Technology in Society*, 1–9. doi:10.1016/j.techsoc.2019.01.003

Siddiqui, M. W., Rahman, M. S., & Wani, A. A. (Eds.). (2018). Innovative Packaging of Fruits and Vegetables: Strategies for Safety and Quality Maintenance. Boca Raton, FL: CRC Press.

Sillanpää, I., & Sillanpää, S. (2014). Supply Chain Strategy: Empirical Case Study in Europe and Asia. *Management*, *9*(2).

Silva, M. R. A., de Amorim, J. C., & Dias, A. L. (2019). Determinants of Employee Retention: A Study of Reality in Brazil. In Strategy and Superior Performance of Micro and Small Businesses in Volatile Economies (pp. 44–56). Hershey, PA: IGI Global. doi:10.4018/978-1-5225-7888-8.ch004

Silverman, D. (2013). Doing qualitative research: A practical handbook. New York, NY: Sage.

Simpson, P. M., Siguaw, J. A., & Enz, C. A. (2006). Innovation orientation outcomes: The good and the bad. *Journal of Business Research*, *59*(10-11), 1133–1141. doi:10.1016/j.jbusres.2006.08.001

Singh, B. (2013). Given the rupee fall, Floriculture of Northeast India facing heat over rising cost of imported planting materials, ET Bureau, Aug. 23.

Singh, A. I., & Kaur, G. (2012). Motion detection method to compensate camera flicker using an algorithm. *International Journal of Computational Engineering Research*, *2*(3), 919–926.

Singh, H., & Mahmood, R. (2013). Determining the effect of export market orientation on export performance of small and medium enterprises in Malaysia: An exploratory study. *Advances in Management and Applied Economics*, *3*(6), 223.

Singh, M. (2018). Adoption of Web 2.0 tools in higher education in India: A study. *Global Knowledge. Memory and Communication*, *67*(4/5), 297–315. doi:10.1108/GKMC-10-2017-0084

Sivarajah, U., Irani, Z., & Weerakkody, V. (2015). Evaluating the use and impact of Web 2.0 technologies in local government. *Government Information Quarterly*, *32*(4), 473–487. doi:10.1016/j.giq.2015.06.004

Skvortsova, T. A., Denisova, I. P., Romanenko, N. G., & Sukhovenko, A. V. (2018). Innovations and Support for Quality in Agriculture: A Case Study. *European Research Studies*, *21*, 423.

Slate, J. R., Manuel, M., & Jr., K. H. B. (2002). The "Digital Divide": Hispanic college students' views of educational uses of the Internet. *Assessment & Evaluation in Higher Education*, *27*(1), 75–93. doi:10.1080/02602930120105081

Slater, S. F., & Narver, J. C. (2000). The positive effect of a market orientation on business profitability: A balanced replication. *Journal of Business Research*, *48*(1), 69–73.

Smircich, L. (1983). Concepts of Culture and Organizational Analysis. *Administrative Science Quarterly*, *28*(3), 339. doi:10.2307/2392246

Smith, P., & Taylor, J. (2004). Marketing Communications: An Integrated Approach. London, UK: Kogan Page.

Smith, T., & Lambert, R. (2014). A systematic review investigating the use of Twitter and Facebook in university-based healthcare education. *Health Education*, *114*(5), 347–366. doi:10.1108/HE-07-2013-0030

Sodexo. (2017). 2017 global workplace trends. Retrieved from http://www. sodexo.com/files/live/sites/sdxcom-global/files/PDF/Media/ Sodexo-2017-workplace-trends-report.pdf

Soni, G., & Kodali, R. (2011). The strategic fit between "competitive strategy" and "supply chain strategy" in Indian manufacturing industry: An empirical approach. *Measuring Business Excellence*, *15*(2), 70–89. doi:10.1108/13683041111131637

Sorensen, L. (2009). *User managed trust in social networking comparing facebook, myspace and linkdin*. Paper presented at the 1st International Conference on Wireless Communication, Vehicular Technology, Information Theory and Aerospace & Electronic System Technology, (Wireless VITAE 09), Denmark.

Souri, M. K., Goodarzizadeh, S., Ahmadi, M., & Hatamian, M. (2018). Characteristics of postharvest quality of chrysanthemum cut flowers under pretreatment with nitrogenous compounds. *Acta Scientiarum Polonorum. Hortorum Cultus*, *17*(3), 83–90. doi:10.24326/asphc.2018.3.8

Souza, J. G. S., Lages, V. A., Sampaio, A. A., Souza, T. C. S., & Martins, A. M. E. D. B. (2019). The absence of functional dentition is associated with the lack of commitment to oral functions among Brazilian adults. *Ciencia & Saude Coletiva*, *24*(1), 253–260. doi:10.1590/1413-81232018241.30432016 PMID:30698258

Spekman, R. E., Kamauff, J. W. Jr, & Myhr, N. (1998). An empirical investigation into supply chain management: A perspective on partnerships. *Supply Chain Management*, *3*(2), 53–67. doi:10.1108/13598549810215379

Spencer, L., Ritchie, J., Ormston, R., O'Connor, W., & Barnard, M. (2014). Analysis: principles and processes. In J. Ritchie, J. Lewis, C. McNaughton Nicholls, & R. Ormston (Eds.), Qualitative Research Practice (pp. 269–293). London, UK: Sage.

Spiggle, S. (1994). Analysis and interpretation of qualitative data in consumer research. *The Journal of Consumer Research*, *21*(3), 491–503. doi:10.1086/209413

Srivastava, A. (2011). How to Calculate Mutual Fund Risk. Retrieved from http://karvy.com/articles/calculatemfrisk.htm

Steijn, B., & Tijdens, K. (2005). Workers and Their Willingness to Learn: Will ICT-Implementation Strategies and HRM Practices Contribute to Innovation? *Creativity and Innovation Management*, *14*(2), 151–159. doi:10.1111/j.1476-8691.2005.00335.x

Steiner, H. J., Alston, P., & Goodman, R. (2008). *International human rights in context: law, politics, morals: text and materials*. USA: Oxford University Press.

Sternad, D., Krenn, M., & Schmid, S. (2017). Business excellence for SMEs: Motives, obstacles, and size-related adaptations. *Total Quality Management & Business Excellence*, 1–18.

Stockdale, M. S., & Crosby, F. (2004). *The Psychology and Management of Workplace Diversity*. Boston, MA: Blackwell.

Study on MSMEs Participation in the Digital Economy in ASEAN. (2018). pp. 1-28.

Sturgess, D. (2013). *A breakthrough in measuring cool*. Forbes. Available at www.forbes.com/sites/onmarketing/2013/03/07/a-breakthrough-in-measuring-cool

Styles, C., Gray, S., Kropp, F., Lindsay, N. J., & Shoham, A. (2006). Entrepreneurial, market, and learning orientations and international entrepreneurial business venture performance in South African firms. *International Marketing Review*, *23*(5), 504–523.

Suárez, E., Calvo-Mora, A., Roldán, J. L., & Periáñez-Cristóbal, R. (2017). Quantitative research on the EFQM excellence model: A systematic literature review (1991–2015). *European Research on Management and Business Economics*, *23*(3), 147–156.

Sundararajan, B., Sheehan, L., & Gilbert, S. (2013). Mediated Discourse in Higher Ed Classrooms Using Text Messaging. In C. Wankel (Ed.), Increasing Student Engagement and Retention Using Classroom Technologies: Classroom Response Systems and Mediated Discourse Technologies (Cutting-edge Technologies in Higher Education, Volume 6 Part E) (pp. 199–232). Emerald Group Publishing. doi:10.1108/S2044-9968(2013)000006E010

Sundar, S. S., & Nass, C. (2001). Conceptualizing sources in online news. *Journal of Communication*, *51*(1), 52–72. doi:10.1111/j.1460-2466.2001.tb02872.x

Sundar, S., Tamul, D., & Wu, M. (2014). Capturing cool: Measures for assessing coolness of technological products. *International Journal of Human-Computer Studies*, *72*(2), 169–180. doi:10.1016/j.ijhcs.2013.09.008

Sune, A., & Gibb, J. (2015). Dynamic capabilities as patterns of organizational change: An empirical study on transforming a firm's resource base. *Journal of Organizational Change Management*, *28*(2), 213–231. doi:10.1108/JOCM-01-2015-0019

Sun, S. (2011). The Internet Effects on Students Communication at Zhengzhou Institute of Aeronautical Industry Management. In S. Lin, & X. Huang (Eds.), *Advances in Computer Science, Environment, Ecoinformatics, and Education. CSEE 2011* (pp. 418–422). Berlin, Germany: Springer; doi:10.1007/978-3-642-23357-9_74

Suppiah, V., & Sandhu, M. S. (2011). Organisational culture's influence on tacit knowledge-sharing behaviour. *Journal of Knowledge Management*, *15*(3), 462–477. doi:10.1108/13673271111137439

Suppliers Directory. (2019a). Timber and Lumber Companies from Bangladesh. Retrieved from https://www.lesprom.com/en/members/countries/Bangladesh/

Suppliers Directory. (2019b). Timber and Lumber Companies from Pakistan. Retrieved from https://www.lesprom.com/en/members/countries/Pakistan/

Suppliers Directory. (2019c). Timber and Lumber Companies from Afghanistan. Retrieved from https://www.lesprom.com/en/members/countries/Afghanistan/

Suppliers Directory. (2019d). Timber and Lumber Companies from Sri Lanka. Retrieved from https://www.lesprom.com/en/members/countries/Sri_Lanka/

Sureka, A. (2011). Mining user comment activity for detecting forum spammers in youtube. *arXiv preprint arXiv*:1103.5044.

Swartz, S. H., Cowan, T. M., & Batista, I. A. (2004). Using claims data to examine patients using practice-based Internet communication: Is there a clinical digital divide? *Journal of Medical Internet Research*, *6*(1), e1. doi:10.2196/jmir.6.1.e1 PMID:15111267

T. R. A. C. Media. (1988). Demographics. Retrieved from http://www.tracmedia.com/Library/Concepts/Demographics/Default.aspx

Tabassum, S. (2015). State of Human Resource Development (HRD) in Pakistan. *International Journal of Science and Research*, *6*(4).

Taha, J., Sharit, J., & Czaja, S. (2009). Use of and satisfaction with sources of health information among older Internet users and nonusers. *The Gerontologist*, *49*(5), 663–673. doi:10.1093/geront/gnp058 PMID:19741112

Tait, L. (2007). 15 Years of Chinese Internet Usage in 13 Pretty Graphs. NanjingMarketingGroup.com. Beijing, China: CNNIC.

Tamura, M., Nagata, H., & Akazawa, K. 2002. Extraction and systems analysis of factors that prevent safety and security by structural models. In *41st SICE Annual Conference*, Osaka, Japan. 10.1109/SICE.2002.1196584

Tandi Lwoga, E. (2014). Integrating Web 2.0 into an academic library in Tanzania. *The Electronic Library*, *32*(2), 183–202. doi:10.1108/EL-06-2012-0058

Tan, J., & Tan, D. (2005). Environment–strategy co-evolution and co-alignment: A staged model of Chinese SOEs under transition. *Strategic Management Journal*, *26*(2), 141–157. doi:10.1002mj.437

Tan, Q., & Sousa, C. M. (2015). Leveraging marketing capabilities into competitive advantage and export performance. *International Marketing Review*, *32*(1), 78–102.

Tarn, J. M., Razi, M. A., Yen, D. C., & Xu, Z. (2002). Linking ERP and SCM systems. *International Journal of Manufacturing Technology and Management*, *4*(5), 420–439. doi:10.1504/IJMTM.2002.001459

Tayebi, M. A., Ester, M., Glasser, U., & Brantingham, P. L. (2014). CRIMETRACER: Activity space based crime location prediction. In *Proceedings of IEEE/ACM International Conference on Advances in Social Networks Analysis and Mining*. Beijing, China: IEEE. 10.1109/ASONAM.2014.6921628

Teece, D. J. (2010). Business models, business strategy and innovation. *Long Range Planning, 43*(2-3), 172–194. doi:10.1016/j.lrp.2009.07.003

Temkin, B. (2016, January 11). *Temkin Group's 2015 Research: Insights For CX Success.* Academic Press.

Teo, T. S. (2001). Demographic and motivation variables associated with Internet usage activities. *Internet Research, 11*(2), 125–137. doi:10.1108/10662240110695089

Terra, D. (2013). How the Internet works. Retrieved from http://www.giantbomb.com/forums/off-topic-31/how-the-internet-works-kinda-1432819/

The Express Tribune. (2018). Pakistan Stock Exchange to launch exchange traded fund this year. Retrieved from https://tribune.com.pk/story/1655461/2-pakistan-stock-exchange-launch-exchange-traded-fund-year/

The World Bank. (2019). GDP per capita growth. Retrieved from https://data.worldbank.org/indicator/NY.GDP.PCAP.KD.ZG

ThichNhatHanh. (1988). *The heart of the Buddha's teaching.* Berkeley, CA: Broadway Books.

Thomas, A., Passaro, R., & Quinto, I. (2019). Developing Entrepreneurship in Digital Economy: The Ecosystem Strategy for Startups Growth. In Strategy and Behaviors in the Digital Economy. IntechOpen.

Thomas, R. E. (2019). Strengthening the Leadership Platform Through Effective Mentoring Programs. *Conflict Resolution & Negotiation Journal, 2019*(1).

Thomson, M., MacInnis, D. J., & Park, C. W. (2005). The ties that bind: Measuring strength of consumers' attachment to brand. *Journal of Consumer Psychology, 15*(1), 77–91. doi:10.120715327663jcp1501_10

Thun, E. (2018). Innovation at the middle of the pyramid: State policy, market segmentation, and the Chinese automotive sector. *Technovation, 70*, 7–19.

Tian, S. W., Yu, A. Y., Vogel, D., & Kwok, R. C. W. (2011). The impact of online social networking on learning: A social integration perspective. *International Journal of Networking and Virtual Organisations, 8*(3), 264–280. doi:10.1504/IJNVO.2011.039999

Tigo, M. (2012). Revisiting the Impact of Integrated Internet Marketing on Firms' Online Performance: European Evidences. *Procedia Technology, 5,* pp. 418-426.

Timberweb(2019b). The Global Timber and Lumber eMarket. Retrieved from http://www.timberweb.com/Members/CountryResults/24/1.html

Timberweb. (2019a). The Global Timber and Lumber eMarket. Retrieved from https://www.timberweb.com/Contacts/CountryResults/210/AA/1.html

Tiruwa, A., Yadav, R., & Suri, P. K. (2018). Modelling Facebook usage for collaborative learning in higher education. *Journal of Applied Research in Higher Education, 10*(3), 357–379. doi:10.1108/JARHE-08-2017-0088

Tiwari, A. K., & Nigam, V. K. (2019). Recent Bio-Processing Technologies for Value Added Horticultural Products. In *Applied Microbiology and Bioengineering* (pp. 57–67). Cambridge, MA: Academic Press. doi:10.1016/B978-0-12-815407-6.00004-6

Tiwari, S., Wee, H. M., & Daryanto, Y. (2018). Big data analytics in supply chain management between 2010 and 2016: Insights to industries. *Computers & Industrial Engineering, 115*, 319–330. doi:10.1016/j.cie.2017.11.017

Tjahjono, B., Esplugues, C., Ares, E., & Pelaez, G. (2017). What does industry 4.0 mean to supply chain? *Procedia Manufacturing, 13*, 1175–1182. doi:10.1016/j.promfg.2017.09.191

Tohidinia, Z., & Mosakhani, M. (2010). Knowledge sharing behavior and its predictors. *Industrial Management & Data Systems, 110*(4), 611–631. doi:10.1108/02635571011039052

Tompkins, B. (2014). Aligning Supply Chains with Business Strategy. Available at http://archive.tompkinsinc.com/wp-content/uploads/2014/04/Supply-Chain-Alignment-with-Business-Strategy.pdf

Toole, J. L., Eagle, N., & Plotkin, J. B. (2011). Spatiotemporal correlations in criminal offense records. *ACM Transactions on Intelligent Systems and Technology*, *2*(4), 38. doi:10.1145/1989734.1989742

Toppr guide, Digital Economy, 2019.

Tosifyan, M., & Tosifyan, S. (2017). A Research on the effect of social media on tendency to entrepreneurship and business establishment (Case Study: Active Iranian Entrepreneurs in Social Media). *Italian Journal of Science & Engineering, 1*(1), pp. 43-48.

Towill, D. R. (1997). The seamless supply chain-the predator's strategic advantage. *International Journal of Technology Management, 13*(1), 37–56. doi:10.1504/IJTM.1997.001649

Towner, T. L., & Lego Muñoz, C. (2011). Facebook and education: A classroom connection? In C. Wankel (Ed.), *Educating Educators with Social Media (Cutting-edge Technologies in Higher Education* (Vol. 1, pp. 33–57). Emerald Group Publishing Limited. doi:10.1108/S2044-9968(2011)0000001005

Trani, J. F., Bakhshi, P., Noor, A. A., Lopez, D., & Mashkoor, A. (2010). Poverty, vulnerability, and provision of healthcare in Afghanistan. *Social Science & Medicine*, *70*(11), 1745–1755. doi:10.1016/j.socscimed.2010.02.007 PMID:20359809

Traunmueller, M., Quattrone, G., & Capra, L. (2014). Mining Mobile Phone Data to Investigate Urban Crime Theories at Scale. In *Proceedings of International Conference on Social Informatics*. Berlin, Germany: Springer. 10.1007/978-3-319-13734-6_29

Trkman, P., McCormack, K., De Oliveira, M. P. V., & Ladeira, M. B. (2010). The impact of business analytics on supply chain performance. *Decision Support Systems*, *49*(3), 318–327. doi:10.1016/j.dss.2010.03.007

Turney, P. D. (2002, July). Thumbs up or thumbs down?: Semantic orientation applied to unsupervised classification of reviews. *In Proceedings of the 40th annual meeting on association for computational linguistics* (pp. 417-424). Association for Computational Linguistics.

Turnley, H., Bolino, M., Lester, S., & Bloodgood, J. (2003). The Impact of Psychological Contract Fulfillment onthe Performance of In-Role and OrganizationalCitizenship Behaviors. *Journal of Management*, *29*(2), 187–206. doi:10.1177/014920630302900204

Tyagi, S. (2012). Adoption of Web 2.0 technology in higher education: A case study of universities in National Capital Region, India. [IJEDICT]. *International Journal of Education and Development Using Information and Communication Technology*, *8*(2), 28–43. Retrieved from https://files.eric.ed.gov/fulltext/EJ1084132.pdf

Tzeng, G. H., Chiang, C. H., & Li, C. W. (2007). Evaluating intertwined effects in e-learning programs: A novel hybrid mcdm model based on factor analysis and DEMATEL. *Expert Systems with Applications*, *32*(4), 1028–1044. doi:10.1016/j.eswa.2006.02.004

U.S. Department of Labor. (2012). American time use survey – 2012 results. Retrieved from http://www.bls.gov/news.release/pdf/atus.pdf

Uittenbogaard, A. C., & Ceccato, V. (2012). Space-time clusters of crime in Stockholm, Sweden. *Review of European Studies*, *4*(5), 148–156. doi:10.5539/res.v4n5p148

Ul Haq, M. (1995). *Reflections on human development*. Oxford, UK: Oxford University Press.

Ulrich, D., Allen, J., Brockbank, W., Younger, J., & Nyman, M. (2009). HR Transformation Building Human Resources from the Outside In. New York, NY: McGraw-Hill.

Umble, E. J., Haft, R. R., & Umble, M. M. (2003). Enterprise resource planning: Implementation procedures and critical success factors. *European Journal of Operational Research*, *146*(2), 241–257. doi:10.1016/S0377-2217(02)00547-7

UNDP. (1995). *Human Development Report-United Nations Development Program*. Retrieved from UNDP. (2015). *Human Development Report 2015*. Retrieved from UNDP. (2016). *Human Development Report 2016-Human Development for everyone*. Retrieved from UNO. (2010). *The Millennium Development Goals Report*. In Y. D. Uslu, Y. Hancıoğlu, & E. Demir (2015), Applicability to green entrepreneurship in Turkey: A situation analysis. *Procedia: Social and Behavioral Sciences*, *195*, 1238–1245.

Uniply Industries. (2018). Annual Report 2017. Retrieved from https://www.uniply.in/pdf-excel/Uniply-Annual-Report-2017.pdf

University of Alberta Students' Union. (2011). *Sustainability Summary Report and Recommendations, University of Alberta*. Retrieved from https://www.su.ualberta.ca/media/uploads/538/2011assessment.pdf

Urde, M. (1999). Brand orientation: A mindset for building brands into strategic resources. *Journal of Marketing Management*, *15*(1-3), 117–133.

US Department of Commerce. (2002). Environmental Technologies Export Market Plan. Retrieved from http://www.fao.org/3/a-am628e.pdf

Uyar, M. (2014). A Research on Total Cost of Ownership and Firm Profitability. *The International Institute for Science, Technology and Education*, *5*(1), 9–16.

Valaskivi, K. (2016). *Cool Nations: Media and the Social Imaginary of the Branded Country*. London: Routledge. doi:10.4324/9781315794662

Valmohammadi, C., & Roshanzamir, S. (2015). The guidelines of improvement: Relations among organizational culture, TQM and performance. *International Journal of Production Economics*, *164*, 167–178.

Van Dijk, J. (2004). Divides in succession: Possession, skills, and use of new media for societal participation. *Media access: Social and psychological dimensions of new technology use*, 233-254.

Van Dijk, J. A. G. M. (2002). A framework for digital divide research. *Electronic Journal of Communication*, *12*(1), 2.

Van Dijk, J., & Hacker, K. (2003). The digital divide as a complex and dynamic phenomenon. *Inf. Soc. Int. J.*, *19*(4), 315–326. doi:10.1080/01972240309487

Vandebosch, H., & Eggermont, S. (2002). Elderly people's media use: At the crossroads of personal and societal developments. *Communications*, *27*(4), 437–455. doi:10.1515/comm.2002.002

Verdon, N. (2002). *Rural women workers in nineteenth-century England: gender, work and wages*. Rochester, NY: Boydell Press.

Verma, S., Bhattacharyya, S. S., & Kumar, S. (2018). An extension of the technology acceptance model in the big data analytics system implementation environment. *Information Processing & Management*, *54*(5), 791–806. doi:10.1016/j.ipm.2018.01.004

Verma, S., & Sekhar Bhattacharyya, S. (2016). Micro-foundation strategies of IOT, BDA, Cloud Computing: Do they really matter in bottom of pyramid? *Strategic Direction*, *32*(8), 36–38. doi:10.1108/SD-06-2015-0093

Viet, B. N., Le Tan, B., Thanh, V. N., & Kim, N. V. (2017). Determinants of export performance: Case of seafood firms in Viet Nam. *Business and Economic Horizons*, *13*(5), 724–737.

Virkus, S., & Bamigbola, A. A. (2011). Students' conceptions and experiences of Web 2.0 tools. *New Library World*, *112*(11/12), 479–489. doi:10.1108/03074801111190473

Vivakaran, M. V., & Maraimalai, N. (2016). Feminist pedagogy and social media: A study on their integration and effectiveness in training budding women entrepreneurs. *Gender and Education*, pp. 1-21. doi:10.1080/09540253.2016.1225008

Vonderembse, M. A., Uppal, M., Huang, S. H., & Dismukes, J. P. (2006). Designing supply chains: Towards theory development. *International Journal of Production Economics*, *100*(2), 223–238. doi:10.1016/j.ijpe.2004.11.014

Vos, J., van Ommen, P., & Mena-Vásconez, P. (2019). 12 To certify or not to certify. *Sustainability Certification Schemes in the Agricultural and Natural Resource Sectors: Outcomes for Society and the Environment*, 259.

Vouzas, F., & Psychogios, A. (2007). Assessing managers' awareness of TQM. *The TQM Magazine*, *19*(1), 62–75.

Waas, T., Hugé, J., Verbruggen, A., & Wright, T. (2011). Sustainable development: A bird's eye view. *Sustainability*, *3*(10), 1637–1661. doi:10.3390u3101637

Walker, C., & Peterson, C. L. (2018). A sociological approach to resilience in health and illness. *Journal of Evaluation in Clinical Practice*, *24*(6), 1285–1290. doi:10.1111/jep.12955 PMID:29901240

Wang, K. Y., & Yip, T. L. (2018). Cold-Chain Systems in China and Value-Chain Analysis. In Finance and Risk Management for International Logistics and the Supply Chain (pp. 217–241). Amsterdam, The Netherlands: Elsevier. doi:10.1016/B978-0-12-813830-4.00009-5

Wang, W. (2001). Impact of ICTs on farm households in China. *ZEF of University Bonn.*

Wang, C.-H., & Chen, T.-M. (2018). Incorporating data analytics into design science to predict user intentions to adopt smart TV with consideration of product features. *Computer Standards & Interfaces*, *59*, 87–95. doi:10.1016/j.csi.2018.02.006

Wang, H., Kifer, D., Graif, C., & Li, Z. (2016). Crime rate inference with big data. In *Proceedings of the 22nd ACM SIGKDD international conference on knowledge discovery and data mining*. San Francisco, CA: ACM Press. 10.1145/2939672.2939736

Wang, H., Lu, Y., & Zhai, C. (2010). Latent aspect rating analysis on review text data: a rating regression approach. *In Proceedings of the 16th ACM SIGKDD international conference on Knowledge discovery and data mining* (pp. 783-792). ACM. 10.1145/1835804.1835903

Wang, L., & Dalton, A. (2016). How and Why Wearing Sunglasses Makes For Cool Consumers. In P. Moreau, & S. Puntoni (Eds.), *NA - Advances in Consumer Research, 44* (pp. 663–664). Duluth, MN: Association for Consumer Research.

Wang, S., & Noe, R. A. (2010). Knowledge sharing: A review and directions for future research. *Human Resource Management Review*, *20*(2), 115–131. doi:10.1016/j.hrmr.2009.10.001

Wang, T., Rudin, C., Wagner, D., & Sevieri, R. (2013). Learning to detect patterns of crime. In *Proceedings of European conference on machine learning and knowledge discovery in databases.* Heidelberg, Germany: Springer.

Wang, X., Gerber, M. S., & Brown, D. E. (2012). Automatic Crime Prediction Using Events Extracted from Twitter Posts. In *Proceedings of Social Computing, Behavioral - Cultural Modeling and Prediction.* New York, NY: Springer. doi:10.1007/978-3-642-29047-3_28

Ward, R., Moule, P., & Lockyer, L. (2008). Adoption of Web 2.0 technologies in education for health professionals in the UK. Where are we and why? In *7th European Conference on E-Learning, Agia Napa, Cyprus*. Retrieved from http://eprints.uwe.ac.uk/65/

Ward, S. (1974). Consumer Socialization. *The Journal of Consumer Research*, *1*(2), 1–14. doi:10.1086/208584

Warren, C., & Campbell, M. (2014). What makes things cool? How autonomy influences perceived coolness. *The Journal of Consumer Research*, *41*(2), 543–563. doi:10.1086/676680

WCED. (1987). *S. W. S.* World Commission on Environment and Development.

Weerahewa, J., & Gunatilake, H. (2010). Timber Market Liberalization in Sri Lanka: Implications for Forest Conservation. *Sri Lankan Journal of Agricultural Economics*, *8*(0), 1–20. doi:10.4038jae.v8i0.1826

Weerawardena, J., & Coote, L. (2001). An empirical investigation into entrepreneurship and organizational innovation-based competitive strategy. *Journal of Research in Marketing and Entrepreneurship*, *3*(1), 51–70.

Wei, J., & Lv, S. (2019, February). Research on the Distribution System of Agricultural Products Cold Chain Logistics Based on Internet of Things. In *IOP Conference Series: Earth and Environmental Science* (Vol. 237, No. 5, p. 052036). Bristol, UK: IOP Publishing 10.1088/1755-1315/237/5/052036

Weisburd, D., Green, L., Gajewski, F., & Bellucci, C. (1994). Defining the Street Level Drug Market. In D. L. MacKenzie, & C. D. Uchida (Eds.), *Drugs and Crime: Evaluating Public Policy Initiatives*. Thousand Oaks, CA: Sage.

Wernerfelt, B. (1984). A resource-based view of the firm. *Strategic Management Journal*, *5*(2), 171–180.

West, D. M., & Kimmons, R. (2019). Twitter in Education. In *The International Encyclopedia of Media Literacy* (pp. 1–6). Hoboken, NJ: Wiley; doi:10.1002/9781118978238.ieml0237

Wieser, D., & Seeler, J.-M. (2018). Online, Not Distance Education: The Merits of Collaborative Learning in Online Education. In A. Altmann, B. Ebersberger, C. Mössenlechner, & D. Wieser (Eds.), *The Disruptive Power of Online Education* (pp. 125–146). Emerald Publishing Limited. doi:10.1108/978-1-78754-325-620181008

Wiles, J., Miskelly, P., Stewart, O., Kerse, N., Rolleston, A., & Gott, M. (2019). Challenged but not threatened: Managing health in advanced age. *Social Science & Medicine*, *227*, 104–110. doi:10.1016/j.socscimed.2018.06.018 PMID:29941204

Wilkins, H. (2008). The Integration of the Pillars of Sustainable Development: A Work in Progress. *McGill International Journal of Sustainable Development Law & Policy*, *4*(2). Available at https://ssrn.com/abstract=2623221

Williams, L., & Cothrell, J. (2000). Four smart ways to run online communities. *Sloan Management Review*, *41*, 81–91.

Wilson, E. (1989). *Hallucinations: Life in the post- modern city*. London, UK: Hutchinson Radius.

Wilson, K. R., Wallin, J. S., & Reiser, C. (2003). Social stratification and the digital divide. *Social Science Computer Review*, *21*(2), 133–143. doi:10.1177/0894439303021002001

Witry, M., Comellas, A., Simmering, J., & Polgreen, P. (2018). The Association between Technology Use and Health Status in a Chronic Obstructive Pulmonary Disease Cohort: Multi-Method Study. *Journal of Medical Internet Research*, *20*(4), e125. doi:10.2196/jmir.9382 PMID:29610113

Wohn, D. Y., & Na, E. K. (2011). Tweeting about TV: Sharing television viewing experiences via social media message streams. *First Monday*, *16*(3). doi:10.5210/fm.v16i3.3368

Wolff, B. (2019). The Truth About Employee Disengagement. *Professional Safety*, *64*(2), 24–24.

Wolf, M. J. (1999). *The entertainment economy—How mega-media forces are transforming our lives*. New York: Times Books, Random House.

Wong, K. (2007). *Emotional Labour of diversity work: Women of colour faculty in predominantly White Institutions*. doi:. doi:10.13140/RG.2.1.2880.6562

Wood, S., & De Menezes, L. (1998). High commitment management in the UK: Evidence from the workplace industrial relations survey, and employers' manpower and skills practices survey. *Human Relations*, *51*(4), 485–515. doi:10.1177/001872679805100403

Wook Kim, S. (2006). Effects of supply chain management practices, integration and competition capability on performance. *Supply Chain Management*, *11*(3), 241–248. doi:10.1108/13598540610662149

World Bank. (2010). ITU, *Info*Dev, IFC, *Telecommunications Regulatory Toolkit*, 10th Anniversary edition. Retrieved from https://openknowledge.worldbank.org/

World Trade Organization. (2013). *E-Commerce in Developing Countries: Opportunities and Challenges for Small and Medium-Sized Enterprises*. Geneva, Switzerland: World Trade Organization.

Wu, G. (2005). The mediating role of perceived interactivity in the effect of actual interactivity on attitude toward the website. *Journal of Interactive Advertising*, *5*(2), 29–39. doi:10.1080/15252019.2005.10722099

Wu, W. W., & Lee, Y. T. (2007). Developing global managers' competencies using the fuzzy DEMATEL method. *Expert Systems with Applications*, *32*(2), 499–507. doi:10.1016/j.eswa.2005.12.005

Xie, Z., & Li, J. (2018). Exporting and innovating among emerging market firms: The moderating role of institutional development. *Journal of International Business Studies*, *49*(2), 222–245.

Xuemei, L. (2015). Surveying the crime analysis in U. S. prediction policing from big data. *Journal of Intelligence*, *34*(12), 16–20.

Xu, L. D., Xu, E. L., & Li, L. (2018). Industry 4.0: State of the art and future trends. *International Journal of Production Research*, *56*(8), 2941–2962. doi:10.1080/00207543.2018.1444806

Yadav, A., Avasthe, R. K., & Dutta, S. K. (2018). Sikkim organic horticulture: Scope, challenges and prospects. *Progressive Horticulture, 50*(1, 2), 82-91.

Yadav, M., & Basera, K. (2013). Status of Forest Products Production and Trade *Centre for Sustainable Forest Management and Forest Certification Working Paper*. Retrieved from http://iifm.ac.in/sites/default/files/working/IIFMWP-13-10-01.pdf

Yahoo finance(2018b). V. R. WoodArt Limited. Retrieved from https://in.finance.yahoo.com/quote/VRWODAR.BO/history?p=VRWODAR.BO

Yahoo finance. (2018a). National Plywood Industries. Retrieved from https://in.finance.yahoo.com/quote/516062.BO/history?p=516062.BO

Yang, W.-H. (1967). Human resource as the key factor of economic development.

Yang, H., Lee, W., & Lee, H. (2018). IoT smart home adoption: The importance of proper level automation. *Journal of Sensors*, *2018*. doi:10.1155/2018/6464036

Yang, J., & Yecies, B. (2016). Mining Chinese social media UGC: A big-data framework for analyzing Douban movie reviews. *Journal of Big Data*, *3*(1), 3. doi:10.118640537-015-0037-9

Yang, Y., Liu, C., Li, C., Hu, Y., Niu, Y., & Li, L. (2014, July). The recommendation systems for smart TV. In *Proceedings of the 2014 International Conference on Computing, Communication and Networking Technologies (ICCCNT)* (pp. 1-6). doi:10.1109/ICCCNT.2014.6963095

Yao, R., & Chen, J. (2013, December). Predicting movie sales revenue using online reviews. *In 2013 IEEE International Conference on Granular Computing (GrC)* (pp. 396-401). IEEE.

Yassine, A., Singh, S., Hossain, M. S., & Muhammad, G. (2019). IoT big data analytics for smart homes with fog and cloud computing. *Future Generation Computer Systems*, *91*, 563–573. doi:10.1016/j.future.2018.08.040

Yazdani, B., Attafar, A., Shahin, A., & Kheradmandnia, M. (2016). The impact of TQM practices on organizational learning case study: Automobile part manufacturing and suppliers of Iran. *International Journal of Quality & Reliability Management*, *33*(5), 574–596.

Yin, S., Wang, G., Qiu, Y., & Zhang, W. (2007). *Research and implement of classification algorithm on web text mining*. Paper presented at the 3rd International Conference on Semantics, Knowledge and Grid, China.

Yoon, H., Jang, Y., Vaughan, P. W., & Garcia, M. (2018). Older adults' Internet use for health information: Digital divide by race/ethnicity and socioeconomic status. *Journal of Applied Gerontology*. PMID:29661052

Yoshida, K., Tsuruoka, Y., Miyao, Y., & Tsujii, J. (2007). *Ambiguous part-of-speech tagging for improving accuracy and domain portability of syntactic parsers*. Paper presented at the 20th International Conference on Artificial Intelligence, China.

Youssef, A. B., Boubaker, S., & Omri, A. (2018). Entrepreneurship and sustainability: The need for innovative and institutional solutions. *Technological Forecasting and Social Change*, *129*, 232–241. doi:10.1016/j.techfore.2017.11.003

Yovovich, B. (1988). What is your brand really worth. *Adweek's Marketing Week*, *8*, 18–24.

Yu, C. H., Ward, M. W., Morabito, M., & Ding, W. (2011). Crime Forecasting Using Data Mining Techniques. In *Proceedings of 2011 IEEE 11th International Conference on Data Mining Workshops*. Washington, DC: IEEE. 10.1109/ICDMW.2011.56

Yusr, M. M. (2016). Innovation capability and its role in enhancing the relationship between TQM practices and innovation performance. *Journal of Open Innovation: Technology, Market, and Complexity*, *2*(1), 1.

Zablith, F., & Osman, I. H. (2019). ReviewModus: Text Classification and Sentiment Prediction of Unstructured Reviews using a Hybrid Combination of Machine Learning and Evaluation Models. *Applied Mathematical Modelling*, *71*, 569–583. doi:10.1016/j.apm.2019.02.032

Zahra, S. A., & Covin, J. G. (1995). Contextual influences on the corporate entrepreneurship-performance relationship: A longitudinal analysis. *Journal of Business Venturing*, *10*(1), 43–58.

Zahra, S. A., Gedajlovic, E., Neubaum, D. O., & Shulman, J. M. (2009). A typology of social entrepreneurs: Motives, search processes and ethical challenges. *Journal of Business Venturing*, *24*(5), 519–532. doi:10.1016/j.jbusvent.2008.04.007

Zapf, D., Vogt, C., Seifert, C., Mertini, H., & Isic, A. (1999). Emotion Work as a Source of Stress: The Concept and Development of an Instrument. *European Journal of Work and Organizational Psychology*, *8*(3), 371–400. doi:10.1080/135943299398230

Zaqout, F., & Abbas, M. (2012). Towards a model for understanding the influence of the factors that stimulate university students' engagement and performance in knowledge sharing. *Library Review*, *61*(5), 345–361. doi:10.1108/00242531211280478

Zavaleta, D., & Tomkinson, J. (2015). *Training Material for Producing National Human Development Reports*. UNDP Human Development Report Office.

Zebrowitz, L., & Rhodes, G. (2004). Sensitivity to "bad genes" and the anomalous face over generalization effect: Cue validity, cue utilization, and accuracy in judging intelligence and health. *Journal of Nonverbal Behavior*, *28*(3), 167–186. doi:10.1023/B:JONB.0000039648.30935.1b

Zeeng, L., Robbie, D., Adams, K. M., & Hutchison, C. (2009). Where's my class? Using Web 2.0 for collaboration in a design environment. In *Ascilite*. Auckland, New Zealand. Retrieved from http://www.ascilite.org/conferences/auckland09/procs/zeeng.pdf

Zezer, M. (1982). *The Dark Ages: Life in the United States 1945-1960*. Cambridge, MA: South End Press.

Zhang, B., Wu, M., Kang, H., Go, E., & Sundar, S. S. (2014). Effects of security warnings and instant gratification cues on attitudes toward mobile websites. In *Proceedings of the 2014 Annual Conference on Human Factors in Computing Systems (CHI'14)* (pp. 111-114). 10.1145/2556288.2557347

Zhong, B. (2013). From smartphones to iPad: Power viewers' disposition toward mobile media devices. *Computers in Human Behavior*, *29*(4), 1742–1748. doi:10.1016/j.chb.2013.02.016

Zhou, K., Liu, T., & Zhou, L. (2015, August). Industry 4.0: Towards future industrial opportunities and challenges. In *2015 12th International Conference on Fuzzy Systems and Knowledge Discovery (FSKD)* (pp. 2147-2152). IEEE. 10.1109/FSKD.2015.7382284

Zhuang, L., Jing, F., & Zhu, X. Y. (2006). Movie review mining and summarization. In *Proceedings of the 15th ACM international conference on Information and knowledge management* (pp. 43-50). ACM.

Zientek, L. R., Werner, J. M., Campuzano, M. V., & Nimon, K. (2018). The Use of Google Scholar for Research and Research Dissemination. *New Horizons in Adult Education and Human Resource Development*, *30*(1), 39–46. doi:10.1002/nha3.20209

Zimmer, M. R., & Golden, L. L. (1988). Impression of retail store: A content analysis of consumer images. *Journal of Retailing*, *64*(Fall), 265–294.

Zimmerman, M. S. (2018). Assessing the reproductive health-related information-seeking behavior of low-income women: Describing a two-step information-seeking process. *Journal of Health Communication*, *23*(1), 72–79. doi:10.1080/10810730.2017.1411996 PMID:29265926

Žižek, S. (2009). *First as Tragedy, Then as Farce,* The RSA, November, 24. Retrieved from https://www.thersa.org/discover/videos/event-videos/2009/11/first-as-tragedy-then-as-farce

Zuckerman, M. (1971). Dimensions of sensation seeking. *Journal of Consulting Psychology*, *36*(1), 45–52. doi:10.1037/h0030478

About the Contributors

Patricia Ordóñez de Pablos is a professor in the Department of Business Administration in the Faculty of Business and Economics at The University of Oviedo (Spain). She completed her education in The London School of Economics (UK). Her teaching and research interests focus on the areas of strategic management, knowledge management, organizational learning, human resource management, intellectual capital, information technologies, with special interest in Asia (Bhutan, China, Laos, Myanmar). She is Editor-in-Chief of the International Journal of Learning and Intellectual Capital (IJLIC) and International Journal of Asian Business and Information Management (IJABIM), respectively. She has edited books for IGI-Global, Routledge, and Springer.

Kwok Tai Chui received the B.Eng. degree in electronic and communication engineering - Business Intelligence Minor from and Ph.D. degree from City University of Hong Kong in 2013 and Feb. 2018 respectively. He had industry experience as Senior Data Scientist in Internet of Things (IoT) company. He joined the Department of Technology, School of Science and Technology, at The Open University of Hong Kong as Research Assistant Professor. His research interests include artificial intelligence, data analytics, wireless communication, pattern recognition, healthcare, machine learning algorithms and optimization. He was the recipient of 2nd Prize Award (Postgraduate Category) of 2014 IEEE Region 10 Student Paper Contest. Also, he received Best Paper Award in IEEE The International Conference on Consumer Electronics-China, in both 2014 and 2015. He has more than 40 research publications: edited books, book chapters, journal papers, and conference papers. His research interests include computational intelligence, wireless communication, pattern recognition, healthcare, machine learning algorithms and optimization. He is the Managing Editor of International Journal on Semantic Web and Information Systems, Associate Editor of International Journal of Asian Business and Information and Guest Editors of various SCIE-listed journals.

Xi Zhang is a professor of College of management and Economics, Tianjin University of China. His research include knowledge management and big data analytics, with several publications in international journals such as JMIS, IEEE TEM, CHB, JKM.

* * *

Jawad Abbas is serving as a lecturer in Near East University, Faculty of Economics and Administrative Sciences. He completed his PhD from the same university. Dr Abbas has published a number of articles in various SSCI, SCIE, ESCI, and Scopus indexed journals.

Neetima Agarwal is Assistant Professor in Faculty of Management Studies, Amity University, Noida. She is a Software Engineer and completed her masters in the area of Organisation Behaviour. She has more than six years of industrial and academic experience at various places. She has published her research work in many journals of high repute. Her areas of expertise are in Training Effectiveness, Training Need Analysis, Human Behaviour in IOT Ecosystem, and Smart Cities.

Mohammad Nabil Almunawar is a senior assistant professor at School of Business and Economics, Universiti of Brunei Darussalam (UBDSBE), Brunei Darussalam. He is a former dean of UBDSBE. He received his bachelor degree in 1983 from Bogor Agricultural University, Indonesia, master Degree (MSc Computer Science) from the Department of Computer Science, University of Western Ontario, London, Canada in 1991, and Ph.D. from the University of New South Wales (School of Computer Science and Engineering, UNSW) in 1998. Dr Almunawar has published more than 100 papers in refereed journals, books, book chapters, and international conferences. He has more than 30 years of teaching experience in the area of computer and information systems. His overall research interests include applications of IT in Management, Electronic Business/Commerce, Digital Platforms, Health Informatics, information security, cloud computing, Big Data, and Internet of Things. He is also interested in object-oriented technology, databases, and multimedia retrieval.

Nurul Afiqah Nor Amin is a PhD candidate at the School of Business and Economics, Universiti Brunei Darussalam. She is also an undergraduate tutor and research assistant in Universiti Brunei Darussalam. Her work has appeared in journals such as Journal of Management Development and International Journal of Asian Business and Information Management. Her main research interest is in Strategic Management, but she is also active in conducting research activities in organizational behavior and knowledge management.

Nurul Nazirah Binti Besar received her first degree in environmental studies from Universiti Brunei Darussalam in 2014 and obtained a Master Degree in Master of Management (Public Management) in the same university in 2015. In 2013, she was awarded the prestigous Dean's list. Her research interests include Strategic Management, Human Resource Management, E-government, and Forestry & Environmental Management. She is working as a teacher in a local private school. She is also actively involved in cultural activities and volunteering works, especially in assisting less fortunate students in their studies.

Som Sekhar Bhattacharyya is a Strategic Management faculty at National Institute of Industrial Engineering (NITIE), Mumbai, India. He has published many research papers and teaching cases in international journals of repute. He was awarded Emerald Literati Awards for Excellence: Outstanding Paper Award – 2011 and Emerald Literati Awards for Excellence: Highly Commended Paper Award – 2014. For teaching, he received NITIE best teacher award 2014 and Business School Award for Best Professor in Strategic Management in India in 2014.

Yanan Cao comes from Hebei, China. She graduated from Dalian University of Technology with a bachelor's degree, majored in Business Administration, and began her postgraduate studies at Tianjin University in September 2018, majoring in Business Administration. She has won many honors, including merit student of Dalian University of Technology, outstanding graduate of Dalian University of Technology, class A Comprehensive Scholarship and so on. Her current research focuses on charitable giving, aiming to find out which factors, relating to the charity or to the donors, affect the potential donors' donation behavior most and to identify the mechanisms under these potential effects.

Kun Cheng earned a Bachelor of Economics, School of Economics and Management, China University of Petroleum (2014-2018), and is working on a Master of Public Administration, College of Management and Economics, Tianjin University (2018-2021). He is researching Energy environment economy, public policy.

Mohammad Ashraful Ferdous Chowdhury is the faculty member of the Department of Business Administration, Shahjalal University of Science & Technology (SUST), Sylhet, Bangladesh. He holds a PhD and MSc degree in Islamic Finance from International Centre for Education in Islamic Finance (INCEIF), The Global University of Islamic Finance, a post-graduate university specializing in Islamic Finance, established by the Central Bank of Malaysia. Previously, Dr. Mohammad Ashraful Ferdous Chowdhury obtained a Bachelor degree in Finance and MBA in Finance & Banking from Shahjalal University of Science & Technology (SUST), Bangladesh. He has written a number of articles in refereed international journals published by Wiley, Taylor & Francis, Elsevier, Inderscience, Emerald, Springer, etc. His main research interests are in the areas of Banking, Financial Economics, Corporate Finance, Development Finance, Institutional Economics, Islamic Economics, Islamic Finance and banking, Sustainable development, Ethical and socially responsible investments, etc.

Rituparna Das the corresponding author teaches at the School of Economics and Commerce of Adamas University as Professor. He previously taught at the University of Petroleum and Energy Studies as Associate Professor. He headed the Centre for Studies in Banking and Finance, established by the Reserve Bank of India corpus and also the Centre of Risk Management and Derivatives. His PhD on Econometrics is preceded by dual Masters in Economics and Management. Earlier, he taught at National Law University, National Institute of Bank Management (established by the Reserve Bank of India), National University of Juridical Sciences and Indian Institute of Social Welfare and Business Management. His areas of preference and international publications include Strategies, Bank Risk Management, Financial Economics, and Policy Research. He is a Life Member of the Ramakrishna Mission Institute of Culture, the Indian Statistical Institute, the Indian Economic Association, and the Indian Econometric Society. He received the Honours of being 'Financial Economics and Risk Educator and Author' from Marquis Who's Who and 'One of the IBC's Leading Educators of the World' from International Biographical Centre. He can be contacted over <dr.rituparnadas@gmail.com>. He is available at www.linkedin.com/in/rpd

Ekta Duggal is Assistant Professor of Commerce, MotiLal Nehru College, University of Delhi. She has done her doctorate from Faculty of Management Studies (FMS), University of Delhi. She also has an Advance Diploma in Computer Software Technology and DOEACC 'O' Level certification. Her areas of teaching and research include business organization and management, e-commerce, services marketing, and retail quality. She has taught courses in masters programs at Department of Commerce, Delhi School of Economics, and Guru Gobind Singh College of Commerce, University of Delhi. She has participated in numerous conventions and seminars. She has published about a dozen research papers in journals of repute including International Journal of Business Ethics in Developing Economies, Journal of Marketing & Business Communication, and South Asian Journal of Global Business Research. Her research papers won best paper awards at national and international conferences. She participated as the Consortium Fellow at 3rd Academy of Indian Marketing-American Marketing Association Sheth Foundation Doctoral Consortium at Dubai. She has coauthored a book titled 'Marketing', published by Oxford University Press.

Nitu Ghosh, Assistant Professor in the School of Management Studies, REVA University, is a dynamic instructor and thought leader. She has been awarded the Doctorate of Philosophy in Management Studies from REVA University and is an alumnus of Alliance Business School, Bangalore with a degree in M.B.A. with HR specialization, P.G.D.M., B.Ed , UGC-NET. She has worked four years in industry and over eight years in teaching. She presented more than 25 papers in International and National conferences and four research papers in International Conferences in IIMs. She is the author of a book on Workplace Ethics. One of her papers on "Emotional Intelligence and entrepreneurship development" has been appreciated and published in the South Asian Journal of Management. She has 15 research publications in international journals and conference publications. She has actively participated in more than 10 Faculty Development Programmes and Workshops. She is also a trainer and consultant in career management, and personality and soft skills development. Her areas of interest in teaching and research lie in the field of Organization Behavior, Human Resources Management, Entrepreneurship, Competency Development, International Business, and Strategic Management.

Lijiao Gong is a student at the college of management and economics in Tianjin University, China. Her research interest is focused on supply chain management.

Ikramul Hasan is a senior lecturer at Independent University, Bangladesh. He holds a Ph.D. in Management, with specialization in Knowledge Management, from Universiti Brunei Darussalam, Brunei. He received his M.B.A. and B.B.A. degrees in HRM and Finance from the School of Business, Independent University, Bangladesh. Dr. Hasan has seven years of working experience in the field of academia, research, and organizational activities (includes teaching, research grant, and R&D activities). He has published several refereed journals in the area of knowledge management and CSR. His research interest lies in Knowledge Management, CSR, HRM, and organizational effectiveness. Amy Suliza binti Hasnan has completed both her undergraduate and post graduate studies at the University of Brunei Darussalam. She earned a Bachelor of Arts in Environmental studies award in 2014. In 2015, she received her Master of Management in public sector. Her areas of interest include human resource management, strategic management, as well as environmental management. She enjoys going to educational field trips and participates in volunteering activities.

Syed Far Abid Hossain is a Ph.D. researcher at Xi'an Jiaotong University. He is a lecturer of College of Business Administration, IUBAT University. He has published numerous papers in well-indexed journals.

Xiaoyang Hu comes from Henan, China. He graduated from Chongqing University with a bachelor's degree, majored in Logistics Engineering, and began his postgraduate studies at Tianjin University in September 2018, majoring in Management Science and Engineering. He has won many honors, including merit student of Chongqing University, outstanding student cadre, outstanding graduate of Chongqing University, class A Comprehensive Scholarship, Meritorious Winner of the Interdisciplinary Contest in

Modeling in 2017, and his undergraduate thesis was rated as excellent. Now, his research focuses on the design and scheduling of bus network, focusing on the combination of big data and machine learning. His main job is to use intelligent algorithms such as genetic algorithms to design new scheduling rules to optimize existing bus networks. He participated in the research training program for college students in Chongqing University and successfully applied for two patents.

Muhammad Imran has completed the PhD in Management from Universiti Utara Malaysia, Sintok, Kedah, Malaysia. Imran has worked as an assistant professor in the faculty of business administration of Iqra University, Karachi, Pakistan, and currently serves as an assistant professor in the department of management sciences in the Islamia University of Bahawalpur. He has published articles in SSCI, SCEI, ESCI and Scopus index journals. His research interest are SMEs, Industry 4.0, Service Innovation, and entrepreneurship.

Jawad Iqbal is working as Associate Professor and Chairman of the Department of Management Sciences, The Islamia University of Bahawalpur, Pakistan. He has published articles in various SSCI, SECI, ESCI and Scopus index journals. His research interests areas are Research Methodology; Organization Behavior; Industrial Revolution (IR 4.0); Entrepreneurship; Service Innovation.

MD. Nazmul Islam is a doctoral researcher in human resource management at University of Malaya. Currently, he is serving as a Graduate Research Assistant at University of Malaya. Prior to this he serves as a Manager, BBA Program, Research and Development Officer and Teaching Assistant at IUB School of Business, Independent University, Bangladesh. His research interest includes Human Resource Management, Organizational Leadership. Organizational Change Management, Employee Workplace Behavior and Organizational Alignment.

Urmila Itam, Assistant Professor, School of Management Studies, REVA University, holds a Doctoral Degree in Management from the Central University of Hyderabad, Hyderabad, and Master's degree in "HRM" from Andhra University, Visakhapatnam. She worked one year in Industry and five years in teaching. Dr. Urmila has been teaching various MBA subjects like Human Resource Management, Business Research Methods, Labour Laws, ITM and Business Analytics, Corporate Environment, and Organizational Behaviour. She has presented her works in international conferences held in prestigious institutes in India. Dr. Urmila has publications in the areas of Diversity Management, Employer Branding, Internal branding, and Employee engagement.

Raunak Jaiswal is a PGDIE scholar at NITIE, Mumbai.

Ajeya Jha has headed the Department of Management Studies at Sikkim Manipal Institute of Technology (SMIT) since February 2000. In addition to having post-graduate and doctoral degrees in Business Administration, he possesses graduation and post-graduation in Pharmaceutical Sciences and is a fellow of International Society of Management. He has a rich and diverse industrial experience of 15 years (includes launching two green-field projects), teaching 18 years and with NGO 8 months. He is the founder Chief coordinator of Entrepreneurship Development Cell (EDC) at SMIT established with support from Department of Science and Technology, Government of India. He is an associate editor of Advances in Management, an International Journal of Management, and International Journal of Biotechnology.

Boxiang Jia majors in management science and engineering in Tianjin University, and researches finance engineer and fintech, etc.

Zhang Kaihua is majoring in information management, supply chain management, at Tianjin University.

Shakeela Kausar is a PhD scholar at the Department of Management Sciences, The Islamia University of Bahawalpur, under the supervision of Dr. Jawad Iqbal.

Farhana Khowaja is a Business Student of the Department of Management Sciences at Isra University (IU), Pakistan. Her main interests are Management and Social Sciences, Health Care Management, with a main focus on Asia. She conducted Research in Health Care Management.

Jitendra Kumar is working as Associate Professor in Department of Management Studies, Sikkim Manipal Institute of Technology, Majitar, Sikkim, India. He graduated in Commerce from the Sikkim Manipal University in 2006. He holds a Master Degree in Business Administration with Finance as major from Sikkim Manipal Institute of Technology, Sikkim, in 2008. Dr. Jitendra Kumar completed his PhD in the 2016 and has 11 years of work experience in academia and two years in the corporate sector. His areas of interest include Behavioral Finance, Portfolio Management and Taxation and Financial Marketing.

Bibhash Laik is a masters-level student of Industrial Engineering specializing in Supply Chain Management and Technology management.

Yi Li is a graduate student in the Department of Management and Economics, Tianjin University. Her research interest is recommender systems.

Diqing Liu received her BS degree in Industrial Engineering from Tianjin University, Tianjin, China, in 2018. She is a master candidate in Management Science and Engineering from Tianjin University, Tianjin.

Xinye Liu is a master degree candidate of Tianjin University. His research field is machine learning and data mining.

Gaofeng Ma from 2014 to 2018 was an undergraduate Mining engineering major of Central South University. Ma is in the second year of a Master's degree program in Management science and engineering at Tianjin University.

Nivedita Mandal is Associate Professor at the Department of Economics, Adamas University, Kolkate. She has specialised on Econometrics and Finance, but she keeps interest in the areas of social sector, and economics of disability, as well. Her recent works mostly focus on interdisciplinary studies. She graduated in Economics from Lady Brabourne College, University of Calcutta, post graduated in Applied Economics from Department of Economics, University of Calcutta, and completed M.Phil in Finance from ICFAI University, Hyderabad, and Ph.D. in Financial Economics from Mizoram University in 2015. She received a Doctoral Fellowship from Indian Council of Social Science Research (ICSSR), New Delhi. She is a passionate teacher and has taught students from a variety of disciplines. There are several reputed conference papers and journal papers to her credit – both national and international. Her research outputs include papers in reputed peer-reviewed journals like Business Sciences, International Research Journal, and conferences hosted by Indian Institute of Management, Kolkata, and Indian Institute of Technology, Chennai. She is on the Editorial Advisory Board of IGI-Global's book project.

Samrat Kumar Mukherjee is working as an assistant professor in the department of management studies, Sikkim Manipal Institute of Technology. He has vast knowledge in subjects like Project Management, and Industrial management, and has published works in national and international peer-reviewed journals.

Mohammad Musa is the Senior Lecturer of the College of Tourism and Hospitality Management. He is studying as a Ph.D. student in the International Business school at Shaanxi Normal University, Xian, China.

Chandra Sekhar Patro is serving as Assistant Professor at Gayatri Vidya Parishad College of Engineering (Autonomous), India. He has a Ph.D in Faculty of Commerce and Management Studies from Andhra University, India. He has a post-graduate degree in Management from JNT University, Commerce from Andhra University, and Master in Financial Management from Pondicherry University. Dr. Patro has over nine years of teaching experience in higher education. Dr. Patro has learned about Human Resource Management and Accountancy/Finance. He has published a number of research papers in reputed National and International Journals and Conferences.

Preethu Rahman is studying at Shaanxi Normal University, Xian, China in the department of international business as a Ph.D. student.

Jaya Rani is currently working as Assistant Professor at Sikkim Manipal Institute of Technology since January 2008. She holds masters degree in Economics and also in Business administration. She has completed her doctoral work on a topic related to Pharmaceutical marketing and awarded with PhD degree. She has 11 papers to her credit and which have been published in peer-reviewed national and international journals. She has co-authored Two books. She is also a co-investigator in a project funded by All India Council for Technical Education. Her interest area include Human Resource Management, Marketing and Training and Development.

Khalil Raza is a sustainable development professional with strong analytical skills in policy analysis, science & technology, and sustainable energy. He serves as the Scientific Officer at the Economic Cooperation Organisation Science Foundation (ECOSF). He plans, coordinates and implements ECOSF's research programs, notably in the area of global environmental challenges with particular focus on Sustainable Energy, Water, and Climate Change. Prior to joining the ECOSF, he worked at the United Nations Development Program (UNDP) as a Coordinator for Sustainable Energy for All (SE4ALL) Program in Pakistan. During his assignment at the UNDP, Khalil co-authored the Country Action Plan on Sustainable Energy for All. Previously, he has worked at the Planning Commission, Government of Pakistan, where he assisted the Energy Wing of Planning Commission in the development of Integrated Energy Model and Planning tool for the Government of Pakistan. He is passionate about the emerging technologies (Blockchain, Machine Learning & Artificial Intelligence, etc.) and their applications to respond to the global challenges of Sustainable Energy, Food Security, and Climate Change. Khalil is a recipient of the prestigious Fulbright Scholarship, and holds Masters in Renewable and Clean Energy from Wright State University, Ohio.

Saibal Kumar Saha (UGC NET Qualified) holds a first class Master's degree in MBA and a first class Bachelor's Degree in Electronics and Communication Engineering. He has over eight years of experience and has worked in MNCs like Cognizant Technology Solutions and Tata Aig Life. He has served National Institute of Technology - Silchar, Jyotirmoy School of Business – Kolkata, University of Technology and Management - Shillong and is presently working as Assistant Professor at Sikkim Manipal Institute of Technology - Majitar, India.

Xu Shan is an administrative secretary at Xi'an Jiaotong University, PRC. Her research interests include entrepreneurship, innovation, and performance management. He has published papers in a well-reputed CSSCI journal.

Bibeth Sharma is an Associate Professor at Sikkim Manipal Institute of Technology - Sikkim Manipal University.

Sharmiladevi J.C. is working as an Assistant Professor in Symbiosis International University, Viman Nagar, Pune. She is a Ph.D. from Bharathiar University, Coimbatore, India. Her Ph.D. title is, "An Analysis of Foreign Direct Investment Inflow in India". Dr. Sharmiladevi is an academician with 14 years of experience in teaching and industry. She teaches subjects like Research Methodology, International Business, International Economics, International Relations, and Statistical Methods. She is a profound researcher and her research areas include Foreign Direct Investment, Macro aspects of Growth, Time Series Econometrics and Modelling, and Tourism Management.

Swati Sisodia is a Research Scholar at G.D Goenka University, Gurgaon. She has completed her masters (PGDM) in Human Resource Management. She has experience in the Education Industry. Her areas of expertise lie in training and development, organization behavior, performance management, and compensation management.

Manzoor Hussain Soomro is a Professor and the Founder President of ECO Science Foundation (ECOSF) since its establishment in December 2011. He previously served as Chairman of Pakistan Science Foundation (PSF), Director General of Pakistan Scientific and Technological Information Centre (PASTIC), Director General of Pakistan Museum of Natural History (PMNH), Professor at the Sindh Agriculture University Tandojam, Chief Scientific Officer in PSF, and Programme Coordinator/Expert of a six-country Regional Programme in the Food and Agriculture Organization (FAO) of the United Nations. The renowned Pakistani Scientist did his Ph.D. from England in 1987. His professional focus has been on research and promotion of Science, Technology, and Innovation (STI), as well as Science Education, especially at schools, over the past 35 years. He is a great proponent of Inquiry Based Science Education (IBSE) at schools and launched IBSE in Pakistan in 2010. In FAO-UN, he led a programme of adult/farmer education about the scientific processes in their crops. Prof. Soomro is founder Fellow and Council Member of the Academy of Engineering and Technology of the developing World (AETDEW) and part of the Digital Belt & Road Double Hundred Universities Cooperation Program

(DHUCP). He is also the founder member of the Asian Innovation Forum & Asian STI Think Tanks Network (ASTN); Vice President of the Belt and Road International Science Education Coordinating Committee (BRISECC) under the patronage of China Association for Science and Technology (CAST). He is Member of Governing Boards of ISTIC Malaysia and IRIS Iran as well as Member of the Global Council of Inter-Academy Partnership Science Education Programme (IAP- SEP). Dr. Soomro has six industrial patents, nine gene sequences, and over 190 publications including books, book chapters, research papers, and popular articles. He serves as the Regional Editor (West & Central Asia) of the Journal of Science, Technology Policy Management, and Editor-in-Chief of the Pakistan Journal of Nematology.

Harsh V. Verma is professor of Faculty of Management Studies, University of Delhi. His areas of academic interest include branding, services marketing, and consumer behavior. He has also taught courses at IIMs. He has published over 30 research papers and five books. One of his books titled "Services Marketing: The Strategies for Success" won DMA-ESCORTS book of the year award in 1993. His last book is titled "Branding Demystified: From Plans to Payoffs" and was published by Response Books (Sage) in 2010. Earlier books include "Services Marketing: Text and Cases," Pearson Education, 2011, "Brand Management: Text and Cases," Excel Books, 2012, "Managing a Service Business Successfully" (Global Business Press, 1995). He is also a corporate trainer and conducted training programs for several Indian and multinational companies in the area of marketing. He also writes a blog, "MarketingCrow" in which he tries to understand and explain business, social, cultural, and political phenomena through a marketing angle.

Xiaoqian Wang graduated with a bachelor's degree in Mechanism Design Manufacturing and Automatization from Chongqing University, and is pursuing a master's degree in Management Science and Engineering in Tianjin University.

Yufei Xia comes from Tianjin, China. She graduated from Tianjin University with a bachelor's degree in Food Science and Engineering, and began her postgraduate studies at Tianjin University in September 2018, majoring in Business Administration. She has won many honors, including merit student of Tianjin University, outstanding student cadre, outstanding graduate of Tianjin University, class A Comprehensive Scholarship, and so on. Her undergraduate thesis was rated as excellent and her graduation thesis was published. Now her research focuses on the influence of equity pledge on corporate performance, finding out the impact of equity pledge rate on corporate innovation and corporate performance.

Tan Xinyun is a student of master's degree at Tianjin University. His research focuses on solving problems in the energy and environment sector, relating to energy economy, energy policy, efficiency evaluation, etc.

Chunfa Xu comes from Tianjin, China. He graduated from Tianjin University with a bachelor's degree in information management, and began his postgraduate studies at Tianjin University in September 2018, majoring in management science and engineering. During his four consecutive years in college, he ranked first in his major and won more than 20 national honors such as national scholarship, mathematical modeling competition, and mathematical competition. Now his research focuses on financial market and big data, and study of the behavior of financial market participants from the perspective of big data. He has published two papers in Chinese core journals in the past year.

Fan Yang is an Assistant Professor in the Department of Communication at University at Albany, SUNY. Her research focuses on new media and strategic communications. She is interested in examining psychological effects of new communication technologies on decision making using methods such as experiment, big-data analysis, and meta-data analysis.

Kinza Yousfani is Assistant Professor of the Department of Management Sciences at Isra University (IU), Pakistan. Her main interests are Management and Social Sciences, Behavioral Finance, International Trade with main focus on South Asia. She conducted Research in the well-known University of Pakistan on "The impact of investors decision making and its impact on Profitability". Ms. Kinza also works in a German government-funded project that focuses on providing training to the local community on Entrepreneurship and Supply Chain Management.

Liping Yuan has been a graduate student of Business Administration Major, College of Management and Economics, Tianjin University, since 2018. The research direction is Enterprise Management and Marketing.

Cailing Zhang comes from Jiang Xi, China. She graduated from Central South University with a bachelor's degree, majored in construction management, and began her postgraduate studies at Tianjin University in September 2018, majoring in management science and engineering. She has won many honors, including merit student of Central South University, outstanding student of college of civil engineering, academic scholarship of Central South University, provincial third prize in National Mathematics Competition for College Students, and so on. Now her research focuses on the organizational behaviors in the project such as transformational leadership, abusive supervision, opportunistic behavior, workplace bullying, social identity, and organizational citizenship behavior.

Yimin Zhang comes from Inner Mongolia, China. He graduated from Tianjin University with a bachelor's degree, majored in Confidentiality management, and began his postgraduate studies at Tianjin University in September 2018, majoring in management science and engineering. He has won many honors, including merit student of Tianjin University, merit student of college of management and economics, outstanding student cadre, outstanding graduate of Tianjin University, class A academic scholarship and so on, and his undergraduate thesis was rated as excellent. Now his research focuses on the use of blockchain technology, and his main work is to utilize blockchain's decentralized, immutable, open, and transparent features to strengthen network security. He has participated in the construction of national virtual simulation experiment teaching project and is responsible for system design.

Bu Zhong is an associate professor of journalism in College of Communications, Pennsylvania State University. His research applies decision-making theories to studying how judgment and decisions may be altered by information processing and the use of information and communication technology.

Xiaodong Zhou is a graduate student from the Department of Management and Economics, Tianjin University, majoring in Management Science and Engineering. The main research direction is data mining.

Index

F

G

H

I

J

K

L

M

N

O

P

R

S

T

V

W

www.ingramcontent.com/pod-product-compliance
Lightning Source LLC
LaVergne TN
LVHW062107300926
849161LV00010B/2392

* 9 7 8 1 7 9 9 8 0 3 5 7 7 *